BASIC ELECTROMAGNETIC THEORY

McGraw-Hill Physical and Quantum Electronics Series

Beam: Electronics of Solids
Collin: Foundations for Microwave Engineering
Elliott: Electromagnetics
Goodman: Introduction to Fourier Optics
Johnson: Field and Wave Electrodynamics
Kirstein, Kino, and Waters: Space-charge Flow
Louisell: Radiation and Noise in Quantum Electronics
Moll: Physics of Semiconductors
Papas: Theory of Electromagnetic Wave Propagation
Paris and Hurd: Basic Electromagnetic Theory
Siegman: Microwave Solid-state Masers
Smith, Janak, and Adler: Electronic Conduction in Solids
Smith and Sorokin: The Laser
Tanenbaum: Plasma Physics
White: Basic Quantum Mechanics

BASIC ELECTROMAGNETIC THEORY

Demetrius T. Paris
Professor of Electrical Engineering
Georgia Institute of Technology

F. Kenneth Hurd
Professor of Electrical Engineering
Georgia Institute of Technology

McGraw-Hill Book Company

New York   St. Louis   San Francisco   London   Sydney   Toronto   Mexico   Panama

BASIC ELECTROMAGNETIC THEORY

Library of Congress Catalog Card Number 68-8775

07 - 0 4 8 4 7 0 - 8

890 KPKP 798765

To Our Families

PREFACE

This book is designed primarily for undergraduate use. However, it can be also used as an intensive review at the beginning graduate level.

The text presumes that the students have a satisfactory background in calculus and in sophomore physics. In particular, it presumes that they have become acquainted with the basic concepts of energy, force, momentum, and power and of electric charge, electric current, and electric and magnetic fields. Furthermore, it presumes that, in their basic physics courses, the students have adequately covered the historical evolution of subject matter through the inductive process of piecing experimental results together to form a coherent theory.

Based on this premise, the presentation begins with Maxwell's equations, the Lorentz force law, and some subsidiary information, which together form a well-established and complete statement of electromagnetic theory in its most general (nonrelativistic) form. The behavior of electromagnetic fields in special cases is deduced from the general formulation. For example, in the electrostatic case all time derivatives and current densities are set equal to zero. This approach has the advantage of requiring an explicit statement at the beginning of the restrictions under which the special case is valid. Any other approach obscures the limits of validity and, furthermore, frequently leads to erroneous generalization.

Aside from the deductive approach, this text treats most of the classical topics of electromagnetic theory in approximately the traditional sequence. However, several topics which are new and several presentations which are not traditional are given. Many topics are treated in somewhat greater depth and from a somewhat more sophisticated point of view than is customary in an undergraduate text. This is deliberate, and is based on the belief that both the student and the curriculum are moving in this direction.

We gratefully acknowledge our appreciation to those who have made significant contributions to the successful completion of this text. Prominent among them is Professor B. J. Dasher, Director of the School of Electrical Engineering at the Georgia Institute of Technology, whose encouragement and full support in adjusting work schedules has been invaluable. On the basis of extensive classroom testing, several of our colleagues, in particular, Professors A. M. Bush, H. A. Ecker, R. D. Hayes, R. W. Larson, and D. C. Ray, provided meaningful critiques of the manuscript.   Professor D. C. Fielder allowed the use of numerous excellent problems.   Several graduate students, including D. G. Bodnar, D. A. Conner, W. T. Mayo, and R. P. Wharton were helpful in eliminating errors from the original manuscript.

We are indebted to Mrs. Betty R. Sims for her skillful typing of the manuscript.

Demetrius T. Paris
F. Kenneth Hurd

CONTENTS

CHAPTER 3.   THE ELECTROSTATIC FIELD   114

CHAPTER 8.   REFLECTION AND REFRACTION OF PLANE WAVES   346

CHAPTER 9.   TRANSMISSION LINES AND WAVEGUIDES   398

CHAPTER 1

VECTOR ANALYSIS

1.1 Introduction. The language of electromagnetic field theory is primarily mathematical, and, in particular, it relies heavily on vector analysis. It is presumed that the students using this book will have had a thorough course in calculus and at least an elementary introduction to differential equations. It is also presumed that the student will have had an introduction to vector analysis. However, since in many cases a mathematician's presentation of vector analysis is rather abstract, and since certain aspects of vector analysis are particularly essential to the study of electromagnetic field theory, this chapter is devoted to a brief presentation of the essential aspects of the subject. For the well-prepared student this chapter will serve as a review and summary. In classes where the student's knowledge of the subject is minimal, it will be desirable for the instructor to supplement the material of this chapter with additional examples, explanations, proofs, and exercises.

1.2 Fields: Vector and Scalar. For our purposes, a *scalar quantity* is one which is completely described by giving its magnitude, and a *vector quantity* is one which requires a magnitude, a direction, and a location for its complete description. (These definitions might not be completely satisfactory from a mathematically rigorous point of view, but they are sufficiently precise and general for our purposes.) Thus, if we say that a certain box is 20 in. high, we have completely specified its height. On the other hand, if we wish to describe a force applied to the box, we need to know its magnitude, its direction, and where it is applied. A little reflection will show that the last two aspects of this vector quantity imply the existence of a reference point and a reference direction. Thus a prescribed coordinate system is implied in the description. This is true for all vector-quantity descriptions. It is a subtlety of vector analysis that vector operations are independent of the coordinate system used but that the existence of a suitable coordinate system is always implied.

The concept of a field requires the concept of a region. We shall not try to give a precise definition of a region. The usual intuitive concept of a region as that part of all space inside (or outside) of a closed boundary surface is usually sufficiently precise. (This is a three-dimensional region. One can extend the concept to two-dimensional, one dimensional, and *n*-dimensional regions.) It is important to realize that the boundary can be at infinity, and thus the region occupies all space, and is then described as an *unbounded region*.

Unfortunately, the word field, as an English-language term, is ambiguous. For example, one commonly used dictionary lists nineteen definitions for the word. Generally, these definitions divide into two major classes. One of these classes defines a field as a region of space devoted to some particular use or having some distinguishing character. Examples are a football field and a gold field. The other major class defines a field as the influence of some agent in a region. Examples here are a gravitational field, an electric field, and a temperature field. Our specialized definition of a field belongs to this class. For our purposes, *a field is defined as the specification of a particular quantity everywhere in a region*. If the quantity specified is a scalar quantity, we have a *scalar field*. If the quantity specified is a vector quantity, we have a *vector field*. The particular quantity which is specified is properly called a *field quantity*. However, it is somewhat unconventional to make an explicit distinction between a field and a field quantity.

Most field quantities of interest in electromagnetic field theory are continuous, and have continuous derivatives of all orders in the regions of interest, although they are frequently discontinuous on the boundary of the region. The continuity of these field quantities allows us to deduce many general properties of the fields, which can then be applied to specific problems.

1.3 Vectors as Directed Line Segments. The mathematics of vector manipulations relies heavily, at least in a conceptual sense, upon the fact that a vector quantity can be mapped in a three-dimensional space as a directed line segment whose length is proportional to the magnitude of the vector quantity, whose direction is the direction of the vector quantity, and whose beginning point is the location of the vector quantity. Thus a vector field has a directed line segment associated with every point in the field region. If we now assign a specific coordinate system to our field region and use the same coordinate system for our directed-line-segment representation of the vector field, we can write expressions in terms of the coordinate system which give the direction and magnitude of directed line segments at every point in the mapping region. Since we have chosen a one-to-one correspondence between the actual vector field and its mapping, we usually discuss the mapping as though it were the actual field. In any event, since the directed line segments and the actual vectors behave mathematically in the same manner, we at least temporarily forget the distinction between them, and call the directed line segments *vectors*, and the manipulations which we perform with them *vector analysis*. Some of the operations of vector analysis deal with vector fields, some deal with the individual vectors, and some are applicable to both. The distinction is sometimes subtle and is, fortunately, seldom necessary. For this reason it is customary not to make the distinction implicitly, and to simply use the term vector to mean either.

In this book we use boldface type to denote a vector. Ordinary type is used, normally, to denote its magnitude. That is, we say a given vector **A** has a magnitude A (or $|A|$). If it is essential to emphasize that we mean a vector field, or that it is a vector field of nonconstant value, we indicate this fact by specifying the chosen coordinate system and saying, for instance, given a vector $\mathbf{A}(x,y,z)$ of magnitude $A(x,y,z)$.

1.4 Vector Addition. All vector manipulations are *defined operations*. The definitions are chosen so as to correspond to the behavior of actual physical vector quantities. It is frequently helpful to interpret the results in terms of a specific physical example.

Consider the two vectors **A** and **B** shown in Fig. 1-1*a*. The dashed lines shown in the figure, together with **A** and **B**, form a parallelogram. The vector sum **A** + **B** is defined as the diagonal vector of this parallelogram, as shown. The definition of vector addition states that any number of vectors can be added together by sequential application of the addition rule in any order. Thus

$$\mathbf{A} + \mathbf{B} + \mathbf{C} + \mathbf{D} = [(\mathbf{A} + \mathbf{B}) + \mathbf{C}] + \mathbf{D} = (\mathbf{A} + \mathbf{B}) + (\mathbf{C} + \mathbf{D})$$
$$= \mathbf{A} + (\mathbf{B} + \mathbf{C}) + \mathbf{D} = [(\mathbf{A} + \mathbf{D}) + \mathbf{C}] + \mathbf{B} = \cdots$$

Also implied in this concept is the possibility of considering any given vector as the sum of two or more other vectors. In particular, consider the vector **A** shown in Fig. 1-1*b*. The figure illustrates the concept of considering the vector **A** as the sum of three orthogonal *components*, $\mathbf{A}_x$, $\mathbf{A}_y$, and $\mathbf{A}_z$. This concept is equally applicable in any other orthogonal coordinate system. We shall be concerned only with the three more common orthogonal coordinate systems, namely, *rectangular* (x,y,z), *cylindrical*, (r,φ,z), and *spherical* (r,θ,φ). Figure 1-2 shows these coordinate systems.

A more useful component representation of a vector is obtained by showing each component as the product of its magnitude times a vector of unit magnitude $\mathbf{a}_i$ in the direction of increasing coordinate. In this representation, we should write

$$\mathbf{A} = A_x\mathbf{a}_x + A_y\mathbf{a}_y + A_z\mathbf{a}_z$$
$$= A_r\mathbf{a}_r + A_\varphi\mathbf{a}_\varphi + A_z\mathbf{a}_z$$
$$= A_r\mathbf{a}_r + A_\theta\mathbf{a}_\theta + A_\varphi\mathbf{a}_\varphi$$

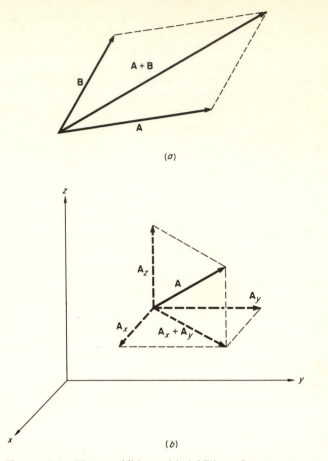

FIGURE 1-1. Vector addition. (a) Addition of two vectors; (b) adding the components of a vector to obtain the vector itself.

depending upon the coordinate system we were using.

This component form of vector representation is the basic representation for all vector manipulations. For example, it can be readily shown that the sum of two vectors, **A** and **B**, is given by

$$\mathbf{A} + \mathbf{B} = (A_x + B_x)\mathbf{a}_x + (A_y + B_y)\mathbf{a}_y + (A_z + B_z)\mathbf{a}_z$$
$$= (A_r + B_r)\mathbf{a}_r + (A_\varphi + B_\varphi)\mathbf{a}_\varphi + (A_z + B_z)\mathbf{a}_z$$
$$= (A_r + B_r)\mathbf{a}_r + (A_\theta + B_\theta)\mathbf{a}_\theta + (A_\varphi + B_\varphi)\mathbf{a}_\varphi$$

Similarly, the difference is given by

$$\mathbf{A} - \mathbf{B} = (A_x - B_x)\mathbf{a}_x + (A_y - B_y)\mathbf{a}_y + (A_z - B_z)\mathbf{a}_z = \cdots$$

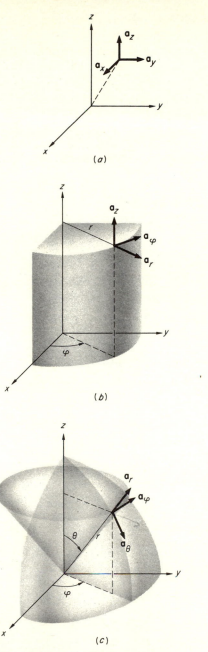

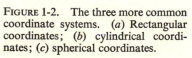

FIGURE 1-2.   The three more common
coordinate systems. (*a*) Rectangular
coordinates; (*b*) cylindrical coordinates; (*c*) spherical coordinates.

1.5   The Dot, or Scalar, Product.   The dot product of two vectors **A** and **B** is written **A · B**. It is defined geometrically as the projection of **A** onto **B** multiplied by B (or vice versa), which is just the scalar quantity

$$\mathbf{A} \cdot \mathbf{B} = AB \cos \theta \tag{1-1}$$

Here A and B denote magnitudes, and θ is the smaller angle between **A** and **B**. The dot product is defined operationally as

$$\mathbf{A} \cdot \mathbf{B} = A_x B_x + A_y B_y + A_z B_z \tag{1-2}$$

which in cylindrical and spherical coordinates becomes

$$\mathbf{A} \cdot \mathbf{B} = A_r B_r + A_\varphi B_\varphi + A_z B_z = A_r B_r + A_\theta B_\theta + A_\varphi B_\varphi \tag{1-3}$$

The two definitions are, of course, equivalent.

The dot product obeys the commutative and distributive laws; thus

$$\mathbf{A} \cdot \mathbf{B} = \mathbf{B} \cdot \mathbf{A} \qquad \mathbf{A} \cdot (\mathbf{B} + \mathbf{C}) = \mathbf{A} \cdot \mathbf{B} + \mathbf{A} \cdot \mathbf{C}$$

It is worth emphasizing that $\mathbf{A} \cdot \mathbf{B} = 0$ says that **A** and **B** are perpendicular to each other, and that $\mathbf{A} \cdot \mathbf{B} = AB$ says that **A** and **B** are parallel to each other. The special case $\mathbf{A} = \mathbf{B}$ yields

$$\mathbf{A} \cdot \mathbf{A} = A^2 = |\mathbf{A}|^2 = A_x{}^2 + A_y{}^2 + A_z{}^2$$

1.6   The Cross, or Vector, Product.   The cross product of two vectors **A** and **B** is written **A × B**. It is defined geometrically as the vector quantity

$$\mathbf{A} \times \mathbf{B} = AB \sin \theta \, \mathbf{n} \tag{1-4}$$

where **n** is a unit vector normal to the plane formed by **A** and **B** and whose sense is positive in the direction of advance of a right-hand screw when **A** is rotated into **B**, through the smaller angle θ, as indicated in Fig. 1-3. Note that this makes the ordering of the terms in the cross product significant. In particular,

$$\mathbf{A} \times \mathbf{B} = -\mathbf{B} \times \mathbf{A} \tag{1-5}$$

Operationally, the cross product is

$$\mathbf{A} \times \mathbf{B} = (A_y B_z - A_z B_y)\mathbf{a}_x + (A_z B_x - A_x B_z)\mathbf{a}_y + (A_x B_y - A_y B_x)\mathbf{a}_z \tag{1-6}$$

which may also be written, in determinant form,

$$\mathbf{A} \times \mathbf{B} = \begin{vmatrix} \mathbf{a}_x & \mathbf{a}_y & \mathbf{a}_z \\ A_x & A_y & A_z \\ B_x & B_y & B_z \end{vmatrix} = \begin{vmatrix} \mathbf{a}_r & \mathbf{a}_\varphi & \mathbf{a}_z \\ A_r & A_\varphi & A_z \\ B_r & B_\varphi & B_z \end{vmatrix} = \begin{vmatrix} \mathbf{a}_r & \mathbf{a}_\theta & \mathbf{a}_\varphi \\ A_r & A_\theta & A_\varphi \\ B_r & B_\theta & B_\varphi \end{vmatrix} \tag{1-7}$$

Like the dot product, the cross product obeys the distributive law

$$\mathbf{A} \times (\mathbf{B} + \mathbf{C}) = \mathbf{A} \times \mathbf{B} + \mathbf{A} \times \mathbf{C}$$

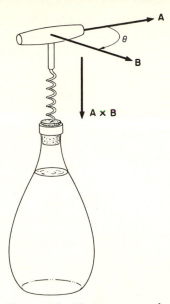

FIGURE 1-3.   Vector cross product.

It is worth mentioning that $\mathbf{A} \times \mathbf{B} = 0$ says that $\mathbf{A}$ and $\mathbf{B}$ are parallel to each other, and that $|\mathbf{A} \times \mathbf{B}| = AB$ says that A and B are perpendicular to each other.

1.7   Vector Coordinate Transformations.   Occasionally, we have need to transform vectors from one coordinate system to another coordinate system.   The problem is similar to scalar-coordinate-system transformations, with the additional necessity of transforming the individual component vectors.   The vector part of the transformation is handled by transforming the unit vectors.

As a specific example, we shall transform from a rectangular system (x,y,z) to an associated cylindrical system (r,φ,z), where the coordinate transformations are

$$x = r \cos \varphi$$

$$y = r \sin \varphi \tag{1-8}$$

$$z = z$$

We wish to transform $\mathbf{A}$ given in the form

$$\mathbf{A} = A_x \mathbf{a}_x + A_y \mathbf{a}_y + A_z \mathbf{a}_z$$

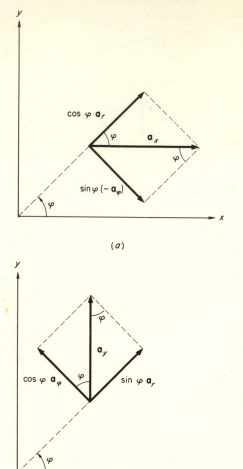

FIGURE 1-4. The geometry of the unit vector transformations from rectangular to cylindrical coordinates. (a) The cylindrical components of $\mathbf{a}_x$; (b) the cylindrical components of $\mathbf{a}_y$.

to the form

$$\mathbf{A} = A_r\mathbf{a}_r + A_\varphi\mathbf{a}_\varphi + A_z\mathbf{a}_z$$

Since the z axis is common to both, we need only deal with $\mathbf{a}_x$, $\mathbf{a}_y$, $\mathbf{a}_r$, and $\mathbf{a}_\varphi$. The relationship between $\mathbf{a}_x$ and $\mathbf{a}_y$ and $\mathbf{a}_r$ and $\mathbf{a}_\varphi$ may be obtained geometrically by inspection of Fig. 1-4, where it will be seen that

$$\mathbf{a}_x = \cos \varphi \, \mathbf{a}_r - \sin \varphi \, \mathbf{a}_\varphi$$

and

$$\mathbf{a}_y = \sin \varphi \, \mathbf{a}_r + \cos \varphi \, \mathbf{a}_\varphi$$

Substituting these results into the expression for **A** in rectangular coordinates and collecting terms gives

$$\mathbf{A} = (A_x \cos \varphi + A_y \sin \varphi)\mathbf{a}_r + (-A_x \sin \varphi + A_y \cos \varphi)\mathbf{a}_\varphi + A_z\mathbf{a}_z \quad (1\text{-}9)$$

or

$$A_r = A_x \cos \varphi + A_y \sin \varphi$$
$$A_\varphi = -A_x \sin \varphi + A_y \cos \varphi \quad (1\text{-}10)$$
$$A_z = A_z$$

The inverse transformation can be obtained from the above relationships:

$$A_x = A_r \cos \varphi - A_\varphi \sin \varphi$$
$$A_y = A_r \sin \varphi + A_\varphi \cos \varphi \quad (1\text{-}11)$$
$$A_z = A_z$$

For the spherical coordinate system, the coordinate transformations are

$$x = r \sin \theta \cos \varphi$$
$$y = r \sin \theta \sin \varphi \quad (1\text{-}12)$$
$$z = r \cos \theta$$

We should similarly find the vector component transformations to be

$$A_r = A_x \cos \varphi \sin \theta + A_y \sin \varphi \sin \theta + A_z \cos \theta$$
$$A_\theta = A_x \cos \varphi \cos \theta + A_y \sin \varphi \cos \theta - A_z \sin \theta \quad (1\text{-}13)$$
$$A_\varphi = -A_x \sin \varphi + A_y \cos \varphi$$

and

$$A_x = A_r \sin \theta \cos \varphi + A_\theta \cos \theta \cos \varphi - A_\varphi \sin \varphi$$
$$A_y = A_r \sin \theta \sin \varphi + A_\theta \cos \theta \sin \varphi + A_\varphi \cos \varphi \quad (1\text{-}14)$$
$$A_z = A_r \cos \theta - A_\theta \sin \theta$$

As an exercise, the student should show that these transformations can be obtained by use of geometry and the definition of dot product. For instance, the first expression in Eq. (1-10) can be obtained by noting that

$$\mathbf{a}_r \cdot (A_r\mathbf{a}_r + A_\varphi\mathbf{a}_\varphi + A_z\mathbf{a}_z) = \mathbf{a}_r \cdot (A_x\mathbf{a}_x + A_y\mathbf{a}_y + A_z\mathbf{a}_z)$$

or since $\mathbf{a}_r \cdot \mathbf{a}_r = 1$ and $\mathbf{a}_r \cdot \mathbf{a}_\varphi = \mathbf{a}_r \cdot \mathbf{a}_z = 0$,

$$A_r = A_x\mathbf{a}_r \cdot \mathbf{a}_x + A_y\mathbf{a}_r \cdot \mathbf{a}_y + A_z\mathbf{a}_r \cdot \mathbf{a}_z$$

Now $\mathbf{a}_r \cdot \mathbf{a}_z = 0$, and from Fig. 1-5,

$$\mathbf{a}_r \cdot \mathbf{a}_x = \cos \varphi \qquad \mathbf{a}_r \cdot \mathbf{a}_y = \sin \varphi$$

Therefore

$$A_r = A_x \cos \varphi + A_y \sin \varphi$$

which is the desired result.

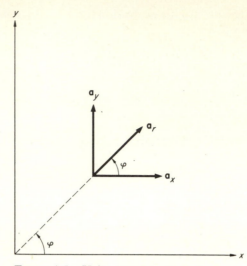

FIGURE 1-5.   Unit vector transformations.

1.8   Line Integrals. Before we discuss a line integral, it is proper to examine the concept of a line.   A line is the path in space along a curve from a beginning point to an ending point.   Notice that this interpretation gives a line a defined positive direction.   In view of the other connotations of the term line, a more acceptable name for the concept given above is a *contour*. We shall use the terms *line, contour, along the path*, and *along the curve*, interchangeably.   Sometimes the path followed by our line is along a closed curve, and if we follow such a curve all the way, we get back to the point where we started.   This line is usually called a *closed contour*.

A curve in space can be specified parametrically by specifying any two of the coordinates as functions of the third.   That is, a curve is specified by equations such as

$$\left.\begin{array}{l} y = g(x) \\ z = h(x) \end{array}\right\} \quad \text{or} \quad \left.\begin{array}{l} \varphi = g(r) \\ z = h(r) \end{array}\right\} \quad \text{or} \quad \left.\begin{array}{l} r = g(\varphi) \\ \theta = h(\varphi) \end{array}\right\}$$

This means that, along the curve, any arbitrary continuous function of position, such as $f_C(x,y,z)$, can be expressed as a function of any one of the three coordinates.   We could write this out explicitly as

$$f_C(x,y,z) = F_C(x) = G_C(y) = H_C(z)$$

Now, if we are given a piecewise continuous function of position, $f(x,y,z)$, a continuous curve C, and two points on C, $P_1 = (x_1,y_1,z_1)$ and $P_2 = (x_2,y_2,z_2)$, we define the integral

$$\int_C f_C(x,y,z)\, dx = \int_{x_1}^{x_2} F_C(x)\, dx \qquad (1\text{-}15)$$

as the line integral of $f(x,y,z)$ along C with respect to x.

Similar integrals with respect to y and with respect to z are defined as

$$\int_C f(x,y,z)\,dy = \int_{y_1}^{y_2} G_C(y)\,dy \qquad (1\text{-}16)$$

and

$$\int_C f(x,y,z)\,dz = \int_{z_1}^{z_2} H_C(z)\,dz \qquad (1\text{-}17)$$

Example 1-1 Line Integrals. Given $f(x,y,z) = 2x + y + z^2$ and the curve specified by $y = 2x$, $z = x$. Let the contour be that part of C beginning at the point $(0,0,0)$ and ending at the point $(2,4,2)$, and calculate the line integral of $f(x,y,z)$ along this contour (a) with respect to x; (b) with respect to y; (c) with respect to z.

SOLUTIONS

$$(a)\quad \int_C f(x,y,z)\,dx = \int_0^2 (4x + x^2)\,dx = \left[2x^2 + \frac{x^3}{3} \right]_0^2 = \frac{32}{3}$$

$$(b)\quad \int_C f(x,y,z)\,dy = \int_0^4 \left(2y + \frac{y^2}{4} \right) dy = \left[y^2 + \frac{y^3}{12} \right]_0^4 = \frac{64}{3}$$

$$(c)\quad \int_C f(x,y,z)\,dz = \int_0^2 (4z + z^2)\,dz = \left[2z^2 + \frac{z^3}{3} \right]_0^2 = \frac{32}{3}$$

A line integral of similar nature to the preceding is the line integral of a function with respect to displacement along the path. The defining process is very similar.

With the beginning point of the path (or some other convenient point of the path) as the origin, define a coordinate l which measures arc length along the path. Now using l as a parameter, write the equations defining the curve in parametric form, and use these equations to express $f_C(x,y,z)$ in parametric form. That is, write

$$f_C(x,y,z) = L_C(l)$$

and define

$$\int_C f(x,y,z)\,dl = \int_{l_1}^{l_2} L_C(l)\,dl \qquad (1\text{-}18)$$

as the line integral of the function f with respect to displacement along the path from the beginning point l_1 to the ending point l_2.

One specific form of Eq. (1-18) in rectangular coordinates is

$$\int_C f(x,y,z) \sqrt{ 1 + \left(\frac{dy}{dx} \right)^2 + \left(\frac{dz}{dx} \right)^2 }\,dx \qquad (1\text{-}19)$$

For most paths, the parametric form is not easy to obtain. There is a special class of line integrals of the type described above which are of extreme

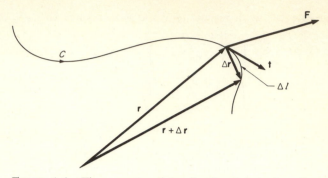

FIGURE 1-6.　The geometry of the directional derivative of an arbitrary radial vector in the direction of a curve.

importance in electromagnetics.　As shown below, they can be evaluated without using the parametric form.　These line integrals are those for which $f_C(x,y,z)$ is the component of some vector field $\mathbf{F}(x,y,z)$ in the direction of the tangent to the line.

Consider the contour C, shown in Fig. 1-6.　Define $\mathbf{t}$ as the unit vector tangent to C.　Let $\mathbf{F}(x,y,z)$ be a vector field which is defined at every point along the path.　Then

$$\int_C \mathbf{F}(x,y,z) \cdot \mathbf{t}(x,y,z)\, dl \tag{1-20}$$

is defined as the line integral of the tangential component of $\mathbf{F}$ along C.

To see how this integral can be evaluated, consider the radial vector $\mathbf{r}$ from an arbitrary origin to a point on C, as shown in Fig. 1-6.　Now form the directional derivative of $\mathbf{r}$ in the direction of l.　That is, form

$$\frac{d\mathbf{r}}{dl} = \lim_{\Delta l \to 0} \frac{\Delta \mathbf{r}}{\Delta l} \tag{1-21}$$

and examine its significance.　Its direction is obviously that of the tangent to the curve C.　Its magnitude is, obviously, unity.　Thus we have

$$\frac{d\mathbf{r}}{dl} = \mathbf{t} \tag{1-22}$$

If we make this substitution for $\mathbf{t}$ in our integrand, we obtain

$$\int_C \mathbf{F} \cdot \mathbf{t}\, dl = \int_C \mathbf{F} \cdot \frac{d\mathbf{r}}{dl}\, dl = \int_C \mathbf{F} \cdot d\mathbf{r} \tag{1-23}$$

The final form shows that we have succeeded in changing from the scalar parameter l to the vector parameter $\mathbf{r}$.　That this results in a simplification of the problem can be seen by examination of dr. Recall that, in rectangular coordinates, the radial vector $\mathbf{r}$ is given by

$$\mathbf{r} = x\mathbf{a}_x + y\mathbf{a}_y + z\mathbf{a}_z$$

and hence

$$d\mathbf{r} = dx\,\mathbf{a}_x + dy\,\mathbf{a}_y + dz\,\mathbf{a}_z$$

Since

$$\mathbf{F} = F_x\mathbf{a}_x + F_y\mathbf{a}_y + F_z\mathbf{a}_z$$

we have

$$\int_C \mathbf{F} \cdot d\mathbf{r} = \int_C (F_x\,dx + F_y\,dy + F_z\,dz)$$

$$= \int_{x_1}^{x_2} F_x\,dx + \int_{y_1}^{y_2} F_y\,dy + \int_{z_1}^{z_2} F_z\,dz \tag{1-24}$$

In other words, we have transformed our original line-integral problem into three much simpler problems.

From the form of the integral (1-23) it will be readily seen that, in cylindrical coordinates, the result will be of the form

$$\int_C \mathbf{F} \cdot d\mathbf{r} = \int_{r_1}^{r_2} F_r\,dr + \int_{\varphi_1}^{\varphi_2} F_\varphi r\,d\varphi + \int_{z_1}^{z_2} F_z\,dz \tag{1-25}$$

and in spherical coordinates,

$$\int_C \mathbf{F} \cdot d\mathbf{r} = \int_{r_1}^{r_2} F_r\,dr + \int_{\theta_1}^{\theta_2} F_\theta r\,d\theta + \int_{\varphi_1}^{\varphi_2} F_\varphi r \sin\theta\,d\varphi \tag{1-26}$$

where, of course, the integrands will have to be evaluated on the curve as functions of the variables of integration.

If the path of integration is completely around a closed curve, we use the notation

$$\oint_C \mathbf{F} \cdot d\mathbf{l} \tag{1-27}$$

and the terminology *closed-contour integration;* we frequently call the result the *circulation of* $\mathbf{F}$ *around C.*

Example 1-2 Line Integral of the Tangential Component of a Vector. Given the vector field

$$\mathbf{F} = xy\mathbf{a}_x + y^2\mathbf{a}_y$$

and the closed (triangular) contour in the xy plane shown in Fig. 1-7, which begins at the origin and goes along the line $x = 0$ to the point $y = 2$, and then along the line $y = 2$ to the point $x = 2$, and back to the origin along the line $x = y$. Calculate

$$\oint_C \mathbf{F} \cdot d\mathbf{l}$$

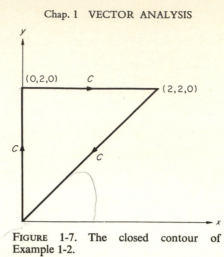

FIGURE   1-7.   The   closed   contour   of
Example 1-2.

(*a*) in rectangular coordinates; (*b*) in cylindrical coordinates.

SOLUTIONS

(*a*) $\oint_C \mathbf{F} \cdot d\mathbf{l} = \int_{\substack{C \\ x=0}} \mathbf{F} \cdot d\mathbf{l} + \int_{\substack{C \\ y=2}} \mathbf{F} \cdot d\mathbf{l} + \int_{\substack{C \\ x=y}} \mathbf{F} \cdot d\mathbf{l}$

$$= \int_{\substack{0 \\ x=0}}^{2} F_y \, dy + \int_{\substack{0 \\ y=2}}^{2} F_x \, dx + \int_{\substack{2 \\ x=y}}^{0} F_x \, dx + \int_{\substack{2 \\ x=y}}^{0} F_y \, dy$$

$$= \int_{0}^{2} y^2 \, dy + \int_{0}^{2} 2x \, dx + \int_{2}^{0} x^2 \, dx + \int_{2}^{0} y^2 \, dy$$

$$= \frac{y^3}{3}\bigg|_0^2 + \frac{2x^2}{2}\bigg|_0^2 + \frac{x^3}{3}\bigg|_2^0 + \frac{y^3}{3}\bigg|_2^0 = \frac{4}{3}$$

(*b*) We transform **F** and get
$$\mathbf{F} = r^2 \sin \varphi \, \mathbf{a}_r$$

We then observe that, in cylindrical coordinates, the contour starts at the origin and goes along the line $\varphi = \pi/2$ to $r = 2$, and then along the line $r \sin \varphi = 2$ to the point $r = 2\sqrt{2}$, $\varphi = \pi/4$, and then back to the origin along the line $\varphi = \pi/4$.   The solution is

$$\oint_C \mathbf{F} \cdot d\mathbf{l} = \int_{\substack{0 \\ \varphi=\pi/2}}^{2} F_r \, dr + \int_{\substack{2 \\ r \sin \varphi=2}}^{2\sqrt{2}} F_r \, dr + \int_{\substack{2\sqrt{2} \\ \varphi=\pi/4}}^{0} F_r \, dr$$

$$= \int_{0}^{2} r^2 \, dr + \int_{2}^{2\sqrt{2}} 2r \, dr + \int_{2\sqrt{2}}^{0} \frac{1}{\sqrt{2}} r^2 \, dr$$

$$= \frac{r^3}{3}\bigg|_0^2 + \frac{2r^2}{2}\bigg|_2^{2\sqrt{2}} + \frac{1}{\sqrt{2}}\frac{r^3}{3}\bigg|_{2\sqrt{2}}^{0} = \frac{4}{3}$$

The student should fill in the details, and he should verify that the contributions from the separate parts of the contour are the same in both coordinate systems.   This property is frequently useful.   *It means that each part of the contour integral may be expressed in terms of the coordinate system which makes that part easiest to handle.* See Prob. 1-23 for an example of the application of this property.

1.9 Surface Integrals.   The definition of a surface integral is as follows: Consider a surface S in three-dimensional space, as shown in Fig. 1-8, and let g be a scalar point function defined at every point on S.   Let S be sub-divided into N contiguous elements of area $\Delta a_1, \Delta a_2, \ldots, \Delta a_N$, and let p_k be any point in the kth element of area.   Denote the value of g at p_k by $g(p_k)$.   If the sum

$$\sum_{k=1}^{N} g(p_k)\, \Delta a_k$$

has a limiting value as $N \to \infty$ and the greatest of the Δa_k's approaches zero, define this limiting value as the *surface integral of the function g over the surface S.*   We emphasize the fact that g is a function of position by writing $g(x,y,z)$ and denote the surface integral by

$$\int_S g(x,y,z)\, da \tag{1-28}$$

If the surface is closed, we use the notation

$$\oint_\Sigma g(x,y,z)\, da \tag{1-29}$$

In most of our work, g is the normal component of some vector field **G**.

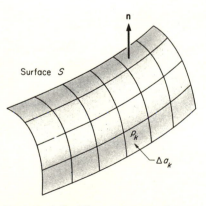

FIGURE 1-8. The geometry for a surface integral.

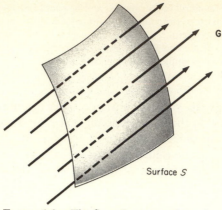

FIGURE 1-9.   The flux of a vector through a
surface.

Thus, if **n** is a unit vector normal to S, we shall be dealing with a function

$$g(x,y,z) = \mathbf{G}(x,y,z) \cdot \mathbf{n}(x,y,z)$$

and we shall denote our surface integral by

$$\int_S \mathbf{G} \cdot \mathbf{n} \, da \tag{1-30}$$

or, for closed surfaces,

$$\oint_\Sigma \mathbf{G} \cdot \mathbf{n} \, da \tag{1-31}$$

We call this surface integral the *flux of* **G** *through* S, or, if **n** is the outward
normal for a closed surface, we call the result the *net outward flux of* **G** *through*
S (Fig. 1-9).   Notice that, for an open surface, we have to make an arbitrary
choice for the positive direction of **n**, and that the sign of the result depends
upon this choice.

The evaluation of the surface integrals (1-30) and (1-31) is relatively
straightforward in those special cases where the surface S is (at least piecewise) specified by constant coordinate surfaces.   In these cases, the normal
to the surface is parallel to a coordinate unit vector.

The following example is given to illustrate the evaluation procedure.

Example 1-3 Surface Integrals.   Given the vector field

$$\mathbf{G} = x^2 \mathbf{a}_x + (y + z)\mathbf{a}_y + xy\mathbf{a}_z$$

We wish, first, to find the flux of **G** through the rectangular surface in the xy plane
bounded by the lines $x = 0$, $x = 3$, $y = 1$, $y = 2$, as shown in Fig. 1-10.

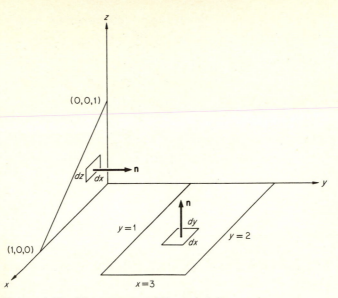

FIGURE 1-10.   The geometry of Example 1-3.

From the figure it will be seen† that $\mathbf{n} = \mathbf{a}_z$ and that $da = dx\,dy$.   Hence

$$\mathbf{G} \cdot \mathbf{n}\,da = G_z\,dx\,dy = xy\,dx\,dy$$

and
$$\int_S \mathbf{G} \cdot \mathbf{n}\,da = \int_1^2 \int_0^3 xy\,dx\,dy$$

For this integral, the order of integration is immaterial.   The student will readily verify that the result is

$$\int_S \mathbf{G} \cdot \mathbf{n}\,da = \frac{27}{4}$$

Next, we calculate the flux of $\mathbf{G}$ through the triangular surface in the xz plane bounded by the x axis, the z axis, and the line $x + z = 1$, as shown in Fig. 1-10. From the figure it will be seen that $\mathbf{n} = \mathbf{a}_y$ and that $da = dx\,dz$.   Hence

$$\mathbf{G} \cdot \mathbf{n}\,da = G_y\,dx\,dz = (y + z)\,dx\,dz$$

But $y = 0$; so
$$\mathbf{G} \cdot \mathbf{n}\,da = z\,dx\,dz$$

Once again the order of integration is immaterial, but for either order the first integration has a variable limit.   The result is

$$\int_0^1 z \left(\int_0^{1-z} dx \right) dz = \frac{1}{6} \quad \text{or} \quad \int_0^1 \left(\int_0^{1-x} z\,dz \right) dx = \frac{1}{6}$$

† A general formula for determining $\mathbf{n}$ is given in Prob. 1-44.

Generally, the surface S is defined by an equation of the form $z = f(x,y)$, where x and y range over a region R in the xy plane. Then

$$\int_S g(x,y,z)\, da = \int_R g(x,y,z) \sec \gamma \, dx \, dy \tag{1-32}$$

where, in the integral on the right, $z = f(x,y)$, and γ is the acute angle between the normal to S at (x,y,z) and the positive z axis. Specifically,

$$\sec \gamma = \left[1 + \left(\frac{\partial z}{\partial x}\right)^2 + \left(\frac{\partial z}{\partial y}\right)^2\right]^{\frac{1}{2}} \tag{1-33}$$

Note that once $\sec \gamma$ is determined, the double integration in Eq. (1-32) can proceed as in Example 1-3.

1.10   Differential Vector Operations.

There are four differential operations which are extremely useful in vector analysis. These are also defined operations. Although it is not the sophisticated approach, we shall define them as manipulations in rectangular coordinates and then indicate what the results mean. The results for other coordinate systems are summarized in Sec. 1.12.

The del operator   Let us first define a differential operator called *del* as

$$\mathbf{\nabla} = \mathbf{a}_x \frac{\partial}{\partial x} + \mathbf{a}_y \frac{\partial}{\partial y} + \mathbf{a}_z \frac{\partial}{\partial z} \tag{1-34}$$

The gradient of a scalar function   Given a scalar function of position $f(x,y,z)$, define the *gradient* of f as

$$\mathbf{\nabla} f = \frac{\partial f}{\partial x} \mathbf{a}_x + \frac{\partial f}{\partial y} \mathbf{a}_y + \frac{\partial f}{\partial z} \mathbf{a}_z \tag{1-35}$$

Thus $\mathbf{\nabla} f$ is a vector whose rectangular components are $\partial f/\partial x$, $\partial f/\partial y$, and $\partial f/\partial z$.

The term gradient is used because a geometrical interpretation of $\mathbf{\nabla} f$ shows that it is a vector perpendicular to the $f(x,y,z) = \text{constant}$ surface and has a magnitude equal to the directional derivative of $f(x,y,z)$ in a direction normal to the $f(x,y,z) = \text{constant}$ surface. As a matter of fact, an equally suitable definition of $\mathbf{\nabla} f$ is

$$\mathbf{\nabla} f = \frac{\partial f}{\partial n} \mathbf{n} \tag{1-36}$$

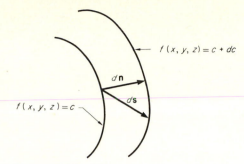

$f(x, y, z) = c + dc$

$d\mathbf{n}$

$d\mathbf{s}$

$f(x, y, z) = c$

FIGURE 1-11. The geometry associated with the discussion of the gradient.

where $\mathbf{n}$ is a unit vector in the normal direction defined above. This geometrical interpretation of $\boldsymbol{\nabla}f$ is important enough that we show a proof.

Consider the surface $f(x,y,z) = c$, as in Fig. 1-11, together with the surface $f(x,y,z) = c + dc$, where c is a constant and dc is an infinitesimal change in c. Define $d\mathbf{s}$ as an infinitesimal displacement vector from an arbitrary point on the surface $f(x,y,z) = c$ to an arbitrary point on $f(x,y,z) = c + dc$. Next, form the scalar product

$$\boldsymbol{\nabla}f \cdot d\mathbf{s} = |\boldsymbol{\nabla}f|\,|d\mathbf{s}|\cos\theta \tag{1-37}$$

where θ is the angle between $\boldsymbol{\nabla}f$ and $d\mathbf{s}$. Obviously, this quantity has its maximum value when $\theta = 0$, that is, when $\boldsymbol{\nabla}f$ is in the direction of $d\mathbf{s}$. Written explicitly in rectangular coordinates, Eq. (1-37) becomes

$$\boldsymbol{\nabla}f \cdot d\mathbf{s} = \left(\frac{\partial f}{\partial x}\,\mathbf{a}_x + \frac{\partial f}{\partial y}\,\mathbf{a}_y + \frac{\partial f}{\partial z}\,\mathbf{a}_z\right) \cdot (dx\,\mathbf{a}_x + dy\,\mathbf{a}_y + dz\,\mathbf{a}_z)$$

$$= \frac{\partial f}{\partial x}\,dx + \frac{\partial f}{\partial y}\,dy + \frac{\partial f}{\partial z}\,dz = df = \frac{\partial f}{\partial s}\,ds$$

where $\partial f/\partial s$ is the directional derivative of f in the direction of $d\mathbf{s}$. Now, in passing from the surface f to the surface $f + df$, the change in f is the same, namely, dc, whatever surface points are chosen. However, the distance ds will be least, and hence $\partial f/\partial s$ greatest, when $d\mathbf{s}$ has the direction of $d\mathbf{n}$. Therefore we have

$$\boldsymbol{\nabla}f \cdot d\mathbf{s}\Big|_{\text{max}} = \frac{\partial f}{\partial s}\Big|_{\text{max}}ds = \frac{\partial f}{\partial n}\,dn = \boldsymbol{\nabla}f \cdot d\mathbf{n} \tag{1-38}$$

where the last equality follows from the definition of the dot product. Hence the vector $\boldsymbol{\nabla}f$ coincides with the local surface normal, in the direction in which the derivative of f has its maximum value.

From Eq. (1-35) it will be seen also that each component of ∇f is equal to the maximum rate of increase of the function f in the local direction of increasing coordinate.

The divergence of a vector Given a vector field $\mathbf{A}$, we then define the *divergence* of $\mathbf{A}$ as the scalar field

$$\nabla \cdot \mathbf{A} = \frac{\partial A_x}{\partial x} + \frac{\partial A_y}{\partial y} + \frac{\partial A_z}{\partial z} \tag{1-39}$$

It can be shown that this definition is equivalent to

$$\nabla \cdot \mathbf{A} = \lim_{V \to 0} \frac{\oint_{\Sigma} \mathbf{A} \cdot \mathbf{n} \, da}{\int_{V} dv} \tag{1-40}$$

where V is the volume contained inside the closed surface Σ. Recalling that the closed-surface integral in Eq. (1-40) is the net outward flux of $\mathbf{A}$ through the surface Σ, this definition says that, at any point (given by $\lim V \to 0$), $\nabla \cdot \mathbf{A}$ is the net outward flux of $\mathbf{A}$ per unit volume. Therefore, if $\nabla \cdot \mathbf{A}$ is nonzero, the implication is that the vector quantity $\mathbf{A}$ originated (has a source) inside of the infinitesimal volume and that this source density is given by $\nabla \cdot \mathbf{A}$. This interpretation of $\nabla \cdot \mathbf{A}$ is frequently taken as its definition.

The curl of a vector Given a vector field $\mathbf{A}$, we then define the *curl* of $\mathbf{A}$ as the vector field

$$\nabla \times \mathbf{A} = \left(\frac{\partial A_z}{\partial y} - \frac{\partial A_y}{\partial z}\right) \mathbf{a}_x + \left(\frac{\partial A_x}{\partial z} - \frac{\partial A_z}{\partial x}\right) \mathbf{a}_y + \left(\frac{\partial A_y}{\partial x} - \frac{\partial A_x}{\partial y}\right) \mathbf{a}_z \tag{1-41}$$

It is advantageous to note that the right-hand side of this equation can be written as the expansion of the determinant

$$\nabla \times \mathbf{A} = \begin{vmatrix} \mathbf{a}_x & \mathbf{a}_y & \mathbf{a}_z \\ \dfrac{\partial}{\partial x} & \dfrac{\partial}{\partial y} & \dfrac{\partial}{\partial z} \\ A_x & A_y & A_z \end{vmatrix} \tag{1-42}$$

It can be shown that the defining equation (1-41) is such that the component of $\nabla \times \mathbf{A}$ in any arbitrary direction $\mathbf{n}$ is given by

$$(\nabla \times \mathbf{A}) \cdot \mathbf{n} = \lim_{S \to 0} \frac{\oint_{C} \mathbf{A} \cdot d\mathbf{l}}{\int_{S} da_n} \tag{1-43}$$

where da_n is an element of area perpendicular to $\mathbf{n}$.

Recall that the contour integral in (1-43) is called the circulation of **A** around the contour C. In hydrodynamic fluid flow, if **A** is the vector velocity field, $|\nabla \times \mathbf{A}|$ is the magnitude of the circulation of the fluid per unit area, and the direction of $\nabla \times \mathbf{A}$ is the axis of this circulation. For other physical vector fields it is rather difficult to visualize the physical meaning of $\nabla \times \mathbf{A}$. However, its interpretation as the vector whose direction is the axis of *circulation* and whose magnitude is the *circulation per unit area* perpendicular to the axis of circulation is still useful. In particular, this concept is frequently given as the definition of the curl.

The Laplacian of a scalar field Several second-order differential vector operations are defined. A particularly important one is called the Laplacian, defined as

$$\nabla^2 f = \nabla \cdot \nabla f = \frac{\partial^2 f}{\partial x^2} + \frac{\partial^2 f}{\partial y^2} + \frac{\partial^2 f}{\partial z^2} \qquad (1\text{-}44)$$

From its definition we see that the Laplacian of $f(x,y,z)$ is just the divergence of the gradient of f. From the interpretation of the divergence we see that $\nabla \cdot \nabla f$ gives the source density of ∇f.

1.11 Integral Vector Theorems. Several integral vector relations which are useful in electromagnetic theory are given in the following paragraphs. Extensive rigorous proofs of the theorems have been omitted deliberately. The proofs may be found in any of several vector-analysis† (or field-theory) texts.

Divergence, or Gauss', theorem Given a vector field **A**, a volume V, and the closed surface Σ enclosing V, Gauss' theorem states that

$$\int_V \nabla \cdot \mathbf{A} \, dv = \oint_\Sigma \mathbf{A} \cdot \mathbf{n} \, da \qquad (1\text{-}45)$$

This can be seen intuitively to follow from the definition of the $\nabla \cdot \mathbf{A}$ as the net outward flux per unit volume for infinitesimal volumes. As a matter of fact, a rigorous proof of the theorem can be built on this definition.

Stokes' theorem Given a vector field **A** and an open surface S whose periphery is the contour C, Stokes' theorem states

$$\int_S (\nabla \times \mathbf{A}) \cdot \mathbf{n} \, da = \oint_C \mathbf{A} \cdot d\mathbf{l} \qquad (1\text{-}46)$$

† H. B. Phillips, "Vector Analysis," John Wiley & Sons, Inc., New York, 1933. R. B. McQuistan, "Scalar and Vector Fields," John Wiley & Sons, Inc., New York, 1965.

This theorem can be seen to be an extension of the definition of the $\mathbf{\nabla} \times \mathbf{A}$ as a vector whose component in any direction $\mathbf{n}$ is given by the circulation of $\mathbf{A}$ around an infinitesimal area da normal to $\mathbf{n}$.

Green's theorem   Given two continuous scalar functions of position $f(x,y,z)$ and $g(x,y,z)$, each of which has continuous derivatives at least to the second order, then for any volume V enclosed by the closed surface Σ, Green's theorem states

$$\oint_{\Sigma} (f\,\mathbf{\nabla} g - g\,\mathbf{\nabla} f) \cdot \mathbf{n}\, da = \int_V (f\,\mathbf{\nabla}^2 g - g\,\mathbf{\nabla}^2 f)\, dv \qquad (1\text{-}47)$$

Green's theorem follows readily from the divergence theorem and the vector identity

$$\mathbf{\nabla} \cdot f\mathbf{A} = \mathbf{\nabla} f \cdot \mathbf{A} + f(\mathbf{\nabla} \cdot \mathbf{A})$$

Green's first identity   To be complete, we should note that the result obtained by applying the divergence theorem to the vectors

$$\mathbf{A} = f\,\mathbf{\nabla} g \qquad \mathbf{B} = g\,\mathbf{\nabla} f$$

is known as *Green's first identity:*

$$\oint_{\Sigma} f\,\mathbf{\nabla} g \cdot \mathbf{n}\, da = \int_V (\mathbf{\nabla} f \cdot \mathbf{\nabla} g + f\,\mathbf{\nabla}^2 g)\, dv \qquad (1\text{-}48)$$

Symmetry yields for $\mathbf{B}$

$$\oint_{\Sigma} g\,\mathbf{\nabla} f \cdot \mathbf{n}\, da = \int_V (\mathbf{\nabla} g \cdot \mathbf{\nabla} f + g\,\mathbf{\nabla}^2 f)\, dv$$

Other identities   If ϕ and $\mathbf{A}$ are continuous functions with at least piecewise continuous first derivatives within a volume V and on the surface Σ enclosing it, and if $\mathbf{n}$ denotes the outward unit normal to Σ, then

$$\int_V \mathbf{\nabla}\phi\, dv = \oint_{\Sigma} \phi\mathbf{n}\, da \qquad (1\text{-}49)$$

$$\int_V \mathbf{\nabla} \times \mathbf{A}\, dv = \oint_{\Sigma} \mathbf{n} \times \mathbf{A}\, da \qquad (1\text{-}50)$$

If ϕ and $\mathbf{A}$ are continuous functions with at least piecewise continuous first derivatives on a surface S and the contour C bounding S, and if $d\mathbf{l}$ is a differential vector tangent to C and related to the unit vector $\mathbf{n}$ normal to S according to the right-hand-screw rule, then

$$\int_S \mathbf{n} \times \mathbf{\nabla}\phi\, da = \int_C \phi\, d\mathbf{l} \qquad (1\text{-}51)$$

Helmholtz' theorem This theorem states in essence that every vector field can be considered to be the sum of two other vector fields, one of which has a zero divergence and the other a zero curl.

1.12 Useful Formulas

Rectangular coordinates

$$\text{grad } \phi = \nabla\phi = \frac{\partial\phi}{\partial x}\mathbf{a}_x + \frac{\partial\phi}{\partial y}\mathbf{a}_y + \frac{\partial\phi}{\partial z}\mathbf{a}_z \tag{1-52}$$

$$\text{div } \mathbf{A} = \nabla\cdot\mathbf{A} = \frac{\partial A_x}{\partial x} + \frac{\partial A_y}{\partial y} + \frac{\partial A_z}{\partial z} \tag{1-53}$$

$$\text{curl } \mathbf{A} = \nabla\times\mathbf{A} = \left(\frac{\partial A_z}{\partial y} - \frac{\partial A_y}{\partial z}\right)\mathbf{a}_x + \left(\frac{\partial A_x}{\partial z} - \frac{\partial A_z}{\partial x}\right)\mathbf{a}_y + \left(\frac{\partial A_y}{\partial x} - \frac{\partial A_x}{\partial y}\right)\mathbf{a}_z \tag{1-54}$$

$$(\text{Laplacian of } \phi) = \nabla\cdot\nabla\phi = \nabla^2\phi = \frac{\partial^2\phi}{\partial x^2} + \frac{\partial^2\phi}{\partial y^2} + \frac{\partial^2\phi}{\partial z^2} \tag{1-55}$$

$$(\text{Vector Laplacian of } \mathbf{A}) = \nabla^2\mathbf{A} = \nabla^2 A_x\mathbf{a}_x + \nabla^2 A_y\mathbf{a}_y + \nabla^2 A_z\mathbf{a}_z \tag{1-56}$$

Cylindrical coordinates

$$\nabla\phi = \frac{\partial\phi}{\partial r}\mathbf{a}_r + \frac{1}{r}\frac{\partial\phi}{\partial\varphi}a_\varphi + \frac{\partial\phi}{\partial z}\mathbf{a}_z \tag{1-57}$$

$$\nabla\cdot\mathbf{A} = \frac{1}{r}\frac{\partial}{\partial r}(rA_r) + \frac{1}{r}\frac{\partial A_\varphi}{\partial\varphi} + \frac{\partial A_z}{\partial z} \tag{1-58}$$

$$\nabla\times\mathbf{A} = \left(\frac{1}{r}\frac{\partial A_z}{\partial\varphi} - \frac{\partial A_\varphi}{\partial z}\right)\mathbf{a}_r + \left(\frac{\partial A_r}{\partial z} - \frac{\partial A_z}{\partial r}\right)\mathbf{a}_\varphi + \left[\frac{1}{r}\frac{\partial(rA_\varphi)}{\partial r} - \frac{1}{r}\frac{\partial A_r}{\partial\varphi}\right]\mathbf{a}_z \tag{1-59}$$

$$\nabla^2\phi = \frac{1}{r}\frac{\partial}{\partial r}\left(r\frac{\partial\phi}{\partial r}\right) + \frac{1}{r^2}\frac{\partial^2\phi}{\partial\varphi^2} + \frac{\partial^2\phi}{\partial z^2} \tag{1-60}$$

$$\nabla^2\mathbf{A} = \nabla(\nabla\cdot\mathbf{A}) - \nabla\times\nabla\times\mathbf{A} \tag{1-61}$$

Note that $\nabla^2\mathbf{A} \neq \nabla^2 A_r\mathbf{a}_r + \nabla^2 A_\varphi\mathbf{a}_\varphi + \nabla^2 A_z\mathbf{a}_z$ since $\nabla^2\mathbf{a}_r A_r \neq \mathbf{a}_r\nabla^2 A_r$, etc., because the orientation of the unit vectors $\mathbf{a}_r$, $\mathbf{a}_\varphi$ varies with the coordinates

r, φ. In fact,

$$\nabla^2 \mathbf{A} = \left(\frac{\partial^2 A_r}{\partial r^2} + \frac{1}{r} \frac{\partial A_r}{\partial r} - \frac{A_r}{r^2} + \frac{1}{r^2} \frac{\partial^2 A_r}{\partial \varphi^2} - \frac{2}{r^2} \frac{\partial A\varphi}{\partial \varphi} + \frac{\partial^2 A_r}{\partial z^2} \right) \mathbf{a}_r$$

$$+ \left(\frac{\partial^2 A_\varphi}{\partial r^2} + \frac{1}{r} \frac{\partial A_\varphi}{\partial r} - \frac{A_\varphi}{r^2} + \frac{1}{r^2} \frac{\partial^2 A_\varphi}{\partial \varphi^2} + \frac{2}{r^2} \frac{\partial A_r}{\partial \varphi} + \frac{\partial^2 A_\varphi}{\partial z^2} \right) \mathbf{a}_\varphi$$

$$+ \left(\frac{\partial^2 A_z}{\partial r^2} + \frac{1}{r} \frac{\partial A_z}{\partial r} + \frac{1}{r^2} \frac{\partial^2 A_z}{\partial \varphi^2} + \frac{\partial^2 A_z}{\partial z^2} \right) \mathbf{a}_z \qquad (1\text{-}62)$$

Spherical coordinates

$$\nabla \phi = \frac{\partial \phi}{\partial r} \mathbf{a}_r + \frac{1}{r} \frac{\partial \phi}{\partial \theta} \mathbf{a}_\theta + \frac{1}{r \sin \theta} \frac{\partial \phi}{\partial \varphi} \mathbf{a}_\varphi \qquad (1\text{-}63)$$

$$\nabla \cdot \mathbf{A} = \frac{1}{r^2} \frac{\partial}{\partial r} (r^2 A_r) + \frac{1}{r \sin \theta} \frac{\partial}{\partial \theta} (\sin \theta \, A_\theta) + \frac{1}{r \sin \theta} \frac{\partial A_\varphi}{\partial \varphi} \qquad (1\text{-}64)$$

$$\nabla \times \mathbf{A} = \frac{\mathbf{a}_r}{r \sin \theta} \left[\frac{\partial}{\partial \theta} (A_\varphi \sin \theta) - \frac{\partial A_\theta}{\partial \varphi} \right] + \frac{\mathbf{a}_\theta}{r} \left[\frac{1}{\sin \theta} \frac{\partial A_r}{\partial \varphi} - \frac{\partial}{\partial r} (r A_\varphi) \right]$$

$$+ \frac{\mathbf{a}_\varphi}{r} \left[\frac{\partial}{\partial r} (r A_\theta) - \frac{\partial A_r}{\partial \theta} \right] \qquad (1\text{-}65)$$

$$\nabla^2 \phi = \frac{1}{r^2} \frac{\partial}{\partial r} \left(r^2 \frac{\partial \phi}{\partial r} \right) + \frac{1}{r^2 \sin \theta} \frac{\partial}{\partial \theta} \left(\sin \theta \frac{\partial \phi}{\partial \theta} \right) + \frac{1}{r^2 \sin^2 \theta} \frac{\partial^2 \phi}{\partial \varphi^2} \qquad (1\text{-}66)$$

$$\nabla^2 \mathbf{A} = \left(\frac{\partial^2 A_r}{\partial r^2} + \frac{2}{r} \frac{\partial A_r}{\partial r} - \frac{2}{r^2} A_r + \frac{1}{r^2} \frac{\partial^2 A_r}{\partial \theta^2} \right.$$

$$+ \frac{\cot \theta}{r^2} \frac{\partial A_r}{\partial \theta} + \frac{1}{r^2 \sin^2 \theta} \frac{\partial^2 A_r}{\partial \phi^2} - \frac{2}{r^2} \frac{\partial A_\theta}{\partial \theta}$$

$$\left. - \frac{2 \cot \theta}{r^2} A_\theta - \frac{2}{r^2 \sin \theta} \frac{\partial A_\varphi}{\partial \varphi} \right) \mathbf{a}_r$$

$$+ \left(\frac{\partial^2 A_\theta}{\partial r^2} + \frac{2}{r} \frac{\partial A_\theta}{\partial r} - \frac{A_\theta}{r^2 \sin^2 \theta} + \frac{1}{r^2} \frac{\partial^2 A_\theta}{\partial \theta^2} \right.$$

$$+ \frac{\cot \theta}{r^2} \frac{\partial A_\theta}{\partial \theta} + \frac{1}{r^2 \sin^2 \theta} \frac{\partial^2 A_\theta}{\partial \varphi^2} + \frac{2}{r^2} \frac{\partial A_r}{\partial \theta}$$

$$\left. - \frac{2 \cot \theta}{r^2 \sin \theta} \frac{\partial A_\varphi}{\partial \varphi} \right) \mathbf{a}_\theta$$

$$+ \left(\frac{\partial^2 A_\varphi}{\partial r^2} + \frac{2}{r} \frac{\partial A_\varphi}{\partial r} - \frac{1}{r^2 \sin^2 \theta} A_\varphi + \frac{1}{r^2} \frac{\partial^2 A_\varphi}{\partial \theta^2} \right.$$

$$+ \frac{\cot \theta}{r^2} \frac{\partial A_\varphi}{\partial \theta} + \frac{1}{r^2 \sin^2 \theta} \frac{\partial^2 A_\varphi}{\partial \varphi^2} + \frac{2}{r^2 \sin \theta} \frac{\partial A_r}{\partial \varphi}$$

$$\left. + \frac{2 \cot \theta}{r^2 \sin \theta} \frac{\partial A_\theta}{\partial \varphi} \right) \mathbf{a}_\varphi \qquad (1\text{-}67)$$

Vector identities If $\mathbf{r} = x\mathbf{a}_x + y\mathbf{a}_y + z\mathbf{a}_z$ is the radius vector drawn from the origin to the point (x,y,z), then

$$\nabla \cdot \mathbf{r} = 3 \qquad \nabla \times \mathbf{r} = 0 \tag{1-68}$$

$$\nabla \cdot (\nabla \times \mathbf{A}) = 0 \tag{1-69}$$

$$\nabla \times \nabla\phi = 0 \tag{1-70}$$

$$\mathbf{A} \cdot \mathbf{B} \times \mathbf{C} = \mathbf{B} \cdot \mathbf{C} \times \mathbf{A} = \mathbf{C} \cdot \mathbf{A} \times \mathbf{B} \tag{1-71}$$

$$\mathbf{A} \times (\mathbf{B} \times \mathbf{C}) = (\mathbf{A} \cdot \mathbf{C})\mathbf{B} - (\mathbf{A} \cdot \mathbf{B})\mathbf{C} \tag{1-72}$$

$$(\mathbf{A} \times \mathbf{B}) \cdot (\mathbf{C} \times \mathbf{D}) = \mathbf{A} \cdot \mathbf{B} \times (\mathbf{C} \times \mathbf{D})$$
$$= \mathbf{A} \cdot (\mathbf{B} \cdot \mathbf{D}\mathbf{C} - \mathbf{B} \cdot \mathbf{C}\mathbf{D})$$
$$= (\mathbf{A} \cdot \mathbf{C})(\mathbf{B} \cdot \mathbf{D}) - (\mathbf{A} \cdot \mathbf{D})(\mathbf{B} \cdot \mathbf{C}) \tag{1-73}$$

$$(\mathbf{A} \times \mathbf{B}) \times (\mathbf{C} \times \mathbf{D}) = (\mathbf{A} \times \mathbf{B} \cdot \mathbf{D})\mathbf{C} - (\mathbf{A} \times \mathbf{B} \cdot \mathbf{C})\mathbf{D} \tag{1-74}$$

$$\nabla(\phi + \psi) = \nabla\phi + \nabla\psi \tag{1-75}$$

$$\nabla(\phi\psi) = \phi\nabla\psi + \psi\nabla\phi \tag{1-76}$$

$$\nabla \cdot (\mathbf{A} + \mathbf{B}) = \nabla \cdot \mathbf{A} + \nabla \cdot \mathbf{B} \tag{1-77}$$

$$\nabla \times (\mathbf{A} + \mathbf{B}) = \nabla \times \mathbf{A} + \nabla \times \mathbf{B} \tag{1-78}$$

$$\nabla \cdot \phi\mathbf{A} = \mathbf{A} \cdot \nabla\phi + \phi\nabla \cdot \mathbf{A} \tag{1-79}$$

$$\nabla \times \phi\mathbf{A} = \nabla\phi \times \mathbf{A} + \phi\nabla \times \mathbf{A} \tag{1-80}$$

$$\nabla(\mathbf{A} \cdot \mathbf{B}) = (\mathbf{A} \cdot \nabla)\mathbf{B} + (\mathbf{B} \cdot \nabla)\mathbf{A} + \mathbf{A} \times (\nabla \times \mathbf{B}) + \mathbf{B} \times (\nabla \times \mathbf{A}) \tag{1-81}$$

$$\nabla \cdot (\mathbf{A} \times \mathbf{B}) = \mathbf{B} \cdot \nabla \times \mathbf{A} - \mathbf{A} \cdot \nabla \times \mathbf{B} \tag{1-82}$$

$$\nabla \times (\mathbf{A} \times \mathbf{B}) = \mathbf{A}\nabla \cdot \mathbf{B} - \mathbf{B}\nabla \cdot \mathbf{A} + (\mathbf{B} \cdot \nabla)\mathbf{A} - (\mathbf{A} \cdot \nabla)\mathbf{B} \tag{1-83}$$

$$\nabla \times \nabla \times \mathbf{A} = \nabla(\nabla \cdot \mathbf{A}) - \nabla^2\mathbf{A} \tag{1-84}$$

1.13 Summary. This chapter has stated, defined, and discussed the various aspects of vector analysis which are important and useful in the study of electromagnetic theory. The objective has been mostly to review, summarize, and explain, rather than to give a rigorous treatment of the subject.

Problems

1-1 Vector Addition, Dot and Cross Products. Given two vector fields $\mathbf{A}$ and $\mathbf{B}$, where

$$\mathbf{A} = x\mathbf{a}_x + y^2\mathbf{a}_y + 3t\mathbf{a}_z$$
$$\mathbf{B} = y\mathbf{a}_x + y^2\mathbf{a}_y + zt\mathbf{a}_z$$

(a) Determine $\mathbf{A}$ and $\mathbf{B}$ at $x = 1$, $y = 2$, $z = 3$, $t = 4$.
(b) Determine $\mathbf{A} + \mathbf{B}$ at $x = 1$, $y = 2$, $z = 3$, $t = 0$.
(c) Determine $\mathbf{A} \cdot \mathbf{B}$.

(d)   Determine $\mathbf{A} \cdot \mathbf{B}$ at $x = 1$, $y = 2$, $z = 3$, $t = 1$.

(e)   Determine the equation for $\mathbf{A} \cdot \mathbf{B}$ for all points in the $z = 3$ plane.

(f)   Using results of part (e), rework part (d) above.

(g)   Determine $\mathbf{A} \times \mathbf{B}$.

(h)   Determine $\mathbf{A} \times \mathbf{B}$ at $x = 1$, $y = 2$, $z = 3$, $t = 1$.

(i)   Determine $\mathbf{A} \times \mathbf{B}$ for all points in the $y = 2$ plane.

1-2 **Vector Addition, Dot and Cross Products.**   Given

$$\mathbf{A} = 3\mathbf{a}_x + x\mathbf{a}_y + y\mathbf{a}_z$$

$$\mathbf{B} = x^2\mathbf{a}_x + 4\mathbf{a}_y$$

$$\mathbf{C} = y\mathbf{a}_x + x\mathbf{a}_y + 4\mathbf{a}_z$$

(a)   Determine $\mathbf{A} \cdot (\mathbf{B} + \mathbf{C})$ at the point $x = 2$, $y = 3$.

(b)   Determine $\mathbf{A} \cdot (\mathbf{B} \times \mathbf{C})$, $\mathbf{C} \cdot (\mathbf{A} \times \mathbf{B})$, $\mathbf{B} \cdot (\mathbf{C} \times \mathbf{A})$, $\mathbf{B} \cdot (\mathbf{A} \times \mathbf{C})$, $(\mathbf{B} \times \mathbf{A}) \cdot \mathbf{C}$, $\mathbf{A} \cdot (\mathbf{C} \times \mathbf{B})$.

Can you interpret these products geometrically?

(c)   Determine $\mathbf{A} \times (\mathbf{B} + \mathbf{C})$.

1-3 **Vector Addition, Dot and Cross Products.**   Using Eqs. (1-1) and (1-4), find $\mathbf{A} \times \mathbf{B}$ if:

(a) $\mathbf{A} \cdot \mathbf{B} = 0$

(b) $\mathbf{A} \cdot \mathbf{B} = AB$

1-4 **Transformation of Coordinates.**

(a)   Express the vectors $\mathbf{A}$ and $\mathbf{B}$ of Prob. 1-1 in cylindrical coordinates.

(b)   Evaluate $\mathbf{A}$ and $\mathbf{B}$ at the point $r = \sqrt{5}$, $\varphi = \tan^{-1} 2$, $z = 3$, at the time $t = 4$. Compare your results with Prob. 1-1a.

1-5 **Dot Product.**   Using the results of Prob. 1-4:

(a)   Express $\mathbf{A} \cdot \mathbf{B}$ in cylindrical coordinates.

(b)   Evaluate this result at the point given in Prob. 1-1d.

1-6 **Cross Product.**   Using the results of Prob. 1-4:

(a)   Express $\mathbf{A} \times \mathbf{B}$ in cylindrical coordinates.

(b)   Evaluate $\mathbf{A} \times \mathbf{B}$ at the point given in 1-1h.

1-7 **Transformation of Coordinates.**   Given

$$\mathbf{A} = \cos \varphi \mathbf{a}_r + \sin \varphi \mathbf{a}_\varphi + r\mathbf{a}_z$$

$$\mathbf{B} = r\mathbf{a}_r + \varphi \mathbf{a}_\varphi + 2\mathbf{a}_z$$

with φ expressed in radians.   The origin of the cylindrical coordinates coincides with the xyz-coordinate origin, and the x axis coincides with the $\varphi = 0$ axis.   Determine $\mathbf{A} \cdot \mathbf{B}$ at the point $x = 2$, $y = 3$.

1-8 **Transformation of Coordinates.**   Express the vector

$$\mathbf{A} = z \cos \varphi \mathbf{a}_r + r^2 \sin \varphi \mathbf{a}_\varphi + 16r\mathbf{a}_z$$

in rectangular coordinates.

1-9 **Transformation of Coordinates.**   A vector lies in the xy plane, and is given by the following equation.

$$\mathbf{B} = x\mathbf{a}_x + y\mathbf{a}_y$$

(a) What is the expression for **B** in cylindrical coordinates?

(b) What are the magnitude and direction of **B** at the point $x = 3$, $y = 4$?

1-10 Transformation of Coordinates. Derive the vector component transformations from spherical to rectangular coordinates, and vice versa.

1-11 Vectors and Analytic Geometry. Two vectors A and D are given by

$$A = 3a_x + 4a_y + 5a_z$$
$$D = a_x + 2a_y + 6a_z$$

Determine the angle between the two vectors.

1-12 Vectors and Analytic Geometry. A vector **B** is given by $B = a_x + 2a_y + 3a_z$. A vector **A** has a magnitude of $\sqrt{3}$ and an x component of 1. Determine **A** (express it in the form $aa_x + ba_y + ca_z$) so that **A** and **B** are at right angles to each other.

1-13 Vectors and Analytic Geometry. A vector A is given by $A = ra_r + ra_\varphi$.

(a) Describe, mathematically and in words, the locus of points in the xy (or $r\varphi$) plane on which the magnitude of **A** is constant.

(b) Find the points in the xy plane where **A** makes an angle of 45° with the x axis and has a magnitude $A = \sqrt{2}$.

1-14 Vectors and Analytic Geometry. A vector field is defined by $A = axa_x + bya_y$, where a and b are constants.

(a) What is the configuration of the lines of constant magnitude of **A** in a constant-z plane?

(b) What is the configuration of the lines of constant magnitude for **A** if a and b are equal?

1-15 Vectors and Analytic Geometry. Consider the vectors

$$A = 5a_x + 2a_y + 3a_z$$
$$B = B_xa_x + 2a_y + B_za_z$$
$$C = 3a_x + C_ya_y + a_z$$

Specify B_x, B_z, and C_y so that **A**, **B**, **C** are mutually orthogonal.

1-16 Line Integral. Compute the value of the line integral $\int_C [(\cos \varphi)/r] \, dy$, where C is the straight line from $(a,0)$ to (a,a).

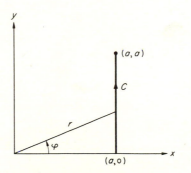

PROBLEM 1-16

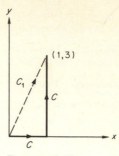

PROBLEM 1-17

1-17 Line Integrals. Evaluate the line integral $\int_C [(x^2 + 2y)\,dx + (x - 5y^2)\,dy]$
(*a*) Along the path C
(*b*) Along the straight line joining (0,0) to (1,3), that is, along C_1

1-18 Line Integral. For the vector $\mathbf{A} = 2x^2y\mathbf{a}_x + (y + z)\mathbf{a}_z$ and the path C_1 in Prob.
1-17, evaluate $\int_{C_1} \mathbf{A} \cdot d\mathbf{l}$.

1-19 Line Integral. Evaluate the line integral of Prob. 1-18 along that portion of the
parabola $y = 4 - x^2$ which lies in the first quadrant. The path traversal is to initiate at
the point (0,4).

1-20 Line Integral. For the path shown, compute

$$\int_C (\sin \varphi \mathbf{a}_r + r \cos \varphi \mathbf{a}_\varphi + \tan \varphi \mathbf{a}_z) \cdot d\mathbf{l}$$

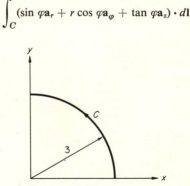

PROBLEM 1-20

1-21 Line Integrals. Given the vector point function

$$\mathbf{A} = 3(x^2 + x)\mathbf{a}_x + 2yx\mathbf{a}_y + 4z\mathbf{a}_z$$

and the path C consisting of three straight-line segments,

First segment: from (0,0,0) to (1,0,0)
Second segment: from (1,0,0) to (1,2,1)
Third segment: from (1,2,1) to (0,0,1)

evaluate $\int_C \mathbf{A} \cdot d\mathbf{l}$.

1-22 Line Integral. Find the line integral of the vector $\mathbf{A} = y\mathbf{a}_x - x\mathbf{a}_y$ around the closed path in the xy plane that follows the parabola $y = x^2$ from the point $x = -1$, $y = 1$ to the point $x = 2$, $y = 4$ and returns along the straight line $y = x + 2$.

1-23 Contour Integration Using Two Coordinate Systems. Given the vector field of Example 1-2, which was $\mathbf{F} = xy\mathbf{a}_x + y^2\mathbf{a}_y$, calculate $\oint_C \mathbf{F} \cdot d\mathbf{l}$, where C is the closed contour in the xy plane which begins at the point $(\sqrt{3},0,0)$ and goes along the line $y = 0$ to the point $(2,0,0)$, and then along the circular arc $x^2 + y^2 = 4$ to the point $(\sqrt{3},1,0)$, and then along the line $x = \sqrt{3}$ to the point of beginning. (*Hint:* See Example 1-2 and note that $F_\varphi = 0$.)

1-24 Surface Integral. A vector field is given by $\mathbf{B} = \mathbf{a}_x + 2\mathbf{a}_y + 3\mathbf{a}_z$. Evaluate $\int_S \mathbf{B} \cdot \mathbf{n} \, da$ over a plane rectangular area bounded by the lines joining the points

$$(0,0,0) \qquad (2,0,0) \qquad (0,2,1) \qquad (2,2,1)$$

1-25 Surface Integral. For a vector field $\mathbf{A} = xy^2\mathbf{a}_x$ evaluate $\oint_\Sigma \mathbf{A} \cdot \mathbf{n} \, da$ over the surface of a box with corners at

$$(0,0,0) \qquad (2,0,0) \qquad (2,2,0) \qquad (0,2,0)$$
$$(0,0,2) \qquad (2,0,2) \qquad (2,2,2) \qquad (0,2,2)$$

1-26 Surface Integral. Determine the total flux of the vector $\mathbf{B} = 3\mathbf{a}_x + 4\mathbf{a}_y + 5\mathbf{a}_z$ passing through the face *abcd* of the prismatic box shown.

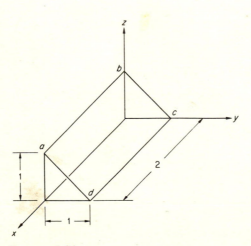

PROBLEM 1-26

1-27 Surface Integral. A vector A is given by $\mathbf{A} = x^2\mathbf{a}_x + yz\mathbf{a}_y + xy\mathbf{a}_z$. Calculate the flux of this vector through the surface in the $z = 4$ plane bounded by the lines $x = 0$, $x = 3$, $y = 1$, $y = 2$.

1-28 Surface Integral. The equation for a field is $\mathbf{B} = x\mathbf{a}_x + y\mathbf{a}_y + z\mathbf{a}_z$. Evaluate $\int_S \mathbf{B} \cdot \mathbf{n}\, da$ over a circular area, of radius 2, which is centered on the z axis and is parallel to the xy plane at $z = 4$.

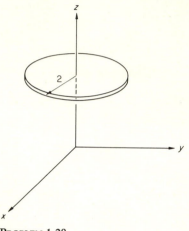

PROBLEM 1-28

1-29 Surface Integral. The equation for a field is $\mathbf{B} = \mathbf{a}_x + xy\mathbf{a}_y$. Evaluate $\int_S \mathbf{B} \cdot \mathbf{n}\, da$ over the surface above the xy plane defined by $z = 1 - (x^2 + y^2)$.

1-30 Surface Integral. If $\mathbf{B} = 2\mathbf{a}_x + 4\mathbf{a}_y + 5\mathbf{a}_z$ describes a vector field in the space occupied by the box shown, determine the net outward flux passing through the side *abcd* plus that through the side *fghk*. The top and bottom of the box are parallel to the xy plane, and the front and the back are parallel to the yz plane.

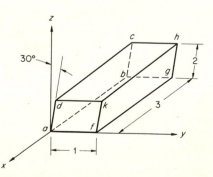

PROBLEM 1-30

1-31 Surface Integral. Evaluate the surface integral $\int_S (x\mathbf{a}_x + y\mathbf{a}_y + z\mathbf{a}_z) \cdot \mathbf{n} \, da$ over the first octant of a spherical surface, radius 1, centered at the origin.

1-32 Surface Integral. Find the surface area above the xy plane for the surface defined by $z = 1 - (x^2 + y^2)$.

1-33 Surface Integral. Given a vector $\mathbf{B} = 10\mathbf{a}_r + 3r\mathbf{a}_\varphi - 2zr\mathbf{a}_z$, find the total flux of this vector emanating from a cylindrical volume enclosed by a cylinder of radius 2, height 4, whose axis is the z axis and whose base lies in the $z = 1$ plane.

1-34 Surface Integral. A vector is given by

$$\mathbf{B} = z\cos\varphi\mathbf{a}_r + r\ln z\mathbf{a}_\varphi + z\mathbf{a}_z$$

Find the net outward flux of $\mathbf{B}$ emanating from a cylinder whose axis is the z axis, whose radius is 1, whose base is at $z = -2$, and whose top is at $z = 2$.

1-35 Evaluation. Given

$$\mathbf{A} \cdot \mathbf{B} = 0 \qquad A_z = B_z = 0 \qquad A = x^2 + y^2 \qquad B = 2A$$

evaluate $\mathbf{A} \times \mathbf{B}$ at $x = 3$, $y = 5$, $z = 4$.

1-36 Evaluation. Show that $\oint_C (x\mathbf{a}_x + x\mathbf{a}_y - y\mathbf{a}_z) \cdot d\mathbf{l} = \int_S da$ if C is a circle of diameter 1 in the xy plane centered at the origin and S is the circular area bounded by C. (*Hint:* Calculate formally and compare with use of Stokes' theorem.)

1-37 Stokes' Theorem. Use Stokes' theorem to transform the contour integral of Prob. 1-17 for C and $-C_1$ into a surface integral, and evaluate the resulting surface integral directly.

1-38 Divergence Theorem. Use the divergence theorem to transform the closed surface integrals of the following problems into volume integrals and evaluate the volume integrals directly.

    (*a*) Prob. 1-25
    (*b*) Prob. 1-34

1-39 Divergence and Curl. Calculate the divergence and curl of the vectors given in Prob. 1-2.

1-40 Gradient. Given a scalar function $f(x,y,z) = x^2 + y^2 + z^2$, find ∇f in (*a*) rectangular coordinates; (*b*) cylindrical coordinates; (*c*) spherical coordinates.

1-41 Gradient. Given the scalar function $f(r,\varphi,z) = r\cos\varphi + r\sin\varphi + z^2$, find ∇f in (*a*) rectangular coordinates; (*b*) cylindrical coordinates; (*c*) spherical coordinates.

1-42 Useful Vector Identities. There are several useful second-order vector operations. All of them are included in Sec. 1.12. Verify, in rectangular coordinates, the two identities given by Eqs. (1-69) and (1-70).

1-43 The Vector Laplacian. A very useful vector quantity, called the *Laplacian of a vector* [Eq. (1-56)] is, in general, defined by the identity

$$\nabla^2\mathbf{A} = \nabla(\nabla \cdot \mathbf{A}) - \nabla \times \nabla \times \mathbf{A}$$

Show that in cylindrical coordinates the z component of $\nabla^2\mathbf{A}$ is given by $\nabla^2 A_z$ (the r and φ components do *not* have this simple form.)

1-44   Surface Normal.   Let $f(x,y,z) = $ constant denote the equation of a surface. Show that the components of the normal **n** at an arbitrary point on the surface are given by

$$n_x = \frac{1}{\Delta}\frac{\partial f}{\partial x} \qquad n_y = \frac{1}{\Delta}\frac{\partial f}{\partial y} \qquad n_z = \frac{1}{\Delta}\frac{\partial f}{\partial z}$$

where

$$\Delta = \left[\left\{\frac{\partial f}{\partial x}\right\}^2 + \left\{\frac{\partial f}{\partial y}\right\}^2 + \left\{\frac{\partial f}{\partial z}\right\}^2 \right]^{\frac{1}{2}}$$

(*Hint:* Use the definition of gradient.)

CHAPTER 2

GENERAL PRINCIPLES

2.1 Introduction. Our interest in electromagnetic fields stems from a desire to understand the behavior of practical devices and systems, to be able to describe such devices and systems mathematically, and to predict their performance. We also desire to understand propagation of energy in space and the means of launching and receiving this energy. Having set this as our ultimate goal, we must make a diligent effort to obtain this understanding from both the theoretical and the practical points of view. The theoretical study is essentially a study of Maxwell's equations. The practical study consists in applying these equations to practical problems. Maxwell's equations, the terminology associated with these equations, and examples illustrating their application are presented in this chapter. Most of the ideas presented in this chapter will be abstract, and the explanations of the theory will be quite general. The special cases considered in later chapters will help clarify the contents of this chapter.

2.2 The Field Concept. We defined a *field* in some detail in Sec. 1.2. We summarize the definition by saying that a field is the spatial distribution of a quantity as a function of time. For a *vector field* both its *magnitude and direction* must be known as a function of the space coordinates and time before it is fully defined. A force field is an example of a vector field. In contrast, a *scalar field* can be completely defined by specifying its magnitude at each point in space and at each instant of time. As an example, the density of the earth's atmosphere is a scalar field.

A field is *static*, or *stationary*, if it is independent of time; a time-varying field is often called *dynamic*. No physical quantity remains constant indefinitely, but over finite periods of time (or when the time variations are small) it is often convenient to consider them static quantities. When time variations are large but slow, we use the term *quasistatic*.

In general, physical fields are three-dimensional fields; that is, they depend on three space variables. The pressure of the earth's atmosphere is a three-dimensional field. Ideally speaking, there are also two-dimensional, and even one-dimensional, fields, examples of which are the density of paint on the surface of a wall (two-dimensional) and the tension at each point along a stretched piece of thin wire (one-dimensional).

Fields generally result from some "cause." That is to say, for most fields of interest a definite cause-and-effect relationship exists between a "resultant field" and its "source field." Often the decision as to which is the source field and which is the resultant field is largely a matter of the point of view. Sometimes the choice is "obvious." For instance, "obviously," the source of temperature field of the earth's atmosphere is the sun's radiation field (or is it the atmospheric-pressure field which moves the air masses which causes the temperature field to be what it is?). When we are able to establish definite physical laws which give the relationship between the source fields and the resultant fields we have what is known as a *field theory*.

In summary, by a field we mean the set of values assumed by a physical quantity at various points in a region of space and at various instants of time. The region of space and the time interval in question may be either finite or infinite in extent. A field theory comprises the physical laws which give the relations between source fields and resultant fields.

2.3 Charge: The Source of Electromagnetic Fields. Charge is a *fundamental quantity*, just as mass, length, and time are fundamental quantities in mechanics. Charge cannot be conveniently defined in terms of these other three quantities. It manifests itself only as the cause of effects whose origin is beyond the range of mechanics. Thus charges at rest and/or in motion exert forces on other charges at rest and/or in motion. These new kinds of forces are called *electromagnetic forces* and the new fields are called *electromagnetic fields*.

Experimental evidence indicates the existence of two kinds of charges—positive and negative. Quantitatively, the smallest known amount of charge is possessed by an electron. In the rationalized mks system of units,† adopted in this text, the charge of an electron is equal to -1.6×10^{-19} coulomb (C). The negative sign is chosen arbitrarily, and is the result of definition.

An allied concept is that of *charge density.* To introduce it, we consider a volume Δv containing a net (positive or negative) charge Δq, and let Δv shrink to a limiting volume δ about a point P in Δv. Then

$$\rho = \lim_{\Delta v \to \delta} \frac{\Delta q}{\Delta v} \qquad C/m^3 \qquad (2\text{-}1)$$

is defined to be the charge density at P. The dimensions of δ are assumed to

† For a discussion on units and dimensions, see Sec. 2.14.

be large compared with atomic dimensions.   Thus ρ is a scalar point function whose integral over a specified volume determines the net charge Q contained within the volume; that is,

$$Q = \int_V \rho \, dv \qquad (2\text{-}2)$$

The process described by Eq. (2-1) implies a subtle mathematical idealization.   We know that charges are discrete entities, which in the real world are separated by distances large compared with the physical dimensions of a charge.   Yet the limiting ratio, Eq. (2-1), defines a smooth and continuous distribution of charge having a finite density everywhere in space.   Charge is thus considered to exist at *every* point in a coarse-grid space, contrary to physical reality.   However, from the so-called *macroscopic* point of view, which is the engineering point of view adopted in this text, this simplified *model* of the physical world fortunately predicts very accurate results.

Two related concepts are those of surface charge density and line charge density.   When charge is distributed on a surface, it gives rise to a surface distribution of so many coulombs per square meter.   Likewise, when charge is distributed along a cylinder very small in diameter, it constitutes a linear distribution of so many coulombs per meter.   Actually, both of these concepts are limiting cases of the general concept of volume charge density, because, to arrive at them, we need only allow the volume to shrink first to a thin sheet and then to a line.   They will both be used often.

Charges in motion constitute a *current*.   Quantitatively speaking, the net (positive or negative) charge which crosses a surface per unit time constitutes the current flowing through that surface:

$$I = \lim_{\Delta t \to 0} \frac{\Delta q}{\Delta t} \qquad \text{A} \qquad (2\text{-}3)$$

Here the symbol I denotes current, and Δq the net amount of charge crossing the given surface in a time interval Δt.   Note that the unit of current is the ampere (A), which is equal to 1 coulomb per second (C/s).†

Current is a scalar quantity.   In contrast, *current density* is a vector quantity which is defined such that the integral of its normal component over a surface gives the current flowing through that surface:

$$I = \int_S \mathbf{J} \cdot \mathbf{n} \, da \qquad (2\text{-}4)$$

---

† The standard sign convention for current will be used: a net positive charge moving in a specified direction constitutes a positive current flowing in that direction.

where da = a differential element of surface S

$\mathbf{n}$ = a unit vector normal to da

I = net current through S

$\mathbf{J}$ = current density, in amperes per meter squared (A/m^2)

The vector function $\mathbf{J}$ specifies at each point not only the intensity of the flow of charge, but also its direction.   The magnitude of $\mathbf{J}$ is equal to the charge which crosses a unit surface area perpendicular to the flow of charge per unit of time.   The direction of $\mathbf{J}$ coincides with the direction of motion of positive charge, consistent with the convention adopted for current.

To summarize, charge distributions are described by charge density functions.   Charge motion is defined as current.   Current distribution is described by a vector current density function such that $\mathbf{J} \cdot \mathbf{n}$ is the net current per unit area flowing in the direction of the normal.

2.4   The Electromagnetic Field Vectors and the Field Equations.   The description of a field, be it a scalar field or a vector field, is conceptually a fairly simple matter.   Conceptually, the description involves measurement of the field quantity, everywhere followed by a reduction of these data to field plots or to a mathematical description of the data.   In contrast, a *field theory* involves the complete mathematical specification of the interdependence of a set of field quantities with each other, with the source fields, and with all other related quantities.   Electromagnetic field theory requires the specification of the interdependence of four vector fields and two source fields (charge density and current density); it also requires the specification of the interaction of these fields with the mechanical world.   The symbols commonly used to represent the so-called *field vectors*, together with the names by which they are known, are

$\mathbf{E}$ = electric field intensity

$\mathbf{B}$ = magnetic flux density

$\mathbf{D}$ = electric flux density (or displacement vector)

$\mathbf{H}$ = magnetic field intensity

Eventually, we shall define the field vectors in terms of physical concepts.   At this point we find it convenient to use them as a means of defining an electromagnetic field.   In abstract language, *an electromagnetic field is defined as the domain of the vectors* $\mathbf{E}$, $\mathbf{B}$, $\mathbf{D}$, $\mathbf{H}$.   It is also defined as *the transmitter of interactions between charges at rest and/or in motion.*

Since charges comprise the sources of electromagnetic fields, we expect them to be related, either explicitly or implicitly, to all four field vectors. Indeed, such is the case.   The interrelationships between charges and the field vectors were first formulated explicitly in the nineteenth century by

James Clerk Maxwell, a Scottish physicist and mathematician. Maxwell's results, written in modern-day mathematical language, together with a short explanation of what the mathematics say, is the subject of this section. In Sec. 2.5 we present a brief history of electromagnetics which outlines the basic experimental phenomena from which Maxwell deduced what is now known as *Maxwell's equations*. In this text we start by assuming that Maxwell's equations, the Lorentz force law, and some other peripheral information which is implicit in Maxwell's formulation specify the behavior of electromagnetic fields in all possible situations. From this general information, we deduce in the following chapters the behavior of electromagnetic fields in special cases. We shall see that each special case is simply the situation which prevails when particular terms in the basic laws are set equal to zero. For example, in the static case all time derivatives are zero. This approach has the advantage of requiring an explicit statement of the restrictions under which the special case is valid at the beginning of the problem. Any other approach results in obscuring the limits of validity of the special case and, furthermore, frequently leads to the erroneous conclusion that the result obtained in a particular special case is the general result which can be applied to a different special case.

Integral form of the electromagnetic field equations Stated in integral form, Maxwell's equations are

Faraday's Law

$$\oint_C \mathbf{E} \cdot d\mathbf{l} = -\frac{d}{dt} \int_S \mathbf{B} \cdot \mathbf{n} \, da \tag{1}$$

Generalized Form of Ampère's Circuital Law

$$\oint_C \mathbf{H} \cdot d\mathbf{l} = \int_S \mathbf{J} \cdot \mathbf{n} \, da + \frac{d}{dt} \int_S \mathbf{D} \cdot \mathbf{n} \, da \tag{2}$$

Gauss' Law For the Magnetic Field

$$\oint_\Sigma \mathbf{B} \cdot \mathbf{n} \, da = 0 \tag{3}$$

Gauss' Law for the Electric Field

$$\oint_\Sigma \mathbf{D} \cdot \mathbf{n} \, da = \int_V \rho \, dv \tag{4}$$

The law which couples electromagnetic fields to mechanical fields is known as the *Lorentz force law*. This law, in integral form, is

$$\int_V \mathbf{f} \, dv = \int_V \rho(\mathbf{E} + \mathbf{v} \times \mathbf{B}) \, dv \tag{5}$$

where **f** is the force per unit volume, in newtons per cubic meter (N/m³), and **v** is the velocity with which the charge density ρ is moving.

It is implicit in these laws (and can be shown explicitly, as in Sec. 2.15) that there is an additional relationship between **D** and **E**, and **B** and **H**, and **J** and **E** which can be written in general functional notation as

$$\mathbf{D} = D(\mathbf{E})$$

$$\mathbf{B} = B(\mathbf{H}) \qquad\qquad (2\text{-}5)$$

$$\mathbf{J} = J(\mathbf{E})$$

These additional relationships specify the electromagnetic properties of the medium, and are called the *constitutive relations of the medium*. Thus Maxwell's equations, the Lorentz force law, and the subsidiary relations which serve as the means for the electromagnetic characterization of the medium are the *complete basic laws* of electromagnetic theory. A study of these laws and their implications is the subject of this text.

As a first task, let us state the laws in words and the restrictions on the integrals and apply them to a few simple examples.

Faraday's law of induction, Eq. (**1**), states: The line integral of the field vector **E** around any closed contour C is equal to the time rate of change of the total *flux* of the vector **B** crossing any surface S bounded by C, provided that (1) the contour C remains fixed with respect to time, and (2) the surface S is *simply connected*. A piece of paper without holes in it is a simply connected surface regardless of its size, shape, or configuration. If it has holes, it is said to be *multiply connected*. Figure 2-1a shows a simply connected surface, and Fig. 2-1b shows a multiply connected surface. A multiply connected surface can be converted into a simply connected region by merely introducing so-called *cuts*, which, as shown in Fig. 2-1c, consist of lines joining any point on the perimeter of the holes to the outside boundary of S. In the evaluation of a contour integral, these cuts are traversed once each way, as shown by the arrows in Fig. 2-1c, giving a net zero contribution to the value of the integral. Note from Fig. 2-1 that the conventional vector symbolism has been adopted: *a right-hand screw will advance in the direction of the unit normal* **n** *if turned in the direction of the arrows* which indicate the positive direction of traversing the path C. Note also that da is an element of surface area perpendicular to **n**, and that the vector $d\mathbf{l}$ is a differential element of length measured along the tangent at every point on C.

The generalized form of Ampère's circuital law, Eq. (**2**), states: The line integral of **H** along C is equal to the sum of two terms. The first term is the *conduction current I* linked by C; the second term, called the *displacement current*, is the time rate of change of the total electric flux crossing the surface S. The current I is expressed in terms of current density by Eq. (2-4). Physically, this current may arise in many ways, including the movement of

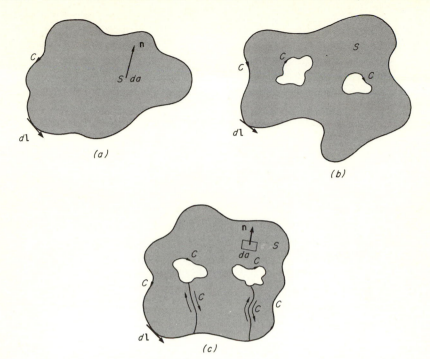

FIGURE 2-1. Simply and multiply connected surfaces. (a) Simply connected; (b) multiply connected; (c) simply connected.

charged particles in free space. Regardless of origin, however, I is the net current flowing in the direction of advance of a right-hand screw as it is turned in the direction indicated on the contour. Various possibilities are shown in Fig. 2-2.

When the electromagnetic field has no time dependence, the displacement current vanishes, and Eq. (2) then reduces to what is known as *Ampère's circuital law*.

Gauss' law for the magnetic field, Eq. (3), states: The net outward flux of the vector **B** through any closed surface (one that completely surrounds some volume) is zero. This implies that the magnetic flux density vector **B** is continuous. Thus, if we start at any point in the region of a magnetic field and move in the direction of the field vector at that point, and keep moving in the direction of the magnetic field, we shall eventually return to our starting point. If we start at a second point which was not passed over by the first path, and move as we did before, we find that we again return to the second point and that nowhere do we cross the first path. [In Eq. (3), **n** is a unit outward vector normal to the surface Σ.]

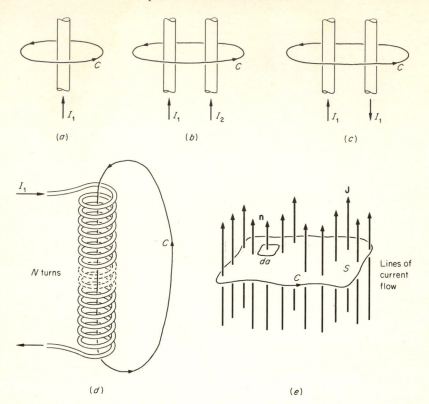

FIGURE 2-2. Net current linked by a closed contour. (a) $I = I_1$; (b) $I = I_1 + I_2$; (c) $I = 0$; (d) $I = NI_1$; (e) $I = \int_S \mathbf{J} \cdot \mathbf{n}\, da$.

Gauss' law for the electric field, Eq. (4), states: The net outward flux of the vector $\mathbf{D}$ through any closed surface Σ is equal to the net charge

$$Q = \int_V \rho\, dv \qquad (2\text{-}6)$$

in the volume V enclosed by the surface Σ. If Q is positive, the net flux of $\mathbf{D}$ will be outward (in the direction of the outward normal vector $\mathbf{n}$); if Q is negative, the net flux of $\mathbf{D}$ will be inward.

The Lorentz force law gives the coupling between electromagnetic fields and the mechanical world. The differential form of this law is more useful and more informative; so at this time we shall simply note the integral form as the field equation which describes this important coupling.

Henceforth, in this text, the four Maxwell equations and the integral form of the Lorentz force law will be referred to by the corresponding boldface Arabic numerals (1), (2), (3), (4), and (5).

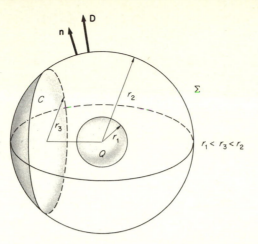

FIGURE 2-3.   Electric flux of a spherical charge distribution.

Now, Eqs. (3) and (4) reveal a fundamental difference in terms of the surface integrals in Eqs. (1) and (2).   In Eq. (1) the net magnetic flux through S is determined solely by the contour C, being completely independent of the shape of the surface S.   By contrast, in Eq. (2), the net electric flux is dependent on both the contour C and the particular shape of S.   This point can be further clarified with the aid of an example.

Example 2-1   Charged Sphere.   Consider a positive charge Q which is uniformly distributed throughout the volume of a sphere of radius r_1.   Let r_2, where $r_2 > r_1$, be the radius of an imaginary spherical surface Σ drawn about the same origin, as shown in Fig. 2-3.   Because of spherical symmetry, the radial component of the electric flux density vector, $D_r = \mathbf{D} \cdot \mathbf{n}$, will be constant on Σ.   Therefore Eq. (4) gives

$$4\pi r_2{}^2 D_r = Q$$

from which we obtain

$$D_r = \frac{Q}{4\pi r_2{}^2}$$

for the radial component of $\mathbf{D}$ at any point on the surface Σ.   The tangential components D_θ and D_φ must be zero.   This can be verified by noting that spherical symmetry demands that they be constant.   Direct application of Eq. (1) will show that the constant value must be zero.

Now, since D_r has the same magnitude everywhere, the $\mathbf{D}$ flux through the upper hemisphere of Σ would be exactly equal to the $\mathbf{D}$ flux through the lower hemisphere. However, the flux through the shaded spherical sector bounded by C would differ appreciably from the flux through the complement of that sector on Σ (not shaded), although both surfaces are obviously bounded by the same contour C.

We thus see that, unlike lines of magnetic flux which have no sources, lines of electric flux emanate and terminate on charges.   In essence, this is the implication of

Eqs. (3) and (4) and the resulting characteristic difference in the surface integrals of Eqs. (1) and (2).

Summarizing, the electromagnetic field equations constitute a set of mathematical statements which summarize past experimental work in electromagnetics, and which therefore constitute the fundamental laws of electromagnetics. Electromagnetic theory *is* the study of these field equations.

Differential form of the electromagnetic field equations The integral form of the electromagnetic field equations has only a limited usefulness in problem solutions because these equations are laws describing the properties of the field over extended regions of space. Consequently, they can be used to solve only those types of problems which exhibit complete symmetry, such as spherical, cylindrical, and one-dimensional rectangular symmetry. To be useful in the general case, the laws must be recast into a form which describes the relations between the field vectors at arbitrary points in space and at arbitrary instants of time. This form is the differential form.

The differential form of Maxwell's equations is

$$\nabla \times \mathbf{E} = -\frac{\partial \mathbf{B}}{\partial t} \tag{I}$$

$$\nabla \times \mathbf{H} = \mathbf{J} + \frac{\partial \mathbf{D}}{\partial t} \tag{II}$$

$$\nabla \cdot \mathbf{B} = 0 \tag{III}$$

$$\nabla \cdot \mathbf{D} = \rho \tag{IV}$$

The differential form of the Lorentz force law is

$$\mathbf{f} = \rho(\mathbf{E} + \mathbf{v} \times \mathbf{B}) \tag{V}$$

Roman numerals **(I)** to **(V)** will denote these equations throughout the text.

The constitutive relations, as before, are

$$\mathbf{D} = D(\mathbf{E}) \qquad \mathbf{B} = B(\mathbf{H}) \qquad \mathbf{J} = J(\mathbf{E}) \tag{2-7}$$

Recasting the basic laws into differential form is accomplished by noting, first, that a differential form of the laws can be valid only if the field vectors are single-valued, bounded, continuous functions of position and time and with continuous derivatives. The field vectors have these properties, except at points where there are abrupt changes in the distribution of current and/or charge. These abrupt changes usually occur at the interface between media of different physical properties, and they determine the discontinuity of the field vectors at these interfaces. The relationships between the discontinuity of the field vectors and the abrupt changes in distribution of currents and/or charges are known as *boundary conditions*. It should be obvious that a *complete* formulation of the basic electromagnetic laws in differential form requires (1) the differential form of the laws (valid everywhere except at points of abrupt change) and (2) the boundary conditions.

Let us now derive the differential laws from the integral laws.   Consider Eq. (1), and let the surface S be a plane surface bounded by the contour C. Since the contour is assumed to be fixed in space, the order of differentiation and integration may be interchanged, and Eq. (1) may be written in the form

$$\oint_C \mathbf{E} \cdot d\mathbf{l} = -\int_S \frac{\partial \mathbf{B}}{\partial t} \cdot \mathbf{n}\, da \qquad (2\text{-}8)$$

Let us divide both sides of Eq. (2-8) by the area of the surface S and allow C to shrink about any point P internal to C.   In the limit, as the surface area approaches zero, the left-hand member of Eq. (2-8) will reduce to the projection of $\nabla \times \mathbf{E}$ in the direction of the unit normal to S, consistent with the definition of curl.   The right-hand member, on the other hand, will reduce to the functional value of the integrand at P, and Eq. (2-8) will then give

$$(\nabla \times \mathbf{E}) \cdot \mathbf{n} = -\frac{\partial \mathbf{B}}{\partial t} \cdot \mathbf{n} \qquad (2\text{-}9)$$

Since $\mathbf{n}$ is arbitrary, it follows immediately from Eq. (2-9) that

$$\nabla \times \mathbf{E} = -\frac{\partial \mathbf{B}}{\partial t} \qquad (\text{I})$$

This is the differential form of Faraday's law of induction.   In any given system of coordinates, it reduces to three scalar equations.   For continuous reference throughout this text, it is being assigned the Roman numeral (I).

Equation (2) may be written

$$\oint_C \mathbf{H} \cdot d\mathbf{l} = \int_S \left(\mathbf{J} + \frac{\partial \mathbf{D}}{\partial t} \right) \cdot \mathbf{n}\, da \qquad (2\text{-}10)$$

Following the previous development, we divide both sides of Eq. (2-10) by the area of S and allow C to shrink about a point interior to C.   In the limit, as the surface area diminishes to zero, we find from the definition of the curl

$$(\nabla \times \mathbf{H}) \cdot \mathbf{n} = \left(\mathbf{J} + \frac{\partial \mathbf{D}}{\partial t} \right) \cdot \mathbf{n} \qquad (2\text{-}11)$$

and by virtue of the arbitrariness of $\mathbf{n}$,

$$\nabla \times \mathbf{H} = \mathbf{J} + \frac{\partial \mathbf{D}}{\partial t} \qquad (\text{II})$$

Equation (II) is the differential form of the second Maxwell equation.

To obtain the differential form of Eq. (3), we divide both sides by the volume V enclosed by the surface Σ. We then let the surface shrink about any interior point P, and observe that, in the limit, when the volume approaches zero, the left-hand member of Eq. (3) reduces to the divergence of $\mathbf{B}$, evaluated at the point P, consistent with the definition of divergence. Therefore we obtain

$$\nabla \cdot \mathbf{B} = 0 \tag{III}$$

as the expression of continuity of $\mathbf{B}$ lines at a point.

The differential expression of Gauss' law for the electric field is obtained in a similar way:

$$\oint_{\Sigma} \mathbf{D} \cdot \mathbf{n} \, da = \int_{V} \rho \, dv \tag{4}$$

Again we divide both sides by V and let Σ shrink about any internal point P. In the limit, as V approaches zero, we obtain

$$\nabla \cdot \mathbf{D} = \rho \tag{IV}$$

This is the differential form of Maxwell's fourth equation.

The differential form of the Lorentz force law follows almost by inspection. If we take the limit of

$$\int_{V} \mathbf{f} \, dv = \int_{V} \rho(\mathbf{E} + \mathbf{v} \times \mathbf{B}) \, dv \tag{5}$$

as $V \to 0$, we obtain

$$\mathbf{f} = \rho(\mathbf{E} + \mathbf{v} \times \mathbf{B}) \tag{V}$$

This is the differential form of the Lorentz force law.

Since the laws of electromagnetic theory are deductions, inferences, and conclusions from experimental measurements, which of necessity are bulk (integral) results, we see that the integral laws are at least historically more fundamental. It is possible, at least by hindsight, to deduce the differential laws as the fundamental laws and to derive the integral laws from them. This derivation is a useful exercise for the student (Prob. 2-15).

Solution of problems by use of the differential laws leads to differential equations. The general solution of every differential equation contains terms (constants, or functions) that can be evaluated only from a known behavior of the variables at space boundaries and/or from initial conditions. Hence we shall need information concerning the behavior of the field vectors at the interface, or boundary, between two different physical media. Logically, this behavior should be discussed next. However, it will be advantageous to first obtain a better understanding of background upon which the laws of electromagnetism are based by examining the macroscopic properties of matter and the nature and types of currents before deriving the boundary conditions, in Sec. 2.9.

2.5 A Brief History of Electromagnetics. "When you can measure what you are speaking about and express it in numbers, you know something about it," said Lord Kelvin, "but when you cannot measure it, when you cannot express it in numbers, your knowledge is of a meager and unsatisfactory kind."

One of the early measurers of electricity was Colonel Charles de Coulomb, a French army engineer. Coulomb invented a device for measuring exceedingly small forces, known as a *torsion balance*, one of the functions of which was to determine the law of electrostatic force. Others had earlier built cruder devices. Volta, in 1775, had built an electrometer, using a pair of dry straws as measuring vanes which would be moved apart by the electric force. However, Volta did not know the magnitude of the force on the straw. It took Charles Coulomb, in 1785, with his torsion balance, to find that the force between two electric charges is proportional to the amount of each of the charges, and inversely proportional to the square of the distance between them. Actually, the inverse square law of electrostatic force was first studied by Cavendish during years of secret research. His manuscripts were not published until 1879, when they were discovered.

To produce electricity in the 1700s, scientists rubbed catskin on glass, used the contact potential from metals, or flew kites to "catch" lightning. Then Alessandro Volta, professor of physics at the University of Pavia, Italy, gave a great impetus to electrical experimentation. In 1800, he invented the *voltaic pillar* (battery), which gave scientists a powerful new tool. It is difficult to appreciate the awe with which men of Volta's time greeted his discovery. Now sparks could be made when desired, wires heated, and new experiments performed.

In a classroom at the University of Copenhagen, Professor H. C. Oersted hoped to show a new effect of electricity by placing a wire near a compass needle. But nothing happened. After class he tried again, and found that the needle swung if he held the current-carrying wire parallel instead of perpendicular to the needle (as he had done in class). Therefore, in 1820, Oersted declared that he had discovered a magnetic effect of voltaic electricity. He found that a magnetic compass needle would deflect when a wire connected to a voltaic pile was brought near the needle. The needle consistently made a right angle with the electric wire. His results were published in the usual manner of his time, in a memoir to the renowned European scientists.

Two months later, news of Oersted's discovery reached France. The French Academy of Science became interested, and requested to see Oersted's experiments. Attending this meeting was André Ampère, who, upon seeing the Oersted experiment successfully demonstrated, decided to extend the results. Professor Ampère was no ordinary man. Quite precocious as a child, he had learned Latin in two weeks at the age of fourteen so that he could read the intricate works of Bernoulli and Euler. Ampère studied impulsively, and was capable of becoming interested in most unexpected subjects; on some occasions he became tired of his studies right at the point of inevitable distinction. Because of his versatility, Ampère dabbled in chemistry, philosophy, electricity, mathematics, psychology, metaphysics, botany, anatomy, animal magnetism, linguistics, and poetry.

Ampère found that a current in one wire exerts a force on another wire even without a magnet present. In a paper which he presented to the Academy, Ampère said, "The electromagnetic action appears in two sorts of effects and I think ought to be distinguished by precise definition.   I will call the first electric tension and the second electric current."   He thus recognized the difference between voltage and current, and used his great mathematical ability to determine that the magnetic field produced by a current-carrying wire was proportional to the current carried, and inversely proportional to the distance from the wire.

The next major contributor to electromagnetic theory was Michael

MICHAEL FARADAY, chemist and electrical experimenter, who discovered electromagnetic induction, the laws of electrolytic action, and magnetic rotation. His discoveries led directly to large-scale electrification and electrical controls in industry. (*Courtesy of the Burndy Library.*)

Faraday. Faraday was born the son of a blacksmith. As a youth he was a bookseller's apprentice, but tired of his work. Luckily, he secured a position at the Royal Institution as a laboratory assistant. After ten years he was ready for scientific work on his own. For several years he experimented with magnetism, obtaining no results. But in August, 1831, he had a great success. He made an iron ring and on it he wound two coils, one on the right side of the ring and one on the left, with no direct connection between them. He connected a battery to one coil and a galvanometer to the other. Nothing happened. But when he disconnected the battery, the galvanometer deflected. When he reconnected the battery, the galvonometer deflected in the opposite direction. The deflections were small and short-lived. Faraday presently found many ways of developing induced currents by changing magnetism. He developed the concept that an induced current is always a result of a change of *lines of magnetic force*. Interestingly, Joseph Henry, of Princeton University, discovered the law of induction by magnetism at the same time that Faraday did. However, Henry failed to publish his results, and therefore Faraday received all the recognition.

Faraday's concept of lines of forces pervading all space was viewed controversially by most physicists of his time. Following a hypothesis of Weber, physicists felt that electric and magnetic forces acted at a distance and in a straight line and, in addition, penetrated all space with infinite velocity. It took Maxwell's genius to verify theoretically Faraday's lines-of-force concept, and Hertz's genius to verify Maxwell's work experimentally.

James Clerk Maxwell was the son of a minor Scottish nobleman. His father had stimulated a curiosity in Clerk by taking him to see the new manufacturing plants and processes that were suddenly being developed in the cities in the 1830s and 1840s. It was while he attended a private high school that Maxwell acquired an interest in mathematics and at the age of fifteen wrote a paper entitled "On the Description of Oval Curves," which was published in the *Proceedings of the Edinburgh Royal Society*. His ability in mathematics grew, and with it his interest in the writings of Faraday. He tried to put Faraday's lines-of-force concept into a mathematical form. In 1857 James Clerk Maxwell wrote to the world-famous Faraday commenting on some of Faraday's writings. Faraday was kind enough to reply, and even encouraged Maxwell with some new ideas. In 1873, after five years of hard work in retreat, Maxwell published his now famous treatise, "Electricity and Magnetism." He unified all knowledge of electricity through a group of simple equations, predicted electromagnetic wave motion, calculated the velocity of light, and explained light propagation as an electromagnetic wave phenomenon. *His work provides the foundation for all electromagnetic devices.* Interestingly, Maxwell believed that his electromagnetic waves were carried by an ether. He had the insight and mathematical intuition to weave 150 years of experimental evidence into a cohesive theory. This new theory radically departed from the rules then accepted by such a wide margin

JAMES CLERK MAXWELL, physicist and electrical experimenter.   He mathematically developed the nature of an electromagnetic field and related it to
the nature of light.   He led to the discovery of electric waves and the
mechanical pressure of light.   (*Courtesy of the Burndy Library.*)

that it took another twenty-five years before Hertz conclusively demonstrated
the accuracy of Maxwell's equations.

Let us now briefly examine the experimental evidence which confronted
Maxwell.†   This evidence can be summed in nine basic experiments, some of
which bear the names of famous men, and others which now seem common-
place, and are easily acceptable to our preeducated minds.   In general, the
experimental evidence is presented as results of briefly defined experiments,
followed by simplifying mathematical statements.   The experimental results
are presented in a logical order, and not in their chronological order of
occurrence.

† This presentation is essentially that given by H. H. Skilling, "Fundamentals of Electric
Waves," John Wiley & Sons, Inc., New York, 1948.

Experiment I   The first experiment established the fact that a field of force exists about any object containing an electric charge, and that this field of force acts upon any other object also containing an electric charge. The field is found to be static and directional.   For two such charged objects the force is found to be given by

$$\mathbf{F} = k\,\frac{Q_1 Q_2}{r^2}\,\mathbf{a}_r$$

where $\mathbf{F}$ = resultant force
  Q_1 = net charge on first object
  Q_2 = net charge on second object
  $\mathbf{a}_r$ = a unit vector in direction joining effective centers of the two charges
  r = separation of effective centers of the two charges

From this it was deduced that each charged object produced a force field $\mathbf{E}_1$ and $\mathbf{E}_2$ such that the force on Q_1 could be written

$$\mathbf{F}_1 = \alpha Q_1 \mathbf{E}_2$$

where α is a constant, and the force on Q_2 could be written

$$\mathbf{F}_2 = \alpha Q_2 \mathbf{E}_1$$

This experiment shows that static charges produce force fields which exert forces on other static charges.

Experiment II   The second experiment established that an electrostatic field is a conservative field.   It demonstrated that a small test charge in an electrostatic field required no total work to be done when moved in any closed path.   As long as the test charge returned to its starting point, the work put into moving the test charge against the field is given back by returning to the starting point.   This experiment showed that the electrostatic field is *conservative*, or *lamellar*, a property which is mathematically stated as

$$\oint_C \mathbf{F} \cdot d\mathbf{l} = 0$$

From the relation established in Experiment I, it follows that

$$\oint_C \mathbf{E} \cdot d\mathbf{l} = 0$$

This mathematical equation also states that the electrostatic field is without curl.

Experiment III   The third experiment established that the total charge Q within a closed surface can be measured.   As in Experiment II, a small test charge is used to measure the field strength over the entire surface; then the total of the field strength normal to each part of the surface, multiplied by the area of each of these surfaces, is a measure of the total charge, Q, enclosed. This experiment showed that the divergence of the electrostatic field is proportional to the charge density, and may be mathematically stated as

$$\oint_{\Sigma} \mathbf{E} \cdot \mathbf{n} \, da = \alpha Q$$

When the experiment is conducted in a vacuum, $\alpha = 1/\epsilon_0$, where ϵ_0 is defined as the *permittivity* of free space.

Experiment IV   The fourth experiment is the same as Experiment III, except that the medium is changed to establish its effect on the field.   When a known charge Q is enclosed by a surface in many different media, a relationship is established between all new media and that for free space.   This experiment established the effect of dielectric substances on electrostatic fields.   This effect is expressed in terms of a quantity known as the *dielectric constant*.

$$\epsilon_r = \frac{\epsilon}{\epsilon_0}$$

or

$$\alpha = \frac{1}{\epsilon} = \frac{1}{\epsilon_r \epsilon_0}$$

Experimentally, the value of ϵ_r for many media has been established and tabulated.

Experiment V   The fifth experiment established Ohm's law.   Experimentally, Ohm kept a constant current flowing through a section of many different conductors, and measured the voltage difference between the ends of these sections, thus establishing the relationship between voltage and current as

$$I = \frac{1}{R} V$$

The resistance R was determined to be a constant dependent upon the composition and geometry of the conductor (there is also a temperature dependence to some extent) and independent of I or V.   Since R is dependent upon geometry and composition of the material, the relationship was determined to be as follows:

$$\frac{1}{R} = \sigma \times \frac{\text{cross-sectional area normal to current flow}}{\text{length of material}}$$

Here σ is defined as the *conductivity* of the material, and has been tabulated for many different substances.   It is at this point that a vector field called current density **J** is introduced, which is related to current flow in a conductor by

$$\int_S \mathbf{J} \cdot \mathbf{n} \, da = I$$

When the differential form of this relationship is applied to Ohm's law, noting that

$$dV = \mathbf{E} \cdot d\mathbf{l}$$

then

$$\mathbf{J} \cdot \mathbf{n} \, da = \sigma \frac{da}{dl} \mathbf{E} \cdot dl$$

and since **n** and dl as vectors have the same direction as **E** and **J**, respectively, we conclude that the current density **J** in a conductor is proportional to the electric field strength **E** at the same point and in the same direction.   Mathematically, this is stated as

$$\mathbf{J} = \sigma \mathbf{E}$$

Experiment VI   The sixth experiment establishes Ampère's law. Ampère mounted a short straight section of conducting wire in such a way that the force exerted upon it can be measured while current is flowing through it.   The result is that the magnetic force on the exploring wire is always normal to the wire, and the force is proportional to the amount of current flowing and to the length, location, and orientation of the exploring wire. This experiment results in a description of a magnetic field and the force it exerts on an exploring wire.   This is mathematically stated as

$$\mathbf{F} = I\mathbf{L} \times \mathbf{B}$$

where **B** is the magnetic flux density.   This is a vector field equation for the magnetostatic force field.

Experiment VII   The seventh experiment was conducted by Faraday in 1831, and established the magnetic and electric field relationship.   When a loop of wire is connected to a ballistic galvanometer, the reading is a measure of the electric charge passing through the loop.   Faraday discovered that whenever the loop was placed in a *changing* magnetic field, a charge flowed through the loop, thus showing that an electric field could be produced magnetically.   By the technique of Experiment VI, the magnetic flux density was measured at the loop, and it was found that the galvanometer readings were proportional to the increase or decrease of the flux passing through the loop.   The galvanometer also indicated that its readings were inversely proportional to the total resistance of the galvanometer.   This information can be used to establish the relationship between the magnetic flux and electric field:

$$Q = -\frac{\Phi}{R} \quad \text{or} \quad RQ = -\Phi$$

Now the time rate of change of this equation expresses Faraday's discovery as

$$R\frac{dQ}{dt} = RI = -\frac{d\Phi}{dt}$$

Since in a closed loop the product of IR is equal to the electromotive force (emf) of the circuit, Faraday's famous law follows immediately.

$$\text{emf} = \oint_C \mathbf{E} \cdot d\mathbf{l} = -\frac{d\Phi}{dt} = -\frac{d}{dt}\int_S \mathbf{B} \cdot \mathbf{n}\, da$$

Experiment VIII The apparatus of this experiment is essentially the same as in Experiment VII, except that, instead of a loop with one turn, there are many turns included in the exploring loop, and the galvanometer is constructed with a long time constant, resulting in a flux meter. This instrument is used to sum the magnetic flux density normal to a closed surface. The resultant sum is always zero, and it is concluded that magnetic flux lines have no beginning or ending, but form closed loops. This is stated mathematically as

$$\oint_\Sigma \mathbf{B} \cdot \mathbf{n}\, da = 0$$

or

$$\nabla \cdot \mathbf{B} = 0$$

and establishes that a magnetostatic field is *sourceless*, or *solenoidal* (without divergence).

Experiment IX This experiment established that the curl of the magnetostatic field is proportional to current density. A flux meter (as described in Experiment VIII) is used to determine the magnetic flux density at every point along a closed path. Then the summation of the products of the tangential components of the magnetic flux density with the corresponding differential lengths along the chosen path in a homogeneous medium is proportional to the amount of the current surrounded by the path. The factor μ, defined as *permeability* of the medium, relates this summation to the enclosed current as follows:

$$\oint_C \frac{\mathbf{B} \cdot d\mathbf{l}}{\mu} = I$$

By introducing a new symbol, $\mathbf{H} = \mathbf{B}/\mu$, which is the magnetic field intensity, this equation becomes

$$\oint_C \mathbf{H} \cdot d\mathbf{l} = I$$

The current I has been defined from Experiment V as the integral of the current density $\mathbf{J}$ over the surface bounded by the closed path. It then follows that

$$\oint_C \mathbf{H} \cdot d\mathbf{l} = \int_S \mathbf{J} \cdot \mathbf{n} \, da$$

or from Stokes' theorem,

$$\nabla \times \mathbf{H} = \mathbf{J}$$

These nine experiments sum the evidence which confronted Maxwell, but there were two more assumptions necessary to complete a basic electromagnetic foundation:

1. It is assumed that the dynamic electric field has divergence proportional to charge density.
2. It is assumed that the dynamic magnetic field has no divergence.

From these assumptions and the experimental evidence, Maxwell derived the now famous equations which include dynamic conditions.

By hindsight we can see how Maxwell's equations follow from the experiments. Experiment II showed that in the static case

$$\oint_C \mathbf{E} \cdot d\mathbf{l} = 0$$

and Experiment VII showed that in the dynamic case

$$\oint_C \mathbf{E} \cdot d\mathbf{l} = -\frac{d\Phi}{dt} = -\frac{d}{dt} \int_S \mathbf{B} \cdot \mathbf{n} \, da \qquad (1)$$

By assuming that the dynamic case is the general case, we have Maxwell's first equation directly. Similarly, Experiment IX showed that in the static case

$$\oint_C \mathbf{H} \cdot d\mathbf{l} = \int_S \mathbf{J} \cdot \mathbf{n} \, da$$

Through a major stroke of genius, Maxwell discovered that the general dynamic case requires the addition of the time rate of change of $\mathbf{D}$. Thus we obtain

$$\oint_C \mathbf{H} \cdot d\mathbf{l} = \int_C \mathbf{J} \cdot \mathbf{n} \, da + \frac{d}{dt} \int_S \mathbf{D} \cdot \mathbf{n} \, da \qquad (2)$$

as Maxwell's second equation.

From Experiments III, IV, and VIII and the assumption that these results are valid in the dynamic case, Maxwell's third and fourth equations,

$$\oint_{\Sigma} \mathbf{B} \cdot \mathbf{n}\, da = 0 \tag{3}$$

$$\oint_{\Sigma} \mathbf{D} \cdot \mathbf{n}\, da = \int_{V} \rho\, dv \tag{4}$$

are obtained.

The constitutive relations

$$\mathbf{D} = D(\mathbf{E}) \qquad \mathbf{B} = B(\mathbf{H}) \qquad \mathbf{J} = J(\mathbf{E})$$

are implied from Experiments IV, IX, and V, respectively, by assuming that their results are general.

The Lorentz force law can be inferred from Coulomb's law (Experiment I) and Ampère's law (Experiment VI). If we assume that a moving charge experiences simultaneously an electric field force and also a magnetic field force, we have

$$\mathbf{F} = \mathbf{F}_e + \mathbf{F}_m = Q\mathbf{E} + I\mathbf{L} \times \mathbf{B}$$

Now if the current I is due to the charge Q moving in the direction $\mathbf{L}$ with a velocity $\mathbf{v}$, we can write $I\mathbf{L} = Q\mathbf{v}$ and recast the equation in the form

$$\mathbf{F} = Q\mathbf{E} + Q\mathbf{v} \times \mathbf{B}$$

If we now assume that this equation holds for distributed sources, we have

$$\mathbf{F} = \int_{V} \mathbf{f}\, dv = \int_{V} \rho(\mathbf{E} + \mathbf{v} \times \mathbf{B})\, dv$$

or in differential form,

$$\mathbf{f} = \rho(\mathbf{E} + \mathbf{v} \times \mathbf{B}) \tag{V}$$

One of the most surprising of Maxwell's results was that he could combine his equations in a manner which predicted a wave solution. This led to the postulation that visible light was merely a form of an electric wave, a truly shocking and doubtful revelation in the mid-nineteenth century. But Maxwell's revelation has stood the test of time; we have but to sit in our living room and look at pictures of the planet Mars transmitted by electromagnetic waves over millions of miles through a widely varying medium to become convinced of the genius of this man.

Further experimental verification of Maxwell's equations was provided later by Heinrich Hertz. The results of Hertz's experiments were the subject of two papers he published in 1888, entitled "On Electromagnetic Waves in Air and Their Reflection" and "On Electric Radiation." Encouraged to study electricity by Professor von Helmholtz of the University of Berlin, Hertz set out to see if electric force travels with infinite velocity, as Weber said, or if it is wavelike, as Maxwell predicted.   He devised a spark-gap generator to be used as a transmitter and used a loop of wire with a very small gap as a receiver.   A spark set off in the transmitter would produce a spark in the receiver loop.   He then set up a fairly large zinc sheet at the other end of his laboratory.   By moving his receiving loop toward and away from the zinc sheet, he found that the intensity of the received spark varied. Therefore standing waves were present.   So electric and magnetic fields were waves, as Maxwell had predicted.   Hertz produced waves at 1 MHz, 100 MHz, and one GHz (the abbreviation for a hertz is Hz).

He also used parabolic cylinders to reflect his waves.   To his amazement, he found that the electric waves passed through wooden doors, that they could be reflected like sunlight, and that they were polarized.   Hertz firmly established Faraday's view of lines of force and Maxwell's view of waves. It was now up to other engineers and scientists to discover the vast implications of Maxwell's equations—implications which Maxwell himself never realized.

2.6 A Physical Interpretation and Units of the Field Vectors.   Maxwell's equations give the relationships between the field vectors $\mathbf{E}$, $\mathbf{B}$, $\mathbf{D}$, $\mathbf{H}$ and the source fields ρ and $\mathbf{J}$.   At this point it will be informative to examine some aspects of the physical nature of the field vectors.

We begin by looking at the Lorentz force law,

$$\mathbf{f} = \rho(\mathbf{E} + \mathbf{v} \times \mathbf{B}) \tag{V}$$

where $\mathbf{v}$ is the velocity of ρ.   The quantity $(\mathbf{E} + \mathbf{v} \times \mathbf{B})$ is seen at once to have the dimension of force per unit charge, that is, newtons per coulomb. From this we see that

$$\text{Units of } \int_C \mathbf{E} \cdot d\mathbf{l} = \text{newton-meters per coulomb} = \text{volts} \tag{2-12}$$

Thus we have that the unit of $\mathbf{E}$ is the newton per coulomb (N/C) or, what is more frequently used, the volt per meter (V/m).

If we now examine $\mathbf{v} \times \mathbf{B}$, we see that it also is in units of newtons per coulomb.   It follows that $\mathbf{B}$ is measured in newton-seconds per coulomb-meter (N · s/C · m), or volt-seconds per square meter (V · s/m²); or since volt-seconds is defined as webers, we usually state $\mathbf{B}$ in webers per square meter (Wb/m²).   It is clear that the Lorentz force law allows us to identify $\mathbf{E}$ and $\mathbf{B}$ in terms of their basic physical units, which are mixed electrical and mechanical units.   These two vector fields provide the explicit connection between electrical and mechanical phenomena.

The units of $\mathbf{D}$ and $\mathbf{H}$ are purely electrical, and hence their mechanical aspects are implicit rather than explicit.   For the electric flux density vector

D we have

$$\oint_{\Sigma} \mathbf{D} \cdot \mathbf{n} \, da = Q \qquad (2\text{-}13)$$

from which it is obvious that the units of **D** are coulombs per square meter.

Similarly, for the magnetic intensity vector **H**, we have, in the static case, Ampère's circuital law,

$$\oint_{C} \mathbf{H} \cdot d\mathbf{l} = I \qquad (2\text{-}14)$$

from which it follows that the units of **H** are amperes per meter.

At this point it will be instructive to give an example of the use of Ampère's circuital law to solve a rather elementary classical problem.

Example 2-2   Current-carrying Long Wire. Consider a long straight conductor as shown in Fig. 2-4. Let the conductor carry a time-independent total current I directed out of the plane of the paper toward the observer. Let this current be uniformly distributed over the area of the conductor, and let the conductor be suspended in air. Because of cylindrical symmetry and because **B** is divergenceless, the magnetic field intensity must be everywhere tangential. The radial component and the axial component must both vanish since they cannot possess complete cylindrical symmetry.

In addition, because of symmetry, the magnitude of the field is constant along all points lying on a circular contour C drawn about the axis of the wire in a plane normal to that axis. Every point on this circular contour is indistinguishable from another. From this analysis we conclude that the magnetic field intensity can be expressed in cylindrical coordinates as

$$\mathbf{H} = H_{\varphi} \mathbf{a}_{\varphi}$$

To evaluate the magnitude H_{φ}, we utilize Ampère's circuital law, which is a special

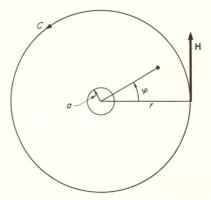

FIGURE 2-4. The field about a long conductor carrying a current I out of the plane of the paper.

case of Maxwell's equation (2), that is,

$$\oint_C \mathbf{H} \cdot d\mathbf{l} = I$$

For the contour C we choose a circular path of radius r along which H has a constant magnitude. Although Ampère's circuital law is true for any path, we choose this path for convenience. Along this path, the vector-displacement element $d\mathbf{l}$ is always in the φ direction. According to our convention, with C traversed in the counterclockwise direction, as indicated, the current I in Ampère's circuital law is the current carried by the conductor. Thus

$$\oint_C \mathbf{H} \cdot d\mathbf{l} = \oint_C (H_\varphi \mathbf{a}_\varphi) \cdot (r \, d\varphi \, \mathbf{a}_\varphi) = \int_0^{2\pi} r H_\varphi \, d\varphi = I$$

This equation is valid for all values of r greater than the radius a of the conductor. Since r and H_φ are constant over all points of the path, they may be moved to the front of the integral sign.

$$r H_\varphi \int_0^{2\pi} d\varphi = 2\pi r H_\varphi = I$$

Therefore, at any radius $r > a$,

$$H_\varphi = \frac{I}{2\pi r}$$

and

$$\mathbf{H} = \frac{I}{2\pi r} \mathbf{a}_\varphi \qquad \text{A/m}$$

It is worth noting that the direction of H can be predicted using the following *right-hand rule:* If the right thumb is pointed along the wire in the direction of current flow, the fingers indicate the direction of the resulting magnetic field (Fig. 2-5).

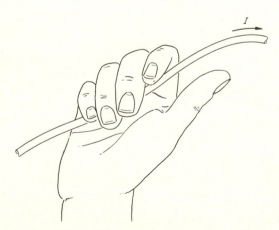

FIGURE 2-5. An illustration of the right-hand rule.

Further insight into the nature of the connection between **B** and **H**, between **D** and **E**, and between **J** and **E**, which were previously given abstractly by

$$\mathbf{B} = B(\mathbf{H}) \qquad \mathbf{D} = D(\mathbf{E}) \qquad \mathbf{J} = J(\mathbf{E}) \tag{2-15}$$

may now be gained by obtaining the physical units of the ratios **B/H**, **D/E**, and **J/E**. We find

$$\text{Units of } \frac{\mathbf{B}}{\mathbf{H}} = \frac{\text{volt-seconds per square meter}}{\text{amperes per meter}} = \frac{\text{volt-seconds}}{\text{ampere-meter}}$$

$$= \frac{\text{volt-seconds squared}}{\text{coulomb-meter}} = \frac{\text{webers}}{\text{ampere-meter}} = \frac{\text{henrys}}{\text{meter}} \tag{2-16}$$

In the last two forms, the weber, as previously defined, is the volt-second, and the henry (H) is defined as the weber per ampere. Obviously, these are *secondary*, or *derived*, units. Their widespread use is our reason for introducing them, rather than insisting on *fundamental* units.

Similarly, we obtain

$$\text{Units of } \frac{\mathbf{D}}{\mathbf{E}} = \frac{\text{coulombs per square meter}}{\text{volts per meter}} = \frac{\text{coulombs}}{\text{volt-meter}} = \frac{\text{farads}}{\text{meter}} \tag{2-17}$$

where the farad (F) is defined as the coulomb per volt.

Combining these results, we find that

$$\text{Units of } \frac{\mathbf{B}}{\mathbf{H}}\frac{\mathbf{D}}{\mathbf{E}} = \frac{\text{volt-seconds squared}}{\text{coulomb-meter}} \times \frac{\text{coulombs}}{\text{volt-meter}}$$

$$= \frac{\text{seconds squared}}{\text{meters squared}} = \frac{1}{(\text{velocity})^2} \tag{2-18}$$

or to show the point more clearly,

$$\text{Units of } \sqrt{\frac{\mathbf{H}}{\mathbf{B}}\frac{\mathbf{E}}{\mathbf{D}}} = \frac{\text{meters}}{\text{second}} = \text{velocity} \tag{2-19}$$

This is a rather surprising result in that it says that this particular, peculiar combination of electromagnetic quantities, even in the most general case, has the units of a purely mechanical quantity, namely, velocity.

Finally, let us examine the units of **J/E**. We find

$$\text{Units of } \frac{\mathbf{J}}{\mathbf{E}} = \frac{\text{amperes per square meter}}{\text{volts per meter}} = \frac{\text{amperes}}{\text{volt-meter}} = \frac{\text{mhos}}{\text{meter}} \tag{2-20}$$

where the final expression has used the circuit-theory definition of amperes per volt = mhos (℧). Thus we see that **J/E** has the units of conductance per meter.

In summary, this section has used the basic electromagnetic laws to obtain a physical interpretation and the basic units of the electromagnetic field vectors. The results are:

E is measured in newtons per coulomb, which may alternatively be stated as volts per meter.
B is measured in newton-seconds per coulomb-meter, which is alternatively stated as volt-seconds per square meter, or as webers per square meter. The weber was defined as the volt-second.
D is measured in coulombs per square meter.
H is measured in amperes per meter.

Next, by taking ratios of units, the units of the general functional relations $\mathbf{B} = B(\mathbf{H})$, $\mathbf{D} = D(\mathbf{E})$, $\mathbf{J} = J(\mathbf{E})$ were obtained. The significance of these ratios is examined in more detail in the next section, in connection with the macroscopic properties of materials.

2.7 Macroscopic Properties of Matter. The behavior of fields in the neighborhood of a single atom is to a large extent the domain of a quantum electrodynamicist. Such fields are called *microscopic fields*. We, on the other hand, are *macroscopic* field theorists, interested only in those fields which vary in space by a discernible amount over distances which are large compared with atomic dimensions. If a material substance is present, we regard it, on a large scale, as an electromagnetic *continuum*, that is, a medium in which a field is continuous and, in addition, constant over a volume containing a large number of atoms   If charges are present, we represent them as continuous, smooth distributions. In short, our main concern is for the average field that prevails within a body, and not in its local values in the immediate surroundings of a specific atom. Hence the continuation of our effort to characterize matter electromagnetically is based on one fundamental premise: our mathematical model must describe adequately only the macroscopic properties of matter. Therefore, in order to determine the constitutive relations of a medium, we need perform our measurements on physically obtainable samples of matter, using ordinary laboratory instruments only. By performing such experiments we obtain the information necessary to express **D** in terms of **E**, **B** in terms of **H**, and **E** in terms of **J**, or conversely. This information may then be displayed in closed mathematical form, or graphically, or in tabular form, depending on the behavior of the particular substance under test.

The simplest relations that can occur are linear relations of the form

$$\mathbf{D} = \epsilon\mathbf{E} \tag{2-21}$$

$$\mathbf{B} = \mu\mathbf{H} \tag{2-22}$$

$$\mathbf{J} = \sigma\mathbf{E} \tag{2-23}$$

where ϵ, μ, and σ are scalar constants of proportionality. Usually, these constants are collectively known as the *parameters* of a medium. However, each parameter has a special name. For example, ϵ is called the *permittivity* of a medium, μ its *permeability*, and σ its *conductivity*. The dimensions (or units) of these parameters were obtained in the preceding section by dimensional analysis.

Free space (vacuum) is a very special case of a medium for which the parameters are linear and constant. (Air behaves electromagnetically as nearly free space, also.) The permittivity of free space is designated ϵ_0. The permeability of free space is designated μ_0. The conductivity of free space is identically zero.

To see how we could measure ϵ_0 and also to obtain a definition of the concept of a *point charge*, let us reexamine the results of Example 2-1.

It was shown there that at any point P outside the charged sphere the electric flux density is

$$\mathbf{D} = \frac{Q}{4\pi r^2}\,\mathbf{a}_r$$

Let us now imagine that the sphere is in free space. According to Eq. (2-21), the electric field intensity outside the sphere would be

$$\mathbf{E} = \frac{1}{\epsilon_0}\,\mathbf{D} = \frac{Q}{4\pi\epsilon_0 r^2}\,\mathbf{a}_r \tag{2-24}$$

Next, let us introduce in the field of the first (source) sphere a very small second (probe) sphere charged uniformly throughout its volume with a density ρ'. Let the radius of the second sphere be r', its volume V', and its total charge Q'. The geometrical arrangement is shown in Fig. 2-6. Let the distance d separating the two spheres be much greater than either r or r', such that $d \pm r'$ is to a very good approximation essentially equal to d. Let us also impose the further condition that the presence of the second sphere does not alter the field of the first sphere. This does not correspond exactly to physical reality, but is nevertheless a very good approximation when Q' is very much smaller than Q. Under these conditions, at every interior point of the second (probe) sphere, the electric field would be. given, to a first approximation, by

$$\mathbf{E} \approx \frac{Q}{4\pi\epsilon_0 d^2}\,\mathbf{a}_r \tag{2-25}$$

If both spheres are fixed in space, the second sphere experiences a force whose density, according to the Lorentz force law, will, to the same approximation, be

$$\mathbf{f} = \rho'\mathbf{E} \approx \frac{\rho' Q}{4\pi\epsilon_0 d^2}\,\mathbf{a}_r$$

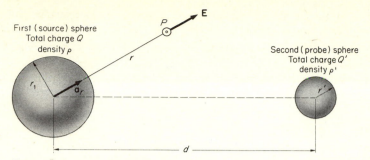

FIGURE 2-6. The interaction between two charged bodies.

The orientation of the unit vector $\mathbf{a}_r$ is evidently different at different points in the probe sphere. However, since by assumption $d \gg r'$, for every point in the second sphere the direction of $\mathbf{a}_r$ is along the dotted line joining the centers of the two spheres. Therefore the total force acting on the second sphere is obtained by integrating over the entire volume of the second sphere the force density $\mathbf{f}$. Thus

$$\mathbf{F} = \int_{V'} \mathbf{f}\, dv = \frac{Q}{4\pi\epsilon_0 d^2}\,\mathbf{a}_r \int_{V'} \rho'\, dv = \frac{QQ'}{4\pi\epsilon_0 d^2}\,\mathbf{a}_r \qquad (2\text{-}26)$$

This is *Coulomb's law*. As a formal mathematical deduction, this result is not surprising, since the first term of the Lorentz force law was deduced from Coulomb's law.

Now, if Q and Q' are both positive or both negative, the magnitude of $\mathbf{F}$ will be positive and the probe sphere will be repelled; otherwise it will be attracted by the source sphere. In either event, the force $\mathbf{F}$ is a quantity that may be measured mechanically. On the other hand, Q and Q' are assumed to be known quantities, and the distance d can be measured easily. Substituting these measurements into Eq. (2-26) permits the evaluation of ϵ_0. The value thus obtained is

$$\epsilon_0 = 8.854 \times 10^{-12} \qquad \text{F/m}$$

An interesting consequence of the foregoing conceptual experiment is that it provides a means for arriving at the notion of a *point charge* as follows.

Consider a region of space such that all linear dimensions are small compared with other macroscopic dimensions. The region is then said to be localized at one of its interior points. In the preceding experiment, the radius of the probe sphere was by assumption small relative to the distance separating the two spheres. For convenience, the probe sphere is said to be located at its center, and when taken together with its charge, it is said to constitute a point charge which is also located at the center. Note, in passing, that because of the large distance of separation, the probe sphere "sees" the source sphere as a point charge. This, in effect, is the meaning of Eq. (2-25).

Now let us return to Eq. (2-19) and specialize it to free space. For this case, it becomes

$$\frac{1}{\sqrt{\mu_0 \epsilon_0}} = c = \text{a velocity} \qquad (2\text{-}27)$$

The question is, the velocity of what? Maxwell combined his equations into a form which resulted in a wave equation. The parameter $1/\sqrt{\epsilon_0 \mu_0}$ appeared in his wave equation for free space as a propagation constant. Hence he postulated that this velocity was the velocity of propagation of electromagnetic waves in free space. A simple extension of this idea suggests that, for every medium, $1/\sqrt{\mu \epsilon}$ is the velocity of propagation of electromagnetic waves in that medium. Thus, to measure this velocity experimentally, we need to launch an electromagnetic wave and to measure its velocity. It was eventually recognized that light waves were electromagnetic. The actual early experiments which measured the free-space propagation velocity were light waves. The measurement has been performed many times and with considerable precision. The results show that $c = 2.997925 \times 10^8$ m/s. For all practical purposes, we use $c = 3 \times 10^8$ m/s as the velocity of light in free space (and in air unless we wish to be extremely precise). Having measured ϵ_0 and c, we may calculate μ_0 from

$$\frac{1}{\sqrt{\mu_0 \epsilon_0}} = 3 \times 10^8 \qquad \text{m/s}$$

The result is

$$\mu_0 = 4\pi \times 10^{-7} \qquad \text{H/m}$$

It is interesting to note that this indirect method of measurement of μ_0 has less experimental uncertainty than a direct measure of the B/H ratio in free space.

Instead of specifying ϵ and μ for a substance, it is frequently advantageous to specify values of *relative permittivity* and *relative permeability*, using as a basis of comparison the values of ϵ_0 and μ_0. Thus, by definition,

$$\text{Relative permittivity, or dielectric constant, } \epsilon_r = \frac{\epsilon}{\epsilon_0}$$

$$\text{Relative permeability } \mu_r = \frac{\mu}{\mu_0}$$

Both ϵ_r and μ_r are dimensionless numbers.

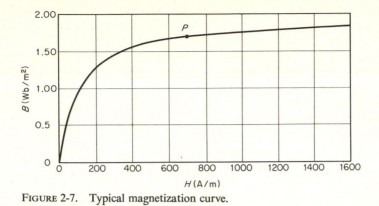

FIGURE 2-7. Typical magnetization curve.

For material media, the parameters ϵ, μ, and σ may differ from the free-space values slightly or by large amounts, may be constant or not, and may in some cases not be single-valued. In some materials the constitutive relations are of such a nature that they must be expressed in *tensor* form [Eq. (2-29)]. In any event, a complete specification of the relationships, no matter how complicated, is necessary for an electromagnetic *characterization* of the medium.

A medium is said to be *linear*, *homogeneous*, and *isotropic* if the complete characterization of the medium is given by specifying single scalar values of ϵ, μ, and σ. Not all media belong to this class. For example, the relation of **H** to **B** for all ferromagnetic materials is nonlinear and nonsingle-valued. Figure 2-7 shows a typical partial *B-H* relation for a ferromagnetic material. The nonsingle valuedness (hysteresis) has been ignored in plotting the curve. Such curves are called *mean magnetization curves*. Obviously, even this approximate curve is nonlinear, and the specification of a single number will not give the magnetic characterization of the medium even after the approximation that the *B-H* relation is single-valued. However, in some cases, variations of *B* resulting from small excursions of *H* about a point on the curve such as *P* may be of interest. It would then be possible to characterize partially such a medium by specifying a value for the *incremental permeability*, defined as

$$\mu_{\text{inc}} = \lim_{\Delta H \to 0} \frac{\Delta B}{\Delta H} = \frac{dB}{dH}\bigg|_P \tag{2-28}$$

This is completely analogous to specifying the incremental, or small-signal, parameters of an active device, such as a transistor, about a specific bias point. This kind of information, of course, does not allow us to predict the behavior of the material, or of the device, in the neighborhood of points other than *P*, or other than at the preselected bias point, as the case may be. Nor would it allow us to draw the magnetization curve of the material or the static characteristics of the device. It would simply provide a partial description of material or device behavior under specific operating conditions. Thus, in case the relation of **H** to **B** is nonlinear, a simple functional relation such as Eq. (2-22) cannot be written except when we are interested in small-signal behavior.

We have been using the terms linear, homogeneous, isotropic, and nonsingle-valued. It may be advantageous to discuss these terms at this time so that we have a more certain knowledge of their meanings.

It is presumed that the meaning of nonsingle-valued is obvious. A substance is said to be linear if, at every point in the medium, the magnitudes of the vectors **D**, **B**, and **J** are directly proportional to the magnitudes of the vectors **E**, **H**, and **E**, respectively. In other words, a medium is linear if the relations of **D** to **E**, **B** to **H**, and **J** to **E** are independent of the level of excitation.

Additionally, a material substance may be classified as homogeneous or inhomogeneous and isotropic or anisotropic. The properties of a homogeneous medium are constant from point to point in space, and the constitutive relations show no *explicit* dependence on the space coordinates. In an isotropic medium, **D** is parallel to **E**, **H** is parallel to **B**, and in a conducting medium, **J** is parallel to **E**. Free space is a linear, homogeneous, and isotropic medium. Although a material substance may be linear, homogeneous, and isotropic, its properties may be subject to changes with environmental conditions, such as temperature, pressure, and atomic radiation, and with changes in the time variations of fields. This kind of behavior deserves special consideration, and falls outside the scope of this book. For the most part, we shall mainly be concerned with fields in linear, homogeneous, and isotropic matter.

Isotropic media exhibit the same properties in all directions. Anisotropic media exhibit a rather complicated behavior. For example, in an anisotropic dielectric each rectangular component of **D** is a linear combination of the three components of **E**; that is,

$$D_x = \epsilon_{11}E_x + \epsilon_{12}E_y + \epsilon_{13}E_z$$
$$D_y = \epsilon_{21}E_x + \epsilon_{22}E_y + \epsilon_{23}E_z \qquad (2\text{-}29)$$
$$D_z = \epsilon_{31}E_x + \epsilon_{32}E_y + \epsilon_{33}E_z$$

or in matrix notation,

$$\begin{bmatrix} D_x \\ D_y \\ D_z \end{bmatrix} = \begin{bmatrix} \epsilon_{11} & \epsilon_{12} & \epsilon_{13} \\ \epsilon_{21} & \epsilon_{22} & \epsilon_{23} \\ \epsilon_{31} & \epsilon_{32} & \epsilon_{33} \end{bmatrix} \begin{bmatrix} E_x \\ E_y \\ E_z \end{bmatrix} \qquad (2\text{-}30)$$

It can be shown that the matrix $[\epsilon]$ is symmetric (Prob. 2-16). Usually, a proper choice of coordinate directions will make some of the off-diagonal terms zero.   An analogous relation may be set up between the vectors **H** and **B** in the case of a material substance such as a ferrite.   The anisotropy of ferrites constitutes the basis of operation of some very important devices used extensively in many applications, such as radar.

Characterization of a medium by specifying ϵ and μ is not the only possible way of giving the desired information.   An alternative characterization is to separate **D** and **B** into two parts,

$$\mathbf{D} = \epsilon_0 \mathbf{E} + \mathbf{P}$$
$$\mathbf{B} = \mu_0(\mathbf{H} + \mathbf{M})$$

(2-31)

and define **P** as the *electric polarization vector*, and **M** as the *magnetic polarization vector*.   If we recall that $\mathbf{D} = \epsilon_0\mathbf{E}$ and $\mathbf{B} = \mu_0\mathbf{H}$ for free space, we see that both **P** and **M** are identically zero for free space.   Thus our arbitrary separation of **D** and **B** into two factors has resulted in defining two new quantities, **P** and **M**, which are intimately related to the presence of material media.   Beyond this, the significance of these two quantities is not at all clear at this point.   However, a dimensional analysis will show that **P** is measured in coulombs per meters squared and that **M** is measured in amperes per meter.   Furthermore, by performing a few manipulations on Eqs. (2-31), we can obtain results which allow us to define a pair of parameters that give the relationship between **P** and **E** and between **H** and **M**.   To do this we write

$$\epsilon_0\mathbf{E} + \mathbf{P} = \epsilon_0\mathbf{E}(1 + \chi_e)$$
$$\mu_0(\mathbf{H} + \mathbf{M}) = \mu_0\mathbf{H}(1 + \chi_m)$$

(2-32)

where we have defined the *electric susceptibility* χ_e,

$$\mathbf{P} = \chi_e\epsilon_0\mathbf{E}$$

(2-33a)

and the *magnetic susceptibility* χ_m,

$$\mathbf{M} = \chi_m\mathbf{H}$$

(2-33b)

Both χ_e and χ_m are dimensionless quantities.   The foregoing development shows that we can also write

$$\epsilon = \epsilon_0(1 + \chi_e)$$
$$\mu = \mu_0(1 + \chi_m)$$
$$\epsilon_r = 1 + \chi_e$$
$$\mu_r = 1 + \chi_m$$

(2-34)

Some further insight can be gained into the physical identity of **P** and **M** by recasting the units for each into volume densities.   In this form we obtain

P in coulomb-meters per cubic meter (C · m/m³)

M in ampere-meters squared per cubic meter (A · m²/m³)

and noting (as we shall show in later chapters) that the coulomb-meter is the unit of *electric dipole moment* and that the ampere-meter squared is the unit of *magnetic dipole moment*.   Thus it appears† that **P** is a measure of the electric dipole moment per unit volume and that **M** is a measure of the magnetic dipole moment per unit volume.

Qualitatively speaking, materials are classified according to their most prominent characteristics.   Thus we have *conductors*, *dielectrics*, and *magnetic* materials.   Glass is a dielectric, iron is a magnetic material, and copper is a conductor.   The dielectric constant of glass is between 5 and 10, and the conductivity is about 10^{-12} ℧/m at 20°C.   The relative permeability is, for all practical purposes, equal to 1.   Copper, on the other hand, has a conductivity on the order of 10^7 to 10^8 ℧/m, while the permittivity and the permeability are essentially equal to those of free space.   So we see that behavioral differences between free space and glass are most significant insofar as the permittivity ϵ is concerned; hence the name dielectric, or *insulator*.   Likewise, copper differs from free space most significantly insofar as conduction processes are concerned; hence the name conductor.

One point is worth mentioning with regard to conductors.   Equation (2-23) shows that as the conductivity of a linear medium increases without limit, and so long as **E** remains finite, **J** approaches infinity.   But an infinitely large current density **J** implies that either an infinite amount of charge is being transported in a finite time or a finite amount of charge is transported in zero time, both events being equally unlikely from the physical point of view.   We conclude that *in a conductor of infinite conductivity the electric field must vanish at all interior points*.   The conductivity of most metals is so large (10^6 to 10^8 ℧/m) that, to a first approximation, it is infinite.   Thus, to the approximation that $\sigma = \infty$, the electric field intensity is zero, even though a current is flowing in the conductor.   Circuit theory uses this concept for the interconnections between circuit elements.   At high frequencies, this leads to some interesting phenomena, which are discussed in Chap. 11.

To summarize, the macroscopic properties of matter are established through experiment.   The constitutive relations which characterize media electromagnetically can be very complicated.   However, a large class of materials exist which are electromagnetically linear, homogeneous, isotropic, and single-valued.   For these materials, ϵ, μ, and σ have fixed scalar values

---

† A rigorous proof of this statement can be found in the following paper: D. T. Paris, A Dynamic Electromagnetic Model for Material Media, *Amer. J. Phys.*, vol. 35, pp. 1125–1127, December, 1967.

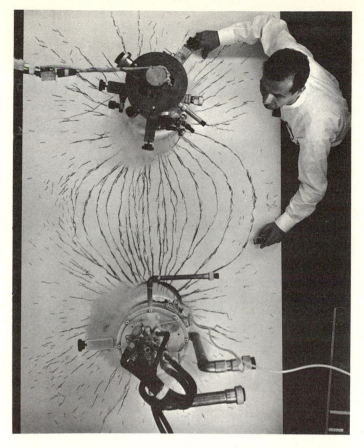

Extremely high magnetic fields (13.2 Wb/m²) are made possible by
superconducting magnets. Here a large-scale iron-filing experiment
is shown, performed by scattering thousands of iron nails on a piece of
white plywood surrounding two superconducting magnets.   When the
board is tapped sharply, the nails align themselves into the classic
picture of magnetic lines of force.   The magnets themselves are
contained within stainless-steel dewar flasks used to maintain the
liquid-helium temperatures of 4.2°K (452°F below zero) which are
required to make the coils superconducting. (*Courtesy of General
Electric Company.*)

regardless of the strength of the field and regardless of position in the medium
and are the same in all directions.   Unless otherwise noted, in this text we
shall be dealing with linear, single-valued, homogeneous, isotropic materials.

2.8 Types of Currents. It is convenient at times to classify currents
according to their physical origin.   Actually, current is always a flow of

charge, but this flow may possess distinguishable characteristics which fall in one of three broad categories, namely, those of (1) convection current, (2) conduction current, and (3) polarization current.

Convection current is the motion of free charged particles produced by some external source. For example, in an ionized medium, such as the earth's ionosphere, both positive ions and negative electrons are present which tend to move under the influence of an electromagnetic field, such as a passing radio wave. (A radio wave is a time-varying electromagnetic field which travels, or *propagates*, carrying with it a finite amount of energy.) As it goes through a medium, a radio wave exerts a force on the charges present in accordance with the Lorentz force law

$$\mathbf{f} = \rho(\mathbf{E} + \mathbf{v} \times \mathbf{B}) \tag{V}$$

In the absence of extraneous binding forces, the charges absorb energy from the field, and are forced to move in such a way that the Lorentz force $\mathbf{f}$ is just balanced by an equal and opposite mechanical force

$$\mathbf{f}_{\text{mech}} = \frac{d}{dt}(m\mathbf{v})$$

Here m is the mass of the charge-carrying bodies per unit of volume. Once in motion, the charged bodies follow specific paths as dictated by the combined effects of both $\mathbf{E}$ and $\mathbf{B}$, and not $\mathbf{E}$ alone. This point is better understood from a consideration of the equation of dynamic equilibrium:

$$\rho(\mathbf{E} + \mathbf{v} \times \mathbf{B}) = \frac{d}{dt}(m\mathbf{v}) \tag{2-35}$$

which applies to both positive ions and negative electrons.

The current density at any point will be determined by both the density of the charges and their velocity. If we denote by ρ_+ and ρ_- the densities of positive and negative charges present, and by $\mathbf{v}_+$ and $\mathbf{v}_-$ their respective velocities, then

$$\mathbf{J} = \rho_+\mathbf{v}_+ + \rho_-\mathbf{v}_- \tag{2-36}$$

is the *convection* current at a point. Equation (2-36) states that the convection current density is specified by a charge density and a corresponding velocity at a point. The current flowing between the cathode and the anode of a vacuum tube is a convection current. The most salient characteristic of convection current flow in a vacuum tube is that the region between cathode and plate is actually charged; that is, a net negative charge is always present when the cathode is heated.

Example 2-3 Convection Current. Let N denote the number of free electrons per unit volume in free space, and v_- the average velocity of the electron cloud motion resulting from the application of an external electromagnetic field. If the charge of an electron is denoted by e, this being algebraically the negative number $e = -1.60 \times 10^{-19}$ C, then the density of the cloud is

$$\rho_- = Ne$$

while the intensity and direction of its flow are, according to Eq. (2-36),

$$\mathbf{J} = \rho_-\mathbf{v}_- = Ne\mathbf{v}_-$$

If we orient a surface ΔA normal to the direction of flow, a total current

$$I = Nev_- \,\Delta A$$

will cross every element of area ΔA on that surface.

A conductor is the name given to a material medium which, although electrically neutral, contains charged particles which are free to move under the influence of an externally applied electromagnetic field. The charged particles, however, move only a short distance between collisions with other particles, and hence the motion is one of migration. The net charge migration is called *conduction current*.

Metals, semiconductors, and ionic solutions are examples of conducting media. Since the gross features of all conduction processes are similar, we shall examine the details only for a metallic conductor.

A metallic conductor is a crystalline solid composed of an ordered array of atoms. The positive ions are fixed, while the electrons are free to move. An externally applied electromagnetic field exerts forces on these free electrons, causing them to move. Once in motion, all electrons occasionally collide with the fixed positive lattice, giving up part or all of their energy. In the process, some of the energy of the electrons is dissipated as heat. In a particular crystalline lattice, an initially stationary electron may be displaced by an electron originally in motion. The average drift velocity of the electrons depends both on the strength of the applied field and on the *mobility* of electrons; mobility, in general, is different in different metals. Thus the average drift velocity of the electrons is

$$\mathbf{v}_d = \mu_d\mathbf{E} \tag{2-37}$$

where μ_d is the mobility of the electrons and is measured in meters squared per volt-second. If N is the number of electrons per unit volume taking part in the motion, and e is the electronic charge, the formula

$$\mathbf{J} = \rho_-\mathbf{v}_d = Ne\mu_d\mathbf{E} \tag{2-38}$$

gives the density of current flow at a point in the conductor. Comparing Eq. (2-37) with Eq. (2-38), we see that

$$\sigma = Ne\mu_d \tag{2-39}$$

As stated previously, the density of total (or net) charge is zero at every interior point of a conductor, even when conduction current is present.

Both convection currents and conduction currents involve the movement of *free* charges. *Polarization* current, by contrast, results from the motion of *bound* charges. In a *perfect* dielectric material, electrons are *bound* to their parent atoms by strong atomic forces, and it takes exceptionally high field intensities to *break* them down, that is, to produce free electrons so that they can move as in a conductor. (No dielectric material is perfect. A conduction current is always present, even though it may be very small in magnitude.)

The current which is of more interest in dielectrics is the polarization current resulting from the relative motion of a bound positive charge and a bound negative charge. The charge may be molecular (two parts of a molecule), ionic (such as the Na^+ and the Cl^- ions of sodium chloride), and/or electronic, where the negative electrons move relative to the nearly stationary positive nucleus. In any event, the electric field will exert a force on the positive charge in one direction and a force on the negative charge in the other direction. Since the charges are bound, no charge motion can be produced by a static electric field other than the initial transient motion which occurred when the field was applied. There can, however, be a static displacement of the positive charge relative to the negative charge. This relative charge displacement is called *electric polarization*, and it contributes to that part of the electric flux density **D** given by the polarization vector **P**. The *polarizability* of the medium, which is related to the electric polarization per unit electric field strength, is accounted for in the permittivity ϵ of the medium by the electric susceptibility χ_e, and is the reason why $\epsilon \neq \epsilon_0$ for polarizable materials. (Almost all dielectrics are polarizable.) If the applied electromagnetic field is time-dependent, the electric polarization will change as the applied field changes. The charge motion expressed as a current flow,

$$\mathbf{J}_p = \rho_+\mathbf{v}_+ + \rho_-\mathbf{v}_- \qquad (2\text{-}40)$$

is called the *polarization current*. *Note carefully that although this polarization current is a true current in the sense that we have defined it, Maxwell's equations, as we have written them, include this current* in $\partial \mathbf{D}/\partial t$, *and not in* **J**. It is accounted for by $\epsilon \neq \epsilon_0$, just as in the static case. However, since the dynamic polarizability, or polarization per unit electric field strength, involves the ability of the charges to *follow* the applied field changes, we should expect the polarizability (and therefore the permittivity ϵ) to be frequency-dependent. Experiment verifies this expectation. Fortunately, for most materials, and for most frequencies of interest, the frequency variation is slow, and ϵ exhibits a monotonically decreasing value.

To summarize, it is convenient to divide currents into three groups, each having its own characteristics. Convection current flows when *free* charges, of known density and velocity, are set in motion. Conduction current flows when *free* charges drift under the influence of an electric field. Polarization current flows when *bound* charges move within an atomic, or molecular, structure. Polarization and polarization current are accounted for in Maxwell's equations by specifying the permittivity ϵ of the medium. The current density **J** appearing in Maxwell's second equation includes *only* the convection and conduction currents.

2.9 Derivation of the Boundary Conditions.

The behavior of the field vectors at surfaces, marking abrupt changes in the constitutive properties of the medium (which usually correspond to surfaces between different material media), were given as an essential part of the differential form of the field equations. The behavior of the field vectors at surfaces of discontinuity in the media are called *boundary conditions.*

Let us restate the necessity for supplementing the differential equations with boundary conditions. From a formal mathematical point of view the differential field equations state relationships for *derivatives* of functions. Obviously, these relationships have no meaning at points of discontinuity of the functions because the derivatives have no meaning at such points. Mathematically, the behavior at such points can be handled by examining the discontinuity of the function. Our electromagnetic boundary conditions do precisely this job. It is the purpose of this section to show the derivation of the boundary conditions.

Physically, the transition of the field vectors across boundaries between dissimilar media is manifested in everyday phenomena. For example, reception by automobile radios is subject to significant changes as the car passes through a tunnel or under a bridge; in some cases reception ceases completely. This example clearly shows the dependence of a field upon the constitution of material media, and also suggests that, generally, the field vectors will change at a boundary between two different media.

Consider Fig. 2-8. The surface S marks the boundary between two arbitrary media, simply labeled 1 and 2. At a point P on S we erect a unit normal vector **n**, which points from medium 1 into medium 2, and a unit vector **t** tangent to S at P. The orientation of the unit tangent vector **t** is immaterial so long as the surface S is smooth. We also construct about the same point P a small rectangular path C_0 in the plane of **n** and **t**, with sides parallel to **n** and **t**, respectively.

For the sake of generality, we consider that charge is distributed on S with a density of ρ_s C/m² and that a current flow resulting from motion of this distribution may exist on S. This kind of current flow does not occur in nature ordinarily; however, it is mathematically a useful concept and a

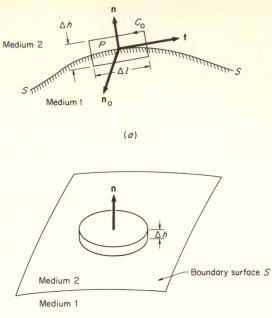

FIGURE 2-8.   Behavior of the field vectors at the boundary between two media.   (a) The transition of the tangential components of **E** and **H**; (b) the transition of the
normal components of **B** and **D**.

good approximation to physical reality, especially in the case of rapidly
changing fields in conducting regions of space, such as copper.   The current
is then confined to a very thin layer on the outer surface of the conductor, in
the form of a surface flow, rather than volume flow, of charge.

The symbol **J** was previously used to denote a volume distribution of
current, the unit of which, we recall, was the ampere per meter squared.   To
distinguish a surface distribution from a volume distribution of current, we
must use a different symbol.   In this text we use the symbol **K**, which we
call the *linear current-density vector*.   The unit of **K** is the ampere per meter.

To illustrate, let us imagine a distribution of surface current **K** on a flat
copper surface.   If we draw a line normal to the direction of current flow,
charge will cross that line at a rate of K coulombs per second per meter length
of line measured at the point of observation.   If the distribution is uniform,
then K C/s will cross every meter length of line per second.

Let us now apply the first Maxwell equation to the configuration of
Fig. 2-8a.   Denoting by S_0 the surface bounded by C_0, we have

$$\oint_{C_0} \mathbf{E} \cdot d\mathbf{l} = -\frac{d}{dt} \int_{S_0} \mathbf{B} \cdot \mathbf{n}_0 \, da \qquad (2\text{-}41)$$

where $n_0 = t \times n$ is the unit normal to S_0 and, like t, is tangent to S at P. Now we let C_0 shrink about P by allowing the height Δh of the rectangle to become progressively smaller. In the process, the area of S_0 will become vanishingly small, and the surface integral on the right-hand side of Eq. (2-41) will approach zero. Bearing in mind that the contribution to the contour integral from the left and right sides of the rectangle will also vanish with Δh, we can write in place of Eq. (2-41) the simpler relation

$$E_{t_1} \Delta l - E_{t_2} \Delta l = 0 \qquad (2\text{-}42)$$

Here E_{t_1} and E_{t_2} are the tangential components of the field vector E, respectively, in media 1 and 2. The minus sign accounts for the reversal in positive sense of traversing C_0 relative to an arbitrary system of coordinates during the transition from medium 1 into medium 2. If Δl is small but not zero, it can be divided out from both sides of Eq. (2-42). Then

$$E_{t_1} - E_{t_2} = 0 \qquad (2\text{-}43)$$

or in formal mathematical notation,

$$n \times (E_2 - E_1) = 0 \qquad (VI)†$$

This is the desired result. It states: *At the interface between any two media the tangential component of the vector E is continuous.*

An identical approach may be used to establish the transition of the tangential component of H at a surface of discontinuity. For this purpose, we use Eq. (II) in the form

$$\oint_{C_0} H \cdot dl = \int_{S_0} J \cdot n\, da + \frac{d}{dt} \int_{S_0} D \cdot n_0\, da \qquad (2\text{-}44)$$

Again, we let C_0 shrink about P by allowing the height Δh to approach zero. In the limit, the last term on the right will vanish because, like all field components, D is bounded; the left side will reduce to $t \cdot (H_1 - H_2) \Delta l$, thus giving

$$t \cdot (H_1 - H_2) = \lim_{\Delta l \to 0} \frac{1}{\Delta l} \int_{S_0} J \cdot n_0\, da \qquad (2\text{-}45)$$

In precise mathematical language, the right-hand member of Eq. (2-45) is the projection of the linear current density vector K in the direction of n_0. Dimensionally, the quantities on the left and right sides of Eq. (2-45) are measured in units of amperes per meter. Now if we rotate the plane of C_0, along with t, $90°$ about n, and repeat the same limiting process, we shall arrive at an expression similar in form to that of Eq. (2-45), but one which shows the relations of the tangential component of H to surface current in a

† Roman numerals designate the formal mathematical statements of the boundary conditions throughout the book.

plane normal to that of the surface S_0; that plane is normal to the plane of the paper in Fig. 2-8a. Adding this new equation to Eq. (2-45) produces the general result

$$\mathbf{n} \times (\mathbf{H}_2 - \mathbf{H}_1) = \mathbf{K} \tag{VII}$$

Equation (VII) states: *At the interface between two media, the tangential component of the vector* **H** *is discontinuous by an amount equal to the value of the linear current density vector* **K** *at the point.* In addition, Eq. (VII) establishes the orientation of **K** relative to the direction of the magnetic field intensity at the surface of discontinuity.

Two special cases of Eq. (VII) which occur frequently should be singled out for special consideration. Suppose the conductivity of medium 1 is infinitely large. Then $E_1 = 0$ inside medium 1, and according to the first differential law, Eq. (I),

$$0 = \nabla \times \mathbf{E}_1 = -\frac{\partial \mathbf{B}_1}{\partial t}$$

We all agree that the generation of all time-varying fields occurred some time in the finite past. Therefore this equation can be satisfied only by the condition $\mathbf{B}_1 = 0$. A finite μ_1 then implies that H_1 is also zero, and we have the final result

$$\mathbf{n} \times \mathbf{H}_2 = \mathbf{K}_1 \qquad \sigma_1 \to \infty \tag{VIIa}$$

For all practical purposes, the requirement $\sigma_1 \to \infty$ can be met only at relatively high frequencies. Then Eq. (VIIa) gives a relation between the current flowing on the surface of a perfect conductor and the tangential component of the magnetic field intensity just outside the conductor. If medium 1 is not a perfect conductor, the field inside cannot vanish, and current flow cannot be confined just on the surface. Then

$$\mathbf{n} \times (\mathbf{H}_2 - \mathbf{H}_1) = 0 \qquad \sigma_1, \sigma_2 \text{ finite} \tag{VIIb}$$

is the condition that must be satisfied in place of (VIIa). Thus, according to Eq. (VIIb), *the transition of the tangential component of* **H** *across the boundary between two media, neither of which is a perfect conductor, is continuous.*

Let us imagine now that the contour C_0 is allowed to turn through a complete revolution about the unit normal **n**, thus forming a cylindrical *pillbox*, as indicated in Fig. 2-8b. Let Σ denote the closed surface so formed. By virtue of the third integral law, we have

$$\oint_\Sigma \mathbf{B} \cdot \mathbf{n} \, da = 0 \tag{3}$$

where the integration extends over the curved wall and flat end surfaces of the cylinder. Let the cylindrical volume shrink about the interior point (which, as before, is a point on S) by allowing Δh once again to approach zero. If the base has a sufficiently small area Δa, Eq. (3) may be approximated thus:

$$(B_{n_2} - B_{n_1}) \Delta a = 0 \qquad (2\text{-}46)$$

We should remember, of course, that since **B** is finite, the contribution of the walls to the surface integral is vanishingly small. The negative sign in Eq. (2-46) comes about as a result of the fact that the surface outward normal must point downward in medium 1, and that **B** must be measured in the same frame of reference in both media. Therefore, as $\Delta h \to 0$, we obtain

$$\mathbf{n} \cdot (\mathbf{B_2} - \mathbf{B_1}) = 0 \qquad (\text{VIII})$$

as a boundary condition for the magnetic field. This condition may be expressed as follows: *At the boundary between two material media the normal component of* **B** *is continuous.*

To establish the transition of the electric flux density vector **D** across the surface S, we consider the field equation

$$\oint_{\Sigma} \mathbf{D} \cdot \mathbf{n} \, da = \int_{V} \rho \, dv \qquad (4)$$

Let Σ denote the surface of the pillbox in Fig. 2-8b, and V its volume. Once again, let the cylinder shrink into the surface S. In the limit, when $\Delta h \to 0$, the contribution from the cylindrical wall to the surface integral on the left side of Eq. (4) becomes vanishingly small, and we have the result

$$\lim_{\Delta h \to 0} \oint_{\Sigma} \mathbf{D} \cdot \mathbf{n} \, da = (D_{n_2} - D_{n_1}) \Delta a \qquad (2\text{-}47)$$

In Eq. (2-47), Δa denotes the area of the base. If any surface charge is present, the right-hand member of Eq. (4) becomes

$$\lim_{\Delta h \to 0} \oint_{V} \rho \, dv = \rho_s \, \Delta a \qquad (2\text{-}48)$$

where ρ_s denotes the density of the surface distribution of charge. Combining Eqs. (2-47) and (2-48) gives

$$D_{n_2} - D_{n_1} = \rho_s \qquad (2\text{-}49)$$

or

$$\mathbf{n} \cdot (\mathbf{D_2} - \mathbf{D_1}) = \rho_s \qquad (\text{IX})$$

Equation (**IX**) states: *The normal component of* **D** *is discontinuous by an amount equal to the surface charge density present at the interface between two media.*

To summarize, we have obtained four relations, called boundary conditions, which complete the differential statement of the laws by establishing the transition of an electromagnetic field from one medium to another.

These conditions are

$$\mathbf{n} \times (\mathbf{E}_2 - \mathbf{E}_1) = 0 \qquad \text{(VI)}$$

$$\mathbf{n} \times (\mathbf{H}_2 - \mathbf{H}_1) = \mathbf{K} \qquad \text{(VII)}$$

$$\mathbf{n} \cdot (\mathbf{B}_2 - \mathbf{B}_1) = 0 \qquad \text{(VIII)}$$

$$\mathbf{n} \cdot (\mathbf{D}_2 - \mathbf{D}_1) = \rho_s \qquad \text{(IX)}$$

and, together with the differential laws, they describe completely the properties of a field.   Here $\mathbf{K}$ and ρ_s represent surface densities of current and charge, respectively.

In instances where it is desirable to express $\mathbf{D}$ and $\mathbf{B}$ in terms of the polarization vectors $\mathbf{P}$ and $\mathbf{M}$, we obtain by direct substitution into Eqs. (VIII) and (IX) the alternative forms

$$\mathbf{n} \cdot (\mathbf{H}_2 - \mathbf{H}_1) = \mathbf{n} \cdot (\mathbf{M}_1 - \mathbf{M}_2) \qquad \text{(VIIIa)}$$

$$\mathbf{n} \cdot \epsilon_0 (\mathbf{E}_2 - \mathbf{E}_1) = \rho_s + \mathbf{n} \cdot (\mathbf{P}_1 - \mathbf{P}_2) \qquad \text{(IXa)}$$

2.10   A Formal Statement of the Principle of the Conservation of Charge.   Maxwell's equations contain implicitly the principle of the conservation of charge.   To display the principle explicitly, we manipulate the equations as follows.   Starting with

$$\mathbf{\nabla} \times \mathbf{H} = \mathbf{J} + \frac{\partial \mathbf{D}}{\partial t} \qquad \text{(II)}$$

$$\mathbf{\nabla} \cdot \mathbf{D} = \rho \qquad \text{(IV)}$$

we take the divergence of Eq. (II) and obtain

$$\mathbf{\nabla} \cdot (\mathbf{\nabla} \times \mathbf{H}) = \mathbf{\nabla} \cdot \mathbf{J} + \mathbf{\nabla} \cdot \frac{\partial \mathbf{D}}{\partial t}$$

Noting that

$$\mathbf{\nabla} \cdot (\mathbf{\nabla} \times \mathbf{H}) \equiv 0$$

and interchanging the order of differentiation in the second term on the right, we obtain

$$0 = \mathbf{\nabla} \cdot \mathbf{J} + \frac{\partial}{\partial t} (\mathbf{\nabla} \cdot \mathbf{D})$$

Substituting Eq. (IV) into this equation gives us the desired result:

$$\mathbf{\nabla} \cdot \mathbf{J} + \frac{\partial \rho}{\partial t} = 0 \qquad (2\text{-}50)$$

This equation formally states that current and charge cannot be specified completely independently of each other.   More precisely, it states mathematically the conservation of charge in the neighborhood of a point.   In integral form, it provides a clearer view of the situation. Let us therefore integrate Eq. (2-50) over a volume V surrounded by a smooth closed surface Σ. We obtain

$$\int_V \mathbf{\nabla} \cdot \mathbf{J} \, dv = -\int_V \frac{\partial \rho}{\partial t} \, dv \qquad (2\text{-}51)$$

Using the divergence theorem, we can transform the left-hand member of Eq. (2-51) into a surface integral. Assuming that Σ is stationary, the order of integration and differentiation on the right may be interchanged, and Eq. (2-51) may then be written

$$\oint_\Sigma \mathbf{J} \cdot \mathbf{n} \, da = -\frac{d}{dt} \int_V \rho \, dv \qquad (2\text{-}52)$$

If we let

$$Q = \int_V \rho \, dv \qquad (2\text{-}53)$$

be the net charge contained in V, we finally have

$$\oint_\Sigma \mathbf{J} \cdot \mathbf{n} \, da = -\frac{dQ}{dt} \qquad (2\text{-}54)$$

as the integral form of the equation of continuity. According to our customary convention, the positive normal $\mathbf{n}$ to the closed surface Σ is drawn outward. Therefore the left-hand member of Eq. (2-54) represents the net amount of charge which in unit time crosses the surface Σ in the outward direction. The right-hand member represents the rate of depletion of charge contained within Σ. Therefore, in words, Eq. (2-54) states:

$$\text{Net outward flow of charge in unit time} = \text{depletion rate of charge contained within the volume}$$

This is equivalent to the statement that charge is conserved. It cannot appear or disappear from a volume without passing in or out through its boundary.

The continuity equation can be used to predict the disposition of charge in a charged conducting region. Let the region be linear, homogeneous, and isotropic, and let ϵ and σ be, respectively, the permittivity and conductivity. Then

$$\mathbf{D} = \epsilon \mathbf{E} \qquad (2\text{-}55)$$

and

$$\mathbf{J} = \sigma \mathbf{E} \qquad (2\text{-}56)$$

at every point in the medium. Let us substitute Eq. (2-55) in the fourth differential law and also substitute Eq. (2-56) into Eq. (2-50). We have

$$\nabla \cdot \epsilon \mathbf{E} = \rho$$

and
$$\nabla \cdot \sigma \mathbf{E} + \frac{\partial \rho}{\partial t} = 0$$

Combining this pair of equations leads to

$$\frac{\partial \rho}{\partial t} + \frac{\sigma}{\epsilon} \rho = 0 \qquad (2\text{-}57)$$

Therefore, at any instant of time, the distribution of charge will be

$$\rho = \rho_0 e^{-(\sigma/\epsilon)t} \qquad (2\text{-}58)$$

where ρ_0 denotes the density at the time $t = 0$. Both ρ and ρ_0 are, in general, functions of position in the region.

From Eq. (2-58) it is apparent that charge cannot exist permanently in the interior of a conducting region. Instead, an initial charge distribution will eventually diminish at a rate fixed by the *relaxation time constant*

$$\tau = \frac{\epsilon}{\sigma} \qquad \text{s} \qquad (2\text{-}59)$$

This is the time in which the initial distribution of charge will diminish to $1/e$ of its original value. (That the unit of τ is the second may easily be deduced from a simple dimensional analysis.)

The physical implication of Eq. (2-58) is that any amount of charge, placed within a conducting body, must ultimately decay to zero at every point *within* the conductor. Since charge is conserved, the initial deployment, no matter how arbitrary, will be followed instantaneously by an outward current flow, which will ultimately deposit the entire charge on the outer surface of the conductor. This process will be entirely independent of an applied field, if one is present. In the absence of an externally applied time-varying field, the transient phenomenon described by Eq. (2-58) will culminate in a surface distribution of charge which will give zero internal field.

The rearrangement of charges within a conductor begins at the very instant of time when the charge is placed within. A measure of the time it takes for the migration of charges to be completed is provided by the value of the relaxation time τ. In copper, this time is 1.5×10^{-19} *second;* in silver, it is 1.3×10^{-19} *second;* in distilled water, it is 10^{-6} *second.* In direct contrast, $\tau = 10$ *days* in fused quartz. In general, the better the conductor, the lower the value of τ for that substance.

Example 2-4 An Initially Charged Spherical Region. Consider a conducting sphere
  of radius R surrounded by free space. The parameters of the sphere are scalar
  constants ϵ, μ, and σ. The parameters of free space are, of course, ϵ_0, μ_0, and $\sigma_0 = 0$.

Assume that at $t = 0$ charge is distributed uniformly throughout the conducting sphere with a charge density ρ_0. We wish to find the current distribution inside the sphere and the electric field intensity vector both inside and outside the sphere as functions of position and time.

First, we find the initial fields $\mathbf{D}_0$, $\mathbf{E}_0$, and $\mathbf{J}_0$, which we shall define as the fields at $t = 0$. Initially, the charge distribution is uniform and spherically symmetric. From the results of Example 2-1 it is clear that, for $r < R$,

$$\mathbf{D}_0 = \frac{\rho_0 r}{3} \mathbf{a}_r$$

$$\mathbf{E}_0 = \frac{\rho_0 r}{3\epsilon} \mathbf{a}_r$$

Consequently, inside the sphere,

$$\mathbf{J}_0 = \sigma \mathbf{E}_0 = \frac{\rho_0 \sigma r}{3\epsilon} \mathbf{a}_r$$

and outside the sphere,

$$\mathbf{D}_0 = \frac{\rho_0 R^3}{3r^2} \mathbf{a}_r$$

$$\mathbf{E}_0 = \frac{\rho_0 R^3}{3\epsilon_0 r^2} \mathbf{a}_r$$

$$\mathbf{J}_0 = 0$$

Applying the equation of continuity for $r < R$, we obtain

$$\left. \frac{\partial \rho}{\partial t} \right|_{t=0} = -\boldsymbol{\nabla} \cdot \mathbf{J}_0 = -\frac{\sigma}{\epsilon} \rho_0 \qquad r < R$$

We find that the rate of change of charge density is independent of r, and hence the charge density remains uniform as required by Eq. (2-58). With this result we can write down the fields. We have, for $r < R$,

$$\mathbf{D} = \frac{\rho_0 r e^{-(\sigma/\epsilon)t}}{3} \mathbf{a}_r$$

$$\mathbf{E} = \frac{\rho_0 r e^{-(\sigma/\epsilon)t}}{3\epsilon} \mathbf{a}_r$$

$$\mathbf{J} = \frac{\rho_0 \sigma r e^{-(\sigma/\epsilon)t}}{3\epsilon} \mathbf{a}_r$$

and for $r > R$,

$$\mathbf{D} = \frac{\rho_0 R^3}{3r^2} \mathbf{a}_r$$

$$\mathbf{E} = \frac{\rho_0 R^3}{3\epsilon_0 r^2} \mathbf{a}_r$$

$$\mathbf{J} = 0$$

It is also interesting to calculate the surface charge density on the sphere. Using the boundary condition, Eq. (IX), and the previous results, we obtain

$$\rho_s = \frac{\rho_0 R}{3} \left(1 - e^{-(\sigma/\epsilon)t}\right)$$

Summarizing, the principle of conservation of charge states that charge can neither be created nor destroyed.   This principle is tacitly implied by the field laws.

Charge and current cannot be specified completely independently of each other.

At every interior point of a linear, homogeneous, and isotropic conducting substance, there can never be a permanent distribution of charge.

2.11 Power and Energy.   One of the more spectacular properties of an electromagnetic field is its ability to transfer energy over long distances even in the absence of an intervening medium.   No other form of energy can be transported even a short distance in the absence of a material medium.   The power and energy aspects of electromagnetic fields are implicitly contained in the field equations.   To obtain explicit relations which display the energy and power behavior, we must manipulate the field equations properly and then examine the significance of the results.

We begin with the vector identity, Eq. (1-82), by applying it to the vectors **E** and **H**.   We obtain

$$\mathbf{\nabla} \cdot (\mathbf{E} \times \mathbf{H}) = \mathbf{H} \cdot (\mathbf{\nabla} \times \mathbf{E}) - \mathbf{E} \cdot (\mathbf{\nabla} \times \mathbf{H})$$

Substituting Maxwell's equations

$$\mathbf{\nabla} \times \mathbf{E} = -\frac{\partial \mathbf{B}}{\partial t} \tag{I}$$

$$\mathbf{\nabla} \times \mathbf{H} = \mathbf{J} + \frac{\partial \mathbf{D}}{\partial t} \tag{II}$$

into this identity, we obtain

$$\mathbf{\nabla} \cdot (\mathbf{E} \times \mathbf{H}) = -\mathbf{H} \cdot \frac{\partial \mathbf{B}}{\partial t} - \mathbf{E} \cdot \mathbf{J} - \mathbf{E} \cdot \frac{\partial \mathbf{D}}{\partial t} \tag{2-60}$$

or

$$-\mathbf{\nabla} \cdot (\mathbf{E} \times \mathbf{H}) = \mathbf{E} \cdot \mathbf{J} + \left(\mathbf{E} \cdot \frac{\partial \mathbf{D}}{\partial t} + \mathbf{H} \cdot \frac{\partial \mathbf{B}}{\partial t}\right) \tag{2-61}$$

This equation is known as the *differential form of Poynting's theorem.* Integrating over a volume V bounded by the closed surface Σ and applying the divergence theorem to the left-hand side of the equation, we find

Total P flowing out

$$-\oint_{\Sigma} (\mathbf{E} \times \mathbf{H}) \cdot \mathbf{n}\, da = \int_{V} \mathbf{E} \cdot \mathbf{J}\, dv + \int_{V} \left(\mathbf{E} \cdot \frac{\partial \mathbf{D}}{\partial t} + \mathbf{H} \cdot \frac{\partial \mathbf{B}}{\partial t}\right) dv \tag{2-62}$$

This equation is known as the *integral form of Poynting's theorem.*   By dimensional analysis it can be seen that each term in Poynting's theorem is measured in newton-meters per second (N $\cdot$ m/s) = joules per second (J/s) = watts (W).   That is to say, Poynting's theorem is a power theorem, and

High-voltage spark ignited in a tank of water creates intense shock waves which can remove automatically the coating of manganese or copper—and possibly other metals—which is electrolytically deposited onto cathodes in metal refineries.   (*Courtesy of General Electric Company.*)

amounts to an electromagnetic statement of the principle of conservation of energy.   We shall now give an interpretation of each term in Poynting's theorem.

The term on the left is the net *inward* flux of a vector,

$$\mathbf{S} = \mathbf{E} \times \mathbf{H} \qquad \text{W/m}^2 \qquad\qquad (2\text{-}63)$$

into the volume V.   Since this is the only such term in the equation, we interpret this term as the net total power delivered to the volume by outside sources or fields.   If the net inward flux is negative, we interpret it to mean that power is delivered *from* the volume *to* outside sources, or fields.   The vector **S** is called the *Poynting vector*, and is usually interpreted as being the *actual* power flow density at each point on the surface Σ.   This gives rise to some subtle philosophical questions, but no practical difficulties.

The volume integral

$$\int_V \mathbf{E} \cdot \mathbf{J} \, dv \tag{2-64}$$

which for convection currents can be written

$$\int_V \mathbf{E} \cdot (\rho_+ \mathbf{v}_+ + \rho_- \mathbf{v}_-) \, dv \tag{2-65}$$

is seen to represent the rate at which the electric field does work on the charges in *V*.   This term may be positive or negative.   If positive, a net amount of work per unit time is being done by the field on the charges.   If negative, a net amount of power is being transferred from the charges to the field.   For example, if the volume contains only conducting material, $\mathbf{J} = \sigma \mathbf{E}$, and

$$\int_V \mathbf{E} \cdot \mathbf{J} \, dv = \int_V \sigma |E|^2 \, dv > 0 \tag{2-66}$$

and the integral represents the ohmic losses in the volume, that is, the conversion of electromagnetic energy into heat.   A simple example of a case where this integral is negative occurs when *V* is the volume occupied by a battery.   In this case the integral represents the rate of conversion of chemical energy into electric energy.   More complex situations, particularly dynamic situations, will be more complicated than indicated by these simple examples, but in all cases the integral represents the *net* rate of energy transfer from the field to the charges in the volume.

Since the integral

$$\int_V \left(\mathbf{E} \cdot \frac{\partial \mathbf{D}}{\partial t} + \mathbf{H} \cdot \frac{\partial \mathbf{B}}{\partial t} \right) dv \tag{2-67}$$

contains only the field vectors and their time derivatives, it is apparent that this term must be interpreted as the net rate at which the energy in the volume is increasing in the volume due to the electromagnetic field.   This becomes even more apparent when the relations of **D** to **E** and of **B** to **H** are both linear and the medium is isotropic.   For, in this case,

$$\int_V \left(\mathbf{E} \cdot \frac{\partial \mathbf{D}}{\partial t} + \mathbf{H} \cdot \frac{\partial \mathbf{B}}{\partial t} \right) dv = \frac{d}{dt} \int_V \left(\frac{1}{2} \epsilon E^2 + \frac{1}{2} \mu H^2 \right) dv \tag{2-68}$$

since

$$\mathbf{E} \cdot \frac{\partial \mathbf{D}}{\partial t} = \mathbf{E} \cdot \frac{\partial (\epsilon \mathbf{E})}{\partial t} = \frac{1}{2} \epsilon \frac{\partial E^2}{\partial t} \tag{2-69}$$

$$\mathbf{H} \cdot \frac{\partial \mathbf{B}}{\partial t} = \mathbf{H} \cdot \frac{\partial (\mu \mathbf{H})}{\partial t} = \frac{1}{2} \mu \frac{\partial H^2}{\partial t} \tag{2-70}$$

From our study of mechanics we know that power is the time rate of change of energy. Hence

$$W = \int_V \left(\frac{1}{2} \epsilon E^2 + \frac{1}{2} \mu H^2 \right) dv \qquad (2\text{-}71)$$

is the total electromagnetic energy stored within the volume. The unit of W is the watt-second, or the joule.

It is customary to conclude from Eq. (2-71) that, throughout a linear and isotropic medium, *electric energy* is distributed with a density $w_e = \frac{1}{2}\epsilon E^2$ and *magnetic energy* is distributed with a density $w_m = \frac{1}{2}\mu H^2$. We hasten to add that the notion of localized energy can be justified only when regarded from a gross point of view. Also, the differential form of Poynting's theorem, Eq. (2-61), may be regarded as an expression of the principle of conservation of energy at a point only from the macroscopic point of view, according to which energy is distributed continuously throughout a volume. These interpretations, like the interpretation of Poynting's vector, lead to some philosophical difficulties, since it is hardly realistic for us to say *where* the energy is when we are not even sure of its identity. However, the application of this interpretation yields correct results for total energy (after integration). If we wish to be careful and precise, we should say: Linear and isotropic media *behave as though* the energy *were* distributed in them with electric and magnetic energy densities of $w_e = \frac{1}{2}\epsilon E^2$ and $w_m = \frac{1}{2}\mu H^2$.

We may summarize by observing that Poynting's theorem follows directly from the field equations and expresses the conservation of electromagnetic energy in the form

Net inward power flow through a closed surface	$=$	power imparted to charges in the volume enclosed by the surface	$+$	rate of change of energy stored in the volume

2.12 Properties and Types of Fields. A substantial part of our effort thus far has been devoted to a discussion of accepted terminology and a general presentation of the laws which govern the behavior of electromagnetic fields. We have developed fundamental ideas in a manner that transcends specific applications. But in order to proceed further, we must first undertake a fairly complete study of special types of fields, beginning with the simplest and advancing to the more complex. We begin this approach in Chap. 3, and continue it systematically throughout the remainder of the text. This purpose can best be served by classifying fields according to their most dominant properties.

Before beginning the field classification, let us observe a few features which are common to all electromagnetic fields.

First, we note that every macroscopic field obeys Maxwell's equations. These equations are linear; hence, if in a linear medium we superimpose two fields, each of which satisfies Maxwell's equations, the combined field will also satisfy these equations. If, in addition, each field satisfies a set of boundary conditions, the sum of the two fields will satisfy the same set of boundary conditions.

Second, in a bounded region, the solution of Maxwell's equations is unique; that is, only one solution exists which satisfies a prescribed set of boundary conditions. This can be shown rigorously.† To obtain a unique solution in the general time-varying case, the following information is required: (1) the initial values of electric and magnetic field vectors throughout the region, and (2) the values of the tangential component of either the electric vector or of the magnetic vector over the boundary of the region at all instants of time $t \geqq 0$. If the properties of the region are linear and isotropic, a solution of Maxwell's equations satisfying these conditions is unique. In our study of special types of fields, we shall encounter several special forms of this general uniqueness theorem.

The primary classification of fields is based on their time dependence. Fields which do not change with time are called *static*. Fields which change with time are called *dynamic*. This definition of dynamic fields, although seemingly precise, in reality is not. One almost immediately asks, how dynamic? Even this question requires a clarification, as indicated by the question, dynamic with respect to what? As a classification, dynamic fields are subdivided and called *quasistatic* if certain features of the field may be analyzed as though the field were static, and are called *time-varying* if the important properties of the field are not given by a static analysis. The distinction is clarified slightly below, and is discussed in detail in Chap. 11. A precise understanding of the distinction will not be required in the chapters preceding Chap 11.

Static fields may be further classified into three types according to the nature of the sources and the properties of the region in which the fields exist.

If the source of field is purely a static distribution of charges, the field is called *electrostatic*. This field has several features which are obvious, or they will become apparent in our discussion in Chap. 3. The outstanding features of an electrostatic field are:

1.   The only nonzero field vectors are **E** and **D**.
2.   The field is *conservative*, that is, for any contour C, $\oint_C \mathbf{E} \cdot d\mathbf{l} \equiv 0$.
3.   Conductors can have no internal electrostatic field, and they are therefore *equipotential* regions.

† J. A. Stratton, "Electromagnetic Theory," chap. 9, McGraw-Hill Book Company, New York, 1941.

If the source of the field is purely a static current distribution, and the region where the current distribution exists has a high enough conductivity so that the electric field intensity in the conducting region can be neglected, the field is called *magnetostatic*.   The outstanding features of a magnetostatic field are:

1.   The only nonzero field vectors are **B** and **H**.
2.   The field is *not* conservative, but is *solenoidal*; that is, for any surface Σ,

$$\oint_{\Sigma} \mathbf{B} \cdot \mathbf{n} \, da \equiv 0.$$

An electrostatic and a magnetostatic field are completely *uncoupled*. That is to say, each one can be determined completely independently of the other.

If the sources of the field are time-independent currents, flowing in a conducting region whose conductivity is low enough so that the electric field intensity in the conducting region *cannot* be neglected, the field is called *electromagnetostatic*.   The outstanding features of an electromagnetostatic field are:

1.   Each field vector is nonzero.
2.   The electric field is electrostatic, and the magnetic field is magnetostatic, *but* the two fields are *coupled* together by the relation $\mathbf{J} = J(\mathbf{E})$ in the conducting region.
3.   The electric field in the conducting region gives rise to a static charge distribution, usually on the surface of the conductors, which are the *sources* for the electrostatic fields.
4.   In the analysis of an electromagnetostatic field, it is necessary, generally, to determine the electrostatic field first (sometimes only in the conducting region).   After we have determined the electrostatic field, we can calculate the current density vector **J**, through $\mathbf{J} = J(\mathbf{E})$, and then we can determine the magnetostatic field.

Time-varying fields have the following important characteristics:

1.   The electric and magnetic fields are always *coupled*.
2.   The coupling through $\mathbf{J} = J(\mathbf{E})$ remains in conducting regions, but the significant additional coupling is through the time derivatives which appear in the field equations.
3.   The fields exhibit wave phenomena and a finite propagation velocity. These effects give rise to a time lag, called *retardation*.
4.   The field produces a power flow which can be intimately associated with the wave phenomena.   This power flow may be *guided* by the material media present, or it may be *radiated*.   Transmission lines of various kinds guide the power flow.   Antennas radiate power.

Quasistatic fields are those time-varying fields for which the wave phenomena, propagation velocity, and time delay can be neglected. On this basis, remembering that the propagation velocity is always finite, it should be obvious that time-varying fields in infinite regions are *never* quasistatic and that *all* time-varying fields are quasistatic at a *single point* in space. Thus quasistatic must be defined on the basis of the size of the region in relation to the rate of change with time and the propagation velocity. Quasistatic fields are discussed extensively in Chap. 11. At that time, the significance of the above definition of quasistatic should become clear.

Example 2-5   Charged Parallel Plates. As an example of a formal application of Maxwell's equations and the boundary conditions, let us determine the field of two equal and opposite distributions of charge on the surfaces of two parallel conducting plates. For simplicity, we shall assume that the plates are in free space.

We shall make the further simplifying assumption that fringing may be neglected. In mathematical language, this means that, with reference to Fig. 2-9a, $\partial/\partial y$ and $\partial/\partial z$ are equal to zero.

Let a and b be the dimensions of each plate, and d their distance of separation. Let ρ_s be the density of surface charge on the top plate, and $-\rho_s$ the density on the lower plate. Since the charges are stationary, the resulting field is independent of time, and $\mathbf{J} = 0$. The differential laws at a point not on the plates, consequently, reduce to the simpler forms

$$\nabla \times \mathbf{E} = 0 \tag{2-72}$$

$$\nabla \times \mathbf{H} = 0 \tag{2-73}$$

$$\nabla \cdot \mathbf{B} = 0 \tag{2-74}$$

$$\nabla \cdot \mathbf{D} = 0 \tag{2-75}$$

We may eliminate two of the four field vectors by introducing the constitutive relations

$$\mathbf{D} = \epsilon_0 \mathbf{E} \qquad \mathbf{B} = \mu_0 \mathbf{H}$$

Since ϵ_0 and μ_0 are both constants, Eqs. (2-74) and (2-75) may be replaced, respectively, by

$$\nabla \cdot \mathbf{H} = 0$$

$$\nabla \cdot \mathbf{E} = 0$$

Thus $\mathbf{E}$ must satisfy the pair

$$\nabla \times \mathbf{E} = 0 \qquad \nabla \cdot \mathbf{E} = 0$$

and $\mathbf{H}$ must satisfy the similar pair

$$\nabla \times \mathbf{H} = 0 \qquad \nabla \cdot \mathbf{H} = 0$$

Since the equations in $\mathbf{E}$ and $\mathbf{H}$ are uncoupled, we shall deal with the $\mathbf{E}$ field first.

Using the definitions of curl and divergence, and remembering the hypothesis $\partial/\partial y = \partial/\partial z = 0$, we expand the field equations in $\mathbf{E}$ to obtain

From $\nabla \times \mathbf{E} = 0$: $\qquad\qquad -\dfrac{\partial E_z}{\partial x}\, \mathbf{a}_y + \dfrac{\partial E_y}{\partial x}\, \mathbf{a}_z = 0 \tag{2-76}$

From $\nabla \cdot \mathbf{E} = 0$: $\qquad\qquad \dfrac{\partial E_x}{\partial x} = 0 \tag{2-77}$

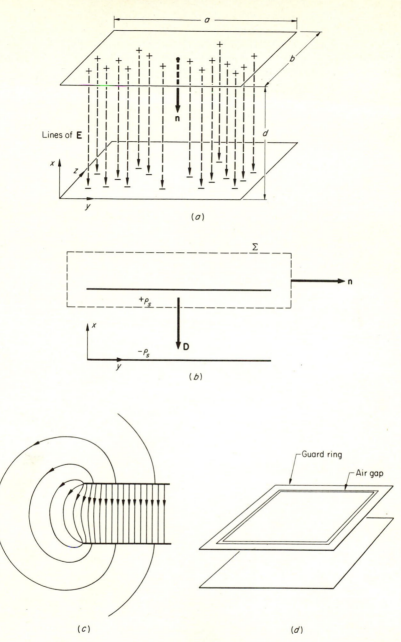

FIGURE 2-9. The field between two charged plates. (*a*) Uniform charge distribution; (*b*) application of Gauss' law; (*c*) nonuniform charge distribution; (*d*) guard ring.

From Eq. (2-76) we conclude that both E_z and E_y must be constants independent of x, y, and z. However, the boundary conditions require that they should both vanish on the conducting plates; otherwise the continuity of tangential $\mathbf{E}$ would require the existence of an $\mathbf{E}$ field within the conducting plates. This $\mathbf{E}$ field in turn would exert a Lorentz force on the charges, causing them to move, contrary to the assumed stationary character of the charge distribution. We see that this condition may be satisfied only if we set

$$E_y = E_z = 0$$

On the other hand, from Eq. (2-77) we conclude that

$$E_x = \text{constant } C$$

To evaluate the constant C, we calculate

$$D_x = \epsilon_0 E_x = \epsilon_0 C$$

and recall the boundary condition

$$\mathbf{n} \cdot (\mathbf{D}_2 - \mathbf{D}_1) = \rho_s \tag{IX}$$

Let us consider the upper plate, which we denote medium 1. If medium 2 is the space between the plates, the unit normal to the upper plate will point downward, as shown in Fig. 2-9a. Since $\mathbf{E}_1$ is zero, $\mathbf{D}_1$ in Eq. (IX) is zero. With $\mathbf{n} = -\mathbf{a}_x$, Eq. (IX) gives

$$-\mathbf{a}_x \cdot D_x \mathbf{a}_x = \rho_s$$

Therefore $$D_x = -\rho_s$$

The same result would be obtained by repeating the procedure for the lower plate. Therefore

$$C = -\frac{\rho_s}{\epsilon_0}$$

and at any point between the plates

$$\mathbf{E} = C\mathbf{a}_x = -\frac{\rho_s}{\epsilon_0}\mathbf{a}_x \tag{2-78}$$

The same formal procedure can be used to obtain the $\mathbf{H}$ field. The student will verify that, since there are no currents anywhere, the result is $\mathbf{H} = 0$, everywhere.

The same problem may be solved by a totally different and much easier approach. Let us imagine that the upper conducting plate is completely surrounded by a surface Σ, in the form of a parallelepiped whose top and bottom sides are parallel to the plates. This is shown in Fig. 2-9b. According to Gauss' law for the electric field,

$$\oint_{\Sigma} \mathbf{D} \cdot \mathbf{n}\, da = \int_V \rho\, dv \tag{4}$$

Because fringing is neglected, the top and vertical sides of the parallelepiped contribute nothing to the left member of this equation. Moreover, since $\partial/\partial y = \partial/\partial z = 0$, $\mathbf{D}$ is the same at every point in the space between plates. We have no a priori knowledge about the direction of $\mathbf{D}$. Let us assume, however, that it is downward, as indicated in Fig. 2-9b. The algebraic sign of $\mathbf{D}$ will ultimately establish the actual direction. Since ρ_s is constant, Eq. (4) now gives

$$-ab\,D_x = ab\,\rho_s$$

from which we obtain $D_x = -\rho_s$. Hence the displacement vector between plates is given by

$$\mathbf{D} = D_x \mathbf{a}_x = -\rho_s \mathbf{a}_x$$

and the electric field intensity vector is given by

$$\mathbf{E} = \frac{1}{\epsilon_0} \mathbf{D} = -\frac{\rho_s}{\epsilon_0} \mathbf{a}_x$$

This is obviously in complete agreement with Eq. (2-78).

It is important to note that when a system of charges is placed on a pair of conducting surfaces, such as those of Fig. 2-9a, they do not distribute themselves uniformly. Fringing will occur, as indicated in Fig. 2-9c, although the field will be much weaker outside the region bounded by the plates. A good approximation to the uniform field depicted in Fig. 2-9a may be obtained by making both a and b much larger than d, or by the *guard-ring* arrangement shown in Fig. 2-9d.

For completeness, let us finally calculate the total energy stored in the field.

Equation (2-78) shows that the magnitude of the electric field is $E = \rho_s/\epsilon_0$. Therefore the electric energy density is

$$w_e = \tfrac{1}{2}\epsilon_0 E^2 = \tfrac{1}{2}\epsilon_0 \left(\frac{\rho_s}{\epsilon_0}\right)^2$$

Since w_e is independent of the space coordinates, the total energy stored in the field is given by

$$W = \int_V w_e \, dv = \int_V \frac{\rho_s^2}{2\epsilon_0} \, dv = \frac{\rho_s^2}{2\epsilon_0} abd$$

We may summarize this section by stating that electromagnetic fields are classified as static or time-dependent and that each of these classifications is further subdivided according to the nature of the sources, the characteristics of the region, and/or other dominant characteristics of situations which prevail. The various types of fields are studied in detail in the remainder of this text.

2.13 Electromagnetic Potentials. Maxwell's equations contain the four vector fields $\mathbf{D}$, $\mathbf{E}$, $\mathbf{B}$, and $\mathbf{H}$ and the two source fields $\mathbf{J}$ and ρ. The four field equations, together with the subsidiary equations $\mathbf{B} = B(\mathbf{H})$, $\mathbf{D} = D(\mathbf{E})$, $\mathbf{J} = J(\mathbf{E})$, appear mathematically to provide sufficient information so that any of the field vectors may be expressed in terms of $\mathbf{J}$ and ρ. This is indeed true. In fact, Maxwell's equations state that, given complete information about $\mathbf{J}$ and ρ, and given the subsidiary relations everywhere, *all* the field vectors are uniquely determined everywhere.

The above problem, although straightforward in concept, is sometimes complicated in application. Searching for alternative (and hopefully, in some cases, less complicated) approaches to the problem has led to the idea of electromagnetic potentials. The question to be answered is this: Can we define a vector field $\mathbf{A}$ and a scalar field ϕ† in such a manner that we can

† This scalar field, ϕ, is to be distinguished from the coordinate φ throughout the text.

first determine $\mathbf{A}$ and ϕ from a knowledge of $\mathbf{J}$ and ρ and then calculate the field vectors $\mathbf{D}$, $\mathbf{E}$, $\mathbf{B}$, and $\mathbf{H}$ from $\mathbf{A}$ and ϕ? The answer to this question, of course, is yes, since it only really asks whether we can separate the original problem into two parts, one of which is to find the part of the vector fields caused by current distributions and, separately, to find the part of the vector fields caused by charge distributions. The linearity of Maxwell's equations allows us to superimpose the two partial solutions in this manner. The functions $\mathbf{A}$ and ϕ are called *electromagnetic potentials*. Use of electromagnetic potentials is of particular advantage in linear, homogeneous, and isotropic regions. For this reason, our derivation and definition of them will be restricted to such regions.

To begin our derivation, we need Maxwell's equations and several items from vector calculus. First, recall that Helmholtz's theorem states that both the curl and the divergence must be given for a vector. Also note that a scalar can be specified to within an additive constant by giving its gradient. Recall also the vector identities

$$\nabla \cdot (\nabla \times \mathbf{A}) \equiv 0 \qquad\qquad (1\text{-}69)$$

$$\nabla \times \nabla\phi \equiv 0 \qquad\qquad (1\text{-}70)$$

Now, from Maxwell's equation,

$$\nabla \cdot \mathbf{B} = 0 \qquad\qquad \textbf{(III)}$$

we see that we can define $\nabla \times \mathbf{A}$ as

$$\mathbf{B} = \nabla \times \mathbf{A} \qquad\qquad (2\text{-}79)$$

and still satisfy $\nabla \cdot \mathbf{B} = \nabla \cdot (\nabla \times \mathbf{A}) = 0$.

Next, we substitute Eq. (2-79) into Eq. (**I**). We obtain

$$\nabla \times \mathbf{E} = -\frac{\partial}{\partial t}(\nabla \times \mathbf{A})$$

Since time differentiation and the curl operation are commutative, this can be rewritten

$$\nabla \times \left(\mathbf{E} + \frac{\partial \mathbf{A}}{\partial t}\right) = 0 \qquad\qquad (2\text{-}80)$$

In view of Eq. (1-70) above, the quantity within parentheses in Eq. (2-80) may be equated to the gradient of some scalar, and be within an additive vector whose curl vanishes; that is, we may write

$$\mathbf{E} + \frac{\partial \mathbf{A}}{\partial t} = -\nabla\phi \qquad\qquad (2\text{-}81)$$

and neglect the additive vector because it can always be lumped in with $\nabla\phi$. The negative sign is arbitrary, and is chosen for convenience.

We have now defined $\nabla \times \mathbf{A}$ and $\nabla\phi$. Also, we have the rules by which $\mathbf{B}$ and $\mathbf{E}$ may be calculated from $\mathbf{A}$ and ϕ. These rules are

$$\mathbf{B} = \nabla \times \mathbf{A} \tag{2-79}$$

$$\mathbf{E} = -\nabla\phi - \frac{\partial \mathbf{A}}{\partial t} \tag{2-82}$$

and since we are restricting ourselves to linear, homogeneous, and isotropic media, we also have

$$\mathbf{H} = \frac{1}{\mu}\,(\nabla \times \mathbf{A})$$

$$\mathbf{D} = -\epsilon\left(\nabla\phi + \frac{\partial \mathbf{A}}{\partial t}\right)$$

We still need to define $\nabla \cdot \mathbf{A}$, and we need to obtain the equations which give the relations between $\mathbf{A}$ and $\mathbf{J}$ and between ϕ and ρ. To accomplish both of these objectives, we write Maxwell's equations (**II**) and (**IV**) in terms of the potentials and obtain

$$\nabla \times \nabla \times \mathbf{A} + \mu\epsilon\nabla\frac{\partial\phi}{\partial t} + \mu\epsilon\frac{\partial^2\mathbf{A}}{\partial t^2} = \mu\mathbf{J} \tag{2-83}$$

and

$$\nabla^2\phi + \nabla \cdot \frac{\partial \mathbf{A}}{\partial t} = -\frac{1}{\epsilon}\,\rho \tag{2-84}$$

We now define $\nabla \cdot \mathbf{A}$ as

$$\nabla \cdot \mathbf{A} = -\mu\epsilon\frac{\partial\phi}{\partial t} \tag{2-85}$$

This is often called the *Lorentz condition,* or the *Lorentz gauge.* Using this definition, we obtain

$$\nabla \times \nabla \times \mathbf{A} - \nabla(\nabla \cdot \mathbf{A}) + \mu\epsilon\frac{\partial^2\mathbf{A}}{\partial t^2} = \mu\mathbf{J}$$

and

$$\nabla^2\phi - \mu\epsilon\frac{\partial^2\phi}{\partial t^2} = -\frac{1}{\epsilon}\,\rho$$

These two equations can be put in a symmetrical form by using the vector identity

$$\nabla \times \nabla \times \mathbf{A} = \nabla(\nabla \cdot \mathbf{A}) - \nabla^2\mathbf{A}$$

to obtain

$$\nabla^2\mathbf{A} - \mu\epsilon\frac{\partial^2\mathbf{A}}{\partial t^2} = -\mu\mathbf{J} \tag{2-86}$$

and

$$\nabla^2\phi - \mu\epsilon\frac{\partial^2\phi}{\partial t^2} = -\frac{1}{\epsilon}\,\rho \tag{2-87}$$

From Eqs. (2-86) and (2-87) it is apparent that ϕ and each of the three rectangular components of **A** are solutions of partial differential equations with identical analytical forms. This facilitates the analysis of electromagnetic problems considerably because the general solution of one equation automatically leads to the general solution of the remaining three.

From the foregoing development it is clear that the problem of obtaining solutions of Maxwell's equations is reduced to an equivalent problem of obtaining solutions of Eqs. (2-86) and (2-87). When the right member is zero, each of the four scalar equations assumes the form of the *wave equation*

$$\nabla^2 \psi - \mu\epsilon \frac{\partial^2 \psi}{\partial t^2} = 0 \qquad (2\text{-}88)$$

where ψ represents either ϕ or any one of the rectangular components of **A**. When the potentials are independent of time, Eq. (2-87) reduces to *Poisson's equation*

$$\nabla^2 \phi = -\frac{1}{\epsilon} \rho \qquad (2\text{-}89)$$

which in turn simplifies to *Laplace's equation*

$$\nabla^2 \phi = 0 \qquad (2\text{-}90)$$

when a region is free of charge. We shall encounter these equations quite frequently in our study of electromagnetic fields.

The functions **A** and ϕ are called the *potentials* of a field, **A** being the *vector potential* and ϕ the *scalar potential*. The unit of **A** is the weber per meter, and the unit of ϕ is the volt, as can easily be deduced from Eqs. (2-79) and (2-82), respectively. The difference in the functional values of ϕ at two points in space is precisely what is defined in circuit theory as potential difference between two points. The limitations involved in this definition are pointed out in Chap. 11.

Example 2-6   Potential Distribution between Two Charged Parallel Plates. Let us illustrate the use of potentials in the solution of the problem considered in Example 2-5. Since the region between the plates is assumed to be free space, Laplace's equation

$$\nabla^2 \phi = \frac{\partial^2 \phi}{\partial x^2} + \frac{\partial^2 \phi}{\partial y^2} + \frac{\partial^2 \phi}{\partial z^2} = 0$$

applies, and under the assumption of negligible fringing, this equation simplifies to

$$\frac{d^2 \phi}{dx^2} = 0$$

The general solution of this ordinary differential equation is given by

$$\phi = C_1 x + C_2$$

where C_1 and C_2 are constants which must be evaluated from a knowledge of boundary conditions. We know that a surface distribution of charge ρ_s exists on the plate at $x = d$, while an equal and opposite distribution of charge exists on the plate at $x = 0$. Since charge marks a discontinuity of the field vector $\mathbf{D}$, an expression for $\mathbf{D}$ must be obtained from the known functional expression for ϕ. Thus, from Eq. (2-82), with $\partial \mathbf{A}/\partial t = 0$, we have

$$\mathbf{E} = -\nabla\phi = -\frac{d\phi}{dx}\mathbf{a}_x = -C_1\mathbf{a}_x$$

Using $\mathbf{D} = \epsilon_0\mathbf{E}$, we obtain

$$\mathbf{D} = -\epsilon_0 C_1 \mathbf{a}_x$$

From the boundary condition

$$\mathbf{n} \cdot (\mathbf{D}_2 - \mathbf{D}_1) = \rho_s \qquad\qquad \textbf{(IX)}$$

noting that, at $x = d$, $\mathbf{D}_1 = 0$ and $\mathbf{n} = -\mathbf{a}_x$, we find

$$-\mathbf{a}_x \cdot (-\epsilon_0 C_1 \mathbf{a}_x) = \rho_s$$

From this it follows that $C_1 = \rho_s/\epsilon_0$ and that

$$\mathbf{D} = -\rho_s\mathbf{a}_x$$

$$\mathbf{E} = -\frac{1}{\epsilon_0}\rho_s\mathbf{a}_x$$

These answers are precisely the same as those obtained in Example 2-5. Oddly enough, the function ϕ is known only to within an additive constant, because the constant C_2 is still unknown. Thus

$$\phi = \frac{\rho_s}{\epsilon_0}x + C_2$$

However, the constant of integration C_2 can be evaluated by choosing *arbitrarily* a reference level for ϕ. This is analogous to choosing the surface of the earth as a reference level, or a datum plane, for specifying the potential energy of a body lifted a certain distance above the surface.

The zero level for ϕ is usually taken to be at the point at infinity. But in this problem it is found more convenient to assign zero as the potential of the lower plate. Both plates must be at constant potentials because a potential which changes from point to point gives rise to a nonzero electric field

$$\mathbf{E} = -\nabla\phi$$

which in turn sets charges in motion in a conductor. This contradicts the assumption of static charge distributions on both plates. Thus the lower plate must be at a certain constant potential, and the upper plate must be at some other fixed potential.

In accordance with our (arbitrary) choice of $\phi = 0$ at the lower plate, we have

$$0 = \phi = \frac{\rho_s}{\epsilon_0}(0) + C_2$$

which gives $C_2 = 0$. Therefore

$$\phi = \frac{\rho_s}{\epsilon_0}x$$

is the required solution. It is seen that at $x = d$,

$$\phi = \frac{\rho_s}{\epsilon_0}d$$

Note that this field can be set up by connecting a battery of $V = \rho_s d/\epsilon_0$ volts across the parallel plates, with the positive terminal tied to the upper plate and the negative terminal connected to the *grounded* lower plate. Thus *grounding* implies fixing a zero reference level for the scalar potential ϕ.

To summarize, the solution of Maxwell's equations is often simplified by formulating an equivalent problem in the vector and scalar potentials. Subject to the Lorentz gauge, these potentials are solutions of a pair of equations, being derivable from Maxwell's equations only when the medium is linear, homogeneous, and isotropic.

2.14 The Development of the Present System of Electromagnetic Units.

Many different physical quantities are governed by the laws of physics. Force, velocity, volume, momentum, and energy are different physical quantities, as are length, mass, and time. It is accepted convention to treat the latter three as fundamental quantities and use the laws of physics to express all other physical quantities in terms of these fundamental quantities. An example is force, expressed as ML/T^2, where M, L, and T represent mass, length, and time, respectively.

We have grown accustomed to learning that mass, length, and time are fundamental quantities, and often do not realize that any other three physical quantities could have been chosen as fundamental. The main reason that mass, length, and time were taken as fundamental was that to do so created a simple, consistent system with accurately reproducible standard units.

The student is invited to pick three quantities, derive his own system, and see how unwieldy the derived units may become (Prob. 2-21).

In the field of mechanics, two systems of units are quite popular, the *cgs* and the *mks* systems. The units in these systems for length, mass, and time are centimeter, gram, and second and meter, kilogram, and second, respectively.

Since the inception of electromagnetics as a discipline within its own right, various systems of units, based upon the three accepted fundamental quantities and one other quantity, which varies with each particular system, have been used to measure the quantities mentioned earlier in this text.

The mksa system   Since 1948 the internationally accepted system of electromagnetic standards has been the meter, the kilogram, the second, and the *absolute* ampere (adopted by the Ninth General Conference on Weights and Measures). The absolute ampere is defined as that current which, when flowing in each of two infinitely long parallel wires of negligible cross-sectional area, separated by a distance of 1 m in vacuum, causes a transverse force per unit length of 2×10^{-7} N/m to act between the wires. Thus the absolute ampere is a derived quantity in terms of mechanical quantities.

The mksa system is an extension of the mechanical mks system to the electromagnetic domain by introducing one more fundamental unit, current. In this system of units the fundamental units are

$$\text{Length } (L) = \text{meter}$$

$$\text{Mass } (M) = \text{kilogram}$$

$$\text{Time } (T) = \text{second}$$

$$\text{Current } (I) = \text{ampere}$$

The ampere is selected as the fourth fundamental unit because it is a convenient size for practical applications and because of the ease in standardization of its measurement.

Next, charge is defined as the integral of the current, and therefore the units of charge are ampere × time, which is defined as coulombs.

The mksc system† In this system, charge is taken as the fourth fundamental quantity. The fundamental unit for charge is the coulomb, and may be defined to be 6.24×10^{19} times the charge of one electron. Current, then, is the rate of flow of charge, and an ampere is equal to 1 C/s.

Rationalized units If Maxwell's equations were expressed in the mksa or the mksc forms, they would have factors of 4π present in various places. The process which eliminates 4π from Maxwell's equations is called rationalization. In a rationalized system, μ_0 has a value of 4π times its value in the unrationalized system, and ϵ_0 has a value of $\frac{1}{4}\pi$ times its value in the unrationalized system. By rationalizing in this manner, the formula relating μ_0, ϵ_0, and the speed of light remains the same in both systems.

It may be noted here that the rmksc system is the *most popular among electrical engineers.*

The cgs esu system The cgs esu (electrostatic unit) system uses the three mechanical units of the cgs system as fundamental units and ϵ_0 as the fourth fundamental unit.

ϵ_0 is set equal to unity (dimensionless), and then all electromagnetic variables are expressed in terms of centimeters, grams, and seconds. It is permissible to set $\epsilon_0 = 1$ since the fundamental units of ϵ_0 are arbitrary.

† The units for physical quantities which are used in this text follow in the main the International System (Appendix 1). Defined and given official status in a resolution of the Eleventh General Conference on Weights and Measures at a 1960 Paris meeting, the International System of Units has also been adopted by the National Bureau of Standards. For a complete discussion, see C. H. Page et al., IEEE Recommended Practice for Units in Published Scientific and Technical Work, *IEEE Spectrum*, vol. 3, no. 3, pp. 169–173, March, 1966.

Thus the fundamental units are

$$\text{Length } (L) = \text{centimeter}$$
$$\text{Mass } (M) = \text{gram}$$
$$\text{Time } (T) = \text{second}$$
$$\epsilon_0 = 1 \text{ (dimensionless)}$$

The selection of the magnitude and units of the electric permittivity ϵ_0 is the principal characteristic of the cgs esu system.

Derived units have the same names as in the rmksc system except for a *stat-* prefixed to these names. For example, voltage is measured in statvolts (299.79 V), and current is measured in statamperes (3.356×10^{-10} A). The unit of charge is the statcoulomb and has the (strange) dimensions $cm^{3/2} \cdot g^{1/2} \cdot s^{-1}$. Interestingly, capacitance has the dimensions of centimeters. The unit of charge is obtained from the following experiment. Two charges of equal magnitude and sign are placed 1 cm apart. If the force on each charge is one dyne (1 dyne = 1 g $\cdot$ cm $\cdot$ s^{-2}), then each charge has a magnitude of 1 statC.

The cgs emu system   The cgs emu (electromagnetic unit) system also uses the three mechanical units of the cgs system as three fundamental units, but μ_0 is chosen as the fourth fundamental unit. For simplicity, the magnetic permeability μ_0 is given a value of unity with no dimensions. The fundamental units are

$$\text{Length } (L) = \text{centimeter}$$
$$\text{Mass } (M) = \text{gram}$$
$$\text{Time } (T) = \text{second}$$
$$\mu_0 = 1 \text{ (dimensionless)}$$

The choice of μ_0 is the distinguishing feature of this system of units. Again, all electromagnetic variables are defined in terms of centimeters, grams, and seconds. Potential is named abvolts (10^{-8} V); current, abamperes (10 A); and resistance abohm. A unit of current is defined as the current that must pass through two infinitely long wires, separated by 1 cm, to produce a force of 1 dyne on each wire if equal currents flow through the wires.

Gaussian system   The gaussian system uses five fundamental units:

$$\text{Length } (L) = \text{centimeter}$$
$$\text{Mass } (M) = \text{gram}$$
$$\text{Time } (T) = \text{second}$$
$$\epsilon_0 = 1 \text{ (dimensionless)}$$
$$\mu_0 = 1 \text{ (dimensionless)}$$

Since μ_0, ϵ_0, and c (the speed of light in free space) cannot all be specified independently, extra constants must be introduced into Maxwell's equations to make them dimensionally correct.

The chief advantage of this system is that fundamental equations do not have ϵ_0's and μ_0's contained in them.

2.15†   Electromagnetic Characterization of a Medium.‡   In stating Maxwell's equations, we asserted that the comparatively simple relations

$$\mathbf{D} = D(\mathbf{E})$$

$$\mathbf{B} = B(\mathbf{H}) \tag{2-91}$$

$$\mathbf{J} = J(\mathbf{E})$$

were valid and were sufficient to complete the description of the interrelations between the field quantities $\mathbf{E}$, $\mathbf{B}$, $\mathbf{D}$, $\mathbf{H}$, $\mathbf{J}$, and ρ. This section presents a heuristic argument showing that this assertion is indeed correct.

We begin by taking the divergence of both sides of Eq. (I).

$$\mathbf{\nabla} \cdot (\mathbf{\nabla} \times \mathbf{E}) = \mathbf{\nabla} \cdot \left(-\frac{\partial \mathbf{B}}{\partial t} \right) \tag{2-92}$$

By virtue of Eq. (1-69), the left-hand side of this equation vanishes identically. The resulting expression

$$\frac{\partial}{\partial t} (\mathbf{\nabla} \cdot \mathbf{B}) = 0 \tag{2-93}$$

implies that the scalar quantity $\mathbf{\nabla} \cdot \mathbf{B}$ is independent of time at every point in space; that is, it depends only on the space coordinates. This means that the field must have always been in existence, is now in existence, and will continue in existence forever, contrary to physical reality, unless of course

$$\mathbf{\nabla} \cdot \mathbf{B} = 0 \tag{III}$$

at every point in space and for every instant of time. We have thus discovered, much to our surprise perhaps, that if we accept the basic premise that every time-varying field currently in existence was first established in the finite past, we can derive Eq. (III) from Eq. (I) using straightforward mathematical techniques. Stated differently, we have discovered that, of the four

† This section may be omitted with no loss in continuity.
‡ D. T. Paris, Maxwell's Equations and Determinate Systems, *Am. J. Phys.*, vol. 34, pp. 618–619, July, 1966.

Maxwell equations, not more than three can possibly be independent when fields are dependent on time.   On the other hand, the statement of an electromagnetic problem in general may not include a complete specification of the sources; instead, the field itself may be specified over a limited region of space, and the problem is then to determine the field everywhere else.   This means that not only **E**, **B**, **D**, and **H** are to be determined from Maxwell's equations, but also **J** and ρ, giving a total of six unknowns to be obtained from a set of three equations.   Evidently, a unique solution is not at all possible.   We conclude that Maxwell's equations must be supplemented by at least three additional relations that result in a system of equations which can be solved uniquely.

Second, the first and third field equations involve only **E** and **B**, and the remaining two involve only **D** and **H**.   It appears, then, that no *coupling* exists between the two pairs of equations and that an electromagnetic field is not one single field but a superposition of two simpler fields.   However, all available experimental evidence suggests that there exists but one field. Therefore, to be consistent with experiment, our mathematical model (field equations) must imply the existence of but a single field.

Third, Maxwell's equations do not show explicit dependence of fields upon the physical constitution of a medium.   This is contrary to our everyday experience.   We saw earlier, for instance, that an electromagnetic field exhibits a definite dependence on the physical properties of a medium. Therefore we must find a means somehow to introduce such properties as may affect a field in the definition of our discipline.   In other words, we must ultimately be able to *characterize*, or *label*, or *identify* media according to their electromagnetic behavior.

Fourth, although the integral laws permit us to establish the transition of certain field components at the interface between two media, they do not enable us to predict the behavior of all of them.   Specifically, they do not enable us to predict the behavior of the normal components of **E** and **H** at a boundary, nor that of the tangential components of **B** and **D**.   It is clear that further conditions must be imposed if we expect to write boundary conditions for E_n, H_n, B_t, and D_t.

Granted, then, that Maxwell's equations do not (and well they should not) show explicit dependence of the field laws upon the physical constitution of a medium, our problem is, from a purely axiomatic point of view, one of identification or characterization: *to arrive at a single model of minimum complexity which reflects completely the electromagnetic behavior of matter under all possible conditions of excitation.*   Granted also that, between the six quantities **E**, **B**, **D**, **H**, **J**, and ρ, Maxwell's equations impose but three independent relations (in the time-varying case at least), this model must enable us to write three additional relations which reflect the local interactions between fields and matter and which, together with three independent field equations, constitute a mathematically determinate system. This same model must also enable us to establish

1. The transition of the normal components of **E** and **H** and of the tangential components of **B** and **D** across the surface separating two media (this information is generally not included under the classical heading of boundary conditions)
2. A coupling between a pair of seemingly disjointed pairs of equations (**E** and **B** appear only in two Maxwell equations, while **D**, **H**, **J**, and ρ appear in the other two)
3. Relations among various field quantities the general validity of which is not contingent upon the character of the field variations with time

It is clear, then, that in order to establish mathematically the constraining effects of a substance upon a field, three constitutive relations of the general form

$$G_1(\mathbf{E},\mathbf{B},\mathbf{D},\mathbf{H},\mathbf{J},\rho) = 0$$
$$G_2(\mathbf{E},\mathbf{B},\mathbf{D},\mathbf{H},\mathbf{J},\rho) = 0 \qquad (2\text{-}94)$$
$$G_3(\mathbf{E},\mathbf{B},\mathbf{D},\mathbf{H},\mathbf{J},\rho) = 0$$

must be derived which express any three field quantities in terms of the other three without resorting to Maxwell's equations, and which, furthermore, satisfy the following requirements:

1. Each relation is strictly local; that is, it does not involve derivatives with respect to the space coordinates or time.
2. Each relation is valid under static as well as dynamic conditions.
3. Each relation is independent of the other two relations and is also independent of Maxwell's equations. Since it is our aim to construct a model possessing maximum simplicity, any duplication of information already contained in Maxwell's equations is to be avoided.

Though not shown explicitly in Eqs. (2-94), the functions G_1, G_2, and G_3 are dependent upon other parameters, such as temperature. However, it is easy to see that the generality of our discussion will not be affected significantly if we disregard altogether all such side dependences. On the other hand, the foregoing triplet of relations is quite general from another point of view, in that it does not preclude the possibility of describing an arbitrary interaction between fields and matter.

Now, a glance at Maxwell's equations reveals at once that the following relations exist:

1. A direct coupling between the vectors **E** and **B** [Eq. (I)]. This coupling

disappears, however, when the derivative of the field quantities with respect to time is identically equal to zero.

2. Explicit relations between the vectors **H**, **J**, and **D** [Eq. (**II**)]. However, when a field is stationary with time, the vector **D** is not coupled either to **H** or to **J**.

3. A direct coupling between **D** and ρ [Eq. (**IV**)] which exists under static as well as dynamic conditions.

In addition, by virtue of the principle of conservation of charge,

$$\mathbf{\nabla \cdot J} + \frac{\partial \rho}{\partial t} = 0 \qquad (2\text{-}95)$$

a relation is seen to exist in the dynamic case between the quantities **J** and ρ. We conclude that the pairs of variables

$$\textbf{(E,B)} \quad \textbf{(D,H)} \quad \textbf{(D,J)} \quad \textbf{(D,}\rho\textbf{)} \quad \textbf{(H,J)} \quad \textbf{(H,}\rho\textbf{)} \quad \textbf{(J,}\rho\textbf{)} \quad (2\text{-}96)$$

ought not to appear simultaneously as independent variables in any one of the functions G_1, G_2, or G_3. This in turn implies that each function can depend at most on two field quantities; in fact, each function can depend only on one of the pairs

$$\textbf{(E,D)} \quad \textbf{(E,H)} \quad \textbf{(E,J)} \quad \textbf{(E,}\rho\textbf{)} \quad \textbf{(B,D)} \quad \textbf{(B,H)} \quad \textbf{(B,J)} \quad \textbf{(B,}\rho\textbf{)}$$

$$(2\text{-}97)$$

Thus we see that eight pairs of independent variables are admissible. Now, to show that only three of these meet all the requirements set forth initially, we must appeal to further evidence.

The Lorentz force density expression, Eq. (**V**), written, with $\mathbf{J} = \rho\mathbf{v}$ (Sec. 2.8),

$$\mathbf{f} = \rho\mathbf{E} + \mathbf{J \times B} \qquad (2\text{-}98)$$

and the behavior of fields in the stationary state provide the clues we need for this purpose. Thus, if charges are stationary, $\mathbf{v} = 0$, then $\mathbf{J} = 0$ and $\mathbf{f} = \rho\mathbf{E}$, and this is true for time-varying as well as for time-independent fields. So **E** and ρ are always related to each other. Since, as noted earlier, **D** and ρ are also related to each other, it follows that **E**, **D**, and ρ always *go together*. Similar reasoning leads us to conclude that **B**, **H**, and **J** always *go together*.

Thus one admissible pair of variables is either **(E,D)** or **(E,**ρ**)**. However, since G_1, G_2, and G_3 should involve no space and time derivatives, it follows that no relation can possibly be written between **E** (effect) and ρ (cause) which does not violate Maxwell's equations. Hence the variables **(E,D)** constitute one appropriate pair of variables which may be used to transform, for example, the first of Eqs. (2-93) to

$$G_1(\mathbf{E},\mathbf{D}) = 0 \qquad (2\text{-}99)$$

or to a so-called constitutive relation

$$\mathbf{D} = D(\mathbf{E}) \qquad (2\text{-}100)$$

For the same reasons, the pair $(\mathbf{B},\rho)$ is also eliminated from further consideration, leaving only

$$(\mathbf{E},\mathbf{H}) \quad (\mathbf{E},\mathbf{J}) \quad (\mathbf{B},\mathbf{D}) \quad (\mathbf{B},\mathbf{H}) \quad (\mathbf{B},\mathbf{J}) \qquad (2\text{-}101)$$

as admissible pairs.

It was suggested earlier that $\mathbf{E}$, $\mathbf{D}$, and ρ as a group are completely decoupled from $\mathbf{B}$, $\mathbf{H}$, and $\mathbf{J}$ under time-independent conditions. One exception does occur, however, in the case of the variables $\mathbf{E}$ and $\mathbf{J}$. According to the Lorentz force density expression, an arbitrary distribution of free charge, initially at rest, will always move under the influence of an applied electric field. Obviously, this motion is affected by the properties of the material substances present. This demands a relation between $\mathbf{E}$ and $\mathbf{J}$ which may be expressed as

$$G_2(\mathbf{E},\mathbf{J}) = 0 \qquad (2\text{-}102)$$

or in accepted notation,

$$\mathbf{J} = J(\mathbf{E}) \qquad (2\text{-}103)$$

This is a second constitutive relation.

Of the remaining pairs

$$(\mathbf{E},\mathbf{H}) \quad (\mathbf{B},\mathbf{D}) \quad (\mathbf{B},\mathbf{H}) \quad (\mathbf{B},\mathbf{J}) \qquad (2\text{-}104)$$

the first and second may be eliminated at once because of the decoupling which occurs in the time-independent case (see Maxwell's equations). The fourth pair needs further discussion.

It is true that the Lorentz force density expression shows that $\mathbf{B}$ and $\mathbf{J}$ *go together*. But in contrast to the electric field case, an arbitrary distribution of free charge, which is initially stationary, cannot be set in motion by an applied $\mathbf{B}$ field. Furthermore, no relation can possibly be written between $\mathbf{B}$ (effect) and $\mathbf{J}$ (cause) which does not violate Maxwell's equations. Therefore, eliminating $(\mathbf{B},\mathbf{J})$ as an admissible pair leads to the final conclusion that $(\mathbf{B},\mathbf{H})$ is the required third pair of variables, in terms of which we write

$$G_3(\mathbf{B},\mathbf{H}) = 0 \qquad (2\text{-}105)$$

or in accepted notation,

$$\mathbf{B} = B(\mathbf{H}) \qquad (2\text{-}106)$$

This is a third (and last) constitutive relation.

We may give further justification for Eqs. (2-100), (2-103), and (2-106) from yet another point of view. We recall that Poynting's theorem

$$-\oint_{\Sigma} (\mathbf{E} \times \mathbf{H}) \cdot \mathbf{n} \, da = \int_{V} \mathbf{E} \cdot \mathbf{J} \, dv + \int_{V} \left(\mathbf{E} \cdot \frac{\partial \mathbf{D}}{\partial t} + \mathbf{H} \cdot \frac{\partial \mathbf{B}}{\partial t} \right) dv \quad (2\text{-}107)$$

is a direct consequence of Maxwell's equations. Figuratively speaking, this equation expresses the *whole*, namely, the total power entering a closed surface Σ, as the *sum of its parts;* that is, it describes mathematically the disposition of the total power

$$P_{\text{in}} = -\oint_{\Sigma} (\mathbf{E} \times \mathbf{H}) \cdot \mathbf{n} \, da \quad (2\text{-}108)$$

within the volume V. In particular, it suggests that in the stationary regime, namely, when $\partial/\partial t = 0$,

$$-\oint_{\Sigma} (\mathbf{E} \times \mathbf{H}) \cdot \mathbf{n} \, da = \int_{V} \mathbf{E} \cdot \mathbf{J} \, dv \quad (2\text{-}109)$$

We see, then, that in this case $\mathbf{E}$ and $\mathbf{J}$ combine to give the total power delivered to the medium surrounded by Σ. Whether it is consumed or generated depends exclusively on the electromagnetic character of the medium. Hence it is reasonable to surmise that, if we agree that in order to characterize a substance electromagnetically we need to write relations among six field quantities, then we must write one such relation between $\mathbf{E}$ and $\mathbf{J}$. A general form of such a relation, which is also independent of time, is the relation expressed by Eq. (2-103), and constitutes an obvious choice.

Carrying on the same sort of reasoning to the other terms in the right member of Eq. (2-107), we soon discover that if we want a simple model, we should write relations between $\mathbf{E}$ and $\mathbf{D}$, $\mathbf{B}$ and $\mathbf{H}$, and $\mathbf{J}$ and $\mathbf{E}$. This is precisely what we have accomplished by writing Eqs. (2-100), (2-103), and (2-106).

Note that the right member of Eq. (2-107) contains exactly three terms. This is an astonishing coincidence because, as we saw earlier, the number of equations we needed to render the system of Maxwell's equations determinate was also three! Thus Poynting's theorem reveals the same kind of *togetherness* between $\mathbf{D}$ and $\mathbf{E}$, $\mathbf{B}$ and $\mathbf{H}$, and $\mathbf{J}$ and $\mathbf{E}$.

To clarify this picture further, let us draw an analogy from mechanics. Let us write the power delivered by a force applied to a moving body:

$$P = \mathbf{F} \cdot \mathbf{v} = F_x \frac{dx}{dt} + F_y \frac{dy}{dt} + F_z \frac{dz}{dt} \quad (2\text{-}110)$$

Employing arguments similar to those used in the electromagnetic case, we might suspect that F_x depends on x, F_y on y, and F_z on z. That this is the case is guaranteed by the relation $\mathbf{F} = m\mathbf{a}$, which implies that

$$F_x = m\frac{d^2x}{dt^2} \qquad F_y = m\frac{d^2y}{dt^2} \qquad F_z = m\frac{d^2z}{dt^2} \qquad (2\text{-}111)$$

and we see at once the analogy of this problem to the electromagnetic problem.

As a parallel illustration of the general modeling problem, let us consider the case of a device such as a transistor. As shown in Chap. 11, network theory is an extreme simplification of electromagnetic field theory. In the case of the transistor, network theory predicts that the circuit external to the transistor can impose at most two independent constraints. Hence only two independent equations can be written in terms of the four terminal transistor variables (two voltages and two currents) if the contents of the *black box*, namely, the transistor itself, are neglected altogether. Therefore, for this problem to be made mathematically determinate, it is obvious that two additional relations must be written that descibe completely the way in which the device itself affects the overall performance of the circuit. Specifically, if the effects of environment and frequency are neglected and the quiescent point of operation is rigidly fixed, transistor behavior in the circuit under small-signal low-frequency conditions is predictable by a set of four parameters that can be measured in the laboratory. It is important to notice that these four parameters suffice to give a complete description of the transistor under the conditions cited, and thus serve as a convenient and practical means of identification. It is also important to notice that although the definition of each of the four parameters is made in terms of three terminal variables only, the overall behavior of the device is affected by all four.

Summarizing, we have shown that the presence of material substances imposes three additional constraints among the six field quantities. Together with three independent Maxwell equations, the so-called constitutive relations of a medium,

$$\mathbf{D} = D(\mathbf{E}) \qquad \mathbf{B} = B(\mathbf{H}) \qquad \mathbf{J} = J(\mathbf{E})$$

comprise a set of six relations among exactly six variables, thus admitting a unique solution to the general problem in electromagnetics.

2.16 Summary. This chapter was devoted to the formulation of electromagnetics as a discipline. We have dealt with semantics, presented the laws of the discipline, and discussed in generalities some aspects of electromagnetic fields and the sequence in which these fields are to be studied. Let us now pause to summarize our presentation.

To begin with, we state that our presentation is *axiomatic*. This, however, presents no particular obstacles. The development of electromagnetics as a discipline has taxed the experimental and intellectual skills of many brilliant pioneers in the field. Acknowledging their remarkable feats, we wish, nevertheless, not to retrace their footsteps, but to draw upon their experience in an effort to learn, to understand, to use, and possibly to advance the discipline to still higher levels. So we have chosen to accept without proof the validity of certain mathematical relations, which we call Maxwell's equations and the Lorentz force law, taking comfort in the fact that these relations do correspond to physical reality, substantiating proof being repeatedly provided by the many experiments which have been performed in the past and are being continued today.

We literally *borrow* from mechanics the fundamental concepts of mass, length, and time. To these fundamental concepts, we add a fourth one— electric charge. We regard the existence of charge as an established fact, although charge has never been seen by the naked eye. In so doing, we do not make an attempt to define charge verbally or to arrive at the unit of charge by some logical procedure. Instead, we observe that charge manifests itself as a force beyond the range of mechanics, and axiomatically assign to it the coulomb (in the mks system) as a unit.

We then observe that the totality of values assumed at different points in a region by the forces attributed to charges at rest and/or in motion constitutes an electromagnetic field; in abstract terms an electromagnetic field may be regarded as the domain of four vectors:

$\mathbf{E}$ = electric field intensity vector (volts per meter)
$\mathbf{B}$ = magnetic flux density vector (webers per meter squared)
$\mathbf{D}$ = electric flux density (or displacement) vector (coulombs per meter squared)
$\mathbf{H}$ = magnetic field intensity (amperes per meter)

Relations between these vectors and the sources (distributions of charges and currents) of the field exist and are expressed by Maxwell's equations (field laws), which as a group constitute our first fundamental postulate:

Faraday's Law

$$\oint_C \mathbf{E} \cdot d\mathbf{l} = -\frac{d}{dt} \int_S \mathbf{B} \cdot \mathbf{n}\, da \tag{1}$$

Generalized Form of Ampère's Circuital Law

$$\oint_C \mathbf{H} \cdot d\mathbf{l} = \int_S \mathbf{J} \cdot \mathbf{n}\, da + \frac{d}{dt} \int_S \mathbf{D} \cdot \mathbf{n}\, da \tag{2}$$

Gauss' Law for the Magnetic Field

$$\int_\Sigma \mathbf{B} \cdot \mathbf{n}\, da = 0 \tag{3}$$

Gauss' Law for the Electric Field

$$\int_{\Sigma} \mathbf{D} \cdot \mathbf{n} \, da = \int_{V} \rho \, dv \qquad (4)$$

where ρ = charge density, C/m^3

$\mathbf{J}$ = current density, A/m^2

The field laws are valid for every surface S bounded by a closed contour C and for every volume V surrounded by a closed surface Σ. On the other hand, Maxwell's equations, while containing all electrodynamics, do not define all four field vectors—they merely define their interrelationships. A complete definition of all four field vectors is made possible only when a second fundamental postulate—the Lorentz force density relation—is introduced to complement Maxwell's equations:

$$\int_{V} \mathbf{f} \, dv = \int_{V} \rho(\mathbf{E} + \mathbf{v} \times \mathbf{B}) \, dv \qquad \text{N} \qquad (5)$$

Between the six field quantities $\mathbf{E}$, $\mathbf{B}$, $\mathbf{D}$, $\mathbf{H}$, $\mathbf{J}$, and ρ, there are but three independent field relations, and further conditions must be imposed if the system is to be made determinate. These conditions are constraints imposed by the medium in the form of macroscopic constitutive relations,

$$\mathbf{D} = D(\mathbf{E}) \qquad \mathbf{B} = B(\mathbf{H}) \qquad \mathbf{J} = J(\mathbf{E})$$

which, for linear and isotropic media, reduce to

$$\mathbf{D} = \epsilon\mathbf{E} \qquad \mathbf{B} = \mu\mathbf{H} \qquad \mathbf{J} = \sigma\mathbf{E}$$

and thus serve to define the macroscopic parameters of the medium ϵ (or χ_e), μ (or χ_m), and σ.

The integral laws describe the properties of fields from the gross point of view. Consequently, they are useful only in predicting fields which have definite symmetry characteristics. In contrast, the differential laws provide interrelationships between the field quantities at a point and are, as a result, best suited for general analysis purposes. In corresponding sequence, Maxwell's equations in differential form are

$$\nabla \times \mathbf{E} = -\frac{\partial \mathbf{B}}{\partial t} \qquad \text{(I)}$$

$$\nabla \times \mathbf{H} = \mathbf{J} + \frac{\partial \mathbf{D}}{\partial t} \qquad \text{(II)}$$

$$\nabla \cdot \mathbf{B} = 0 \qquad \text{(III)}$$

$$\nabla \cdot \mathbf{D} = \rho \qquad \text{(IV)}$$

and the Lorentz force density is

$$\mathbf{f} = \rho(\mathbf{E} + \mathbf{v} \times \mathbf{B}) \qquad\qquad \text{(V)}$$

The differential laws are valid only at points where the field vectors are differentiable.   At surfaces separating media with different electromagneitic properties, the field vectors experience discontinuities, as specified by the following boundary conditions:

$$\mathbf{n} \times (\mathbf{E}_2 - \mathbf{E}_1) = 0 \qquad\qquad \text{(VI)}$$

$$\mathbf{n} \times (\mathbf{H}_2 - \mathbf{H}_1) = \mathbf{K} \qquad\qquad \text{(VII)}$$

$$\mathbf{n} \cdot (\mathbf{B}_2 - \mathbf{B}_1) = 0 \qquad\qquad \text{(VIII)}$$

$$\mathbf{n} \cdot (\mathbf{D}_2 - \mathbf{D}_1) = \rho_s \qquad\qquad \text{(IX)}$$

where $\mathbf{K}$ and ρ_s denote surface densities of current and charge, respectively. In each of these, $\mathbf{n}$ is a unit vector drawn normal to the boundary from medium 1 into medium 2.

The differential laws, when supplemented by the boundary conditions, are entirely equivalent to the integral laws.   In either form, the laws tacitly imply the conservation of charge, a principle which is expressed by the equation of continuity,

$$\mathbf{\nabla} \cdot \mathbf{J} + \frac{\partial \rho}{\partial t} = 0$$

or, in integral form, by

$$\int_{\Sigma} \mathbf{J} \cdot \mathbf{n} \, da = -\int_{V} \frac{\partial \rho}{\partial t} \, dv$$

Thus charge cannot be created or destroyed.

In many situations, it is convenient and expedient to introduce the scalar and vector potentials ϕ and $\mathbf{A}$.   Subject to the Lorentz condition

$$\mathbf{\nabla} \cdot \mathbf{A} + \mu\epsilon \frac{\partial \phi}{\partial t} = 0$$

the electromagnetic potentials are solutions of

$$\mathbf{\nabla}^2\phi - \mu\epsilon \frac{\partial^2 \phi}{\partial t^2} = -\frac{1}{\epsilon} \rho$$

$$\mathbf{\nabla}^2\mathbf{A} - \mu\epsilon \frac{\partial^2 \mathbf{A}}{\partial t^2} = -\mu\mathbf{J}$$

these being derivable from Maxwell's equations only when the region is linear, homogeneous, and isotropic.   The relations of the field vectors to the potentials are

$$\mathbf{E} = -\nabla\phi - \frac{\partial \mathbf{A}}{\partial t}$$

$$\mathbf{B} = \nabla \times \mathbf{A}$$

In addition to Maxwell's equations and the Lorentz force, we include as a third fundamental postulate the familiar principle of conservation of energy, with the aid of which the Poynting relations

$$-\nabla \cdot (\mathbf{E} \times \mathbf{H}) = \mathbf{E} \cdot \mathbf{J} + \left(\mathbf{E} \cdot \frac{\partial \mathbf{D}}{\partial t} + \mathbf{H} \cdot \frac{\partial \mathbf{B}}{\partial t} \right)$$

$$-\oint_{\Sigma} (\mathbf{E} \times \mathbf{H}) \cdot \mathbf{n} \, da = \int_{V} \mathbf{E} \cdot \mathbf{J} \, dv + \int_{V} \left(\mathbf{E} \cdot \frac{\partial \mathbf{D}}{\partial t} + \mathbf{H} \cdot \frac{\partial \mathbf{B}}{\partial t} \right) dv$$

state that the *whole* equals the sum of its *parts*, namely, that total power entering equals the sum of the power imparted to charged bodies plus the rate of change of stored energy. The Poynting vector

$$\mathbf{S} = \mathbf{E} \times \mathbf{H}$$

specifies the direction of power flow at a point, and its magnitude per unit area normal to flow.

In conclusion, we have presented what is probably the most difficult, and yet the most important, chapter of the text, the remaining chapters being extensions and clarifications of fundamental ideas and applications of general theory. To derive full benefits from this chapter, it is advisable to return to it for repeated study and for tying to general concepts and laws the more specialized topics treated in the rest of the book. Good luck!

Problems

2-1 Field Concept. Plot

$$\mathbf{E} = E_0 \sin x \sin t \, \mathbf{a_y}$$

as a function of x holding t constant, and as a function of t holding x constant. E_0 is a scalar constant. Does E change in direction with time?

2-2 Concept of an Electromagnetic Field. Suppose the parallel-plate structure shown in Fig. 2-9a is suspended in free space. In the region between the plates, a static electric field exists due to the charged plates. There is also present a time-independent magnetic field which is due, say, to the magnetic field of the earth. Does this combination of fields constitute an electromagnetic field?

2-3 Charge. Under proper discharge conditions, a battery of 50 A · hr capacity can deliver 1 A for 50 hr. How much transfer of charge does this correspond to?

2-4 Current. A current is comprised of electrons moving in the positive x direction with velocity v_- and heavier ions moving with velocity v_+ in the negative x direction. If ρ_- and ρ_+ are the respective densities of electron- and ion-charge distributions, determine the current density measured by an observer moving at velocity v_0 in the positive x direction.

2-5   Maxwell's Equations.   Show whether the field

$$\mathbf{E} = -E_0 \sin x \sin t \, \mathbf{a}_x$$

$$\mathbf{H} = \frac{1}{\mu_0} E_0 \sin x \cos t \, \mathbf{a}_z$$

is realizable in free space. E_0 is a scalar constant.

2-6   Lorentz Force.   A potential difference of 10 V is applied between two large parallel plates separated by 2 m.   An electron is released at the positive plate with an initial velocity of 2×10^6 m/s at an angle of 30° with respect to the plate.   Describe the electron's motion, and determine its final resting point.

2-7   Lorentz Force.   Suppose that a radio wave is propagating through the ionosphere in the direction of the positive z axis of a rectangular coordinate system.   Assume that there are N particles per unit volume of charge e and mass m and that the geomagnetic field is given by

$$\mathbf{B} = B_0 \mathbf{a}_z = \mu_0 H_0 \mathbf{a}_z$$

The field vectors of the incident wave have x and y components only.   Show that the equations of motion of a particle are

$$\ddot{x} = \frac{e}{m} E_x + \frac{e\mu_0}{m} (\dot{y}H_0 - \dot{z}H_y)$$

$$\ddot{y} = \frac{e}{m} E_y + \frac{e\mu_0}{m} (\dot{z}H_x - \dot{x}H_0)$$

$$\ddot{z} = \frac{e\mu_0}{m} (\dot{x}H_y - \dot{y}H_x)$$

where the dot over letters denotes differentiation with respect to time.

2-8   Force and Energy.   How much energy is acquired by an electron in falling through a static potential difference of 10 V?

2-9   Electromagnetic Properties of Material Media.   If there had been a uniform temperature T_0, greater than ambient, in the spherical structure of Example 2-4 at $t = 0$, the heat would have flowed out into the surrounding medium instead of having been effectively *blocked* at the outer surface.   How do you account for this difference between the two phenomena?

2-10   Electromagnetic Characterization of a Medium.   Given that the constitutive relation for a magnetic substance is

$$\mathbf{B} = \mu_0 \left(1 + \frac{M_0}{H}\right) \mathbf{H}$$

where M_0 is a constant, and H is the magnitude of the magnetic field intensity vector. State whether or not the medium is (a) linear, (b) homogeneous, and (c) isotropic.   Give reasons.

2-11   Orthogonality of the Field Vectors.   Prove that the vectors $\mathbf{E}$ and $\mathbf{B}$ of a time-varying field are, in general, not normal to each other in space.

2-12   Application of the Integral Laws.   Charge, with a uniform density, is fixed on the surface of a nonconducting balloon.   The balloon is in free space, and is being inflated and deflated periodically, thus giving rise to a time-dependent radius

$$r = r_0 + r_1 \cos t \qquad 0 < r_1 < r_0$$

Discuss the field which exists in the region $r > (r_0 + r_1)$.

2-13 Constitutive Relations. A homogeneous and isotropic conducting rod, whose permittivity and permeability are equal to those of free space, behaves according to

$$J = E + 0.1E^3$$

insofar as its conduction properties are concerned, over the range $0 < E < 5$ V/m. Determine the current density vector at points within the rod when a uniform static electric field, of 3 V/m, is applied parallel to the axis of the rod simultaneously with a uniform static magnetic field, of 4 Wb/m², applied normal to the axis of the rod.

2-14 Scalar Potential. Consider two conducting spherical shells, of radii a and b $(a < b)$, sharing a common center. The spherical shells are separated by a dielectric of permittivity ϵ and carry equal but opposite charge totaling Q C. The charge carried by the outer shell is positive, and both distributions are uniform.

(a) Using Laplace's equation, determine **E** and **D** in the region between the two spheres.

(b) If the inner sphere is grounded, what is the potential of the outer sphere?

2-15 Derivation of the Integral Laws from the Differential Laws. Using Stokes' theorem, derive Eqs. (1) and (2) from Eqs. (I) and (II). Also, using the divergence theorem, derive Eq. (3) from Eq. (III) and Eq. (4) from Eq. (IV).

2-16 Macroscopic Properties of Matter. Show that the matrix $[\epsilon]$ of a linear but anisotropic medium is Hermitian. Consider only the intrinsic properties of the medium.

2-17 Conservation of Charge. The space between the parallel plates of Fig. 2-9a is filled with a lossy dielectric having a *resistivity* $(1/\sigma)$ of $10^6 \, \Omega \cdot$ m, and a dielectric constant of unity. The spacing between plates is 0.005 m, and each plate has a length of 0.1 m and width of 0.1 m. A charge of 10^{-6} C is distributed uniformly on the upper plate at the time instant $t = 0$ s. The lower plate is grounded. Assume that fringing is negligibly small and that the electric field between the plates is uniform.

(a) Determine the current density as a function of position within the dielectric for $t > 0$.

(b) Determine **H** also as a function of the space coordinates and time.

2-18 Conservation of Charge. Derive the integral form of the principle of conservation of charge, using the integral laws as a point of departure.

2-19 Boundary Conditions. It is known that the electric field intensity at the interface of two dielectrics is $E_1 = 10$ V/m and makes an angle $\theta_1 = 30°$ with the normal. If $\epsilon_2 = \epsilon_1/2$, calculate E_2 and θ_2.

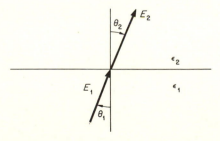

PROBLEM 2-19

2-20 Boundary Conditions. Let S be a surface which bounds medium 1 from medium 2, and let **n** be a unit normal to S, pointing from medium 1 to medium 2.   Prove that conservation of charge requires that

$$(\mathbf{J}_2 - \mathbf{J}_1) \cdot \mathbf{n} + \nabla \cdot \mathbf{K} = -\frac{\partial \rho_s}{\partial t}$$

where $\mathbf{J}_1$ and $\mathbf{J}_2$ = volume current densities
$\mathbf{K}$ = surface current density
ρ_s = surface charge density

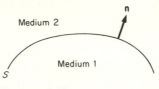

PROBLEM 2-20

2-21 Units. Using voltage V, current I, length L, and time T as fundamental quantities, express the dimensions of mass, charge, force, energy, and permittivity in terms of V, I, L, and T.

2-22 Magnetization Parameters. Calculate the magnetic flux density in a material for which $M = 3$ A/m when a field intensity $H = 10$ A/m is impressed.   Find the relative permeability and susceptibility of the material.

2-23 Conducting Properties of Metals. It is found that conduction through a metal is strongly dependent on internal temperature.   Although over wide variations this temperature dependence is nonlinear in character, much work in practice is conducted on the basis of a linear relation for resisitivity,

$$\rho = \rho_{20}[1 + \alpha_{20}(T - 20)] \qquad \Omega \cdot m$$

which is valid for moderate excursions of temperature T about 20°C.   The constant α_{20}, called the *temperature coefficient*, is the slope of the curve, and ρ_{20} the resistivity at 20°C. The resistivities and temperature coefficients of some common metals are given in the following table:

Material	Resistivity at 20°C, $\Omega \cdot m$	Temperature coefficient/°C at 20°C
Aluminum	2.83×10^{-8}	0.0039
Annealed copper	1.72×10^{-8}	0.00393
Hand-drawn copper	1.77×10^{-8}	0.00382
Gold	2.44×10^{-8}	0.0034
Pure iron	10.00×10^{-8}	0.005
Pure silver	1.64×10^{-8}	0.0038

A wide variety of electrical and electronic equipment, especially that used in spacecraft, is subject to extreme environmental changes in temperature.   For example, the avionics equipment on aircraft must be designed to operate effectively between −60 and +150°F.

Find the percent change in resistivity of annealed copper over a temperature excursion of $\pm 15°$ C to about $20°$ C.

Note: Although the temperature coefficient of metals is positive, certain other types of materials, notably *semiconductors*, exhibit negative temperature coefficients. Semiconductor materials find extensive usage in practice, as in the fabrication of *thermistors* (resistors with negative temperature coefficient) and *transistors*.

2-24 Electrets and Biomedical Electronics. The physical significance of the polarization vector **P** as a quantity describing the state of polarization in a medium is nowhere clearer than in a medium which, through special treatment, gives rise to a static electric field although a distribution of *free* charge is altogether absent. Such a medium constitutes an *electret*.

An electret is a dielectric material, such as plexiglass or a mixture of carnauba wax and beeswax, which has been so treated as to exhibit a permanent positive charge on one surface and negative charge on the other. The material to be made into an electret is subjected to a high temperature in a strong electric field. The electret is then allowed to cool, leaving the substance permanently polarized on solidification. The net effect is an object surrounded by electrostatic lines of force, just as a magnet is surrounded by magnetic lines of force.

Now, medical research has shown that human blood tends to coagulate around a platinum electrode connected to the positive terminal of a battery but not around the negative electrode. Could an electret be used effectively to impede the formation of blood clots in a given artery?

2-25 Diamagnetism, Paramagnetism, and Ferromagnetism. An important difference between the electric and magnetic properties of matter is that, in contrast to the relative permittivity, the relative permeability is less than 1 in some materials. Then χ_m is negative, and the flux density is weakened in the presence of the material. In that case the material is said to be *diamagnetic*. If χ_m is positive but very nearly equal to unity, the material is called *paramagnetic*. The following table gives the susceptibilities of a few substances:

Material	X_m
Aluminum	2.3×10^{-5}
Copper	-0.98×10^{-5}
Gold	-3.6×10^{-5}
Magnesium	1.2×10^{-5}
Mercury	-3.2×10^{-5}
Silver	-2.6×10^{-5}
Tungsten	6.8×10^{-5}

State whether each substance is diamagnetic or paramagnetic.

Note: If the relative permeability is much greater than unity, the material is said to be *ferromagnetic*. Strictly speaking, ferromagnetic materials are not linear, and therefore relative permeability here refers to an incremental quantity (slope of the magnetization curve). Maximum values of relative permeability for some common materials are:

Material	μ_r(max)
Iron (annealed)	5,500
Iron-silicon	7,000
Permalloy	25,000
Mumetal	100,000

2-26 Equivalent Densities of Current and Charge.   Let us eliminate the vectors **D** and **H** from the field equations by means of the constitutive relations for free space and Eqs. (2-31).   We obtain the following relations:

For free space	*For material substances*
$\nabla \times \mathbf{E} = -\dfrac{\partial \mathbf{B}}{\partial t}$	$\nabla \times \mathbf{E} = -\dfrac{\partial \mathbf{B}}{\partial t}$
$\nabla \times \dfrac{\mathbf{B}}{\mu_0} = \mathbf{J} + \dfrac{\partial(\epsilon_0 \mathbf{E})}{\partial t}$	$\nabla \times \dfrac{\mathbf{B}}{\mu_0} = \left(\mathbf{J} + \dfrac{\partial \mathbf{P}}{\partial t} + \nabla \times \mathbf{M}\right) + \dfrac{\partial(\epsilon_0 \mathbf{E})}{\partial t}$
$\nabla \cdot \mathbf{B} = 0$	$\nabla \cdot \mathbf{B} = 0$
$\nabla \cdot \epsilon_0 \mathbf{E} = \rho$	$\nabla \cdot \epsilon_0 \mathbf{E} = \rho - \nabla \cdot \mathbf{P}$

A comparison of corresponding equations shows at once that if *total* current and *total* charge densities are defined as

$$\mathbf{J}_T = \mathbf{J} + \frac{\partial \mathbf{P}}{\partial t} + \nabla \times \mathbf{M}$$

$$\rho_T = \rho - \nabla \cdot \mathbf{P}$$

then the two sets of equations become exactly alike.   Moreover, if $\mathbf{P} = \mathbf{M} = 0$, the set on the right reduces to the set on the left, as expected.   Prove that

$$\nabla \cdot \mathbf{J}_T + \frac{\partial \rho_T}{\partial t} = 0$$

and that charge is thus conserved in the more general sense of total current and total charge.

2-27 Relaxation.   At time $t = 0$, charge is distributed uniformly, with a constant density ρ_0, throughout the region of space occupied by two concentric spherical conductors, as shown in the figure.   Determine the current distribution in both conducting regions as a function of time and of radius r.   Also, find the displacement vector in the free space surrounding the spherical-conductor configuration.

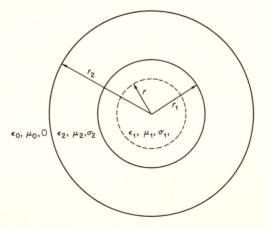

PROBLEM 2-27

2-28 Review of Theory

(a) Identify the source(s) of the electromagnetic field.

(b) Write
   (i)   Maxwell's equations in both differential and integral form, identifying the name of the law associated with each
   (ii) The Lorentz force expression, identifying each term
   (iii) The boundary conditions on the electromagnetic field

(c) Prove or disprove: Maxwell's equations are linear relations.

(d) Together with the Lorentz force expression, Maxwell's set of four equations form the basis for macroscopic electromagnetic theory.   However, given certain additional information, only two of the four equations are required for time-varying fields.   Identify the additional information, and show that both formulations are equivalent.

(e) Define ϵ, μ, and σ and give the mksc units for each.

(f) Explain the meaning of the terms homogeneous, linear, and isotropic when used to describe the electromagnetic properties of matter.

(g) What is a field?

CHAPTER 3

THE ELECTROSTATIC FIELD

3.1   Introduction.   Everything in nature changes with time.   Most of the interesting and useful applications of electromagnetics involve time-varying, that is, dynamic, fields.   For this reason our major objective in this text is to acquire a sound physical feeling for the dynamic behavior of fields, which will enable us to understand the role of electromagnetic fields in devices and systems of practical or theoretical interest.

A valuable and informative intermediate step in the study of dynamic fields is to study static fields.   This chapter is the first of three chapters concerned with the study of those fields for which the charges and/or currents (motion of charges) which produce the fields are static.   This chapter deals with static electric fields, that is, those fields whose sources are stationary charges.

Electrostatics, from our point of view, is a very special case of the application of Maxwell's equations.   Hence we begin by writing Maxwell's equations for the case where the current density **J** and the time derivatives of all field quantities are identically zero everywhere.

The problem divides itself into three fundamental types of problems:

1. Given *all* the sources, find the field.
2. Given the field *everywhere*, find the sources.
3. Given a *region*, usually not infinite in extent, with conductors, dielectrics, and *source* charges, in the region or on its boundary, find the resultant field, and any resultant spatial distribution of charges.

Each of the three problems requires a different approach.   We shall show that the first problem is conceptually quite straightforward, although it may lead to troublesome integrals.   The second problem is a synthesis problem, and can be quite difficult.   A brief presentation of the nature of this problem will be given.   The third problem is the so-called *boundary-value problem*.   It involves the formal solution of Laplace's and Poisson's equations by many different techniques.   The solution of these equations is an extensive subject.   A complete treatment is therefore impossible and, what is more important, not desirable in a volume of this sort. The formal mathematics will be developed to an extent that will allow us to attain our most immediate objectives.

3.2 Electrostatic Field Equations.

The relations which describe the behavior of static electric fields are extreme simplifications of the general field laws. Setting the time derivative equal to zero in the field equations results in the following relations:

Integral form		*Differential form*	
$\oint_C \mathbf{E} \cdot d\mathbf{l} = 0$	(3-1)	$\nabla \times \mathbf{E} = 0$	(3-2)
$\oint_\Sigma \mathbf{D} \cdot \mathbf{n} \, da = \int_V \rho \, dv$	(4)	$\nabla \cdot \mathbf{D} = \rho$	(IV)
$\mathbf{F} = \int_V \rho \mathbf{E} \, dv$	(3-3)	$\mathbf{f} = \rho \mathbf{E}$	(3-4)

Discontinuities in the field vectors $\mathbf{E}$ and $\mathbf{D}$ are expressed by the general boundary conditions

$$\mathbf{n} \times (\mathbf{E}_2 - \mathbf{E}_1) = 0 \qquad \text{(VI)}$$

$$\mathbf{n} \cdot (\mathbf{D}_2 - \mathbf{D}_1) = \rho_s \qquad \text{(IX)}$$

The electrostatic properties of material substances are described by a single relation connecting the vectors $\mathbf{E}$ and $\mathbf{D}$. As in the general case,

$$\mathbf{D} = D(\mathbf{E})$$

expresses a constitutive relation of a medium which reduces to

$$\mathbf{D} = \epsilon \mathbf{E}$$

when the medium is linear and isotropic.

Equation (3-2) implies that the vector $\mathbf{E}$ is derivable from a scalar potential, in accordance with the vector identity $\nabla \times \nabla \phi = 0$. Thus

$$\mathbf{E} = -\nabla \phi \qquad (3-5)$$

a result which is easily confirmed by setting the time derivative equal to zero in Eq. (2-82). The negative sign is again arbitrary, and is chosen for convenience.

In a linear, homogeneous, and isotropic medium, Eq. (IV) reduces to

$$\nabla \cdot \mathbf{E} = \frac{\rho}{\epsilon} \qquad (3-6)$$

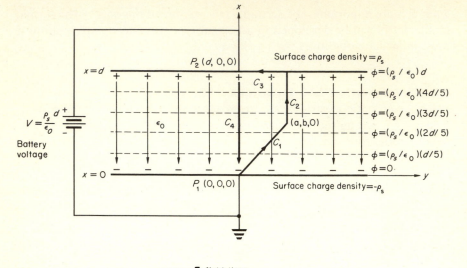

FIGURE 3-1.   The electrostatic field between two parallel plates.

which, with the aid of Eq. (3-5), transforms to

$$\nabla^2\phi = -\frac{\rho}{\epsilon} \qquad (3\text{-}7)$$

This is Poisson's equation; it was previously obtained in Sec. 2.13 as a special case of Eq. (2-87).

   In a source-free region ($\rho = 0$), Poisson's equation reduces to Laplace's equation

$$\nabla^2\phi = 0 \qquad (3\text{-}8)$$

This equation plays an extremely important role in the study of electrostatics.

3.3   Electrostatic Potential.   Let us consider now some of the basic properties of the scalar potential† function ϕ.   It is convenient to begin our discussion by returning to Example 2-6.

   Figure 3-1 summarizes the most important features of the parallel-plate configuration considered in Examples 2-5 and 2-6.   It will be remembered

† For a thorough discussion of potentials see O. D. Kellogg, "Foundations of Potential Theory," Dover Publications, Inc., New York, 1953.

that at every point in the region between the plates, the field is specified by

$$\mathbf{E} = -\frac{\rho_s}{\epsilon_0}\,\mathbf{a}_x \qquad (3\text{-}9)$$

$$\phi = \frac{\rho_s}{\epsilon_0}\,x \qquad (3\text{-}10)$$

where ρ_s is the surface charge density on the upper plate. Recall that ϕ was the solution to a simple boundary-value problem. It should also be observed that, according to Eq. (3.5),

$$\mathbf{E} = -\nabla\phi = -\nabla\left(\frac{\rho_s}{\epsilon_0}\,x\right) = -\frac{\rho_s}{\epsilon_0}\,\mathbf{a}_x \qquad (3\text{-}11)$$

Thus Eqs. (3-9) and (3-10) form a self-consistent pair of relations.

Let us next evaluate the line integral of the electrostatic field intensity around a completely closed contour. According to Eq. (3-1), this integral should vanish. For a closed contour, we choose the path C, shown in Fig. 3-1, which consists of four straight-line segments, C_1, C_2, C_3, and C_4, joining the points $(0,0,0)$, $(a,b,0)$, $(d,b,0)$, and $(d,0,0)$. Along C_1, the line integral of $\mathbf{E}$ is given by

$$\int_{C_1} \mathbf{E}\cdot d\mathbf{l} = \int_0^a E_x\,dx + \int_0^b E_y\,dy = \int_0^a\left(-\frac{\rho_s}{\epsilon_0}\right)dx = a\left(-\frac{\rho_s}{\epsilon_0}\right)$$

Along C_2,

$$\int_{C_2} \mathbf{E}\cdot d\mathbf{l} = \int_a^d E_x\,dx = \int_a^d\left(-\frac{\rho_s}{\epsilon_0}\right)dx = (d-a)\left(-\frac{\rho_s}{\epsilon_0}\right)$$

Along C_3, $\mathbf{E} = 0$; hence

$$\int_{C_3} \mathbf{E}\cdot d\mathbf{l} = 0$$

Finally, along C_4,

$$\int_{C_4} \mathbf{E}\cdot d\mathbf{l} = \int_d^0 E_x\,dx = \int_d^0\left(-\frac{\rho_s}{\epsilon_0}\right)dx = -d\left(-\frac{\rho_s}{\epsilon_0}\right)$$

By combining the four integrals, we find that the right members add to zero; this agrees fully with Eq. (3-1). Exactly the same result will be obtained for any other closed contour. Stated differently, when a small exploring body, carrying a unit charge of positive electricity, is taken completely around a closed path against the forces of an electrostatic field, the net energy expended is zero, regardless of path. Precisely the same thought is conveyed by saying that when the small exploring body is carried from one point in the field to another, the amount of work done is independent of path. In particular, when the probe charge follows the path formed by the line segments C_1, C_2, C_3 from $(0,0,0)$ to $(d,0,0)$, the work done is exactly the same as when it follows the path C_4^- from $(0,0,0)$ to $(d,0,0)$, simply because, according to what was just proved,

$$\int_{C_1+C_2+C_3} \mathbf{E} \cdot d\mathbf{l} = -\int_{C_4} \mathbf{E} \cdot d\mathbf{l} = \int_{C_4^-} \mathbf{E} \cdot d\mathbf{l} \tag{3-12}$$

Here, C_4^- denotes the path C_4 when the latter is traversed in the opposite direction.

By virtue of this property, electrostatic fields are said to be *conservative*.

The conservative nature of an electrostatic field can also be expressed in terms of ϕ, the scalar potential.   For instance, by making the substitution

$$\mathbf{E} = -\nabla\phi \tag{3-13}$$

in Eq. (3-12), we obtain

$$\int_{C_1+C_2+C_3} (-\nabla\phi) \cdot d\mathbf{l} = \int_{C_4^-} (-\nabla\phi) \cdot d\mathbf{l} \tag{3-14}$$

The point of departure P_1 and the terminal point P_2 are the same for both paths in Eq. (3-14).   Now, for any path C, connecting two arbitrary points a and b, the definition of $\nabla\phi$ requires that

$$-\int_C \nabla\phi \cdot d\mathbf{l} = \phi(a) - \phi(b) \tag{3-15}$$

where the contour C is traversed from point a to point b.   By virtue of this expression, Eq. (3-14) reduces to the identity

$$\phi(P_1) - \phi(P_2) = \phi(P_1) - \phi(P_2) \tag{3-16}$$

which confirms that the potential difference between any two points is independent of path.   Therefore *the value of the electrostatic potential at a point is unique.*   In contrast, the time-varying potential difference defined as the line integral of $\mathbf{E}$ is not unique.

The foregoing results may be summarized as follows.   If C_1 and C_2 are any two smooth paths, as shown in Fig. 3-2, then

$$\int_{C_1} \mathbf{E} \cdot d\mathbf{l} = \int_{C_2} \mathbf{E} \cdot d\mathbf{l} \tag{3-17}$$

Another distinguishing characteristic of the electrostatic potential is illustrated in Fig. 3-1, where it will be seen that the $\mathbf{E}$-field lines are normal to the equipotentials.   Actually, this property of the electrostatic field is general, and follows directly from $\mathbf{E} = -\nabla\phi$.   By definition, the gradient of a scalar function is a vector, whose magnitude is a measure of the maximum rate of *increase* of the function at a point and whose direction coincides with

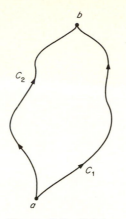

FIGURE 3-2. Points a
and b joined by two
paths.

the direction of maximum increase of the function. (This direction is normal to surfaces over which the value of the scalar function is a constant.)
Thus the vector **E** specifies the maximum rate of *decrease* in ϕ, and is normal
to the equipotential surface passing through a point. (Note that, as laymen,
we know that maximum rate of climb is straight up, normal to the equi-
altitude surface passing through a point.)

We have previously observed that the electrostatic field at any interior
point of a conductor is zero, because charges would move if a field were
present. This means that the potential at every point in a conducting region,
or on its surface, must be the same, and that the surface must be an equipotential. As a result, the electrostatic field vector is always normal to a
conducting surface, and its tangential component is always equal to zero.
Figure 3-1 shows an example of such behavior: the field intensity lines are
normal to both conducting plates, consistent with the requirements imposed
by the boundary conditions, Eqs. (**VI**) and (**IX**). At the interface between
a conductor (medium 1) and a dielectric (medium 2), these become

$$\mathbf{n} \times \mathbf{E}_2 = 0 \qquad\qquad (3\text{-}18)$$

$$\mathbf{n} \cdot \mathbf{D}_2 = \rho_s \qquad\qquad (3\text{-}19)$$

since both E_1 and D_1 must vanish in the conducting region.

It is informative to examine the boundary conditions in terms of the
potential. Thus Eq. (3-18) becomes

$$\mathbf{n} \times (-\nabla \phi_2) = 0 \qquad\qquad (3\text{-}20)$$

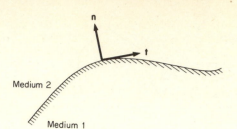

FIGURE 3-3.   Boundary between two media.

and if the dielectric is linear and isotropic, Eq. (3-19) yields

$$\mathbf{n} \cdot (-\epsilon_2 \boldsymbol{\nabla} \phi_2) = \rho_s \tag{3-21}$$

or

$$\mathbf{n} \cdot \boldsymbol{\nabla} \phi_2 = -\frac{\rho_s}{\epsilon_2} \tag{3-22}$$

Physically, $\mathbf{n} \times (-\boldsymbol{\nabla} \phi_2)$ represents the rate of decrease of ϕ_2 in the direction of the unit tangent vector $\mathbf{t}$ shown in Fig. 3-3. Therefore Eq. (3-20) can be written

$$\frac{\partial \phi_2}{\partial t} = 0 \tag{3-23}$$

Likewise, $\mathbf{n} \cdot (-\boldsymbol{\nabla} \phi_2)$ represents the rate of decrease of ϕ_2 in the direction of the normal vector. Hence

$$\frac{\partial \phi_2}{\partial n} = -\frac{1}{\epsilon_2} \rho_s \tag{3-24}$$

If neither medium is a conductor, the general boundary condition, Eq. (VI), may be written

$$\frac{\partial \phi_2}{\partial t} - \frac{\partial \phi_1}{\partial t} = 0 \tag{3-25}$$

while if both media are linear and isotropic, the condition (IX) may be expressed as

$$\epsilon_2 \frac{\partial \phi_2}{\partial n} - \epsilon_1 \frac{\partial \phi_1}{\partial n} = -\rho_s \tag{3-26}$$

From the conservative nature of the field, it follows that ϕ itself must be continuous across the interface between two media.

$$\phi_1 = \phi_2 \tag{3-27}$$

In other words, since the potential at a point is unique, the value of ϕ at two adjacent points on opposite sides of the boundary must be the same.

Example 3-1   Potential Distribution about a Charged Sphere.   Let us suppose that an amount Q of positive charge is uniformly distributed on the surface of a conducting sphere of radius r_1.   It is required to determine the potential and the electric flux density at every point in the free space surrounding the sphere.

A similar problem was solved in Example 2-1 by using the integral form of Maxwell's equations.   This time, however, we shall apply Laplace's equation.

Because of spherical symmetry, the Laplacian in spherical coordinates simplifies to

$$\nabla^2\phi = \frac{1}{r^2}\frac{d}{dr}\left(r^2\frac{d\phi}{dr}\right) \tag{3-28}$$

and the differential equation to be solved reduces to

$$\frac{d}{dr}\left(r^2\frac{d\phi}{dr}\right) = 0 \tag{3-29}$$

The boundary conditions to be satisfied are:

1.   The conducting surface at $r = r_1$ must be an equipotential surface.
2.   The potential ϕ must vanish as $r \rightarrow \infty$.   In other words, the point at infinity is to serve as the *ground*.

A general solution to Eq. (3-29) is

$$\phi = \frac{C_1}{r} + C_2$$

where C_1 and C_2 are two arbitrary constants of integration.   The boundary condition at infinity requires that $C_2 = 0$.   Letting $\epsilon_2 = \epsilon_0$ and $\rho_s = Q/4\pi r_1^2$ in Eq. (3-24), and noting that $\mathbf{n}$ is a unit vector in the direction of increasing radial coordinate, we find

$$\frac{\partial}{\partial r}\left(\frac{C_1}{r}\right)_{r=r_1} = -\frac{Q}{4\pi\epsilon_0 r_1^2}$$

from which we obtain $C_1 = Q/4\pi\epsilon_0$.   Thus

$$\phi = \frac{Q}{4\pi\epsilon_0 r} \tag{3-30}$$

and

$$\mathbf{E} = -\nabla\phi = \frac{Q}{4\pi\epsilon_0 r^2}\mathbf{a}_r \qquad \mathbf{D} = \epsilon_0\mathbf{E} = \frac{Q}{4\pi r^2}\mathbf{a}_r$$

The surfaces of constant potential are spherical surfaces which share a common center with the conducting sphere, while the $\mathbf{E}$ and $\mathbf{D}$ lines are radial lines, as indicated in Fig. 3-4.

The potential of the sphere

$$\phi|_{r=r_1} = \frac{Q}{4\pi\epsilon_0 r_1} \tag{3-31}$$

represents the amount of work required to transfer a unit positive charge from infinity to a point on the surface of the conducting sphere.   The transfer of charge is assumed to occur so slowly as to be equivalent to a sequence of stationary states.

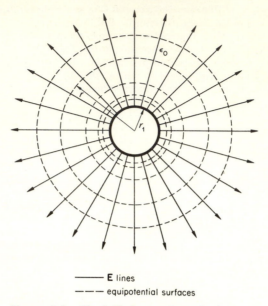

——— **E** lines

——— equipotential surfaces

FIGURE 3-4.   The field about a charged sphere.

Summarizing, the electrostatic field is a *conservative* field.   The potential is continuous at a boundary while the direction of **E** is always perpendicular to surfaces of constant potential.

3.4   The Concept of a Point Charge and Coulomb's Law.   A comparison of the results obtained in Examples 2-1 and 3-1 shows that, at points exterior to a sphere, the displacement vector **D** is always the same no matter whether the total charge is uniformly distributed on the surface of the sphere or throughout its volume.   It is therefore reasonable to expect that, as the radius of the charged sphere is allowed to become very small, the resultant field at all exterior points will be dependent only (1) on the total charge Q, (2) on the properties of the surrounding medium, and (3) on the distance r from the center of the sphere.   In the limit when $r \to 0$, the charge will appear "at" the center of the sphere, and the resulting configuration, according to accepted terminology, will become a *point charge*.

In Chap. 2, a point charge was defined as a finite amount of charge concentrated in a region so small relative to the distance from the point of observation that it appears to be at a point, in much the same way that celestial bodies appear to be at points to earth-bound observers.

According to previous developments, the field of a point charge q, located at the origin of a spherical coordinate system, is given by

$$\mathbf{D} = \frac{q}{4\pi r^2} \mathbf{a}_r \tag{3-32}$$

$$\mathbf{E} = \frac{q}{4\pi \epsilon r^2} \mathbf{a}_r \tag{3-33}$$

$$\phi = \frac{q}{4\pi \epsilon r} \tag{3-34}$$

A second point charge q' in the field of q would experience a force

$$\mathbf{F} = q'\mathbf{E} = \frac{qq'}{4\pi \epsilon r^2} \mathbf{a}_r \tag{3-35}$$

Equation (3-35) expresses Coulomb's law. This law assumes that r is very much larger than the physical dimensions of the charged bodies, and implies that an equal but opposite force will be exerted by q' on q.

It was Coulomb who, in the latter part of the eighteenth century, first established experimentally the validity of the law which is named after him. Although, to a large extent, the analytic study of electrostatics may be deduced from it, this law is seen to be just a special case of the Lorentz force law in the framework of a general theory of electromagnetics.

3.5 Field of Fixed Point-charge Configurations and the Electric Dipole. In Sec. 2.12 we observed that since Maxwell's equations are linear, we are free to superimpose any two solutions which satisfy the same boundary conditions. Thus, if two charged spheres are in a linear, homogeneous, and isotropic medium and are far enough apart so that neither disturbs the distribution of charge on the surface of the other, the potential at every point exterior to both may be expressed as the superposition of the potentials produced by each sphere acting alone. In particular, the potential can be calculated by applying Eq. (3-30) twice and then adding the results. Let us illustrate the point with a concrete example.

Example 3-2 Potential of Two Oppositely Charged Spheres. A positive charge is uniformly distributed on the surface of a conducting sphere centered at the point $(0, 0, d/2)$ of a rectangular coordinate system. An equal, but negative, charge $-q$ is uniformly distributed on the surface of a second conducting sphere centered at $(0, 0, -d/2)$. The radius of each sphere is the same, but much smaller than the spacing d. The surrounding medium is free space.

Without loss in generality, a point of observation may be any point P on the yz plane exterior to the spheres.   Let r_+ and r_- be the distance from P to the centers of the spheres, as indicated in Fig. 3-5a.   According to Eq. (3-30),

$$\phi_+ = \frac{q}{4\pi\epsilon_0 r_+}$$

is the potential at P due to the positive charge, and

$$\phi_- = \frac{-q}{4\pi\epsilon_0 r_-}$$

is the potential at P due to the negative charge.   Since free space is a linear, homogeneous, and isotropic medium, superposition applies, and the total potential is just the sum of ϕ_+ and ϕ_-:

$$\phi = \phi_+ + \phi_- = \frac{q}{4\pi\epsilon_0}\left(\frac{1}{r_+} - \frac{1}{r_-}\right)$$

Noting that

$$r_+ = \sqrt{\left(z - \frac{d}{2}\right)^2 + y^2} \qquad r_- = \sqrt{\left(z + \frac{d}{2}\right)^2 + y^2}$$

we have

$$\phi = \frac{q}{4\pi\epsilon_0}\left[\frac{1}{\sqrt{(z - d/2)^2 + y^2}} - \frac{1}{\sqrt{(z + d/2)^2 + y^2}}\right]$$

The field intensity at P is given by

$$\mathbf{E} = -\nabla\phi = \frac{q}{4\pi\epsilon_0}\left\{\frac{y}{[(z - d/2)^2 + y^2]^{3/2}} - \frac{y}{[(z + d/2)^2 + y^2]^{3/2}}\right\}\mathbf{a}_y$$

$$+ \frac{q}{4\pi\epsilon_0}\left\{\frac{(z - d/2)}{[(z - d/2)^2 + y^2]^{3/2}} - \frac{(z + d/2)}{[(z + d/2)^2 + y^2]^{3/2}}\right\}\mathbf{a}_z$$

The field in the exterior of the two-sphere configuration is plotted in Fig. 3-5b; it is of the same form in any plane containing the z axis.

Equations (3-32) to (3-34) show that the field of a point charge is infinitely large at $r = 0$.   Thus a point charge gives rise to a *zero-order singularity*.   Other charge configurations give rise to higher-order singularities. A first-order singularity, for example, is caused by an *electric dipole*, discussed below.   Higher-order singularities are caused by *multipoles* (ordered arrays of point charges) of yet higher order.

The concept of an electric dipole is intimately tied in with the two-sphere configuration shown in Fig. 3-5a.   Let r denote the distance from the origin to P, and θ be the polar angle measured in the clockwise direction from the positive z axis.   Let us also assume that P is so far away from the two-sphere

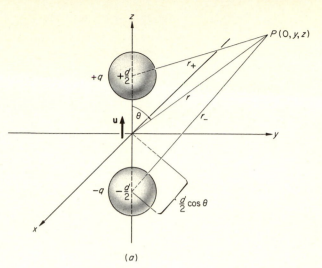

(a)

(b)

——— **E** lines

- - - - equipotential surfaces

FIGURE 3-5.   The geometry and field of two charged spheres.
(a) Geometry; (b) field distribution.

configuration that $r \gg d/2$; to a first approximation, then,

$$r_- \approx r + \frac{d}{2}\cos\theta \tag{3-36}$$

$$r_+ \approx r - \frac{d}{2}\cos\theta \tag{3-37}$$

so that

$$\frac{1}{r_+} - \frac{1}{r_-} \approx \frac{1}{r - \dfrac{d}{2}\cos\theta} - \frac{1}{r + \dfrac{d}{2}\cos\theta} = \frac{d\cos\theta}{r^2 - \left(\dfrac{d}{2}\cos\theta\right)^2} \tag{3-38}$$

Since $d/2 \ll r$,

$$\phi \approx \frac{qd\cos\theta}{4\pi\epsilon_0 r^2} \tag{3-39}$$

We see now that ϕ is a function of the product

$$p = qd \tag{3-40}$$

and varies inversely as the square of the radial distance; in contrast, the potential due to a single point charge varies merely as the inverse first power of the radial distance. It is customary to define *dipole moment* as the vector whose magnitude is p and whose direction is that of the unit vector **u**, shown in Fig. 3-5a. Thus

$$\mathbf{p} = p\mathbf{u} = qd\mathbf{u} \tag{3-41}$$

The field of a dipole is cylindrically symmetrical about the axis of the dipole, namely, about **u**. Using spherical coordinates, we find from Eq. (3-39) that

$$\mathbf{E} = -\boldsymbol{\nabla}\phi = -\left(\frac{\partial\phi}{\partial r}\mathbf{a}_r + \frac{1}{r}\frac{\partial\phi}{\partial\theta}\mathbf{a}_\theta\right) = \frac{p}{4\pi\epsilon_0 r^3}(2\cos\theta\,\mathbf{a}_r + \sin\theta\,\mathbf{a}_\theta) \tag{3-42}$$

In any plane containing the axis of the dipole, ϕ and **E** are specified by Eqs. (3-39) and (3-42), respectively. A plot of these expressions is shown in Fig. 3-6.

In a linear, homogeneous, and isotropic medium, the superposition of fields due to individual point sources can be continued indefinitely. Thus, if $q_1, q_2, q_3, \ldots, q_n$ are point charges, randomly distributed throughout space, the potential at every point of observation will be

$$\phi = \phi_1 + \phi_2 + \phi_3 + \cdots + \phi_n = \sum_{i=1}^{n} \phi_i \tag{3-43}$$

where

$$\phi_i = \frac{q_i}{4\pi\epsilon r_i} \tag{3-44}$$

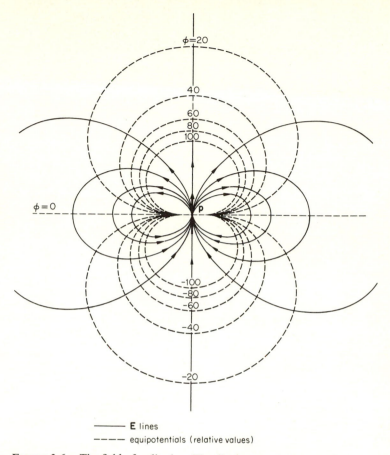

FIGURE 3-6.   The field of a dipole.   The dipole moment is $\mathbf{p} = pd\mathbf{u}$.

Here r_i is the distance from the ith point charge to the point of observation. As more and more charges are added to a region, the distances of separation gradually diminish, and the distribution of charge becomes continuous.   In the limit, the sum in Eq. (3-43) transforms to an integral.   We surmise that, if the charge is distributed in space with density represented by ρ, the potential takes the form of an integral,

$$\phi(x_0, y_0, z_0) = \frac{1}{4\pi\epsilon} \int_V \frac{\rho}{r} \, dv \qquad (3\text{-}45)$$

where

$$r = [(x_0 - x)^2 + (y_0 - y)^2 + (z_0 - z)^2]^{1/2} \qquad (3\text{-}46)$$

is the distance from a source point (x,y,z) to a point of observation (x_0,y_0,z_0). The general validity of Eq. (3-45) will be shown in the next section.

Summarizing, the potential of a system of charges is obtained by summing the potentials from each point or infinitesimal charge. The validity of superposition rests on the linearity of the medium.

3.6 Integration of Poisson's Equation. We now present a formal proof† of superposition by deriving a general solution to Poisson's equation.

Consider Fig. 3-7. A finite amount of charge is distributed with an arbitrary density in certain (shaded) regions of space. Let $\rho(x,y,z)$ denote the density of charge. Also, let (x_0,y_0,z_0) be an interior point of a closed surface Σ bounding a linear, homogeneous, and isotropic volume V. If any conductors are present in V, then Σ is formed in part by the surfaces of these conductors, and the volume V excludes all conducting regions. We wish to determine the potential at a point of observation (x_0,y_0,z_0) due to the bounded, but otherwise arbitrary, distribution of charge.

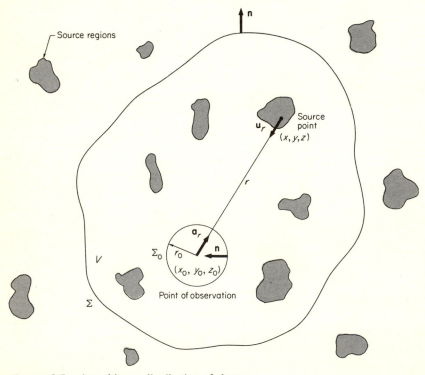

FIGURE 3-7. An arbitrary distribution of charge.

† This proof is rather involved, and may be omitted without loss of continuity. The interpretation of the results, Eqs. (3-58) and (3-59), is the important part of this section.

We begin with Green's theorem,

$$\int_V (\psi \, \nabla^2 \phi - \phi \, \nabla^2 \psi) \, dv = \oint_\Sigma (\psi \, \nabla \phi - \phi \, \nabla \psi) \cdot \mathbf{n} \, da \qquad (3\text{-}47)$$

where ψ and ϕ are assumed to be nonsingular, continuous, and twice differentiable scalar functions of position with continuous first partial derivatives. Remembering that we intend to obtain a general solution to Poisson's equation

$$\nabla^2 \phi = -\frac{\rho}{\epsilon} \qquad (3\text{-}48)$$

it behooves us to pick for the function ϕ, in Eq. (3-47), just the unknown potential function ϕ. As for ψ, experience tells us to pick the scalar function

$$\psi = \frac{1}{r} \qquad (3\text{-}49)$$

where
$$r = \sqrt{(x_0 - x)^2 + (y_0 - y)^2 + (z_0 - z)^2} \qquad (3\text{-}50)$$

is the distance from the variable source point (x,y,z) to the fixed point of observation (x_0, y_0, z_0). Our choice of the so-called *Green's function* ψ is motivated by the simplification which results from the relation

$$\nabla^2 \psi = \nabla^2 \left(\frac{1}{r} \right) = 0 \qquad \text{when } r \neq 0 \qquad (3\text{-}51)$$

This relation can easily be verified by direct substitution in the expression

$$\nabla^2 \psi = \frac{1}{r^2} \frac{\partial}{\partial r} \left(r^2 \frac{\partial \psi}{\partial r} \right) + \frac{1}{r^2 \sin \theta} \frac{\partial}{\partial \theta} \left(\sin \theta \frac{\partial \psi}{\partial \theta} \right) + \frac{1}{r^2 \sin^2 \theta} \frac{\partial^2 \psi}{\partial \varphi^2} \qquad (3\text{-}52)$$

considering, of course, the point (x_0, y_0, z_0) as the origin of a system of spherical coordinates.

Thus chosen, the functions ϕ and ψ reduce the left side of Eq. (3-47) to

$$\int_V (\psi \, \nabla^2 \phi - \phi \, \nabla^2 \psi) \, dv = \int_V \psi \, \nabla^2 \phi \, dv = -\frac{1}{\epsilon} \int_V \frac{\rho(x,y,z)}{r} \, dv \qquad (3\text{-}53)$$

The integrand on the right of Eq. (3-47) is seen to include the unknown function ϕ. Let us observe, however, that ψ fails to satisfy the nonsingularity requirement at $r = 0$, for as $r \to 0$, $\psi \to \infty$.

To exclude this singularity, let us surround the point (x_0, y_0, z_0) with a small sphere of radius r_0 and bounding surface Σ_0, so that V is now the volume bounded by Σ and Σ_0; also, the surface integral in Eq. (3-47) is broken up into two integrals, one of which is to be evaluated over the surface Σ and the other over Σ_0. Taking into account Eq. (3-53), we can write Eq. (3-47) as

$$-\frac{1}{\epsilon}\int_V \frac{\rho}{r}\,dv = \oint_\Sigma (\psi\,\nabla\phi - \phi\,\nabla\psi)\cdot\mathbf{n}\,da + \oint_{\Sigma_0}(\psi\,\nabla\phi - \phi\,\nabla\psi)\cdot\mathbf{n}\,da \quad (3\text{-}54)$$

Now

$$\oint_{\Sigma_0}(\psi\,\nabla\phi - \phi\,\nabla\psi)\cdot\mathbf{n}\,da = \oint_{\Sigma_0}\left[\frac{1}{r_0}\nabla\phi - \phi(\nabla\psi)_{r=r_0}\right]\cdot\mathbf{n}\,da$$

$$= -\frac{1}{r_0}\oint_{\Sigma_0}\frac{\partial\phi}{\partial r}\,da - \frac{1}{r_0{}^2}\oint_{\Sigma_0}\phi\,da \quad (3\text{-}55)$$

We are free to choose r_0 so small that both ϕ and $\partial\phi/\partial r$ are essentially constant at every point on the surface Σ_0. In particular, as we make r_0 progressively smaller, the potential and its normal derivative on Σ_0 approach their limiting values at the center, since by hypothesis they are both defined and continuous functions of position. Hence we can take them outside the integral signs, and in the limit as $r_0 \to 0$, we can write

$$\oint_{\Sigma_0}(\psi\,\nabla\phi - \phi\,\nabla\psi)\cdot\mathbf{n}\,da = -4\pi\phi(x_0,y_0,z_0) \quad (3\text{-}56)$$

noting, of course, that

$$\lim_{r_0\to 0}\frac{1}{r_0}\oint_{\Sigma_0}\frac{\partial\phi}{\partial r}\,da = \lim_{r_0\to 0}\frac{1}{r_0}\left(\frac{\partial\phi}{\partial r}\right)_{r=r_0}\oint_{\Sigma_0}da$$

$$= \lim_{r_0\to 0}\frac{1}{r_0}\left(\frac{\partial\phi}{\partial r}\right)_{r=r_0}4\pi r_0{}^2 = 0 \quad (3\text{-}57)$$

Finally, Eqs. (3-54) and (3-56) combine to give

$$\phi(x_0,y_0,z_0) = \frac{1}{4\pi\epsilon}\int_V \frac{\rho}{r}\,dv + \frac{1}{4\pi}\oint_\Sigma\left[\frac{1}{r}\nabla\phi - \phi\nabla\left(\frac{1}{r}\right)\right]\cdot\mathbf{n}\,da \quad (3\text{-}58)$$

Inside Σ   *outside Σ*

as an expression for the potential at any point (x_0,y_0,z_0) of a linear, homogeneous, and isotropic medium due to a stationary, bounded, but otherwise arbitrary distribution of charge. The volume integral in Eq. (3-58) is obviously the same as the volume integral in Eq. (3-45). The surface integral needs further examination.

If the volume V contains no charges, Poisson's equation reduces to Laplace's equation since ρ is zero at every interior point of Σ. Duplication of the previous development then produces a result similar to Eq. (3-58), except that there is no volume integral on the right side. Of course, the same result can be obtained by setting $\rho = 0$ in the volume integrand of Eq. (3-58). The point is this: The first term on the right of Eq. (3-58) accounts for those charges which are located inside the surface Σ, while the second term accounts for all charges outside the surface Σ. Thus, if Σ were to be expanded so as to enclose all the charges, the surface integral would vanish and, in this case, ϕ would be expressed by

$$\phi = \frac{1}{4\pi\epsilon} \int_V \frac{\rho}{r}\, dv \tag{3-59}$$

This equation shows that potential is a quantity which is linearly related to charge; doubling the charge density ρ doubles the potential.

In the absence of exterior sources, the surface integral in Eq. (3-58) is zero. For, if there are no charges present anywhere, either inside or outside the surface Σ, Laplace's equation would lead to a nonzero potential, contrary to physical reality, which precludes the existence of an *effect* in the absence of a *cause*.

Let us state the conclusions once again: The volume integral in Eq. (3-58) represents the contribution to the potential at (x_0, y_0, z_0) of all interior charges, while the surface integral represents the contribution of all exterior charges.

It can be shown (Prob. 3-28) that Eq. (3-58) converges for interior as well as exterior points of a bounded charge distribution. Thus, on a charged surface, the potential ϕ is defined and is continuous provided the total charge is finite.

In summary, we have derived a general formula which determines the field of a specified system of sources and which provides important information regarding the first fundamental electrostatic problem listed in the introduction to this chapter.

3.7 Field of Continuous Charge Distributions.

In applying Eq. (3-59) we should take into account all charges present, whether they are isolated point charges or continuously distributed in space. To demonstrate how this is done, we shall treat several examples in which the charge distribution is known and the field is to be found.

Example 3-3 Field about a Finite Charged Line. Let us consider first a uniform distribution of charge along a line which extends from $x = a$ to $x = b$, as in Fig. 3-8a. Let ρ_l be the density of distribution measured in coulombs per meter. For the sake of simplicity, let ρ_l be constant at every point $(x, 0)$ in the interval $[a, b]$. The line charge is supposed to be suspended in free space, and the field is to be evaluated at a point $P(x_0, y_0)$, which is not on the line, but which is arbitrarily situated relative to it.

Strictly speaking, a line charge is really an approximation of a distribution of charge along a wire so small in diameter that the sources become condensed practically into a line. Since discontinuous distributions are easier to handle mathematically, it is convenient to idealize the problem by dealing with a discontinuous distribution which is nearly equivalent to the continuous one. Apart from introducing slight errors in the final results, we do not depart drastically from physical reality.

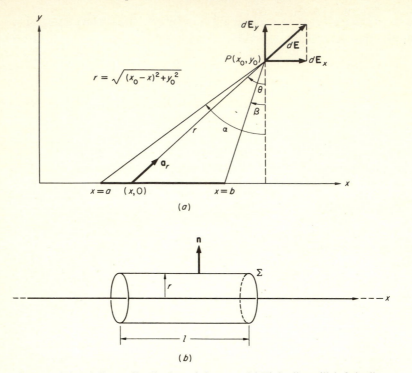

FIGURE 3-8.   A linear distribution of charge.   (a) Finite line; (b) infinite line.

It may help some readers to visualize a line charge as a mere arrangement of mathematical point charges placed side by side along a straight line.

To dispense with the preliminaries, let us, lastly, observe that by locating P on the xy plane we do not destroy the generality of the problem since the configuration shown in Fig. 3-8a obviously exhibits cylindrical symmetry about the x axis.

Our point of departure is Eq. (3-59), in which we now set

$$r = \sqrt{y_0^2 + (x_0 - x)^2}$$

and $\epsilon = \epsilon_0$; in the present case, the volume integration extends over a single straight-line segment, so that

$$\rho \, dv = \rho_l \, dx \tag{3-60}$$

Hence the potential at (x_0, y_0) is given by

$$\phi(x_0, y_0) = \frac{1}{4\pi\epsilon_0} \int_a^b \frac{\rho_l}{\sqrt{y_0^2 + (x_0 - x)^2}} \, dx$$

Remembering that ρ_l is independent of x, we have

$$\phi(x_0, y_0) = \frac{\rho_l}{4\pi\epsilon_0} \int_a^b \frac{dx}{\sqrt{y_0^2 + (x_0 - x)^2}}$$

$$= \frac{\rho_l}{4\pi\epsilon_0} \ln \frac{(x_0 - a) + \sqrt{y_0^2 + (x_0 - a)^2}}{(x_0 - b) + \sqrt{y_0^2 + (x_0 - b)^2}} \tag{3-61}$$

where the symbol ln denotes the natural logarithm.   This is the required result.

As $a \to b$, the fraction on the right of Eq. (3-61) approaches unity. Since the natural logarithm of 1 is zero, the potential at (x_0, y_0) approaches zero, as expected, since the total charge approaches zero.

If we set $a = -b$, the charge distribution becomes symmetric about the y axis; in this case

$$\phi(x_0, y_0) = \frac{\rho_l}{4\pi\epsilon_0} \ln \frac{(x_0 + b) + \sqrt{y_0^2 + (x_0 + b)^2}}{(x_0 - b) + \sqrt{y_0^2 + (x_0 - b)^2}} \tag{3-62}$$

As b is allowed to increase without limit, keeping the point (x_0, y_0) a finite distance away from the origin, the ratio within the logarithm tends to increase without limit, and the potential becomes undefined. Actually, in the limit as $b \to \infty$, it is impossible to expand the surface Σ in Eq. (3-58) to include all the charges, and hence Eq. (3-59) does not apply.

For completeness, the solution of the infinite line charge will be shown. This problem is best treated by application of the integral form of Gauss' law.

With reference to Fig. 3-8b, let Σ be a cylindrical surface length l and radius r. Then

$$\oint_\Sigma \mathbf{D} \cdot \mathbf{n} \, da = \text{total charge enclosed by } \Sigma \tag{3-63}$$

By symmetry

$$D_r(2\pi r)(l) = (\rho_l)(l)$$

so that

$$D_r = \frac{\rho_l}{2\pi r} \tag{3-64}$$

Hence

$$\mathbf{D} = \frac{\rho_l}{2\pi r} \mathbf{a}_r \tag{3-65}$$

and

$$\mathbf{E} = \frac{\rho_l}{2\pi\epsilon_0 r} \mathbf{a}_r \tag{3-66}$$

Since in cylindrical coordinates

$$\mathbf{E} = -\nabla\phi = -\left(\frac{\partial\phi}{\partial r} \mathbf{a}_r + \frac{1}{r} \frac{\partial\phi}{\partial\varphi} \mathbf{a}_\varphi + \frac{\partial\phi}{\partial z} \mathbf{a}_z \right)$$

and the field is cylindrically symmetric, the required potential may be written in the form

$$\phi = -\frac{\rho_l}{2\pi\epsilon_0} \ln r + C$$

The constant C may be determined by establishing an arbitrary cylindrical surface $r = r_0$ as a reference. Thus, setting

$$\phi = 0 = -\frac{\rho_l}{2\pi\epsilon_0} \ln r_0 + C$$

gives the final result for the infinite line charge

$$\phi = \frac{\rho_l}{2\pi\epsilon_0} \ln \frac{r_0}{r} \tag{3-67}$$

Returning to the finite line charge, the field intensity may be determined from $\mathbf{E} = -\nabla_0\phi$, in which the subscript zero attached to the operator del denotes differentiations with respect to x_0, y_0, z_0. Taking ϕ from Eq. (3-61), we find

$$E_x = -\frac{\rho_l}{4\pi\epsilon_0}\left[\frac{1}{\sqrt{y_0^2 + (x_0 - a)^2}} - \frac{1}{\sqrt{y_0^2 + (x_0 - b)^2}}\right] \qquad (3\text{-}68)$$

$$E_y = -\frac{\rho_l}{4\pi\epsilon_0}\left[\frac{\dfrac{y_0}{\sqrt{y_0^2 + (x_0 - a)^2}}}{(x_0 - a) + \sqrt{y_0^2 + (x_0 - a)^2}} - \frac{\dfrac{y_0}{\sqrt{y_0^2 + (x_0 - b)^2}}}{(x_0 - b) + \sqrt{y_0^2 + (x_0 - b)^2}}\right] \qquad (3\text{-}69)$$

Equations (3-68) and (3-69) may be expressed in terms of the angles α and β, shown in Fig. 3-8a. The details are moderately involved, but the results are simple:

$$E_x = \frac{\rho_l}{4\pi\epsilon_0 y_0}\,(\cos\beta - \cos\alpha) \qquad (3\text{-}70)$$

$$E_y = \frac{\rho_l}{4\pi\epsilon_0 y_0}\,(\sin\alpha - \sin\beta) \qquad (3\text{-}71)$$

An alternative approach to Eqs. (3-70) and (3-71) is suggested in Prob. 3-9.

The second point of this section is to show that the electric field intensity of a continuous distribution of charge may be calculated by summing the increments of electric field intensities resulting from a system of point charges of magnitude $\rho\,dv$. The following discussion establishes the general validity of this technique.

We begin with Eq. (3-58). If all charges are within Σ, the surface integral vanishes, and the potential is given by

$$\phi(x_0, y_0, z_0) = \frac{1}{4\pi\epsilon}\int_V \frac{\rho(x, y, z)}{r}\,dv \qquad (3\text{-}72)$$

where r denotes distance from the variable point of integration (x, y, z) to the fixed point of observation (x_0, y_0, z_0). The field intensity at (x_0, y_0, z_0) is

$$\mathbf{E}(x_0, y_0, z_0) = -\boldsymbol{\nabla}_0\phi = -\frac{1}{4\pi\epsilon}\int_V \rho\boldsymbol{\nabla}_0\left(\frac{1}{r}\right)dv = \int_V \frac{\rho\,dv}{4\pi\epsilon r^2}\,\mathbf{u}_r \qquad (3\text{-}73)$$

where $\mathbf{u}_r$ represents a unit vector directed radially outward from the source point. Thus, with reference to Fig. 3-7, $\mathbf{u}_r = -\mathbf{a}_r$. The contribution of a charge element $dq = \rho\,dv$ in V is therefore

$$d\mathbf{E} = \frac{dq}{4\pi\epsilon r^2}\,\mathbf{u}_r \qquad (3\text{-}74)$$

which fully agrees with Eq. (3-33), as expected.

Example 3-4   The Field of a Charged Disk. As an illustration of a two-dimensional distribution of charge, let us examine the case of a uniformly charged disk of radius R. Figure 3-9 shows the pertinent geometry. For simplicity, we assume that the

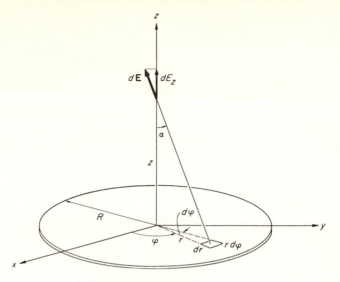

FIGURE 3-9. A uniformly charged disk.

density ρ_s of the charge† on the surface of the disk is independent of the coordinates r and φ and that the permittivity of the medium is ϵ. The electrostatic field intensity at a point on the z axis is to be determined.

Because of cylindrical symmetry, the component of **E** which is parallel to the plane of the disk is always zero at every point on the z axis. This means that we need only calculate the z component of the electric field intensity. The contribution to this component from an element of charge $dq = \rho_s r \, dr \, d\varphi$ is, by virtue of Eq. (3-74),

$$dE_z = dE \cos \alpha = \frac{z \rho_s r \, dr \, d\varphi}{4\pi\epsilon(r^2 + z^2)^{3/2}} \tag{3-75}$$

and the total field is

$$E_z = \int_0^R \int_0^{2\pi} dE_z = \frac{z\rho_s}{4\pi\epsilon} \int_0^{2\pi} d\varphi \int_0^R \frac{r}{(r^2 + z^2)^{3/2}} \, dr = \frac{\rho_s}{2\epsilon}\left(1 - \frac{z}{\sqrt{z^2 + R^2}}\right) \tag{3-76}$$

Example 3-5 Charged Sphere. As the number of independent variables becomes larger, the analysis problem becomes progressively harder. To gain some appreciation of attendant difficulties, we consider a uniform three-dimensional distribution of charge in a spherical volume of radius R. Let Q be the total charge of the distribution, and let it be required to determine the electric flux density vector at a point exterior to the region. The sphere is suspended in free space.

From the treatment of a similar problem in Example 2-1, we know that the answer is

$$\mathbf{D} = \frac{Q}{4\pi r^2} \mathbf{a}_r \tag{3-77}$$

† The charge density would not be constant if the disk were conducting. Refer to Example 3-16.

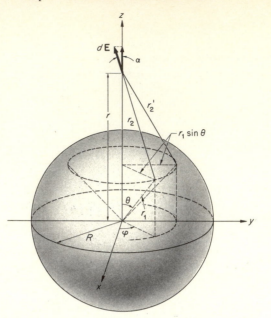

FIGURE 3-10. A spherical distribution of charge.
The angle α is measured in the yz plane.

where r denotes radial distance from the center of the sphere, and $\mathbf{a}_r$ is a unit vector
in the direction of increasing coordinate r.   In this section we should arrive at the
same result by a process of superposition.

An element of spherical volume contains a total charge of $dq = \rho r_1^2 \sin \theta \, dr \, d\theta \, d\varphi$.
where ρ denotes the constant value of the spherical distribution of charge (Fig. 3-10).
From Eq. (3-74) it is clear that the magnitude of $d\mathbf{E}$ is

$$dE = \frac{\rho r_1^2 \sin \theta \, dr_1 \, d\theta \, d\varphi}{4\pi\epsilon_0 r_2^2} \tag{3-78}$$

In performing the integration with respect to the variable φ, we obtain the field of a
differential ring of charge centered about the z axis with radius $r_1 \sin \theta$.   The resultant
$\mathbf{E}$ vector is then contained entirely in the yz plane and is directed radially outward
along r_2'.   Since pairs of differential elements of charge in the ring cancel each other,
this integration yields an intermediate z-directed differential field of magnitude

$$dE_z = \frac{\rho r_1^2 \sin \theta \, dr_1 \, d\theta}{4\pi\epsilon_0 (r_2')^2} 2\pi \cos \alpha \tag{3-79}$$

where
$$r_2'^2 = r^2 + r_1^2 - 2rr_1 \cos \theta \tag{3-80}$$

and
$$\cos \alpha = \frac{r - r_1 \cos \theta}{r_2'} \tag{3-81}$$

Keeping r and r_1 fixed, we have from Eq. (3-80)

$$2r_2' \, dr_2' = 2rr_1 \sin \theta \, d\theta \tag{3-82}$$

Also, from Eq. (3-80), we find

$$-r_1 \cos \theta = \frac{r_2'^2 - r^2 - r_1^2}{2r}$$

and substituting this expression into Eq. (3-81), we obtain

$$\cos \alpha = \frac{r^2 - r_1^2 + r_2'^2}{2rr_2'} \tag{3-83}$$

Combining Eqs. (3-79), (3-82), and (3-83), we find

$$dE_z = \frac{\rho r_1 (r^2 - r_1^2 + r_2'^2)\, dr_2'\, dr_1}{4\epsilon_0 r_2'^2 r^2} \tag{3-84}$$

We now integrate this expression with respect to r_2', over the interval from $r - r_1$ to $r + r_1$, and obtain the contribution to E_z from a charged spherical shell of radius r_1 and thickness dr_1.   Calling this new differential field dE_z', we find

$$dE_z' = \frac{\rho r_1\, dr_1}{4\epsilon_0 r^2} \int_{r-r_1}^{r+r_1} \left[(r^2 - r_1^2)\frac{1}{r_2'^2} + 1 \right] dr_2'$$

$$= \frac{\rho r_1\, dr_1}{4\epsilon_0 r^2} \left[(r^2 - r_1^2)\left(-\frac{1}{r_2'} \right) + r_2' \right]_{r-r_1}^{r+r_1} = \frac{\rho r_1\, dr_1}{4\epsilon_0 r^2}\, 4r_1 \tag{3-85}$$

Finally, integrating dE_z' with respect to r_1 from zero to R, we obtain

$$E_z = \frac{\rho}{\epsilon_0 r^2} \int_0^R r_1^2\, dr_1 = \frac{\rho}{\epsilon_0 r^2} [\tfrac{1}{3} r_1^3]_0^R = \frac{\rho R^3}{3\epsilon_0 r^2} \tag{3-86}$$

Since the total charge in the sphere is

$$Q = \tfrac{4}{3}\pi R^3 \rho$$

and since, moreover, $\mathbf{D} = \epsilon_0 \mathbf{E}$, it follows that

$$\mathbf{D} = \frac{Q}{4\pi r^2}\, \mathbf{a}_z \tag{3-87}$$

Because of spherical symmetry, $\mathbf{a}_z$ may be replaced by the radial unit vector $\mathbf{a}_r$; the resulting expression

$$\mathbf{D} = \frac{Q}{4\pi r^2}\, \mathbf{a}_r \tag{3-88}$$

is clearly in full agreement with Eq. (3-77).

Aside from illustrating a general method for finding the field of a three-dimensional distribution of charge, this example serves to point out the mathematical difficulties encountered in problems of this type, in clear contrast to the ease by which the same result may be obtained, as in Example 2-1, by the general integral laws. Therefore, when symmetry prevails, it is always advisable to use the integral laws.

We saw in Sec. 3-4 that Eq. (3·88) is an expression for the field of a mathematical point charge Q located at the origin.   We shall now show that, at sufficiently remote points of observation, every bounded distribution of charge, no matter how arbitrary, appears as a point charge.   This is clearly

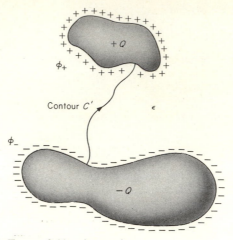

FIGURE 3-11.   A capacitor.

evident from either Eq. (3-72) or Eq. (3-73), both of which require that all charges be located within Σ.   For if we choose a point in V as an arbitrary origin, we can approximate the variable distance r from any source point within Σ by the fixed distance R from the origin to the point of observation. Equation (3-72) then gives

$$\lim_{R \to \infty} \phi(x_0, y_0, z_0) = \lim_{R \to \infty} \int_V \frac{\rho \, dv}{4\pi\epsilon R} = \frac{1}{4\pi\epsilon R} \int_V \rho \, dv = \frac{Q}{4\pi\epsilon R} \qquad (3\text{-}89)$$

which is just the potential of a point charge Q located at the origin.

3.8   The Capacitance Concept.   Consider two conductors in a dielectric of permittivity ϵ.   One conductor carries a positive charge Q, and the second conductor carries an equal but negative charge $-Q$.   The shapes and relative orientations of both conductors are wholly arbitrary (Fig. 3-11).

From previous work we know that charges placed on a system of conductors generate a static field such that the tangential component of the electric field intensity vanishes on the outer surfaces of all conducting bodies. Thus, in Fig. 3-11, both surfaces will be equipotentials.   Moreover, the difference

$$\phi_+ - \phi_- = -\int_{C'} \mathbf{E} \cdot d\mathbf{l} \qquad (3\text{-}90)$$

in the potentials between the two surfaces will in no way depend on the path of integration.

Any pair of conductors, separated by a dielectric medium, constitutes a *capacitor*.   A capacitor may be charged by connecting a battery between the two conducting bodies. In the process, a charge is transferred from one body to the other; the magnitude Q of this charge depends on the battery voltage

$$V = \phi_+ - \phi_- \tag{3-91}$$

and on the *capacitance* C of the capacitor, which is defined by the relation

$$C = \frac{Q}{V} \tag{3-92}$$

Now, according to Eq. (3-59), potential and potential difference are both linearly dependent on charge. Thus, if a charge Q is transferred from one conductor of a capacitor to the other, the potentials ϕ_+ and ϕ_-, as well as the difference $(\phi_+ - \phi_-)$, will vary linearly with Q. In other words, increasing Q by a factor k will have the effect of increasing $V = \phi_+ - \phi_-$ by exactly the same factor k. Therefore, in Eq. (3-92) the ratio Q/V is independent of Q, and C is dependent only on the geometry of the conducting bodies and on the properties of the surrounding medium.

Example 3-6 Parallel-plate Capacitor. The structure considered in Example 2-5 and again in Sec. 3.3 is a parallel-plate capacitor in which a total charge

$$Q = \rho_s ab$$

is transferred from one plate to the other by connecting them to a battery of

$$V = \rho_s \frac{d}{\epsilon_0} \quad \text{V}$$

The capacitance

$$C = \frac{Q}{V} = \epsilon_0 \frac{ab}{d} \tag{3-93}$$

of a parallel-plate configuration is clearly seen to depend on the area of the plates ab, on the distance of separation d, and on the permittivity of the medium. When the plates are separated by a dielectric of permittivity ϵ, the capacitance becomes

$$C = \epsilon \frac{ab}{d} \tag{3-94}$$

Capacitance is measured in coulombs per volt, or coulombs squared per newton-meter. The unit of capacitance is the farad (F). This is confirmed by Eqs. (3-93) and (3-94), since ϵ_0 and ϵ are measured in farads per meter. If the parallel plates measure 1 m on a side and are separated by a distance of 1 cm, then

$$C = 8.854 \times 10^{-12} \frac{(1)\,(1)}{0.01} = 8.854 \times 10^{-10} \quad \text{F}$$

Obviously, the farad is inconveniently large, and for this reason the microfarad (1 μF = 10^{-6} F) and the picofarad (pF), or micromicrofarad (1 $\mu\mu$F = 10^{-12} F), are units most often used in practice.

To summarize, a capacitor is a structure which consists of two† conductors embedded in a dielectric medium. The capacitance of a condenser is defined to be charge divided by voltage; it is dependent only on the geometry and on the medium.

3.9 Formulation of the Boundary-value Problem in Electrostatics.
Mathematically, the electromagnetic field problem is concerned with the solution of a set of differential equations (Maxwell's equations) which meets certain prescribed conditions at the boundaries of the region under consideration. In other words, the analysis problem in electromagnetics is a *boundary-value problem*. If the distribution of sources is completely specified, the field is uniquely determinate; this was demonstrated for static electric fields in the first part of this chapter. Conversely, if the field is specified at every point within a region, the source distribution is determinate, but not necessarily uniquely; this property will be demonstrated for electrostatic fields in the last part of this chapter.

Ordinarily, the electromagnetic field problem is a very difficult one. Generally speaking, neither the charges nor the field itself are known everywhere in a region. Instead, only the *primary*, or *driving*, sources (batteries and other types of *generators* of electromagnetic energy) are specified, and the problem is then to calculate the resultant field, as well as the *secondary* sources *induced* by the specified primary sources in various regions of space.

To illustrate, let us imagine that a third conducting body is brought near the capacitor of Fig. 3-11. The surface of the third body, and the surfaces of of the first two as well, are always true equipotentials. Generally speaking, the shape of the third body will be arbitrary and, most likely, will not conform to an equipotential surface of the primary field. Therefore the presence of the third body will cause a disturbance of the primary field, but the resultant potential will still be constant over each of the three surfaces. This disturbance will be accompanied by a rearrangement of the primary charge Q on the surfaces of both capacitor bodies. Also, a secondary system of charges will be induced on the surface of the third body without altering, of course, its electrically neutral state of charge. In other words, the forced migration of surface charge will keep the net charge on the third body always equal to zero.

To obtain an analytical solution to the three-body problem of a capacitor in the presence of a third body, Laplace's equation must be solved subject to the condition of a vanishing tangential **E** field on the boundaries of the three conducting bodies. This is easily said, but the task is a very difficult one to

---

† The capacitance of a single isolated conductor may be defined if the second conductor is considered to be a spherical shell of infinite radius. The numerical value of C would then specify the degree of coupling between the conductor and other remote objects such as the earth (ground).

perform and, more frequently, even impossible, unless the surfaces of all three conducting bodies are in perfect coincidence with three coordinate surfaces.   This is a basic limitation of boundary-value-problem solutions. To obtain *formal solutions* of Laplace's equation in rectangular coordinates, the boundaries must be plane; in cylindrical coordinates, the boundaries must be cylindrical; in spherical coordinates, the boundaries must be spherical; and so on.   If they are not, the respective solutions will be in the form of infinite series of functions for which the evaluation of the coefficients is more often than not a frustrating, and very frequently an impossible, task.

Fortunately, an electrostatic field problem can be solved by one of several methods.   In order of presentation, these methods are

1.   The method of separation of variables† (formal solutions of static field equations)
2.   The numerical method (computer solutions of static field equations)
3.   The method of adapting a known solution to a new problem
4.   The method of images
5.   The method of conformal transformations
6.   The graphical method
7.   The electrolytic tank method (an experimental method)

The subject of all seven methods cannot be considered as a whole in this work.   Instead, each method is discussed only to the extent required by a first exposition to the general principles of electromagnetics.   The first six methods are examined in this chapter; the treatment of the last is deferred to Chap. 5.

In a great number of cases, the solution of the electrostatic boundary-value problem reduces to a solution of Laplace's equation which fits a prescribed set of boundary conditions; in other words, all field sources are considered to exist either exterior to, or on the boundaries of, the region under consideration; a field is then to be found at every point in the source-free region within, such as to satisfy the boundary conditions over all surfaces of discontinuity.   Every field of this sort is characterized by the following properties:

1.   The potential ϕ will satisfy Laplace's equation at all points where there are no charges.

---

† There are only 11 coordinate systems in which Laplace's equation can be solved by the method of separation of variables.   These so-called *coordinate systems of Eisenhart* are (1) rectangular, (2) circular cylindrical, (3) elliptic cylindrical, (4) parabolic cylindrical, (5) spherical, (6) prolate spheroidal, (7) oblate spheroidal, (8) parabolic, (9) conical, (10) ellipsoidal, and (11) paraboloidal.

2. The potential ϕ will be continuous everywhere except at surfaces bearing an isolated distribution of electric dipoles, called *dipole layers*. Such layers are found principally on the electrode surfaces of electrochemical devices, such as batteries.

3. The potential ϕ will be finite everywhere except at points occupied by mathematical point charges, or line charges.

4. When all sources are located within a finite distance from the origin, ϕ will vanish at infinity.

5. The potential ϕ will assume constant values ($E_{tan} = 0$) over the surfaces of conductors. Conductors are equipotential regions.

6. **E** will always be normal to surfaces of constant ϕ.

7. The tangential component of **E** will be continuous across all surfaces.

8. **D** lines will emanate from positive charges and will terminate on negative charges.

9. At the interface between two dielectrics, the normal component of **D** will always be continuous if no surface charge is present.

10. At the interface between a conductor and a dielectric, the normal component of **D** will be discontinuous by an amount equal to the density of the charge on the surface of the conductor.

Actually, we do not concern ourselves with each of these properties every time we solve an electrostatic problem, because a solution of Laplace's equation that satisfies the prescribed boundary conditions is unique (Sec. 3.11), and thus exhibits each of these properties automatically.

In summary, this section discussed the third part of the problem stated in the introduction of Chap. 2, namely: Given a system of conductors, dielectrics, and driving sources, find the resultant field and the spatial distribution of charges. This, in essence, constitutes the boundary-value problem in electrostatics.

3.10 Formal Solutions of Laplace's Equation in Rectangular Coordinates. Finding closed form solutions of Laplace's equation is not altogether new to us; this method was illustrated once in Example 2-6, using rectangular coordinates, and again in Example 3-1, using spherical coordinates. Though geometrically different, the parallel-plate-capacitor problem of Example 2-6 and the isolated-charged-sphere problem of Example 3-1 were similar in one respect: they both were one-dimensional problems, because the potential function depended on only one coordinate. In both cases a simple solution in closed form was obtained. Two-dimensional and three-dimensional solutions of Laplace's equations are normally more involved.

We shall eventually illustrate the solution of Laplace's equation in the three most widely used systems of coordinates, namely, rectangular, cylindrical, and spherical. This section deals with solutions of Laplace's equation in rectangular coordinates. Solutions in other systems of coordinates are to be found in the literature.†

Example 3-7 Covered Trough. As an illustration of a two-dimensional solution of Laplace's equation in rectangular coordinates, let us consider a very long hollow metal tube of rectangular cross section. The pertinent geometry is shown in Fig. 3-12a. Three sides of the tube are held at zero potential. The fourth side, insulated from the other three, is held at the variable potential $\phi = V \sin (\pi x/a)$, where V is a constant. We wish to determine the electric potential at all points in the free space within the pipe.

Our problem is to find a solution of Laplace's equation

$$\nabla^2\phi = 0 \tag{3-95}$$

which satisfies the following boundary conditions:

$$\phi(0,y) = 0$$
$$\phi(x,0) = 0$$
$$\phi(a,y) = 0$$
$$\phi(x,b) = V \sin \frac{\pi x}{a}$$

Since the pipe is very long, the field inside the pipe will be independent of z. Therefore, in cartesian coordinates, Laplace's equation is

$$\frac{\partial^2\phi}{\partial x^2} + \frac{\partial^2\phi}{\partial y^2} = 0 \tag{3-96}$$

This is a linear second-order partial differential equation. Equations of this type may be solved by assuming a product solution of the form

$$\phi = X(x)Y(y) \tag{3-97}$$

where X is a function of x alone, and Y is a function of y alone. Substituting this expression in Eq. (3-96), dividing by the product XY, and transposing yields

$$\frac{1}{X}\frac{d^2X}{dx^2} = -\frac{1}{Y}\frac{d^2Y}{dy^2} \tag{3-98}$$

The left member of this equation is dependent on x alone; the right member is dependent on y alone. Thus, in Eq. (3-98), the independent variables are *separated*. As a result, changes in y do not affect the left member; similarly, changes in x do not affect the right member. Therefore the only way in which the equality can hold is for both members to be independent of both x and y; in other words, the equality holds if both members are equal to some real constant $-k^2$ (where the negative sign is chosen deliberately to satisfy the boundary conditions):

$$\frac{1}{X}\frac{d^2X}{dx^2} = -\frac{1}{Y}\frac{d^2Y}{dy^2} = -k^2 \tag{3-99}$$

† P. Moon and D. E. Spencer, "Field Theory for Engineers," D. Van Nostrand Company, Inc., Princeton, N.J., 1961.

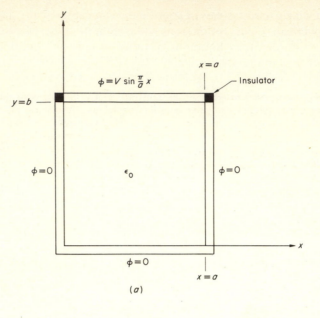

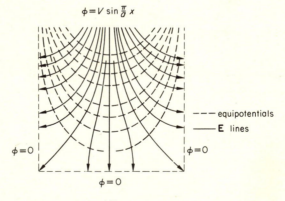

FIGURE 3-12.   The field within a covered trough.   (*a*) Geometry; (*b*) field distribution.

In terms of the *constant of separation* $-k^2$, we now have

$$\frac{d^2X}{dx^2} + k^2X = 0 \qquad (3\text{-}100)$$

$$\frac{d^2Y}{dy^2} - k^2Y = 0 \qquad (3\text{-}101)$$

Thus the so-called *method of separation of variables* has reduced a partial differential equation to a pair of ordinary differential equations. The general solutions of Eqs. (3-100) and (3-101) are, respectively,

$$X = A_1 \cos kx + A_2 \sin kx \qquad (3\text{-}102)$$

$$Y = B_1 e^{ky} + B_2 e^{-ky} \qquad (3\text{-}103)$$

where A_1, A_2, B_1, and B_2 are arbitrary constants whose value will be determined by the boundary conditions. Substitution of these in Eq. (3-97) now gives

$$\phi = (A_1 \cos kx + A_2 \sin kx)(B_1 e^{ky} + B_2 e^{-ky}) \qquad (3\text{-}104)$$

This solution must be adjusted to fit all four boundary conditions simultaneously. Applying the first boundary condition gives

$$0 = A_1(B_1 e^{ky} + B_2 e^{-ky})$$

The only way that this equality can hold identically for all y, without leading to the trivial solution, is for A_1 to be equal to zero. Then

$$\phi = \sin kx \ (B_1 e^{ky} + B_2 e^{-ky}) \qquad (3\text{-}105)$$

where A_2 was absorbed in B_1 and B_2. Applying the second boundary condition to Eq. (3-105) gives

$$0 = \sin kx \ (B_1 + B_2)$$

This requires that $B_1 + B_2 = 0$; hence Eq. (3-105) reduces to

$$\phi = C_k \sin kx \sinh ky \qquad (3\text{-}106)$$

where $C_k = 2B_1$. Applying the third boundary condition now gives

$$0 = C_k \sin ka \sinh ky$$

which in turn demands that $\sin ka = 0$ and that, consequently,

$$ka = n\pi \qquad n = \pm 1, \pm 2, \pm 3, \ldots$$

(Note that the case $n = 0$ was purposely omitted because it leads to the trivial solution.) This means that the most general form of Eq. (3-97) is a superposition of simple solutions† of the form $C_n \sin (n\pi/a)x \sinh (n\pi/a)y$. Thus

$$\phi = \sum_{n=1}^{\infty} C_n \sin \frac{n\pi}{a} x \sinh \frac{n\pi}{a} y \qquad (3\text{-}107)$$

Note that the index n covers only the set of positive integers. Negative values of n merely change the algebraic sign of C_n, without in any way affecting the form of the sum. This is because the constants C_n associated with pairs of terms corresponding to each combination of positive and negative values of n can always be combined into a single constant.

Of the infinitely many terms on the right side of Eq. (3-107), *we shall now retain only as many terms as we need to satisfy* the fourth and last boundary condition. In this case we can satisfy this condition by retaining a single term, namely, $n = 1$. Thus we have

$$C_1 \sin \frac{\pi}{a} x \sinh \frac{\pi}{a} b = V \sin \frac{\pi}{a} x \qquad (3\text{-}108)$$

---

† In mathematical language, the particular values $k = n\pi/a$ are called *eigenvalues*. The corresponding solutions $C_n \sin (n\pi/a)x \sinh (n\pi/a)y$ are in turn called *eigenfunctions*.

from which we obtain

$$\phi = \frac{V}{\sinh (\pi/a)b} \sin \frac{\pi}{a} x \sinh \frac{\pi}{a} y \qquad (3\text{-}109)$$

This is the desired result. The function ϕ is shown plotted, together with the electric field intensity, in Fig. 3-12*b*.

The following observations will serve to complement the discussion of the present example.

1. Substituting k^2 for $-k^2$ interchanges x and y in every expression for ϕ, including Eq. (3-107). However, in this example, a positive k^2 would make it impossible to match the boundary condition at $y = b$. The choice of sign is always dictated by the boundary conditions of the problem.
2. A closely related problem arises when a potential other than zero is specified along two or even three sides of the pipe (Prob. 3-7). Such a problem can always be solved by superposition.
3. A solution of the rectangular-pipe problem is to be found with $\phi = 0$ on all surfaces, except that $\phi = V$, where $V = $ constant, at the top. The solution in this case develops as before, except that the boundary condition at $y = b$ now requires

$$\sum_{n=1}^{\infty} C_n \sin \frac{n\pi}{a} x \sinh \frac{n\pi}{a} b = V \qquad (3\text{-}110)$$

It will be an accident if the scalar potential ϕ is exactly equal to V at $y = b$ when the sum includes only a finite number of terms. On the other hand, even if the number of terms is large, the series may not converge rapidly, and ϕ may not represent anything.

Be that as it may, let us multiply both sides of Eq. (3-110) by $\sin (m\pi/a)x$, where m is a real positive integer. We have

$$\sum_{n=1}^{\infty} C_n \sinh \frac{n\pi}{a} b \sin \frac{n\pi}{a} x \sin \frac{m\pi}{a} x = V \sin \frac{m\pi}{a} x$$

Let us next integrate both sides of this equation from $x = 0$ to $x = a$, interchanging the order of integration and summation in advance.

$$\sum_{n=1}^{\infty} C_n \sinh \frac{n\pi}{a} b \int_0^a \sin \frac{n\pi}{a} x \sin \frac{m\pi}{a} x \, dx = \int_0^a V \sin \frac{m\pi}{a} x \, dx \qquad (3\text{-}111)$$

Using the trigonometric identity

$$\sin A \sin B = \tfrac{1}{2}[\cos (A - B) - \cos (A + B)]$$

transforms the integral on the left side of Eq. (3-111) to

$$\int_0^a \sin \frac{n\pi}{a} x \sin \frac{m\pi}{a} x \, dx$$

$$= \frac{1}{2} \int_0^a \left[\cos (n - m) \frac{\pi}{a} x - \cos (n + m) \frac{\pi}{a} x \right] dx = \begin{cases} \dfrac{a}{2} & m = n \\ 0 & m \neq n \end{cases} \qquad (3\text{-}112)$$

Therefore, after integration, we have

$$C_n \frac{a}{2} \sinh \frac{n\pi}{a} b = 2V \frac{a}{n\pi} \qquad n \text{ odd}$$

or

$$C_n = \frac{4V}{n\pi \sinh (n\pi/a)b}$$

and

$$\phi = \sum_{n \text{ odd}} \frac{4V}{n\pi \sinh (n\pi/a)b} \sin \frac{n\pi}{a} x \sinh \frac{n\pi}{a} y \qquad (3\text{-}113)$$

Equation (3-113) expresses the potential distribution within the pipe when a battery of V volts is connected between the top and the bottom three sides of the pipe. Obviously, it is much more complicated than Eq. (3-109).

It should be noted, in passing, that in order to arrive at these results, we made use of what is called the *orthogonality property* of the trigonometric function, expressed in this case by Eq. (3-112). Generally speaking, two functions f_1 and f_2 are said to be *simply orthogonal* on the closed interval $[a,b]$ if

$$\int_a^b f_1(x) f_2(x) \, dx = 0 \qquad (3\text{-}114)$$

A set of functions, $f_1, f_2, f_3, \ldots$, is said to be an *orthogonal set* on the closed interval $[a,b]$ provided

$$\int_a^b f_i(x) f_j(x) \, dx = \begin{cases} N_i & i = j \\ 0 & i \neq j \end{cases} \qquad (3\text{-}115)$$

where N_i is a constant, usually dependent upon i. In Eq. (3-112), it is seen that $N_i = a/2$ for all i.

4. Further complications arise in a three-dimensional problem. If the pipe is terminated at $z = 0$ and $z = d$, so as to form a parallelepiped with dimensions $a \times b \times d$, the potential is expressed by a doubly infinite sum of the form

$$\phi = \sum_{n=1}^{\infty} \sum_{m=1}^{\infty} C_{mn} \sin \frac{n\pi}{a} x \sinh \tau y \sin \frac{m\pi}{d} z \qquad (3\text{-}116)$$

where

$$\tau = \pi \sqrt{\frac{n^2}{a^2} + \frac{m^2}{d^2}}$$

and

$$C_{mn} = \frac{4}{ad \sinh b\tau} \frac{1}{} \int_0^d \int_0^a f(x,z) \sin \frac{n\pi}{d} x \sin \frac{m\pi}{d} z \, dx \, dz$$

In this expression, the function f represents an arbitrary potential distribution on the top of the pipe. All four sides and the bottom are held at zero potential.

5. The important question of whether Eqs. (3-109), (3-113), and (3-116) are the only possible solutions is answered next, in Sec. 3.11.

Let us now summarize the main points of this discussion. In a source-free region the electrostatic problem reduces to a solution of Laplace's equation, which satisfies the prescribed boundary conditions. In all but the most elementary cases, solutions of Laplace's equations involve fairly complicated expressions, which often may become unmanageable.

The following equations are the most frequently encountered solutions of Laplace's equations in rectangular coordinates:

1. ϕ independent of y and z:

$$\phi = Ax + B \tag{3-117}$$

where A and B are constants.

2. ϕ independent of z:

$$\phi = (A_1 \sin kx + A_2 \cos kx)(B_1 e^{ky} + B_2 e^{-ky}) \tag{3-118}$$

$$\phi = (A_1 e^{kx} + A_2 e^{-kx})(B_1 \sin ky + B_2 \cos ky) \tag{3-119}$$

$$\phi = (A_1 \sin kx + A_2 \cos kx)(B_1 \sinh ky + B_2 \cosh ky) \tag{3-120}$$

$$\phi = (A_1 \sinh kx + A_2 \cosh kx)(B_1 \sin ky + B_2 \cos ky) \tag{3-121}$$

where k is a constant of separation.

In addition, any linear combination of the products xy, yz, zx is also an admissible solution of Laplace's equation.   Three-dimensional solutions are rarely encountered.

3.11 Uniqueness.   In this section we prove a uniqueness theorem which applies to every solution of Poisson's and Laplace's equations.

Generally speaking, the purpose of a uniqueness theorem is to establish conditions under which a particular solution is the only solution of a problem. Specifically, a uniqueness theorem establishes the boundary conditions which are sufficient to specify a solution uniquely, and in addition specifies necessary and sufficient conditions for a one-to-one correspondence between a field and its sources.

THEOREM

Within a region V which is bounded by a closed surface Σ, a solution of Poisson's or Laplace's equation is specified uniquely by the potential over the boundary, or by the normal derivative of the potential over the boundary, or the former over part of the boundary and the latter over the rest.

In accepted terminology, *Dirichlet boundary conditions* are those for which the scalar potential is specified at every point on the boundary.   This occurs when conductors are held at specified potentials, as in Example 3-7. On the other hand, *Neumann boundary conditions* are those for which the normal derivative of the potential is specified everywhere on the boundary. Finally, *mixed boundary conditions* are those for which the potential is specified over a portion of the boundary, and its normal derivative is specified over the rest of the boundary (Example 2-5).

We want to show that both Poisson's equation

$$\mathbf{\nabla}^2\phi = -\frac{\rho}{\epsilon} \tag{3-122}$$

and Laplace's equation

$$\mathbf{\nabla}^2\phi = 0 \tag{3-123}$$

have a unique solution within a volume V when either Dirichlet, Neumann, or mixed boundary conditions are specified on the boundary surface Σ.

To prove this statement, we suppose to the contrary that there exist two solutions, ϕ_1 and ϕ_2, which satisfy either Poisson's or Laplace's equation in V and identical boundary conditions on Σ. Then the difference function

$$\phi = \phi_1 - \phi_2 \tag{3-124}$$

satisfies Laplace's equation in V, regardless of whether or not any sources are present in V. Additionally, either $\phi = 0$ or $\partial\phi/\partial n = 0$ over the bounding surface or $\phi = 0$ over part of Σ and $\partial\phi/\partial n = 0$ over the rest. Now, in Green's first identity,

$$\int_V (\psi\,\mathbf{\nabla}^2\phi + \mathbf{\nabla}\psi \cdot \mathbf{\nabla}\phi)\,dv = \oint_\Sigma \psi\,\frac{\partial\phi}{\partial n}\,da \tag{3-125}$$

we let $\psi = \phi$, and we find

$$\int_V (\phi\,\mathbf{\nabla}^2\phi + |\mathbf{\nabla}\phi|^2)\,dv = \oint_\Sigma \phi\,\frac{\partial\phi}{\partial n}\,da \tag{3-126}$$

For any one of the three types of boundary conditions, the right side of this equation vanishes identically; and since $\mathbf{\nabla}^2\phi = 0$, it reduces to

$$\int_V |\mathbf{\nabla}\phi|^2\,dv = 0 \tag{3-127}$$

The integrand is an essentially positive quantity; hence

$$|\mathbf{\nabla}\phi| = 0 \tag{3-128}$$

and ϕ is constant throughout V; that is,

$$\phi = \phi_1 - \phi_2 = \text{constant} \tag{3-129}$$

For Dirichlet or mixed boundary conditions, $\phi = 0$ on the boundary or on some part of it; hence this constant must be zero, $\phi_1 = \phi_2$, and the solution is unique. When Neumann conditions are specified, the solution is also unique, but only within an arbitrary additive constant. However, this constant has little or no practical significance since, as noted earlier, a zero reference for the static potential can always be chosen arbitrarily.

In closing, it is important to note that the proof just completed is not dependent on the coordinate system.

3.12   Formal Solutions of Laplace's Equation in Cylindrical Coordinates.

Following our brief discussion of uniqueness, let us next seek solutions to Laplace's equations in cylindrical coordinates.

In right circular cylindrical coordinates, we have

$$\nabla^2\phi = \frac{1}{r}\frac{\partial}{\partial r}\left(r\frac{\partial\phi}{\partial r}\right) + \frac{1}{r^2}\frac{\partial^2\phi}{\partial\varphi^2} + \frac{\partial^2\phi}{\partial z^2} = 0 \tag{3-130}$$

As in the rectangular case, using a product solution will reduce Laplace's equation to three interdependent differential equations of a single variable. Thus, substituting

$$\phi = R(r)\Phi(\varphi)Z(z) \tag{3-131}$$

into Eq. (3-130) gives

$$\Phi Z\frac{d^2R}{dr^2} + \frac{1}{r}\Phi Z\frac{dR}{dr} + \frac{1}{r^2}RZ\frac{d^2\Phi}{d\varphi^2} + R\Phi\frac{d^2Z}{dz^2} = 0 \tag{3-132}$$

Dividing this result by the product $R\Phi Z$ and transposing gives

$$\frac{1}{R}\frac{d^2R}{dr^2} + \frac{1}{rR}\frac{dR}{dr} + \frac{1}{r^2\Phi}\frac{d^2\Phi}{d\varphi^2} = -\frac{1}{Z}\frac{d^2Z}{dz^2} = -\lambda^2 \qquad \lambda^2 \geqq 0 \tag{3-133}$$

where λ^2 is the separation constant. From Eq. (3-133) we obtain two differential equations, namely,

$$\frac{d^2Z}{dz^2} - \lambda^2 Z = 0 \tag{3-134}$$

and

$$\frac{r^2}{R}\frac{d^2R}{dr^2} + \frac{r}{R}\frac{dR}{dr} + \lambda^2 r^2 = -\frac{1}{\Phi}\frac{d^2\Phi}{d\varphi^2} = n^2 \tag{3-135}$$

Here n^2 is a second separation constant whose value is yet to be determined. Equation (3-135) may be broken into two ordinary differential equations:

$$\frac{d^2\Phi}{d\varphi^2} + n^2\Phi = 0 \tag{3-136}$$

$$\frac{d^2R}{dr^2} + \frac{1}{r}\frac{dR}{dr} + \left(\lambda^2 - \frac{n^2}{r^2}\right)R = 0 \tag{3-137}$$

Now Eqs. (3-134) and (3-136) can be solved easily, giving

$$Z = A_1 e^{\lambda z} + A_2 e^{-\lambda z} \tag{3-138}$$

$$\Phi = B_1 \cos n\varphi + B_2 \sin n\varphi \tag{3-139}$$

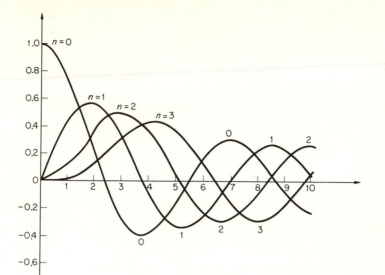

(a)

n	First three zeros		
0	2.405	5.520	8.654
1	0	3.832	7.016
2	0	5.136	8.417
3	0	6.380	9.761

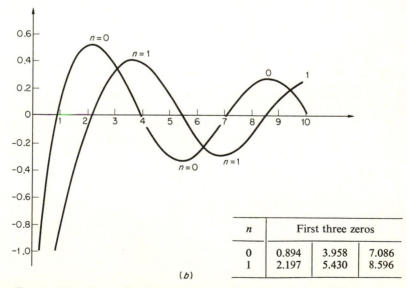

(b)

n	First three zeros		
0	0.894	3.958	7.086
1	2.197	5.430	8.596

FIGURE 3-13.   Curves for Bessel functions of the first and second kind.
(a) First kind; (b) second kind (Neumann functions).

On the other hand, Eq. (3-137), known as *Bessel's equation*, leads to solutions, called *Bessel functions*, which have the forms of infinite series in powers of r. For example, when $n = 0$, the solution is

$$R_0 = C_1 J_0(\lambda r) + C_2 Y_0(\lambda r) \tag{3-140}$$

where

$$J_0(\lambda r) = \sum_{k=0}^{\infty} (-1)^k \frac{(\lambda r/2)^{2k}}{(k!)^2} \tag{3-141}$$

is the Bessel function of the *first kind* and *order zero*, and with

$$\psi(k) = \sum_{m=1}^{k} \frac{1}{m}$$

$$Y_0(\lambda r) = \frac{2}{\pi}\left(\ln\frac{\lambda r}{2} + 0.5772\right) J_0(\lambda r) - \frac{2}{\pi} \sum_{k=1}^{\infty} (-1)^k \psi(k) \frac{\left(\frac{\lambda r}{2}\right)^{2k}}{(k!)^2} \tag{3-142}$$

is the Bessel function of the *second kind*, or *Neumann function*, and *order zero*. Bessel functions of various values of n are plotted in Fig. 3-13.

The general solution of Laplace's equations in cylindrical coordinates is the product of Eqs. (3-138) and (3-139) and Bessel functions similar in form to Eq. (3-140). However, many problems in cylindrical coordinates have solutions which are independent of z. In such cases

$$\phi = R(r)\Phi(\varphi) \tag{3-143}$$

and Laplace's equation reduces to

$$\frac{d^2\Phi}{d\varphi^2} + n^2\Phi = 0 \tag{3-13 6}$$

$$r^2\frac{d^2R}{dr^2} + r\frac{dR}{dr} - n^2R = 0 \tag{3-144}$$

It can be shown that the general solution of the second equation is

$$R = \begin{cases} C_1 \ln r + C_2 & n = 0 \\ C_1 r^n + C_2 r^{-n} & n \neq 0 \end{cases} \tag{3-145}$$

which in combination with Eq. (3-139) gives

$$\phi = \begin{cases} C_1 \ln r + C_2 & n = 0 \\ (B_1 \cos n\varphi + B_2 \sin n\varphi)(C_1 r^n + C_2 r^{-n}) & n \neq 0 \end{cases} \tag{3-146}$$

The arbitrary constants of integration, B_1, B_2, C_1, and C_2, together with all possible values of the constant n, must be obtained from the boundary conditions.

Example 3-8   Coaxial Line. The coaxial line of Fig. 3-14a is a good example to which Laplace's equation in cylindrical coordinates may be applied. The solution to this problem has practical significance because it holds even for time-varying fields.

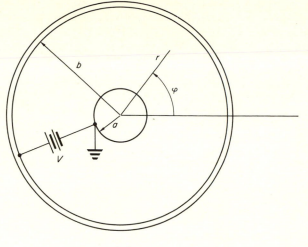

(*a*)

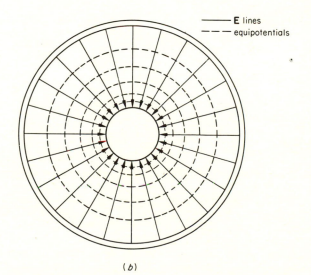

——— **E** lines
– – – equipotentials

(*b*)

FIGURE 3-14. A coaxial line. (*a*) Structure; (*b*) field configuration.

For purposes of discussion we assume that:

1. The coaxial cable is very long, so that $\partial/\partial z = 0$.
2. The radius of the inner conductor is a, and the inner radius of the outer conductor is b (the outer radius is of no concern to us here).
3. The inner conductor (equipotential) is grounded, so that $\phi(a) = 0$.

4. The outer conductor (also an equipotential) is held at a fixed potential by connecting a battery between the two conductors; so $\phi(b) = V$.
5. The interelectrode space, $a < r < b$, is filled with a linear, homogeneous, and isotropic dielectric of permittivity ϵ.

We wish to determine the potential distribution within the dielectric region, the electric field intensity, and the capacitance per unit length of line.

Stated mathematically, we wish to find a solution of Laplace's equation, not dependent on z, which satisfies the (Dirichlet) boundary conditions $\phi(a) = 0$ and $\phi(b) = V$. By uniqueness, only one such solution exists.

A good rule to remember is that the number of arbitrary constants arising in a general solution must be equal to or greater than the number of boundary conditions to be satisfied. Here the number of boundary conditions to be satisfied is two. Hence the general solution must include at least two arbitrary constants. Since, in addition, the solution must be independent of z and φ, the obvious choice dictated from Eq. (3-146) is

$$\phi = C_1 \ln r + C_2 \tag{3-147}$$

Applying the first boundary condition, $\phi(a) = 0$, then gives $C_2 = -C_1 \ln a$, so that

$$\phi = C_1 \ln \frac{r}{a}$$

Next, applying the second boundary condition, $\phi(b) = V$, yields

$$C_1 = \frac{V}{\ln (b/a)}$$

Therefore the potential distribution is

$$\phi = \frac{V}{\ln (b/a)} \ln \frac{r}{a} \qquad a \leq r \leq b \tag{3-148}$$

To obtain an expression for the electric field intensity, we make use of the relation $\mathbf{E} = -\nabla\phi$ in cylindrical coordinates. We find

$$\mathbf{E} = \frac{-V}{\ln (b/a)} \frac{\mathbf{a}_r}{r} \tag{3-149}$$

Both ϕ and $\mathbf{E}$ are plotted in Fig. 3-14b. Equipotential surfaces are concentric, but not equally spaced along a radius. Electric field intensity lines originate on the outer cylinder and terminate on the inner cylinder where the potential is lowest. $\mathbf{E}$ lines become more dense near the surface of the inner conductor, consistent with Eq. (3-149), which shows the amplitude of $\mathbf{E}$ to increase with decreasing r. Thus $r = a$ determines the most critical area from an insulation point of view, because when the magnitude of $\mathbf{E}$ exceeds the *dielectric strength* of a substance, the insulation *breaks down* and arcing (conduction) occurs between conductors. Typically, the dielectric strength of air is 3000 kV/m, while that of paper is 15,000 kV/m. Therefore dielectric breakdown will occur if the quantity $V/[a \ln (b/a)]$ exceeds the dielectric strength of the substance between conductors. Note the linear dependence on V.

Our final objective is to find the capacitance per unit length of coaxial line. To this end, we compute

$$\mathbf{D} = \epsilon\mathbf{E} = -\frac{\epsilon V}{\ln (b/a)} \frac{1}{r} \mathbf{a}_r \tag{3-150}$$

Next, we calculate the density of charge on both cylinders.

Outer cylinder:
$$\rho_s = \frac{\epsilon V}{b \ln (b/a)}$$

(3-151)

Inner cylinder:
$$\rho_s = \frac{-\epsilon V}{a \ln (b/a)}$$

On a section of either cylinder, 1 m in length, the magnitude of the total charge is

$$q = (2\pi)(\text{radius})(1)(\rho_s) = \frac{2\pi \epsilon V}{\ln (b/a)}$$

(3-152)

Therefore, using the relation $C = q/V$, we find that the capacitance per unit length of coaxial line is

$$C = \frac{2\pi \epsilon}{\ln (b/a)}$$

(3-153)

From this formula it is seen that (quite properly) the unit of C is that of ϵ, namely, the farad per meter.

Because of symmetry, the coaxial-cable problem can also be solved by using the integral laws. This is left as an exercise for the student.

In summary, this section was devoted to solutions of Laplace's equation in cylindrical coordinates. The simplest and most frequently used solutions are

1. ϕ independent of φ and z:

$$\phi = C_1 \ln r + C_2$$

(3-154)

where C_1 and C_2 are constants.

2. ϕ independent of r and z:

$$\phi = C_1 \varphi + C_2$$

(3-155)

3. ϕ independent of z:

$$\phi = (B_1 \cos n\varphi + B_2 \sin n\varphi)(C_1 r^n + C_2 r^{-n})$$

(3-156)

where n, B_1, and B_2 are constants, also.

4. ϕ independent of φ:

$$\phi = (B_1 e^{\lambda z} + B_2 e^{-\lambda z})[C_1 J_0(\lambda r) + C_2 Y_0(\lambda r)]$$

(3-157)

where Y_0 is not to be included if the origin ($r = 0$) is a part of the region under consideration.

3.13 Formal Solutions of Laplace's Equation in Spherical Coordinates.

In spherical coordinates Laplace's equation

$$\frac{1}{r^2} \frac{\partial}{\partial r}\left(r^2 \frac{\partial \phi}{\partial r}\right) + \frac{1}{r^2 \sin \theta} \frac{\partial}{\partial \theta}\left(\sin \theta \frac{\partial \phi}{\partial \theta}\right) + \frac{1}{r^2 \sin^2 \theta} \frac{\partial^2 \phi}{\partial \varphi^2} = 0 \quad (3\text{-}158)$$

may be solved again, assuming a product solution of the form

$$\phi = R(r)P(\theta)\Phi(\varphi) \tag{3-159}$$

The details of the solution for the functions R, P, and Φ are quite involved, and are omitted. The simplest and most frequently used solutions of Laplace's equation in spherical coordinates are

1. ϕ independent of θ and φ:

$$\phi = A_1 + \frac{A_2}{r} \tag{3-160}$$

     where A_1 and A_2 are arbitrary constants.

2. ϕ independent of r and φ:

$$\phi = C_1 + C_2 \ln\left(\cot\frac{\theta}{2}\right) \tag{3-161}$$

     where C_1 and C_2 are also arbitrary constants.

3. ϕ independent of φ:

$$\phi = \sum_{n=0}^{\infty} (A_{1n}r^n + A_{2n}r^{-(n+1)})[C_{1n}P_n(\theta) + C_{2n}Q_n(\theta)] \tag{3-162}$$

where the functions P_n are the *Legendre polynomials of the first kind:*

$$P_0(\cos\theta) = 1 \tag{3-163}$$

$$P_1(\cos\theta) = \cos\theta \tag{3-164}$$

$$P_2(\cos\theta) = \tfrac{1}{2}(3\cos^2\theta - 1) \tag{3-165}$$

$$P_3(\cos\theta) = \tfrac{1}{2}(5\cos^3\theta - 3\cos\theta) \tag{3-166}$$

$$P_4(\cos\theta) = \tfrac{1}{8}(35\cos^4\theta - 30\cos^2\theta + 3) \tag{3-167}$$

· ·

The *degree* of these polynomials is signified by the subscripts 0, 1, 2, etc. The functions Q_n in (3-162) are known as the *Legendre functions of the second kind:*

$$Q_0(\cos\theta) = \frac{1}{2}\ln\frac{1+\cos\theta}{1-\cos\theta} \tag{3-168}$$

$$Q_1(\cos\theta) = \frac{1}{2}\cos\theta\ln\frac{1-\cos\theta}{1+\cos\theta} - 1 \tag{3-169}$$

· ·

Every function Q_n is undefined at the north pole ($\theta = 0$) and at the south pole ($\theta = \pi$). Therefore the Q_n's are always excluded when the north and south poles are included in the region under consideration, and Eq. (3-162) is replaced by

$$\phi = \sum_{n=0}^{\infty} (A_{1n}r^n + A_{2n}r^{-(n+1)})P_n(\theta) \qquad (3\text{-}170)$$

Example 3-9   Metal Sphere in a Uniform Field.   Consider a metal sphere of radius a in a dielectric medium of permittivity ϵ.   We suppose that a uniform field $\mathbf{E}_0$ exists everywhere in the dielectric region initially, and wish to determine the field distortion caused by the sphere.

The geometry of the problem is shown in Fig. 3-15a.   The center of the sphere coincides with the origin of the coordinates, and the primary field $\mathbf{E}_0$ has the direction of the negative z axis.

The presence of the metal sphere in the primary field $\mathbf{E}_0$ will generate a secondary field, which, combined with $\mathbf{E}_0$, will render the metal sphere a region of constant potential.   With no loss in generality, the potential of the sphere can always be chosen as zero.   Thus boundary conditions for this problem are

$$\phi = 0 \qquad \text{at} \qquad r = a$$
$$\mathbf{E} \to \mathbf{E}_0 \qquad \text{that is,} \ \phi \to E_0 r \cos \theta \qquad \text{for} \qquad r \gg a \qquad (3\text{-}171)$$

Because of axial symmetry $(\partial/\partial\varphi = 0)$, Eq. (3-170) applies:

$$\phi = \sum_{n=0}^{\infty} (A_{1n}r^n + A_{2n}r^{-(n+1)})P_n(\theta) \qquad a \leq r \qquad (3\text{-}172)$$

Again, we retain only as many terms as we need to satisfy all boundary conditions. Accordingly, let us retain only the first two terms in the infinite sum:

$$\phi = \left(A_{10} + \frac{A_{20}}{r} \right) 1 + \left(A_{11}r + \frac{A_{21}}{r^2} \right) \cos \theta \qquad (3\text{-}173)$$

From the second boundary condition, we have $A_{10} = 0$ and $A_{11} = E_0$.   Thus

$$\phi = \frac{A_{20}}{r} + \left(E_0 r + \frac{A_{21}}{r^2} \right) \cos \theta$$

Now, from the first boundary condition, we obtain $A_{20} = 0$ and $A_{21} = -E_0 a^3$. Hence the potential is

$$\phi = E_0 \left[1 - \left(\frac{a}{r} \right)^3 \right] r \cos \theta \qquad (3\text{-}174)$$

and the electric field intensity is

$$\mathbf{E} = -E_0 \left[1 + 2 \left(\frac{a}{r} \right)^3 \right] \cos \theta \, \mathbf{a}_r + E_0 \left[1 - \left(\frac{a}{r} \right)^3 \right] \sin \theta \, \mathbf{a}_\theta \qquad (3\text{-}175)$$

For r very large,

$$\lim_{r \to \infty} \mathbf{E} = -E_0 \cos \theta \, \mathbf{a}_r + E_0 \sin \theta \, \mathbf{a}_\theta = -E_0 \mathbf{a}_z = \mathbf{E}_0 \qquad (3\text{-}176)$$

as required.   On the other hand, when $r = a$,

$$\mathbf{E} = -3E_0 \cos \theta \mathbf{a}_r \qquad (3\text{-}177)$$

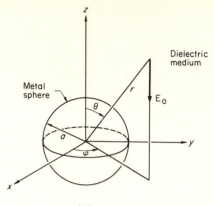

(a)

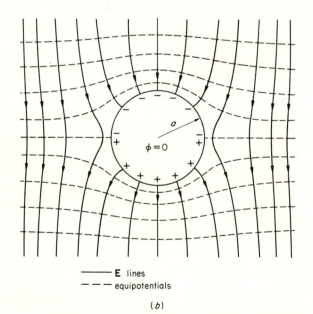

——— **E** lines
- - - equipotentials

(b)

FIGURE 3-15.   Metal sphere suspended in a uniform field.
(a) Geometry; (b) field configuration.

which shows that the electric field intensity vector is normal to the surface of the sphere. Maximum stress occurs at the top and bottom of the sphere, where $|\mathbf{E}| = 3E_0$.

A plot of the field is shown in Fig. 3-15b. Note that, although the sphere is electrically neutral, a distribution of charge,

$$\rho_s = -3\epsilon E_0 \cos \theta \tag{3-178}$$

appears at all points on the outer surface.

3.14 Properties of Laplace's Equation.

Every solution of Laplace's equation is characterized by a number of useful properties. One of these provides the theoretical basis for the numerical solution of Laplace's equation, a topic which is discussed in the next section.

PROPERTY 1

A solution ϕ of Laplace's equation within a region V, completely enclosed by a surface Σ, cannot attain a maximum value or a minimum value at any point within V. Instead, ϕ must reach its maximum and minimum values on the bounding surface Σ.

To prove this assertion, we assume, to the contrary, that ϕ is maximum at some point (x,y,z) within V. At that point, the derivative of ϕ with respect to each space coordinate must be zero, and the second derivative must be negative. Thus

$$\frac{\partial^2 \phi}{\partial x^2} < 0 \qquad \frac{\partial^2 \phi}{\partial y^2} < 0 \qquad \frac{\partial^2 \phi}{\partial z^2} < 0$$

and therefore

$$\frac{\partial^2 \phi}{\partial x^2} + \frac{\partial^2 \phi}{\partial y^2} + \frac{\partial^2 \phi}{\partial z^2} < 0 \tag{3-179}$$

This, however, contradicts the assumption that ϕ is a solution of Laplace's equation

$$\nabla^2 \phi = \frac{\partial^2 \phi}{\partial x^2} + \frac{\partial^2 \phi}{\partial y^2} + \frac{\partial^2 \phi}{\partial z^2} = 0$$

Therefore ϕ cannot assume a maximum value within V.

On the other hand, ϕ cannot assume a minimum value within V because the condition

$$\frac{\partial^2 \phi}{\partial x^2} + \frac{\partial^2 \phi}{\partial y^2} + \frac{\partial^2 \phi}{\partial z^2} > 0 \tag{3-180}$$

would also contradict Laplace's equation.

PROPERTY 2

All partial derivatives of ϕ computed with respect to its rectangular independent variables satisfy Laplace's equation. This statement is generally not true for other types of coordinates, such as cylindrical or spherical.

This follows easily because, for example,

$$\nabla^2\left(\frac{\partial\phi}{\partial x}\right) = \frac{\partial^3\phi}{\partial x^3} + \frac{\partial^3\phi}{\partial x\,\partial y^2} + \frac{\partial^3\phi}{\partial x\,\partial z^2} = \frac{\partial}{\partial x}\left(\nabla^2\phi\right) = 0$$

A conclusion that can be drawn directly from this property is that a set of solutions to Laplace's equation can be obtained from a single solution by differentiation with respect to the rectangular coordinates.

It follows from Property 1 now that all partial derivatives of ϕ with respect to x, y, or z reach their respective maxima and minima on the bounding surface Σ.

PROPERTY 3

A function ϕ which satisfies Laplace's equation at all interior points of a sphere has at the center of the sphere a value equal to the average of its values either on the surface or in the sphere.   (This is often called the *average-value theorem*.)

The assumption that at all interior points of a sphere ϕ satisfies Laplace's equation implies that all continuous sources of the static electric field must be located outside the spherical region.   According to Eq. (3-58), the relation of ϕ to its sources is given by

$$\phi = \frac{1}{4\pi}\oint_\Sigma\left[\frac{1}{r}\nabla\phi - \phi\nabla\left(\frac{1}{r}\right)\right]\cdot\mathbf{n}\,da \tag{3-181}$$

Since r is constant on the spherical surface Σ, the first integral yields

$$\frac{1}{4\pi r_0}\oint_\Sigma\nabla\phi\cdot\mathbf{n}\,da = \frac{1}{4\pi r_0}\int_V\nabla^2\phi\,dv = 0 \tag{3-182}$$

where r_0 denotes the radius of the sphere; the first equality follows directly from the divergence theorem.   On the other hand,

$$\nabla\left(\frac{1}{r}\right)_{r=r_0}\cdot\mathbf{n} = -\frac{1}{r_0{}^2}$$

Therefore the potential at the center is

$$\phi_0 = \frac{1}{4\pi r_0{}^2}\oint_\Sigma\phi\,da = \frac{\oint_\Sigma\phi\,da}{A} \tag{3-183}$$

and this proves the first part of our assertion, since $A = 4\pi r_0{}^2$ is just the area of the spherical surface.

To prove the second part of the assertion, we write Eq. (3-183) for an arbitrary sphere of radius $r \le r_0$ as

$$4\pi r^2\phi_0 = \oint_\Sigma\phi\,da \tag{3-184}$$

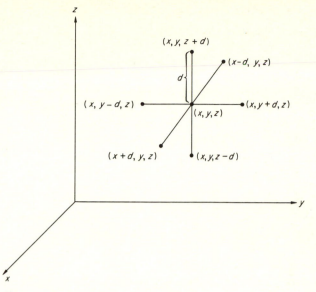

FIGURE 3-16.   A three-dimensional grid.

and subsequently integrate both sides with respect to r:

$$\int_0^{r_0} 4\pi r^2 \phi_0 \, dr = \int_0^{r_0} \left[\oint_\Sigma \phi \, da \right] dr$$

The left member integrates to $\frac{4}{3}\pi r_0^3 \phi_0$.   Hence

$$\phi_0 = \frac{1}{\frac{4}{3}\pi r_0^3} \int_V \phi \, dv = \frac{\int_V \phi \, dv}{V} \tag{3-185}$$

and this proves the second part of our assertion, since $V = \frac{4}{3}\pi r_0^3$ is the volume of the sphere.

Another way of looking at the same property is through the notion of the derivative.   Consider a three-dimensional grid, as shown in Fig. 3-16.   An arbitrary point (x,y,z) is shown at the intersection of three mutually orthogonal lines oriented parallel to the directions of the coordinate axes.   The length $2d$ of each line is very small.

According to the definition of a partial derivative, the *rate of increase* of the potential at (x,y,z) is given by

$$\left. \frac{\partial \phi}{\partial x} \right]_+ \approx \frac{1}{d} [\phi(x + d, y, z) - \phi(x,y,z)]$$

and by

$$\left. \frac{\partial \phi}{\partial x} \right]_- \approx \frac{1}{d} [\phi(x,y,z) - \phi(x - d, y, z)]$$

Therefore

$$\frac{\partial^2 \phi}{\partial x^2} \approx \frac{\left.\dfrac{\partial \phi}{\partial x}\right]_+ - \left.\dfrac{\partial \phi}{\partial x}\right]_-}{d} \approx \frac{1}{d^2}\,[\phi(x + d, y, z) - 2\phi(x,y,z) + \phi(x - d, y, z)]$$

Similarly,

$$\frac{\partial^2 \phi}{\partial y^2} \approx \frac{1}{d^2}\,[\phi(x, y + d, z) - 2\phi(x,y,z) + \phi(x, y - d, z)]$$

$$\frac{\partial^2 \phi}{\partial z^2} \approx \frac{1}{d^2}\,[\phi(x, y, z + d) - 2\phi(x,y,z) + \phi(x, y, z - d)]$$

Adding the last three equations gives

$$\frac{\partial^2 \phi}{\partial x^2} + \frac{\partial^2 \phi}{\partial y^2} + \frac{\partial^2 \phi}{\partial z^2} \approx \frac{1}{d^2}\,[\phi(x + d, y, z) + \phi(x, y + d, z)$$

$$+ \phi(x, y, z + d) - 6\phi(x,y,z) + \phi(x - d, y, z)$$

$$+ \phi(x, y - d, z) + \phi(x, y, z - d)] \quad \text{(3-186)}$$

By virtue of Laplace's equation, the sum of the partial derivatives on the left is zero.   Hence

$$\phi(x,y,z) \approx \tfrac{1}{6}[\phi(x + d, y, z) + \phi(x, y + d, z) + \phi(x, y, z + d)$$

$$+ \phi(x - d, y, z) + \phi(x, y - d, z) + \phi(x, y, z - d)] \quad \text{(3-187)}$$

and this shows that the potential at the center point (x,y,z) is approximately equal to the average of its values on the six corners of the grid.

PROPERTY 4[†]

The field which results from an interchange of the equipotential lines and the lines of force (streamlines) of any two-dimensional solution of Laplace's equation is also a solution of the same equation.

To prove this theorem we note that, if ϕ is a two-dimensional solution of Laplace's equation, then

$$\nabla \phi = \frac{\partial \phi}{\partial x}\,\mathbf{a}_x + \frac{\partial \phi}{\partial y}\,\mathbf{a}_y$$

is a vector perpendicular to the lines of constant potential.   At any point in the field, the ratio $\nabla \phi / |\nabla \phi|$ therefore defines a unit vector tangent to the local streamline,[‡] the slope dy/dx of which is just the slope of the unit tangent

[†] D. T. Paris, Conjugate Solutions of Laplace's Equation, *Proc. IEEE*, vol. 55, pp. 104–105, January, 1967.

[‡] In a conductor, streamlines coincide with lines of current flow.

vector.   Put mathematically, this leads to

$$\frac{dy}{dx} = \frac{\partial \phi / \partial y}{\partial \phi / \partial x} \tag{3-188}$$

for the differential equation of the streamlines.   Integration yields

$$\int \frac{\partial \phi}{\partial x} \, dy = \int \frac{\partial \phi}{\partial y} \, dx + \text{constant} \tag{3-189}$$

which is the equation of the streamlines.

On the basis of the foregoing discussion, we define†

$$\phi' = A \left(\int^y \frac{\partial \phi}{\partial x} \, dy - \int^x \frac{\partial \phi}{\partial y} \, dx \right) + B \tag{3-190}$$

where A and B are constants, and assert that ϕ' is a potential function which satisfies Laplace's equation as well as the condition of orthogonality, $\nabla \phi \cdot \nabla \phi' = 0$.

Proof: We calculate

$$\frac{\partial \phi'}{\partial x} = A \left(\int^y \frac{\partial^2 \phi}{\partial x^2} \, dy - \frac{\partial \phi}{\partial y} \right)$$

and note that, since $\nabla^2 \phi = 0$, $\partial^2 \phi / \partial x^2 = - \partial^2 \phi / \partial y^2$, and

$$\int^y \frac{\partial^2 \phi}{\partial x^2} \, dy = - \int^y \frac{\partial^2 \phi}{\partial y^2} \, dy = - \frac{\partial \phi}{\partial y}$$

hence

$$\frac{\partial \phi'}{\partial x} = -2A \frac{\partial \phi}{\partial y}$$

Now

$$\frac{\partial^2 \phi'}{\partial x^2} = -2A \frac{\partial^2 \phi}{\partial x \, \partial y}$$

Similarly,

$$\frac{\partial^2 \phi'}{\partial y^2} = 2A \frac{\partial^2 \phi}{\partial y \, \partial x}$$

Therefore

$$\nabla^2 \phi' = \frac{\partial^2 \phi'}{\partial x^2} + \frac{\partial^2 \phi'}{\partial y^2} = -2A \frac{\partial^2 \phi}{\partial x \, \partial y} + 2A \frac{\partial^2 \phi}{\partial y \, \partial x} \equiv 0 \tag{3-191}$$

---

† Note that integrals of the type shown in parentheses in Eq. (3-190) give rise to arbitrary additive functions of x or y, alone.   Our definition requires that these functions be zero.

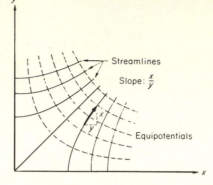

$$- - - \ xy = \text{constant}$$
$$\overline{\qquad} \ x^2 - y^2 = \text{constant}$$

FIGURE 3-17.   Plot of the two-dimensional potential function $\phi = xy$ and its conjugate $\phi' = x^2 - y^2$.

To prove the condition of orthogonality, we calculate

$$\boldsymbol{\nabla}\phi \cdot \boldsymbol{\nabla}\phi' = \left(\frac{\partial \phi}{\partial x}\,\mathbf{a}_x + \frac{\partial \phi}{\partial y}\,\mathbf{a}_y\right) \cdot \left(\frac{\partial \phi'}{\partial x}\,\mathbf{a}_x + \frac{\partial \phi'}{\partial y}\,\mathbf{a}_y\right)$$

$$= \left(\frac{\partial \phi}{\partial x}\,\mathbf{a}_x + \frac{\partial \phi}{\partial y}\,\mathbf{a}_y\right) \cdot \left(-2A\frac{\partial \phi}{\partial y}\,\mathbf{a}_x + 2A\frac{\partial \phi}{\partial x}\,\mathbf{a}_y\right)$$

$$= 2A\left(-\frac{\partial \phi}{\partial x}\frac{\partial \phi}{\partial y} + \frac{\partial \phi}{\partial y}\frac{\partial \phi}{\partial x}\right) \equiv 0 \qquad (3\text{-}192)$$

which proves the second part of our assertion.

Example 3-10   Interchange of Hyperbolic Potentials.   For a specific example, let us consider the problem of interchanging the equipotential lines and the lines of force associated with the hyperbolic potential function $\phi = xy$ (Fig. 3-17).

From the gradient expression

$$\boldsymbol{\nabla}\phi = y\mathbf{a}_x + x\mathbf{a}_y$$

it follows that the streamlines are defined by the differential equation

$$\frac{dy}{dx} = \frac{x}{y}$$

This equation can be solved by the method of separation of variables, giving

$$\tfrac{1}{2}y^2 = \tfrac{1}{2}x^2 + \text{constant}$$

The condition that $y = 0$ when $x = 0$ (45° line in Fig. 3-17) requires that the value of the constant of integration be zero. Hence

$$x^2 - y^2 = \text{constant}$$

defines the family of streamlines associated with $\phi = xy$, and

$$\phi' = x^2 - y^2$$

defines the conjugate potential for the special case $A = 1$. The proof that ϕ' satisfies (a) Laplace's equation, and (b) the orthogonality condition $\nabla\phi \cdot \nabla\phi' = 0$ is an elementary task.

PROPERTY 5

Let ϕ be a function which satisfies Laplace's equation within a region V and which is constant on the bounding surface Σ. Then ϕ must be constant throughout V.

By virtue of its continuity, the function ϕ must take on (in the volume V) its maximum value, its minimum value, and every value in between. But this is contrary to the conditions expressed by Property 1. The only way in which all requirements can be satisfied simultaneously is for ϕ to have a constant value throughout V equal to its value on the surface Σ. In particular, if $\phi = 0$ on Σ, then $\phi = 0$ throughout V.

This, by the way, explains the electrostatic shielding properties of hollow metal cavities.

PROPERTY 6

Let ϕ be a function which satisfies Laplace's equation within a region V, and the normal derivative of which is zero ($\partial\phi/\partial n = 0$) on the bounding surface Σ. Then ϕ must be constant throughout V.

To prove the validity of this statement, we apply the divergence theorem to the function $\phi \nabla\phi$. We have

$$\int_V \nabla \cdot \phi \nabla\phi \, dv = \oint_\Sigma \phi \nabla\phi \cdot \mathbf{n} \, da \qquad (3\text{-}193)$$

and since

$$\nabla \cdot \phi \nabla\phi \equiv \phi \nabla^2\phi + \nabla\phi \cdot \nabla\phi = |\nabla\phi|^2$$

Eq. (3-193) transforms to

$$\int_V |\nabla\phi|^2 \, dv = \oint_\Sigma \phi \frac{\partial\phi}{\partial n} \, da \qquad (3\text{-}194)$$

Moreover, $\partial\phi/\partial n = 0$ everywhere on Σ; hence

$$\int_V |\nabla\phi|^2 \, dv = 0 \qquad (3\text{-}195)$$

The theorem is proved since the integrand is an essentially positive quantity, from which it follows at once that $|\nabla\phi|^2 = 0$, $\nabla\phi = 0$, $\phi = $ constant, in that sequence.

3.15 The Numerical Method.

Property 3 of Laplace's equation provides a most useful basis for solving Laplace's equation by the numerical method. This is an extremely important method because it allows us to proceed to the solution of problems beyond the range of purely formal analytical techniques. Keep in mind that *most problems that can be solved formally have already been solved!*

The method can best be introduced with the aid of an example, one in particular which can be solved analytically, to compare answers.

Example 3-11 Covered Trough. Let us consider the problem of the long hollow rectangular metal tube of Example 3-7. We recall that the bottom three sides of the pipe are grounded, that the top side is held at the variable potential

$$\phi(x,b) = V \sin \frac{\pi}{a} x$$

and that the potential distribution within the pipe is

$$\phi = \frac{V}{\sinh (\pi/a)b} \sin \frac{\pi}{a} x \sinh \frac{\pi}{a} y \qquad (3\text{-}196)$$

For simplicity, we set $V = 1$ and $a = b = \pi/2$. Then, within the pipe,

$$\phi = \frac{\sin 2x \sinh 2y}{\sinh \pi} \qquad (3\text{-}197)$$

and the conditions on the boundaries are those indicated in Fig. 3-18a.

The numerical method seeks to determine the potential at every point on a cross sectional representation of the geometry in the form of a grid by systematically adjusting all unknown potentials until they become equal (within allowable limits) to the average of the values of the potentials at all immediately adjacent points on the grid. The procedure by which the systematic adjustment is made is called *iteration*. An outline of the detailed calculations follows.

First, all known potentials are calculated at various points on the grid. In the present example, potentials are known for all points on the boundary. Thus, from left to right on the top boundary, in Fig. 3-18b, the potentials are:

0.0000   0.5000   0.8660   1.0000   0.8660   0.5000   0.0000

consistent with the requirement $\phi(x, \pi/2) = \sin 2x$. At every point on the remaining three sides of the boundary, the potential is zero.

Next, an initial distribution of potentials is assumed at all remaining points. A reasonably educated initial estimate will normally suffice. However, fewer iterations are required if the assumed distribution approximates the average of the surrounding

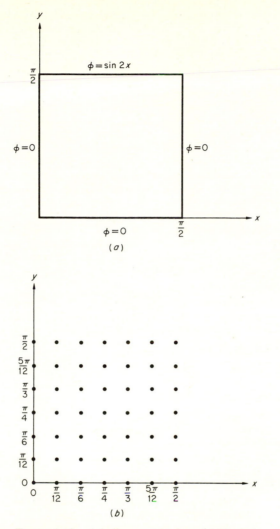

FIGURE 3-18. A hollow metal tube. (a) Physical arrangement; (b) grid of points representing tube cross section.

points.   An initial guess for the problem at hand is tabulated below:

0.0000	0.5000	0.8660	1.0000	0.8660	0.5000	0.0000
0.0000	0.3000	0.7000	0.9000	0.7000	0.3000	0.0000
0.0000	0.2000	0.5000	0.7000	0.5000	0.2000	0.0000
0.0000	0.1500	0.2000	0.5000	0.2000	0.1500	0.0000
0.0000	0.1000	0.1500	0.3000	0.1500	0.1000	0.0000
0.0000	0.0500	0.1200	0.1000	0.1200	0.0500	0.0000
0.0000	0.0000	0.0000	0.0000	0.0000	0.0000	0.0000

This array of numbers is treated as input information.   At every interior point of the array the computer is programmed to calculate the average of the potential at the four adjacent points.   For example, at the location of the second row, second column, the first calculation would be

$$\text{Average potential} = \frac{0.5000 + 0.7000 + 0.2000 + 0.0000}{4} = 0.3500 \qquad (3\text{-}198)$$

This value differs from the assumed value, 0.3000, by 0.0500 unit.   If, arbitrarily, we establish 0.0005 as a difference we are willing to tolerate, the potential at that first point in the field does not satisfy this requirement and must therefore be adjusted.   A suitable adjustment is simply to choose the average of 0.3000 and 0.3500, that is, 0.3250.   Identical operations are then carried out at every interior point of the array. If the average potential, as computed in Eq. (3-198), differs from the assumed potential at any interior point by more than the allowed tolerance (0.0005 in this case), a second iteration is required.   During the second iteration the adjusted values produced in the first iteration become the assumed values for calculating the new adjustment.   This process is continued until no adjustments are necessary.   In general, many iterations are required to satisfy any respectable allowed tolerance.

The answers obtained to the problem of the covered trough are

0.0000	0.5000	0.8660	1.0000	0.8660	0.5000	0.0000
0.0000	0.2998	0.5191	0.5993	0.5191	0.2998	0.0000
0.0000	0.1791	0.3098	0.3575	0.3098	0.1791	0.0000
0.0000	0.1052	0.1817	0.2096	0.1817	0.1051	0.0000
0.0000	0.0584	0.1008	0.1162	0.1008	0.0584	0.0000
0.0000	0.0263	0.0451	0.0520	0.0452	0.0263	0.0000
0.0000	0.0000	0.0000	0.0000	0.0000	0.0000	0.0000

Forty-eight iterations were required to produce the desired tolerance limit of 0.0005 unit.   These answers compare favorably with the exact values of the potential computed from Eq. (3-197):

0.0000	0.5000	0.8660	1.0000	0.8660	0.5000	0.0000
0.0000	0.2951	0.5112	0.5903	0.5112	0.2951	0.0000
0.0000	0.1731	0.2998	0.3462	0.2998	0.1731	0.0000
0.0000	0.0996	0.1725	0.1992	0.1725	0.0996	0.0000
0.0000	0.0540	0.0936	0.1081	0.0936	0.0540	0.0000
0.0000	0.0237	0.0410	0.0474	0.0410	0.0237	0.0000
0.0000	0.0000	0.0000	0.0000	0.0000	0.0000	0.0000

The numerical method always produces approximate results.   Several reasons account for this behavior.   One is that computers inevitably introduce errors due to truncation.   Second, in the formulation of every problem, a continuous region is always represented by a finite set of points.   In addition, the tolerance limit in the averaging process influences the final accuracy.   In most cases the desired accuracy will be dictated by the ultimate use of the final results (for example, the calculation of resistance or capacitance), and can usually be obtained through proper adjustment of grid spacing and tolerance limit.

To summarize, Laplace's equation can always be solved by numerical techniques *even when the physical boundaries of the problem do not coincide with coordinate surfaces.*

3.16 Adapting a Known Solution to a New Problem.

In certain cases, the known solution of one problem helps expedite the solution of a new, more difficult problem.

Example 3-12 Lecher Two-wire Line. To illustrate this method, let us determine the capacitance per unit length of a line formed by two very long parallel cylindrical wires, each of radius a, separated by a distance b. It is possible to solve this problem as a boundary-value problem. However, this approach requires use of the unfamiliar *bipolar coordinates.* Instead, we begin with a much simpler problem.

We consider a pair of parallel line charges (Fig. 3-19a) and proceed to determine the potential distribution when the left line is charged to a uniform density ρ_l, while the right line is charged to a density $-\rho_l$.

As a field point, we choose $P(x,y,0)$. No generality is lost by choosing a point on the xy plane since the line charges extend the same distance, l, above and below the plane. The pair of infinitesimal line charges, $+\rho_l \, dz$ and $-\rho_l \, dz$, each located a distance z away from the xy plane, account for a contribution to the potential at P of magnitude

$$d\phi = \frac{\rho_l \, dz}{4\pi\epsilon}\left(\frac{1}{r_1'} - \frac{1}{r_2'}\right) = \frac{\rho_l \, dz}{4\pi\epsilon}\left(\frac{1}{\sqrt{r_1^2 + z^2}} - \frac{1}{\sqrt{r_2^2 + z^2}}\right)$$

Taking advantage of symmetry, we obtain by integration

$$\phi = \frac{\rho_l}{2\pi\epsilon}\int_0^l \left(\frac{1}{\sqrt{r_1^2 + z^2}} - \frac{1}{\sqrt{r_2^2 + z^2}}\right) dz = \frac{\rho_l}{2\pi\epsilon}\left(\ln\frac{l + \sqrt{r_1^2 + l^2}}{l + \sqrt{r_2^2 + l^2}} - \ln\frac{r_1}{r_2}\right)$$

$$(3\text{-}199)$$

The limit of this expression, as $l \to \infty$, is

$$\phi = -\frac{\rho_l}{2\pi\epsilon}\ln\frac{r_1}{r_2} \qquad (3\text{-}200)$$

Now the generic equation of all equipotential surfaces is obtained by setting ϕ equal to a constant. Thus, for a line infinite in extent, all surfaces of constant potentials are solutions of

$$k_1 = \frac{\rho_l}{2\pi\epsilon}\ln\frac{r_2}{r_1} \qquad (3\text{-}201)$$

or

$$\left(\frac{r_2}{r_1}\right)^2 = k^2 \qquad (3\text{-}202)$$

where now

$$k = \ln^{-1}\frac{2\pi\epsilon k_1}{\rho_l} \qquad (3\text{-}203)$$

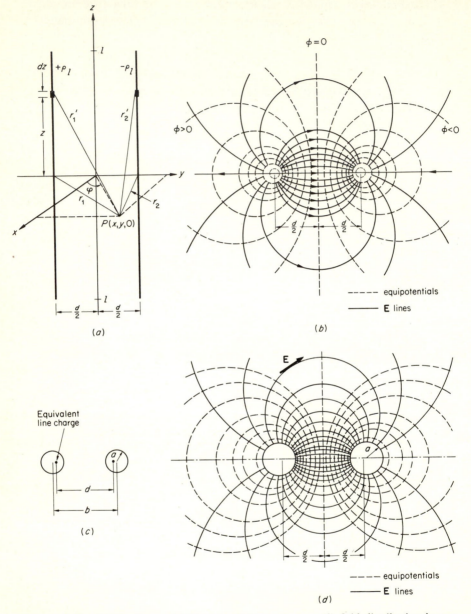

Figure 3-19. Infinite lines. (*a*) Two parallel line charges; (*b*) field distribution in any plane parallel to the *xy* plane for two parallel line charges; (*c*) two-wire line; (*d*) field about a two-wire line.

Substituting the expressions

$$r_1^2 = x^2 + \left(y + \frac{d}{2}\right)^2 \tag{3-204}$$

$$r_2^2 = x^2 + \left(y - \frac{d}{2}\right)^2 \tag{3-205}$$

into Eq. (3-202) and collecting terms, we obtain

$$x^2 + \left(y + \frac{d}{2}\frac{k^2 + 1}{k^2 - 1}\right)^2 = \frac{k^2 d^2}{(k^2 - 1)^2} \qquad \text{for all } z \tag{3-206}$$

From Eq. (3-203) it is clear that the parameter k is a function of the medium, of the density of charge, and of the constant potential selected. So each value of k_1 establishes a different value for k^2, and this in turn, together with d, establishes values for the numerical terms in Eq. (3-206). The form of this equation is easily recognized as a pair of cylinders with equal radii $kd/(k^2 - 1)$ centered about $x = 0$, $y = \pm d(k^2 + 1)/2(k^2 - 1)$.

Figure 3-19b shows a plot of the static field generated by a pair of parallel line charges. The equipotential surfaces are seen to be nonconcentric cylinders completely surrounding the parallel line charges. For very small values of potential (small k_1 and k), the corresponding radii are large. For zero potential ($k = \pm 1$), the radius is infinitely large. Thus, in the xz plane, the potential is zero. On the other hand, for very large values of potential (large k_1 and k), the cylindrical surfaces become diminishingly small and nearly concentric.

Finally, note that all surfaces with positive potential surround the positively charged line on the left; all surfaces with negative potential surround the negatively charged line on the right. This follows from Eq. (3-200), because, for example, around the left line, $r_1 < r_2$ and $\ln (r_1/r_2) < 0$.

Now we are ready to solve the electrostatic problem of the Lecher two-wire line (Fig. 3-19c). We suppose the line to be infinitely long and each wire to have a radius a. The distance of separation between their axes is b.

It is conceivable that we might be able to adjust the distance d shown in Fig. 3-19b in such a way as to make the separation between a corresponding pair of cylindrical equipotential surfaces equal to b. This condition requires that

$$b = d\frac{k^2 + 1}{k^2 - 1} \tag{3-207}$$

At the same time, through proper adjustment of ρ_l, we are also able to adjust k so that the conductor surfaces and the same pair of cylindrical equipotential surfaces coincide exactly. This second condition requires that

$$a = \frac{kd}{k^2 - 1} \tag{3-208}$$

Equations (3-207) and (3-208) can be solved simultaneously for d and k. However, in the present example, we are interested primarily in the value of k. Dividing the latter into the former eliminates d, yielding

$$\frac{b}{a} = \frac{k^2 + 1}{k}$$

from which it follows that

$$k = \frac{b \pm \sqrt{b^2 - 4a^2}}{2a} \tag{3-209}$$

Combining this result with Eqs. (3-202) and (3-200) yields

Left conductor: $\phi_+ = \frac{\rho_l}{2\pi\epsilon} \ln \frac{b + \sqrt{b^2 - 4a^2}}{2a}$

$$\tag{3-210}$$

Right conductor: $\phi_- = \frac{\rho_l}{2\pi\epsilon} \ln \frac{b - \sqrt{b^2 - 4a^2}}{2a}$

Since the quantities within the logarithms are the reciprocals of each other, $\phi_+ = -\phi_-$. Therefore, if a potential difference V is maintained between wires, then

$$V = \phi_+ - \phi_- = \frac{\rho_l}{\pi\epsilon} \ln \frac{b + \sqrt{b^2 - 4a^2}}{2a} \tag{3-211}$$

and the capacitance per unit length is

$$C = \frac{\rho_l}{V} = \frac{\pi\epsilon}{\ln \left[(b + \sqrt{b^2 - 4a^2})/2a \right]} \tag{3-212}$$

This is the required result. When the separation is large compared with wire radius $(b \gg a)$,

$$C \approx \frac{\pi\epsilon}{\ln (b/a)} \tag{3-213}$$

This expression should be compared with Eq. (3-153).

Before terminating this discussion, it is well to note that at every point exterior to the wires the field is identical with that of the parallel line charges; within the wires themselves, however, the fields are different. Each wire, whether hollow or solid, is an equipotential region. The field about a two-wire line is shown in Fig. 3-19d.

Finally, we should note that the distribution of charge on the surface of the two conductors is not uniform. It is more dense on the side near the other wire than it is elsewhere. This behavior is called the *proximity effect* (Prob. 3-37).

3.17 The Method of Images.
The method discussed in the preceding section bears a strong resemblance to a special technique known as the *method of images*.

Example 3-13 Infinite Line Charge Parallel to Axis of a Conductor. To introduce this method, let us suppose that the line charge on the right of Fig. 3-19a is removed and that a wire of finite radius is put in its place. Let us also suppose that the wire carries no charge and that both the line charge and the wire are very long and parallel to each other. The potential distribution about the hybrid structure is desired.

Essentially, the problem is to find a solution of Laplace's equation which becomes undefined at every point along the line charge and which, in the wire, is everywhere constant.

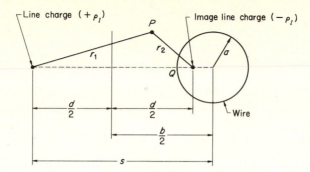

FIGURE 3-20. Imaging an infinite line charge.

The method of images asserts that all requirements can be met by a scalar potential field which arises from the line charge itself and a suitable *image* line charge. The image charge, opposite in sign, is located at some distance d which satisfies Eqs. (3-207) and (3-208) simultaneously. Thus

$$d = b \left[1 - \frac{4a^2}{b(b \pm \sqrt{b^2 - 4a^2})} \right]$$

However, only the positive sign is of interest since the negative sign could make d negative; thus

$$d = b \left[1 - \frac{4a^2}{b(b + \sqrt{b^2 - 4a^2})} \right] \tag{3-214}$$

From Fig. 3-20 it is clear that

$$\frac{b + d}{2} = s \tag{3-215}$$

where s is the distance from the line charge to the axis of the wire. Therefore, with s known, b and d can be determined from Eqs. (3-214) and (3-215). This is enough information to locate the image charge at its proper place.

The potential external to the wire is given by

$$\phi = -\frac{\rho_l}{2\pi\epsilon} \ln \frac{r_1}{r_2} \tag{3-20 0}$$

The constant potential of the wire can be evaluated by allowing the point P, shown in Fig. 3-20, to approach Q; at that point

$$r_1 = \frac{b + d}{2} - a \tag{3-216}$$

$$r_2 = a - \frac{b - d}{2} \tag{3-217}$$

Since b and d are both known in terms of s, the constant potential at the surface of the conductor can be determined by substituting Eqs. (3-216) and (3-217) in Eq. (3-200).

It is clear that some a priori information is needed to solve a problem by the method of images. In fact, the solution is seldom straightforward, and must be found by experience and, possibly, intuition.

The preceding example brings out a second important point about the general types of problems that can be solved. Normally, a distribution of charges and conductors will be given, and the proper (sometimes geometric) image charges are to be found by making an educated guess such that, together with the real charges, the image charges make all conductor surfaces equipotentials. The potential at points exterior to all conductors can then be obtained by summing the effects of both real and image charges, pretending meanwhile that the conductors themselves are all absent. The induced surface charge on all conductors can be calculated as usual from the normal derivative of the potential.

One point should be emphasized: The field within the conductors *cannot* be calculated simply by summing the field of the real charges and that of the image charges.

A simple theoretical proof will now be given to show the general validity of the method of images.

We begin by referring to Fig. 3-7, and suppose that the distribution of charge is such that the boundary Σ is an *equipotential surface*. Then, in Eq. (3-58), which is

$$\phi(x_0,y_0,z_0) = \frac{1}{4\pi\epsilon} \int_V \frac{\rho}{r}\, dv + \frac{1}{4\pi} \oint_\Sigma \left[\frac{1}{r}\nabla\phi - \phi\nabla\left(\frac{1}{r}\right) \right] \cdot \mathbf{n}\, da \qquad (3\text{-}58)$$

the last term on the right may be transformed, using the divergence theorem to give

$$-\frac{1}{4\pi} \oint_\Sigma \phi\nabla\left(\frac{1}{r}\right) \cdot \mathbf{n}\, da = -\frac{\phi}{4\pi} \oint_\Sigma \nabla\left(\frac{1}{r}\right) \cdot \mathbf{n}\, da = -\frac{\phi}{4\pi} \int_V \nabla^2\left(\frac{1}{r}\right) dv$$

$$(3\text{-}218)$$

remembering, of course, that $\phi = $ constant on Σ. But according to Eq. (3-51),

$$\nabla^2\left(\frac{1}{r}\right) = 0 \qquad \text{when } r \neq 0$$

Therefore

$$\phi(x_0,y_0,z_0) = \frac{1}{4\pi\epsilon} \int_V \frac{\rho}{r}\, dv + \frac{1}{4\pi} \oint_\Sigma \frac{1}{r}\nabla\phi \cdot \mathbf{n}\, da \qquad (3\text{-}219)$$

As in Sec. 3.6, the volume integral gives the contribution to the potential due to charges inside of the volume, and the surface integral gives the contribution to the potential due to charges outside of the volume. In terms of the actual charges,

$$\phi(x_0,y_0,z_0) = \frac{1}{4\pi\epsilon} \int_V \frac{\rho}{r}\, dv + \frac{1}{4\pi\epsilon} \int_{V'} \frac{\rho'}{r}\, dv \qquad (3\text{-}220)$$

where V' is the complement of V (assumed to have the same ϵ's) and ρ' is the distribution of charge in V'. It is now clear from a comparison of Eqs. (3-219) and (3-220) that, insofar as the field at an interior point of Σ is concerned, the presence of ρ' in V' can be completely accounted for by an equivalent surface distribution of charge, $\epsilon \, \nabla \phi \cdot \mathbf{n} = \epsilon \, \partial\phi/\partial n$, at every point on Σ, and conversely. Thus, if $\phi = $ constant on Σ, the field at all interior points remains exactly the same when all exterior charges are removed and replaced by a charge density $\epsilon \, \partial\phi/\partial n$ on the equipotential boundary. Conversely, if Σ is a conducting boundary, the field at all interior points remains the same when the conducting boundary and any surface charges on it are removed and replaced by an *image* charge which makes the boundary an equipotential surface. The actual distribution of charge induced on the conducting boundary can then be determined from the normal derivative of the potential so obtained.

Example 3-14   A Point Charge above a Conducting Plane. As a second example, let us consider the case of a point charge q located a distance h above a conducting plane of infinite extent. Figure 3-21a shows the pertinent geometry. We wish to determine the potential distribution and the electric field intensity in the free space above the conducting plane.

Boundary conditions require that the conducting plane be an equipotential surface. In particular, the potential of the plane must be zero since, by assumption, the plane extends to infinity, where the potential must be zero.

The solution of this problem by the method of images proceeds as follows. First, a charge $-q$ is located at the geometric image of the real charge $+q$. Next, we conceptually remove the conducting plane, and for any point P above the plane we calculate the potential by superposition:

$$\phi = \frac{q}{4\pi\epsilon_0 r_+} + \frac{-q}{4\pi\epsilon_0 r_-} \tag{3-221}$$

This expression reduces to zero at $z = 0$ because, at every point on the conducting plane, $r_+ = r_- = r$. The potential and the electric field intensity below the plane are zero. At points above the plane, an expression for $\mathbf{E}$ may be obtained from $\mathbf{E} = -\nabla\phi$ by first establishing a suitable system of coordinates. Thus, in the notation of Fig. 3-21a, the z component of the electric field intensity is

$$E_z = \frac{-q}{4\pi\epsilon_0} \left\{ \frac{h-z}{[x^2 + y^2 + (z-h)^2]^{3/2}} + \frac{h+z}{[x^2 + y^2 + (z+h)^2]^{3/2}} \right\} \tag{3-222}$$

where x and y are measured in the plane of symmetry and z is normal to it. At the surface of the conducting plane, this expression reduces to

$$E_z = \frac{-qh}{2\pi\epsilon_0(x^2 + y^2 + h^2)^{3/2}} = \frac{-qh}{2\pi\epsilon_0 r^3} \tag{3-223}$$

The density of the surface charge

$$\rho_s = D_z = \epsilon_0 E_z = \frac{-qh}{2\pi r^3} \tag{3-224}$$

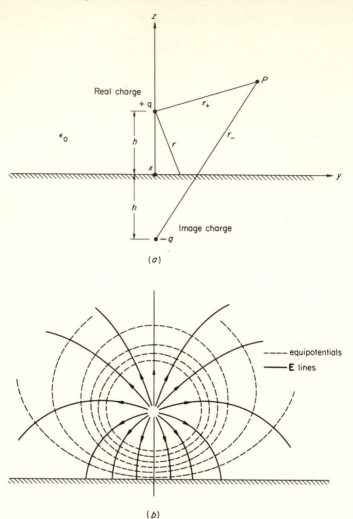

FIGURE 3-21. A point charge over a conducting plane. (*a*) Geometry; (*b*) field distribution.

induced on the conducting plane is seen to fall off inversely as the cube of the distance from the point charge and reaches a maximum immediately below the point charge. The total charge on the plane is

$$\oint_{\Sigma} \rho_s \, da = \int_{\Sigma} \frac{-qh}{2\pi r^3} \, da = -\frac{qh}{2\pi} \int_0^\infty \int_0^{2\pi} \frac{R \, d\varphi \, dR}{(R^2 + h^2)^{3/2}} = -q \qquad (3\text{-}225)$$

consistent with the basic requirements imposed by Gauss' law.

Summarizing, the method of images, though powerful, can be applied only to problems that involve simple geometrical shapes, such as planes, cylinders, and spheres.

3.18   The Method of Conformal Transformations.

This method transforms the geometry of one problem into the simpler geometry of a second problem for which the static field is known.   In the process, the shape of all infinitesimal areas is preserved, a property which accounts for the term *conformal transformation*.   Applications of this method to Laplace's equation are limited to two-dimensional fields (uniform geometry in one of three mutually orthogonal directions).

The subject matter of conformal transformations is an extensive one. However, because of space limitations, the following treatment will be confined to a discussion of a simple illustrative example which will merely serve as an introduction.   For more extensive treatments, the student is referred to the literature.†

Example 3-15   Conducting Wedge.   Let us consider the conducting wedge formed by two semi-infinite planes, as in Fig. 3-22a.   A solution of Laplace's equation is desired in the 45° sector when both planes are grounded ($\phi = 0$).

From Sec. 3.10 we recall that a possible solution of Laplace's equation in rectangular coordinates is Axy, where A is an arbitrary constant dependent on the relative level of excitation.   This function vanishes on the planes $x = 0$ and $y = 0$. Therefore the solution of Laplace's equation which vanishes on the walls of a 90° corner is

$$\phi = Axy \tag{3-226}$$

Figure 3-22b shows a plot of this field in a 90° corner with both walls extending to infinity and grounded.   The method of conformal transformations seeks to adapt this known field to the problem of the wedge by transforming coordinates in such a way as to retain the characteristic orthogonality between field and potential lines.

Let any point in the two-dimensional region of the corner be denoted by a complex‡ variable

$$Z = x + jy \tag{3-227}$$

where j, as usual, denotes the imaginary number $\sqrt{-1}$.   Let any point in the wedge be denoted by another complex variable.

$$W = u + jv \tag{3-228}$$

Then consider the relation

$$W = Z^{1/2} \tag{3-229}$$

---

† K. J. Binns and P. J. Lawrenson, "Analysis and Computation of Electric and Magnetic Fields," The Macmillan Company, New York, 1963.
‡ The student who is unfamiliar with complex numbers and functions of a complex variable may find it helpful to associate the notion of a complex variable with that of a *phasor*, a term sometimes used in circuit theory.   Impedance is a complex variable.

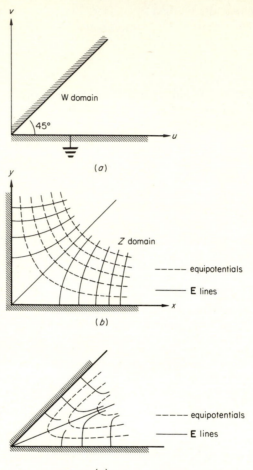

FIGURE 3-22.   Conformal   transformation.   (a)
Geometry of the wedge; (b) geometry and field of
a 90° corner; (c) field distribution.

This formula *maps* points in one domain to points in the other.   To determine the
specific relations between coordinates, let us introduce in Eq. (3-229) the defining
equations (3-227) and (3-228).   We obtain

$$u + jv = (x + jy)^{1/2} \tag{3-230}$$

Squaring both sides of this equation and separating real and imaginary parts, we find

$$u^2 - v^2 = x \tag{3-231}$$

$$2uv = y \tag{3-232}$$

from which we deduce

$$u = \sqrt{\frac{x \pm \sqrt{x^2 + y^2}}{2}} \qquad (3\text{-}233)$$

$$v = \frac{y}{2\sqrt{(x \pm \sqrt{x^2 + y^2})/2}} \qquad (3\text{-}234)$$

Now, substituting x and y from Eqs. (3-231) and (3-232) into Eq. (3-226), we obtain

$$\phi' = 2Auv(u^2 - v^2) \qquad (3\text{-}235)$$

Although the rigor of this substitution is unquestionable, the usefulness of ϕ' as a representation of a physically admissible field remains to be shown. To this end, we note first that

$$\nabla^2 \phi' = 2A \frac{\partial^2}{\partial u^2} (u^3v - uv^3) + 2A \frac{\partial^2}{\partial v^2} (u^3v - uv^3)$$

$$= 12Auv - 12Auv = 0$$

Thus ϕ' is a solution of Laplace's equation. This means that the field lines are normal to surfaces of constant ϕ'. Under the transformation (3-229), the form of all infinitesimal areas is seen to have been preserved, and as a consequence the transformation itself can rightfully be classified as conformal.

Now when $x = 0$, $u = v$ (45° plane) and $\phi' = 0$, as required. By uniqueness, Eq. (3-235) expresses (to within a multiplicative constant) the only solution to Laplace's equation which vanishes at every boundary on the wedge. A plot of this field is shown in Fig. 3-22c.

In the preceding example, the transformation (3-229) is seen to compress, literally, the region of the 90° corner into the region of the wedge, without, however, in any way affecting the orthogonality between field and potential lines. This is a characteristic property of all conformal transformations. Typical among these are:

1. $W = Z^{1/4}$ compresses by a factor of 4.
2. $W = Z^{3/2}$ expands by a factor of $3/2$.
3. $W = e^Z$ transforms, for example, a rectangle in the Z domain to one-fourth of an annular region (Fig. 3-23a).
4. $W = \tan Z$ transforms, for example, a coaxial line into a line consisting of two flat parallel sheets of infinite extent, which together serve as the outer conductor, and an inner conductor of elliptical cross section (Fig. 3-23b).

5. $\displaystyle W = A \int [(Z - x_1)^{k_1}(Z - x_2)^{k_2} \cdots (Z - x_n)^{k_n}]\, dZ \qquad (3\text{-}236)$

is the so-called *Schwarz-Christoffel transformation* and maps the real axis in the Z plane into a polygon in the W plane. The entire upper half of the Z plane is mapped into the region interior to the polygon, $x_1, x_2, \ldots,$

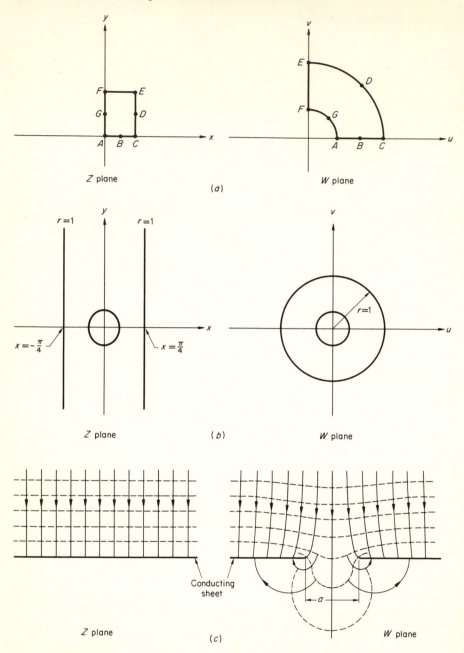

FIGURE 3-23. Various conformal transformations.   (a) $W = e^Z$; (b) $W = \tan Z$; (c) $W = Z + (a/4)^2/Z$ [obtained from Eq. (3-236) by setting $\alpha_1 = \alpha_3 = 2\pi$, $\alpha_2 = -\pi$, $x_1 = -x_3$, $x_2 = 0$].

x_n being the breakpoints, or corners, of the polygon (Fig. 3-23c).   The angles α_i of the polygon are related to the exponents k_i by $k_i = \alpha_i/\pi - 1$.

In summary, from the brief presentation just concluded, it is clear that the method of conformal transformations is limited in many ways.   For one thing, it requires a two-dimensional geometry.   Second, it requires intuition and a priori knowledge of solutions to other problems.   Then, too, some exposure to the theory of functions of a complex variable is a virtual "must" for a complete mastery of the subject.   Yet this technique is often found very valuable.

3.19   The Graphical Method.   This method, known by the alternative name of *flux plotting*, requires a little drafting experience, much patience, and practically no knowledge of formal mathematics.   In the past, many highly skilled field plotters have applied their talents in the design of all types of devices.   Today, the digital computer offers a much more versatile and faster means of solving similar potential problems.   For this reason, we do not attempt to give even a short presentation; instead, we refer the interested reader to the literature.†

3.20   The Natural Distribution of Charge.   The formula

$$\phi = \frac{1}{4\pi\epsilon} \int_V \frac{\rho}{r} \, dv \tag{3-237}$$

relates charge distribution directly to the potential field.   As shown in Sec. 3.6, the limits of integration in this formula are such as to include all the region containing charge.

In this section we consider the natural distribution of charge on the surface of a conductor.   It is assumed that the potential of the surface is known and that the charge distribution is to be found.   Equation (3-237) then becomes a complicated *integral* equation, where the unknown function (the charge distribution function in this case) appears under the integral sign; it may be found in closed form only in a few very simple cases.   However, most field problems are not simple, and an approximate method of solution must be chosen.   An approximation useful for field problems is the *method of subsections*.‡   An elementary presentation of this method is best given for

† W. B. Boast, "Vector Fields," Harper & Row, Publishers, Incorporated, New York, 1964.
‡ R. F. Harrington, Matrix Methods for Field Problems, *Proc. IEEE*, vol. 55, pp. 136–149, February, 1967.

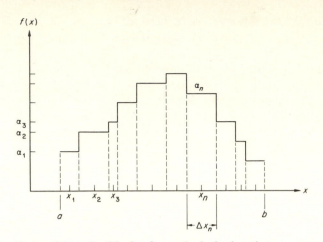

FIGURE 3-24.   Partitioning for method of subsections.

integral equations of a single independent variable of the form

$$f(x) = g(x) + \lambda \int_a^b K(x,y)f(y)\,dy \tag{3-238}$$

where λ is a constant, g is a known function, and K is the *kernel* of the integral equation and is always known [the kernel is the ratio $1/4\pi\epsilon r$ in Eq. (3-237), and the potential ϕ is represented by $-g$].

Consider the interval $[a,b]$ divided into N (not necessarily equal) subintervals Δx, as in Fig. 3-24.   Define functions

$$f_n = \begin{cases} 1 & \text{on } n\text{th subinterval} \\ 0 & \text{on all other subintervals} \end{cases}$$

and let f be represented by

$$f(x) = \sum_{n=1}^{N} \alpha_n f_n \tag{3-239}$$

where $\alpha_1, \alpha_2, \alpha_3, \ldots, \alpha_N$ need to be determined.   The integral equation (3-238) becomes

$$f(x) = g(x) + \sum_{n=1}^{N} \alpha_n \left[\lambda \int_{x_n - \Delta x_n/2}^{x_n + \Delta x_n/2} K(x,y)\,dy \right]$$

where x_n is chosen as the midpoint of the nth subinterval, and Δx_n is the width of the subinterval.   Let x take on the values $x_1, x_2, x_3, \ldots, x_N$, sequentially. The last equation then transforms to

$$\sum_{n=1}^{N} \alpha_n A_{in} = -g(x_i) \qquad i = 1, 2, 3, \ldots, N$$

$$A_{in} = \begin{cases} \lambda \int_{x_n-\Delta x_n/2}^{x_n+\Delta x_n/2} K(x_i, y) \, dy & i \neq n \\[2em] -1 + \lambda \int_{x_n-\Delta x_n/2}^{x_n+\Delta x_n/2} K(x_i, y) \, dy & i = n \end{cases}$$

or in matrix form,

$$\begin{bmatrix} A_{11} & A_{12} & A_{13} & \cdots & A_{1N} \\ A_{21} & A_{22} & A_{23} & \cdots & A_{2N} \\ \multicolumn{5}{c}{\cdots\cdots\cdots\cdots\cdots\cdots} \\ A_{N1} & A_{N2} & A_{N3} & \cdots & A_{N} \end{bmatrix} \begin{bmatrix} \alpha_1 \\ \alpha_2 \\ \cdot \\ \cdot \\ \cdot \\ \alpha_N \end{bmatrix} = \begin{bmatrix} -g(x_1) \\ -g(x_2) \\ \cdot \\ \cdot \\ \cdot \\ -g(x_N) \end{bmatrix} \qquad (3\text{-}240)$$

The linear problem (3-240) can now be solved for the α's. An approximate solution to the integral equation (3-238) is then given by Eq. (3-239).

Example 3-16 Flat Conducting Disk. Consider a charged conducting disk in free space, of radius a, coincident with the xy plane, as in Fig. 3-25. The surface of the disk is at a fixed potential ϕ_0. It is desired to find the surface charge density (which, by symmetry, is only a function of radius).

Since charge on the conducting disk is distributed over the surface, the amount of charge in a differential element of area da will be $\rho_s r \, d\varphi \, dr$, where ρ_s is the unknown surface charge distribution. The distance in Eq. (3-237) from the source point $(r, \varphi, 0)$

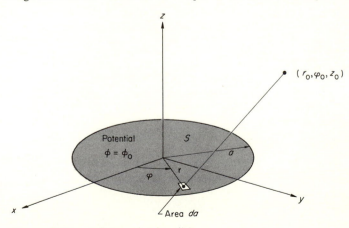

FIGURE 3-25. A flat conducting disk.

to the point of observation (r_0, φ_0, z_0) is simply $[z_0{}^2 + r^2 + r_0{}^2 - 2rr_0 \cos(\varphi - \varphi_0)]^{1/2}$. Therefore the potential $\phi(r_0, \varphi_0, z_0)$ can be expressed as

$$\phi(r_0, \varphi_0, z_0) = \frac{1}{4\pi\epsilon_0} \int_S \frac{\rho_s r \, d\varphi \, dr}{[z_0{}^2 + r^2 + r_0{}^2 - 2rr_0 \cos(\varphi - \varphi_0)]^{1/2}}$$

where S represents the surface of the disk. If the coordinates of the point at which ϕ is being determined are $(r_0, 0, 0)$, the potential will be ϕ_0, and the last expression will reduce to

$$\phi_0 = \frac{1}{2\pi\epsilon_0} \int_0^a \int_0^\pi \frac{\rho_s r \, d\varphi \, dr}{(r^2 + r_0{}^2 - 2rr_0 \cos\varphi)^{1/2}}$$

for all r_0 in the interval from zero to a. The last equation is the desired integral equation by which $\rho_s(r)$ is to be determined.

To obtain the corresponding system of linear algebraic equations, let us subdivide the interval $[0, a]$ into N (not necessarily equal) subintervals, and let r_n fall in the middle of the nth subinterval. According to the method of subsections, the charge density is a constant, ρ_{s_n}, in the nth interval, so that

$$\phi_0 = \sum_{n=1}^N \rho_{s_n} \left[\frac{1}{2\pi\epsilon_0} \int_{r_n - \Delta r_n/2}^{r_n + \Delta r_n/2} \int_0^\pi \frac{r \, d\varphi \, dr}{(r^2 + r_0{}^2 - 2rr_0 \cos\varphi)^{1/2}} \right]$$

This equation is valid at any point $(r_0, 0, 0)$, and specifically at all N points $r_1, r_2, r_3, \ldots, r_N$ of the partition. It can thus be written N times, giving

$$\phi_0 = \sum_{n=1}^N \rho_{s_n} \left[\frac{1}{2\pi\epsilon_0} \int_{r_n - \Delta r_n/2}^{r_n + \Delta r_n/2} \int_0^\pi \frac{r \, d\varphi \, dr}{(r^2 + r_i{}^2 - 2rr_i \cos\varphi)^{1/2}} \right]$$

where $i = 1, 2, 3, \ldots, N$. Denoting the bracketed expression in this equation by A_{in} gives the matrix equation

$$[A_{in}][\rho_{s_n}] = [\phi_0]$$

which is a required intermediate result.

The numerical evaluation of the matrix elements A_{in} is moderately involved, though for $i \neq n$ it can be readily accomplished on the digital computer by any standard integration technique (trapezoidal, Simpson's rule, etc.). When $i = n$, the integrand becomes undefined at $r = r_i$ and $\varphi = 0$. However, even this singularity can be handled successfully, for example, by the method of gaussian quadratures.

Figure 3-26 shows a plot[†] of the approximate charge density for a disk of 1-m radius and 1-V potential. Two cases are displayed: $N = 5$ and $N = 15$. It will be seen that even in the case of five subintervals, agreement with the exact answer, which is known to be $(1.1 \times 10^{-11})/(1 - r^2)^{1/2}$, is fairly good, especially at points near the center of the disk. *The large concentration near the edge is typical. Charge tends to accumulate on sharp corners and conducting-body extremities.*

3.21 Solution of Poisson's Equation. Laplace's equation applies only when the region of interest is free of charge. For a region containing charge, Poisson's equation applies:

$$\nabla^2 \phi = -\frac{\rho}{\epsilon} \tag{3-241}$$

† Numerical data courtesy of T. D. Slagh.

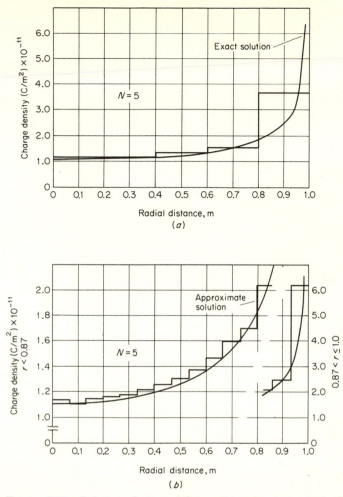

FIGURE 3-26. Solution of charged-disk problem by the method of subsections.

A solution of Poisson's equation which satisfies known boundary conditions is unique (Sec. 3.11). Hence, once a solution of Eq. (3-241) is obtained, expressions for electric field intensity and other field parameters may be found as in preceding sections.

It should be noted that certain special techniques, such as conformal transformations, do not apply to Poisson's equation.

Example 3-17 Charge in Region between Two Parallel Plates. As an example, suppose that the region between plates of the parallel-plate capacitor shown in

Fig. 3-1 is filled with charge continuously distributed with a density $\rho(x) = \rho_0 x$, where ρ_0 is a constant. We wish to find the resulting electric field, neglecting fringing effects.

Since the problem is one-dimensional, Poisson's equation reduces to

$$\frac{d^2\phi}{dx^2} = -\frac{\rho_0}{\epsilon_0} x \qquad (3\text{-}242)$$

The general solution is of the form

$$\phi = -\frac{\rho_0}{6\epsilon_0} x^3 + Bx + C$$

Applying the boundary conditions

$$\phi = \begin{cases} 0 & \text{at} \quad x = 0 \\ V & \text{at} \quad x = d \end{cases}$$

gives

$$\phi = \frac{V}{d} x + \frac{\rho_0}{6\epsilon_0} (xd^2 - x^3) \qquad (3\text{-}243)$$

The electric field intensity calculated from $\mathbf{E} = -\boldsymbol{\nabla}\phi$ is

$$\mathbf{E} = \left[-\frac{V}{d} + \frac{\rho_0}{6\epsilon_0} (3x^2 - d^2) \right] \mathbf{a}_x \qquad (3\text{-}244)$$

The distribution of charge on the bottom plate $(x = 0)$

$$\rho_s = -\epsilon_0 \left(\frac{V}{d} + \frac{\rho_0}{6\epsilon_0} d^2 \right) \qquad (3\text{-}245)$$

is seen to differ from the distribution on the top plate $(x = d)$:

$$\rho_s = \epsilon_0 \left(\frac{V}{d} - \frac{\rho_0}{3\epsilon_0} d^2 \right) \qquad (3\text{-}246)$$

Therefore the total charges on the top and bottom plates are obviously different from each other. This suggests that the term *capacitance* is meaningless here because it can no longer be defined uniquely.

3.22 Synthesis of Electrostatic Fields. So far our treatment of electrostatics has been concerned with the general problem of analysis. That is, for a given distribution of charge, we have determined the resulting coulomb field. Or for a given geometry of dielectrics and conductors and a known deployment of driving sources, we have obtained solutions by several techniques for what is, fundamentally, a boundary-value problem. Now we are ready to attempt a solution of the inverse problem of synthesis; namely, we are to find *that* system of sources which will generate a given field.

Suppose that a mathematical expression is given for the potential in a region.   A systematic procedure for finding its sources is as follows:

1. The volume density of the required continuous charge is first determined by substituting the given expression in Poisson's equation:

$$\rho = -\epsilon \, \nabla^2 \phi \qquad (3\text{-}247)$$

2. The contribution from mathematical point charges, line charges, and other types of charge singularities is next established by locating all points in space where the potential becomes undefined.   For example, a $1/r$ variation, according to Eq. (3-34), will signify the presence of a mathematical point charge; a $1/r^2$ variation, according to Eq. (3-39), will signify the presence of a dipole; and so forth.

3. Finally, the contribution from a surface layer of dipoles is determined by examining the regions of space where the given potential function is discontinuous.   It can be shown rigorously† that a surface distribution of dipoles marks a discontinuity in the potential function, and that the difference in potential on opposite sides of the surface is

$$\Delta\phi = \frac{1}{\epsilon} \, p_s \qquad (3\text{-}248)$$

where p_s denotes the dipole moment per unit area.

Example 3-18   Electrolytic Battery.   Let us synthesize the potential function

$$\phi = \begin{cases} 0 & 0 \le r < a \\ \dfrac{1}{\ln (b/a)} \left(\phi_1 \ln \dfrac{r}{a} + \phi_2 \ln \dfrac{r}{b} \right) & a < r < b \\ 0 & b \le r \end{cases} \qquad (3\text{-}249)$$

in the region of a coaxial line having an inner radius a and an outer radius b.   In Eq. (3-249), r denotes radial distance measured from the axis of the line, and ϕ_1 and ϕ_2 are constants (Fig. 3-14).

It can be shown readily that $\nabla^2 \phi = 0$, and therefore that $\rho = 0$ everywhere.   Moreover, ϕ is nowhere singular.   Therefore, no point sources are present.   However, ϕ is discontinuous both at $r = a$ and at $r = b$.   Hence the surfaces of both conductors are covered by dipole layers.   Specifically, at the surface of the inner conductor, the dipole moment is, from Eq. (3-248),

$$p_{s_i} = \epsilon \, \Delta\phi = \epsilon\phi_2 \qquad (3\text{-}250)$$

the positive layer being the nearest one to the axis.   At the surface of the outer conductor

$$p_{s_0} = \epsilon \, \Delta\phi = \epsilon\phi_1 \qquad (3\text{-}251)$$

the positive layer being also the nearest one to the axis.

The potential, Eq. (3-249), therefore arises from a pair of surface distributions of dipoles—one on each cylindrical surface.   This situation exists, for instance, in short-circuited electrolytic batteries.

† J. A. Stratton, "Electromagnetic Theory," pp. 188–192, McGraw-Hill Book Company, New York, 1941.

3.23 Summary. This chapter has dealt with electrostatic fields. Every electrostatic field is a conservative field.

We showed that a scalar potential can be associated with every electrostatic field and that in linear, homogeneous, and isotropic media this potential satisfies Poisson's equation or Laplace's equation. Both equations follow directly from Maxwell's equations. Excepting surface distributions of dipoles, the value of the electrostatic potential at a point is always unique. This property allows us to define the difference in potential between any two points uniquely.

We indicated that Coulomb's law, as deduced from experimental facts or from the general theory of Maxwell's equation, is a law which applies to complexes of charged bodies whose physical dimensions are small compared with other macroscopic dimensions of the configuration.

We next derived a general solution to Poisson's equation and proved that the potential in a closed region can always be represented by the superposition of a potential arising from interior sources and a second potential arising from exterior sources. If the region is source-free, the scalar potential is a solution of Laplace's equation. In linear, homogeneous, and isotropic media, the relation between potential (effect) and charge (cause) is always a linear one.

Then we defined capacitance as charge divided by voltage. Capacitance provides a measure of electrostatic coupling between conductors.

After formulating the boundary-value problem in electrostatics, we showed that a potential which satisfies either Poisson's equation or Laplace's equation in a region V and which on its boundary meets the prescribed conditions is the only possible solution. Thus the solution of any boundary-value problem, obtained either by intuition or by formal methods, is the correct potential.

The solution of the electrostatic boundary-value problem can be achieved by one of several formal methods. Many of these were elaborated upon to some degree in this chapter. The experimental method is discussed in Chap. 5.

The method of separation of variables provides a tool for obtaining closed-form solutions of Laplace's equation. Only 11 coordinate systems are known for which the separation of variables can be achieved. Even in these it is impossible to apply boundary conditions unless the physical boundaries coincide with coordinate surfaces. Furthermore, the solution often includes an infinite number of terms, and as such it is unwieldly, except when a good approximation is given by the first few terms in the series.

The analytical difficulties encountered when boundaries do not coincide with coordinate surfaces can be overcome by numerical methods.

Following a brief exposition of the solution of Poisson's equation, a short discussion of the problem of synthesis was presented.

Problems

3-1 Coulomb Field. Three point charges, arranged as in the figure, constitute an *axial quadrupole.* Show that, if $d \ll r$, the potential at P is given approximately by

$$\phi \approx \frac{qd^2}{4\pi\epsilon_0 r^3}(3\cos^2\theta - 1)$$

(*Hint:* Use the binomial theorem and retain the first three terms.)

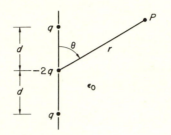

PROBLEM 3-1

3-2 Two Simple Problems.

(*a*) In the accompanying diagram, $q_1 = 5 \times 10^{-6}$ C and $q_2 = -10 \times 10^{-6}$ C. Find the total flux of the vector **D** through the square $ABCD$.

(*b*) Capacitor C_1 is initially charged to 10 V, with the polarity as shown. Find the voltage across C_2 when the switch is closed.

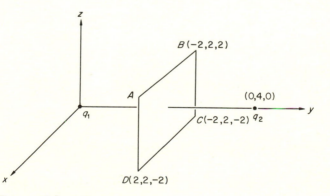

PROBLEM 3-2*a*

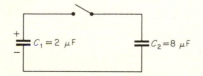

PROBLEM 3-2*b*

3-3 Capacitance. Two parallel metal plates, 1 mm apart, are connected to a battery of 10 V. The battery is then removed, and the separation is increased to 1 cm. Neglecting fringing, find the new potential difference between plates.

3-4 Capacitance. A 9-V battery is connected across a pair of identical parallel-plate air capacitors. The battery is next removed, and a block of solid dielectric ($\epsilon_r = 2$) is inserted between the plates of one of the capacitors. Neglecting fringing, what is the final voltage across the pair?

3-5 Capacitance—Dielectric Strength.
  (*a*) Find the capacitance of a metal sphere in air ($\epsilon \approx \epsilon_0$). The radius of the sphere is 0.1 m.
  (*b*) Find the greatest charge which can be carried by the sphere without breaking down the surrounding air (the dielectric strength of air is 3000 kV/m).

3-6 Capacitance—Formal Solution. A hemispherical indentation is made accidentally in one of two large circular plates forming a parallel plate capacitor. If the indentation is inward at a point remote from the edges, and the radius of the boss resulting from the accident is very small compared with the dimensions and separation of the plates, what is the change in the capacitance of the condenser?

3-7 Solution of Laplace's Equation in Rectangular Coordinates. Refer to Fig. 3-12*a* and suppose that the potential along the conducting wall at $x = a$ is $V \sin(\pi y/b)$, instead of zero. Find the new potential distribution within the rectangular cavity.

3-8 Solution of Laplace's Equation in Rectangular Coordinates. A dielectric slab of permittivity ϵ and thickness d partially fills the interelectrode space between two parallel conducting plates. The plates are separated by a distance l, and they are held at a potential difference V. By matching proper solutions of Laplace's equation in the two regions, find expressions for the potential in the dielectric and in the air space between the plates.

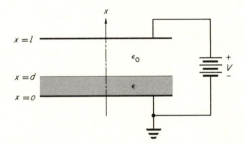

PROBLEM 3-8

3-9 Coulomb Field. Derive Eqs. (3-70) and (3-71) by integrating Eq. (3-74) over the interval [*a*,*b*].

3-10 Dielectric Strength. Consider an infinitely long coaxial cable. The radius of the center conductor is a m, and the inner radius of the outer conductor is b m. If the insulation between conductors has a breakdown strength of K V/m, find the minimum potential difference between the two conductors which causes breakdown. Your answer should be expressed in terms of a, b, and K.

3-11 Solution of Laplace's Equation in Cylindrical Coordinates. A long conducting cylinder of radius a is placed in a field which, far from the cylinder, is given by

$\phi = -E_0 r \cos \varphi$, where r and φ are the usual cylindrical coordinates and E_0 is a constant. The z axis is so oriented as to coincide with the axis of the cylinder.

   (a) Determine the field distribution in the region about the cylinder, assuming the potential of the cylinder is zero.
   (b) Determine the magnitude and direction of the electrostatic field intensity at distant points from the cylinder ($r \gg a$).

3-12 Solution of Laplace's Equation in Cylindrical Coordinates. A potential distribution $\phi = E_0 r \sin \varphi$ exists in an extended dielectric region of permittivity ϵ. A z-directed, infinitely long cylindrical cavity is cut out of the substance. Determine the resulting potential in the dielectric and in the cavity.

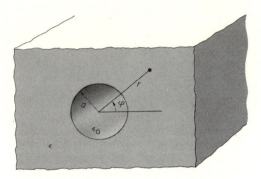

PROBLEM 3-12

3-13 Coulomb Field. A charge Q_1 is distributed uniformly on the surface of a hemisphere. An infinite number of point charges (two of which are shown in the figure) are placed along an axis passing through the center of the hemisphere, perpendicular to the plane of its base. The distance to the kth point charge from the center is $2^{k-1}R$, where $k = 1, 2, 3, \ldots$. Find Q_2 in terms of Q_1 if the potential at the center of the hemisphere is zero and the surrounding medium is free space.

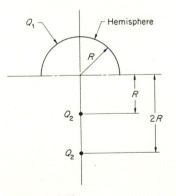

PROBLEM 3-13

3-14 Solution of Laplace's Equation in Spherical Coordinates. Let a sphere of permittivity ϵ_2 be immersed in a medium of permittivity ϵ_1, as in Fig. 3-15. Show that in this case the potential outside and inside the sphere is given, respectively, by

$$\phi_1 = E_0 \left[1 + \left(\frac{a}{r} \right)^3 \frac{\epsilon_1 - \epsilon_2}{2\epsilon_1 + \epsilon_2} \right] r \cos \theta$$

and

$$\phi_2 = E_0 \frac{3\epsilon_1}{2\epsilon_1 + \epsilon_2} r \cos \theta$$

3-15 Solution of Laplace's Equation in Spherical Coordinates. A spherical capacitor is formed by two concentric conducting thin spherical shells of radii a and b ($b > a$), separated by a dielectric of permittivity ϵ. The outer sphere is grounded, and the potential of the inner sphere is held fixed at V volts.

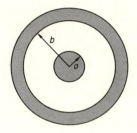

PROBLEM 3-15

    (a) Derive expressions for potential and electric field intensity in the three regions defined by (i) $0 \leq r < a$, (ii) $a < r < b$, and (iii) $b \leq r$.

    (b) Determine the surface charge distributions on both spheres.

    (c) Show that the capacitance of the spherical capacitor is

$$C = 4\pi\epsilon \frac{ab}{b - a}$$

3-16 Capacitance. Find the capacitance of the spherical capacitor of Prob. 3-15 if the space surrounding the inner sphere is filled with a dielectric of permittivity ϵ up to a radius equal to $\frac{1}{2}(a + b)$. The rest of the space has permittivity ϵ_0.

3-17 Adapting a Known Solution to a New Problem

    (a) Two opposite charges, q_1 and q_2, such that $|q_1| > |q_2|$, are separated by a distance d. Show that the equipotential surface $\phi = 0$ is a sphere of radius $q_1 q_2 d/(q_1^2 - q_2^2)$ centered along the line joining the two charges at a point distant $d(q_1^2 + q_2^2)/2(q_1^2 - q_2^2)$ from the midpoint of d measured toward q_2.

    (b) The point charge q_2 is next removed, and an equal amount of charge is placed upon the surface of a conducting sphere of radius a, separated from q_1 by a distance b ($b > a$). Determine the potential at every point within and outside the sphere. (*Hint:* Place two point charges, one at the center and another along the line joining the charge q_1 to the center of the sphere, at a distance R from the center so that $a^2 = Rb$ and so as to match the boundary conditions while maintaining the correct total charge in the sphere.)

3-18 Image. A long thin wire carrying a charge of ρ_l C per unit length stretches parallel to a very large flat conducting plane.   The distance of the wire from the ground plane is h. Obtain an expression for the potential at any point P.

3-19 Coulomb Field. Given the three charges in a line as shown.   The charges are in microcoulombs.   Find all points on the x axis where the potential is zero.

PROBLEM 3-19

3-20 Conformal Mapping. Discuss the way in which the Z plane is mapped on the W plane by the function

$$W = \cosh^{-1} Z$$

3-21 Conformal Transformations. Confirm the transformation shown in Fig. 3-23a.

3-22 Poisson's Equation. The interelectrode space of the coaxial line shown in Fig. 3-14 is filled with a charge which is distributed with a density $\rho = \rho_0/r$, where ρ_0 is a constant.   Determine the potential distribution in the dielectric between cylindrical conductors.

3-23 Synthesis. Determine the sources of the potential $\phi = \cos r/r$, where r is the radial distance from the origin to any point within the region bounded by a metal sphere of radius R.

3-24 Synthesis. Consider the following two-dimensional solution to Laplace's equation in cylindrical coordinates:

$$\phi = r^{-n} \sin n\varphi$$

(a) Show analytically that $\partial\phi/\partial x$ is also a solution to Laplace's equation.

(b) For $n = 1$, synthesize a distribution of sources which will give rise to the potential $\partial\phi/\partial x$ in terms of the sources of ϕ.

3-25 General. The algebraic sum of the charges on a system of conductors is known to be positive.   Show that on the surface of at least one conductor the charge density is everywhere positive.

3-26 Capacitance. Figure 5-5 shows several groups of measured points on a sequence of equipotential surfaces established in the region between conductors of a rectangular coaxial cable.   Assuming that the figure is drawn to scale, determine approximately the capacitance per meter of the air-filled cable.   (*Hint:* Draw the equipotential lines and, normal to them, several streamlines.   Then determine the field intensity near the inner conductor.)

3-27 Experimental Proof of the Law of the Inverse Square. Historically, the following experiment has been regarded as proof that the law of electrostatic force must be that of the inverse square [Eq. (3-35)].

Two spherical shells are concentric and insulated from one another; they can be connected electrically by lowering a piece of wire connected to the outer shell through a small hole in the outer shell.   The experiment is conducted by charging the outer shell and then removing the wire, making electric contact between shells.   If the outer shell is discharged, no charge is found on the inner sphere.

Prove the validity of the aforementioned deduction.

PROBLEM 3-27

3-28 Convergence. Prove that, if the charge density ρ is finite at every point in the volume V, the integral

$$\phi = \frac{1}{4\pi\epsilon} \int_V \frac{\rho}{r}\, dv$$

converges for all points in the region.

3-29 Two-dimensional Dipole. This problem deals with the asymptotic behavior of the configuration considered in Example 3-12. As $d \to 0$ while $\rho_l \to \infty$ in such a way that the product $\rho_l d$ remains constant, the two-wire line approaches a *two-dimensional dipole*. Show that, in cylindrical coordinates, the electric field associated with such a dipole is given by

$$\mathbf{E} = \frac{-p_l}{2\pi\epsilon r^2}\,(\sin\varphi\,\mathbf{a}_r - \cos\varphi\,\mathbf{a}_\varphi)$$

where $p_l = \rho_l d$ is the magnitude of the dipole moment.

3-30 Synthesis. Find the distribution of sources which gives rise to a static electric field defined by $\phi = e^{-ar}/r$ throughout space. In this expression, a is a constant and r is the radial distance from the origin of a spherical system of coordinates.

3-31 Formal Solution. Given a hollow dielectric sphere of dielectric constant ϵ_r and internal and external radii a and b, in a uniform external field $\mathbf{E}_0$. Show that the electric field intensity in the cavity is given by

$$\mathbf{E} = 9\mathbf{E}_0\,\frac{\epsilon_r}{(2\epsilon_r + 1)(\epsilon_r + 2) - 2(\epsilon_r - 1)^2(a/b)^3}$$

3-32 Solution of Laplace's Equation in Cylindrical Coordinates. The solid conducting cylinder of Prob. 3-11 is replaced by an infinitely long hollow conducting cylinder of inner and outer radii a and b $(a < b)$. The externally applied field remains the same. Find the electric field everywhere.

3-33 Solution of Laplace's Equation in Spherical Coordinates. Three concentric thin metal spheres form an electrostatic system. The radius of the inner sphere is R_1 m, the radius of the intermediate sphere is R_2, and the radius of the outer sphere is R_3. The medium between spheres is free space. The inner and outer spheres are grounded. A charge of Q_2 C is placed on the intermediate sphere.

(a) Determine the value of induced charge Q_1 on the inner sphere.

(b) Determine the value of induced charge Q_3 on the outer sphere.

(c) Determine the value of potential ϕ_2 of the intermediate sphere.

3-34 Reciprocity Theorem. Charges $Q_1, Q_2, Q_3, \ldots, Q_n$ are placed on a system of N conductors, all embedded in a linear and isotropic dielectric. Let $\phi_1, \phi_2, \phi_3, \ldots, \phi_n$ be the corresponding potentials on the conductors. A second set of charges, $Q'_1, Q'_2, Q'_3, \ldots, Q'_n$, gives rise to a second set of potentials, $\phi'_1, \phi'_2, \phi'_3, \ldots, \phi'_n$, on the same system of conductors. Prove that

$$\sum_{k=1}^{N} Q_k \phi'_k = \sum_{k=1}^{N} Q'_k \phi_k$$

3-35 Reciprocity. Prove that the charge Q induced on a grounded conductor by a point charge q placed at some external point P is equal to $-q\phi_1/\phi_2$, where ϕ_1 is the potential at the point P when q is absent, and the conductor is not grounded but charged in such a way as to assume a potential ϕ_2.

3-36 Uniqueness. Prove that a charge placed upon an isolated metal sphere distributes itself uniformly.

3-37 Proximity Effect. Determine the distribution of charge on the surfaces of the two conductors forming the two-wire line of Example 3-12.

3-38 Method of Subsections. A conducting wire of radius a and length L stretches along the x axis from $x = 0$ to $x = L$; the axis of the wire and the x axis coincide along the entire length of wire. It is desired to compute the distribution of a quantity of charge, Q, placed upon the wire by the method of subsections, neglecting the effects of end surfaces.

Assume that the charge density can be approximated by a linear charge density ρ_s on the wire axis, and consider the case $L = 100a$. Use a total of N equal subintervals, and do a parametric study versus N.

3-39 Image. A point charge q is located within the right angle formed by a pair of conducting planes intersecting normally. Show by the method of images that the potential

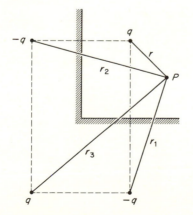

PROBLEM 3-39

at any point P within the right angle is

$$\phi = \frac{q}{4\pi\epsilon_0}\left(\frac{1}{r} - \frac{1}{r_1} - \frac{1}{r_2} + \frac{1}{r_3}\right)$$

3-40 Image. A point charge q is located within the 60° angle formed by a pair of conducting planes. Confirm that image charges are formed as indicated in the accompanying figure.

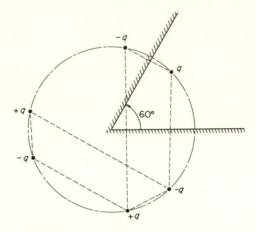

PROBLEM 3-40

3-41 Uniqueness. A point charge q is completely enclosed by a metal sphere, as shown in the figure. Both inside and outside the sphere the medium is free space. Show that outside the spherical shell the potential is

$$\phi = \frac{q}{4\pi\epsilon_0 r}$$

independent of the position of q within the sphere.

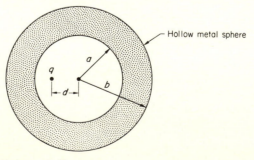

PROBLEM 3-41

3-42 Adapting a Known Solution to a New Problem. Two parallel infinite conducting cylinders of radii a and b are separated by a distance d (where $d > a + b$).

(a) Prove that the capacitance per unit length is given by

$$C = \frac{2\pi\epsilon}{\cosh^{-1}\left[(d^2 - a^2 - b^2)/2ab\right]}$$

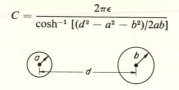

PROBLEM 3-42*a*

(b) If one cylinder is placed completely inside the other and their axes are at a distance d (where $d < b - a$), show that

$$C = \frac{2\pi\epsilon}{\cosh^{-1}\left[(a^2 + b^2 - d^2)/2ab\right]}$$

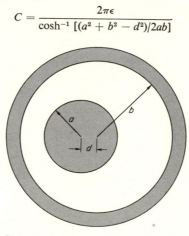

PROBLEM 3-42*b*

CHAPTER 4

THE MAGNETOSTATIC FIELD

4.1 Introduction. We shall begin our study of the phenomenon of the magnetic field assuming that we have a general idea of some of the manifestations of such fields from our experience and from our physics courses. Our objective is to list the properties of magnetic fields and to formulate these properties into precise mathematical expressions, sufficiently general to permit their use in all situations likely to be encountered. These properties are discussed in detail from the viewpoint of Maxwell's equations.

A magnetic field (whether time-varying or not) has a unique magnitude and a unique direction, both of which may vary from point to point in space. (Time-varying magnetic fields also vary both in direction and magnitude at one point; non-time-varying fields are constant with time at one point.) The study of magnetic fields is principally a study of the evaluation of the variations in magnitude and direction and of the effects which they produce.

The two most common sources of macroscopic magnetic fields are permanent magnets and current-carrying conductors. Present-day experimental and theoretical evidence indicates that there is no fundamental difference between the fields produced by these sources, and that, in fact, the ultimate mechanism for the production of magnetic fields is the same for the two types of sources. It should be emphasized in this connection that unless the region in which current flows is electrically neutral at the macroscopic level, an electrostatic field will exist in the region, in addition to the magnetostatic field.

The analytical techniques developed in the preceding chapter apply largely in this chapter as well. However, aside from possessing a purely formal equivalence, a magnetostatic field is fundamentally different from an electrostatic field because, as experimental evidence indicates, there is no *free* magnetic charge, only magnetic dipoles and higher-order singularities.

4.2 Magnetostatic Field Equations. The relations which describe the behavior of stationary magnetic fields are obtained by setting the time

198

derivatives equal to zero in the field equations. We obtain the following relations.

Integral form		Differential form	
$\oint_C \mathbf{H} \cdot d\mathbf{l} = \int_S \mathbf{J} \cdot \mathbf{n}\, da = I$	(4-1)	$\nabla \times \mathbf{H} = \mathbf{J}$	(4-2)
$\oint_\Sigma \mathbf{B} \cdot \mathbf{n}\, da = 0$	(3)	$\nabla \cdot \mathbf{B} = 0$	(III)
$\mathbf{F} = \int_V \rho(\mathbf{v} \times \mathbf{B})\, dv = \int_V \mathbf{J} \times \mathbf{B}\, dv$	(4-3)	$\mathbf{f} = \rho(\mathbf{v} \times \mathbf{B}) = \mathbf{J} \times \mathbf{B}$	(4-4)

As stated in Chap. 2, Eq. (4-1), known as *Ampère's circuital law*, implies that the line integral of the tangential component of **H** around a closed path is equal to the net current enclosed by that path (Fig. 2-2). The right side of Eq. (4-1) or of Eq. (4-2) is generally not equal to zero. It follows, then, that the magnetostatic field is not *conservative*. However, it is *solenoidal*, because Eq. (3) or Eq. (III) requires that all lines of magnetic flux close upon themselves (or start and terminate at infinity), thus giving the appearance of a tube, or pipe ($\sigma\omega\lambda\acute{\eta}\nu$, pronounced sol-een, in Greek).

As in the electrostatic case, the differential forms of the magnetostatic equations are appropriate for most problems, and the integral form of the equations yields closed-form solutions only where the problem has a high degree of symmetry.

Example 4-1 Field within a Toroid. As an example of the use of the integral laws, let us determine the field inside the toroid shown in Fig. 4-1. A *toroid* is a volume of revolution obtained by rotating any planar geometric area about an axis which is external to the area and in the plane of the area. (An automobile tire and a doughnut are examples of toroids.) We assume that the toroid is homogeneous, so that μ is

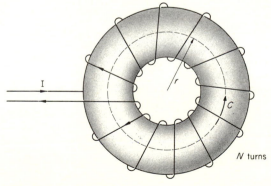

FIGURE 4-1. A toroid.

the same at every point. Let the toroid be wound with N turns of wire which are uniformly and tightly distributed around the circumference, and let this wire carry I A.

Consider the circular path of radius r, as shown. From symmetry we expect that the magnetic field will have only a tangential component at all points on this path. And we also conclude that the magnitude of the magnetic field will be constant along such a path. Hence the left side of Eq. (4-1) becomes

$$\oint_C \mathbf{H} \cdot d\mathbf{l} = \int_0^{2\pi} H_\varphi r \, d\varphi$$

where H_φ is the magnitude of the field intensity at all points on the path. Since H_φ and r are both constant along the path, Eq. (4-1) becomes

$$H_\varphi r \int_0^{2\pi} d\varphi = NI$$

so that inside the toroid,

$$H_\varphi = \frac{NI}{2\pi r} \tag{4-5}$$

The direction of the magnetic field intensity vector in the toroid is counterclockwise, in keeping with the convention adopted in Chap. 2 relative to the vector symbolism of Eq. (4-1). The same direction is predicted by the right-hand rule.

Some additional mathematical relations pertinent to magnetostatic fields are as given below.

Setting $\partial/\partial t = 0$ in the equation of continuity, namely,

$$\mathbf{\nabla} \cdot \mathbf{J} + \frac{\partial \rho}{\partial t} = 0 \tag{4-6}$$

we obtain

$$\mathbf{\nabla} \cdot \mathbf{J} = 0 \tag{4-7}$$

This equation shows that a stationary distribution of current is also solenoidal. *In practical terms, this equation and the definition of divergence tell us why a direct current will not flow through a capacitor; the lines of flow are prevented (blocked) from closing upon themselves.*

Discontinuities in the field vectors $\mathbf{B}$ and $\mathbf{H}$ are expressed by the general boundary conditions

$$\mathbf{n} \times (\mathbf{H_2} - \mathbf{H_1}) = \mathbf{K} \tag{VII}$$

$$\mathbf{n} \cdot (\mathbf{B_2} - \mathbf{B_1}) = 0 \tag{VIII}$$

In linear and isotropic media, the general constitutive relation

$$\mathbf{B} = B(\mathbf{H}) \tag{4-8}$$

reduces to

$$\mathbf{B} = \mu \mathbf{H} \tag{4-9}$$

It is frequently convenient to represent the magnetic flux density vector in terms of the vector potential $\mathbf{A}$, discussed in Sec. 2.13. If the medium is linear, homogeneous, and isotropic, μ is a scalar constant and the relation

$\mathbf{B} = \nabla \times \mathbf{A}$ combines with Eq. (4-2) to give

$$\nabla \times \nabla \times \mathbf{A} = \mu \mathbf{J} \tag{4-10}$$

On the other hand, the Lorentz condition, Eq. (2-85), now becomes

$$\nabla \cdot \mathbf{A} = 0 \tag{4-11}$$

Therefore, when use is made of the vector identity

$$\nabla \times \nabla \times \mathbf{A} = \nabla(\nabla \cdot \mathbf{A}) - \nabla^2 \mathbf{A} = -\nabla^2 \mathbf{A} \tag{4-12}$$

then Eq. (4-10) gives

$$\nabla^2 \mathbf{A} = -\mu \mathbf{J} \tag{4-13}$$

which is just the result obtained by setting the time derivative equal to zero in Eq. (2-86).

4.3 Magnetic Field of a Stationary Current Distribution.

The study of stationary magnetic fields is in many respects similar to the study of stationary electric fields.   Aside from questions pertaining to force, energy, and power, which are discussed in Chap. 6, magnetostatics, like electrostatics, is generally concerned with three types of problems:

1. Given *all* the source current distributions, find the field.
2. Given the field *everywhere*, find the sources.
3. Given a system of conductors, permeable materials, and *source* currents, find the resultant field.

The first problem is the analysis problem, pure and simple.   A function of one, two, or three variables representing the spatial distribution of currents is the given information.   This function is then substituted into formulas, derived in this chapter.   The rest is a routine application of formal mathematics, and we have an expression for the field.

In a practical problem, we seldom know the spatial distribution of currents; these must be found as solutions to a boundary-value problem for a given geometry of current-carrying conductors, permeable materials, and driving current sources.   A certain idealization may even be necessary to reduce a problem to the class of pure magnetostatic field problems.   For example, the conductivity of current-carrying conductors may have to be assumed infinite.   In short, we often have to solve a boundary-value problem.   Beyond that, in a real-life situation, we may be called upon to synthesize a static magnetic field by finding a distribution of source currents which will produce a field of a given description.   This calls for a lot of experience, intuition, and above all, compromise with what is deemed practical.

The first part of this chapter is devoted to the pure analysis problem. Two somewhat different methods are developed for determining the field of a known static current distribution.

The first method of analysis, discussed in this section, is the indirect method. The vector potential $\mathbf{A}$ is determined first, and then the field is found from $\mathbf{B} = \nabla \times \mathbf{A}$.

The second method proceeds directly to the field from the sources by using a relation, known as the *Biot-Savart law*, that follows directly from Maxwell's equations. This method is discussed in the next section.

At this point, let us focus our attention on the first method.

From Eq. (4-13), we have

$$\nabla^2 A_x = -\mu J_x \qquad (4\text{-}14)$$

$$\nabla^2 A_y = -\mu J_y \qquad (4\text{-}15)$$

$$\nabla^2 A_z = -\mu J_z \qquad (4\text{-}16)$$

where each equation bears an obvious resemblance to Poisson's equation. The theory developed in Sec. 3.6 may now be applied directly to each of the three equations. If the current distribution is bounded, and if all the currents are taken into account, then by analogy with Eq. (3-59) we can write the solutions

$$A_x = \frac{\mu}{4\pi} \int_V \frac{J_x}{r} \, dv \qquad (4\text{-}17)$$

$$A_y = \frac{\mu}{4\pi} \int_V \frac{J_y}{r} \, dv \qquad (4\text{-}18)$$

$$A_z = \frac{\mu}{4\pi} \int_V \frac{J_z}{r} \, dv \qquad (4\text{-}19)$$

or in condensed form,

$$\mathbf{A}(x_0, y_0, z_0) = \frac{\mu}{4\pi} \int_V \frac{\mathbf{J}(x,y,z)}{r} \, dv \qquad (4\text{-}20)$$

The geometry of Fig. 3-7 applies here, as well. It is evident from Eq. (4-20) that the relation between the vectors $\mathbf{A}$ and $\mathbf{J}$ is linear and that, consequently, superposition applies. As in the case of the scalar electric potential, Eq. (4-20) gives the correct answer (converges) for interior as well as exterior points of a bounded current distribution.

Example 4-2 Straight-line Segment. By way of an example, let us consider a circuit (closed conducting loop) which contains a straight section of finite length. This circuit carries a current *I*.

Let us find *that* part of the magnetic field produced by the straight conducting segment which, as shown in Fig. 4-2, extends in free space from *a* to *b* along the *x* axis.

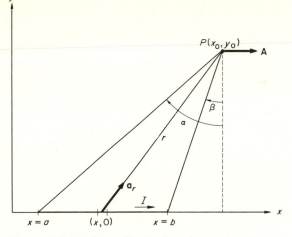

FIGURE 4-2. A linear distribution of current.

With the current flowing from left to right, the vector potential $\mathbf{A}$ at P is directed from left to right, consistent with Eq. (4-17) or Eq. (4-20). Since $J_x \, dv = I \, dx$,

$$A_x(x_0, y_0) = \frac{\mu_0}{4\pi} \int_a^b \frac{I}{\sqrt{y_0^2 + (x_0 - x)^2}} \, dx \tag{4-21}$$

By analogy with Eq. (3-61), integration leads to

$$A_x(x_0, y_0) = \frac{\mu_0 I}{4\pi} \ln \frac{(x_0 - a) + \sqrt{y_0^2 + (x_0 - a)^2}}{(x_0 - b) + \sqrt{y_0^2 + (x_0 - b)^2}} \tag{4-22}$$

The magnetic field is given by $\mathbf{B} = \nabla_0 \times \mathbf{A}$, where the space derivatives are taken with respect to the zero-subscripted variables (coordinates of the point of observation). We obtain

$$\mathbf{B} = -\frac{\partial A_x}{\partial y_0} \mathbf{a}_z \tag{4-23}$$

and, as in Sec. 3.7,

$$B_z = \frac{\mu_0 I}{4\pi y_0} (\sin \alpha - \sin \beta) \tag{4-24}$$

This indicates that in Fig. 4-2 the magnetic flux density vector is directed out of the plane of the paper. Because of symmetry, the vector $\mathbf{B}$ is normal to the plane defined by the wire and the point of observation. The sense of $\mathbf{B}$ is given by the right-hand rule: If the fingers of the right hand are wrapped around the wire with the thumb pointing in the direction of the flow of current, then the fingers will point in the direction of the vector $\mathbf{B}$. Thus, looking down the positive x axis, an observer will see the lines of $\mathbf{B}$ directed clockwise about the wire. If, in particular, the wire extends to infinity in both directions, the observer will see the lines of $\mathbf{B}$ as concentric circles (cylinders), because when $\alpha = -\beta = \pi/2$, Eq. (4-24) transforms to

$$B_z = \frac{\mu_0 I}{2\pi y_0} \tag{4-25}$$

Essentially the same result was deduced in Example 2-2. Thus the field about a straight wire of infinite length, obtained either from Eq. (4-20) or directly from Ampère's circuital law, is seen to vary inversely with the perpendicular distance from the wire to the point of observation.

Other degenerate forms of Eq. (4-24) which deserve special consideration are

$$\beta = 0: \qquad\qquad B_z = \frac{\mu_0 I}{4\pi y_0} \sin \alpha$$

$$\beta = -\alpha: \qquad\qquad B_z = \frac{\mu_0 I}{2\pi y_0} \sin \alpha$$

$$\beta \text{ negative, but } \beta \neq -\alpha: \quad B_z = \frac{\mu_0 I}{4\pi y_0} (\sin \alpha + \sin \beta)$$

$$\alpha = \frac{\pi}{2}, \quad \beta = 0: \qquad\qquad B_z = \frac{\mu_0 I}{4\pi y_0}$$

Example 4-3   Parallel-wire Line.   Figure 4-3 shows two infinitely long wires carrying equal and opposite currents. We wish to determine the magnetic field intensity at any point in the free space surrounding the two-wire line.

Without loss in generality, any point $P(x,y,0)$ on the xy plane may be chosen as a point of observation. An expeditious approach to this problem is to start with a finite-length wire and to let $a = -b$ and $x_0 = 0$ in Eq. (4-22). The resulting expression then gives the magnetic potential at any point on a plane bisecting the

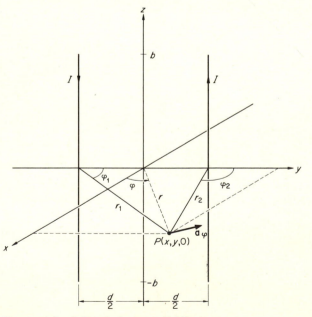

FIGURE 4-3.   Infinitely long parallel lines.

current-carrying wire. Thus, with reference to Fig. 4-2,

$$A_x(0,y_0) = \frac{\mu_0 I}{4\pi} \ln \frac{b + \sqrt{y_0^2 + b^2}}{-b + \sqrt{y_0^2 + b^2}} \tag{4-26}$$

Referring now to the geometry of Fig. 4-3, we see immediately that the contributions to the magnetic potential from the two wires are, respectively,

$$A_z^+(x,y,0) = \frac{\mu_0 I}{4\pi} \ln \frac{b + \sqrt{r_2^2 + b^2}}{-b + \sqrt{r_2^2 + b^2}}$$

$$A_z^-(x,y,0) = \frac{\mu_0(-I)}{4\pi} \ln \frac{b + \sqrt{r_1^2 + b^2}}{-b + \sqrt{r_1^2 + b^2}}$$

Simple addition gives

$$A_z(x,y,0) = \frac{\mu_0 I}{4\pi} \ln \left(\frac{r_1}{r_2} \frac{b + \sqrt{r_2^2 + b^2}}{b + \sqrt{r_1^2 + b^2}} \right)^2$$

Since

$$\lim_{b \to \infty} \frac{b + \sqrt{r_2^2 + b^2}}{b + \sqrt{r_1^2 + b^2}} = 1$$

it follows that, for a very long transmission line,

$$A_z(x,y,0) = \frac{\mu_0 I}{2\pi} \ln \frac{r_1}{r_2} = \frac{\mu_0 I}{2\pi} \ln \frac{\sqrt{(y + d/2)^2 + x^2}}{\sqrt{(y - d/2)^2 + x^2}} \tag{4-27}$$

The corresponding components of the magnetic field intensity are

$$H_x = \frac{I}{2\pi} \left[\frac{y + d/2}{(y + d/2)^2 + x^2} - \frac{y - d/2}{(y - d/2)^2 + x^2} \right]$$

$$= \frac{I}{2\pi} \left(\frac{\cos \varphi_1}{r_1} - \frac{\cos \varphi_2}{r_2} \right) \tag{4-28}$$

$$H_y = \frac{I}{2\pi} \left[\frac{x}{(y - d/2)^2 + x^2} - \frac{x}{(y + d/2)^2 + x^2} \right]$$

$$= \frac{I}{2\pi} \left(\frac{\sin \varphi_2}{r_2} - \frac{\sin \varphi_1}{r_1} \right) \tag{4-29}$$

Equations (4-28) and (4-29) are much simpler in form at distant points of observation. Since

$$r_1 \approx r + \frac{d}{2} \sin \varphi \qquad r_2 \approx r - \frac{d}{2} \sin \varphi$$

Eq. (4-27) becomes

$$A_z = \frac{\mu_0 I}{2\pi} \ln \frac{(r + d \sin \varphi/2)}{(r - d \sin \varphi/2)} \tag{4-30}$$

Since, for $x \ll 1$, $\ln (1 + x) \approx x$,

$$A_z = \frac{\mu_0 I}{2\pi} \ln \frac{(1 + d \sin \varphi/2r)}{(1 - d \sin \varphi/2r)}$$

$$= \frac{\mu_0 I}{2\pi} [\ln (1 + d \sin \varphi/2r) - \ln (1 - d \sin \varphi/2r)]$$

$$\approx \frac{\mu_0 I}{2\pi} \frac{d \sin \varphi}{r} \tag{4-31}$$

Using $\mathbf{H} = (1/\mu_0)(\nabla \times \mathbf{A})$, we finally find

$$\mathbf{H} = \frac{Id}{2\pi r^2}(\cos \varphi \, \mathbf{a}_r + \sin \varphi \, \mathbf{a}_\varphi) \qquad (4\text{-}32)$$

This is the field of a so-called *two-dimensional magnetic dipole*; it should be compared with the two-dimensional electric dipole of Prob. 3-29.

Example 4-4   The Circular Loop.   From large distances of observation, a small circular loop (Fig. 4-4) carrying a current I behaves like a *three-dimensional magnetic dipole*.

To see this, let us calculate the magnetic field at some point P, which, because of symmetry, may be taken in the yz plane.   In the notation of Fig. 4-4,

$$d\mathbf{l} = dl \, \mathbf{a}_\varphi = a \, d\varphi \, \mathbf{a}_\varphi = a \, d\varphi(-\sin \varphi \, \mathbf{a}_x + \cos \varphi \, \mathbf{a}_y)$$

So Eq. (4-20) gives

$$\mathbf{A} = \frac{\mu_0 I}{4\pi} \oint_C \frac{d\mathbf{l}}{R} = \frac{\mu_0 I}{4\pi} \int_0^{2\pi} \frac{a \, d\varphi}{R}(-\sin \varphi \, \mathbf{a}_x + \cos \varphi \, \mathbf{a}_y)$$

The superposition of elementary contributions to the total vector potential is implied by the integral on the right.   However, only a net $-x$-directed component will remain after summation, because the contribution to the y component of $\mathbf{A}$ from any two current elements symmetrically located about the y axis will cancel.   Since, at $P(0,y,z)$, $\mathbf{a}_\varphi = -\mathbf{a}_x$,

$$A_\varphi = \frac{\mu_0 I a}{4\pi} \int_0^{2\pi} \frac{\sin \varphi}{R} \, d\varphi$$

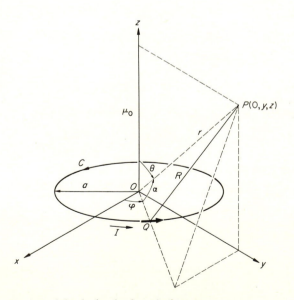

FIGURE. 4-4   A circular loop in free space.

The distance R can be expressed in terms of the radius r from the origin to P as follows.   From the triangle POQ we have

$$R^2 = a^2 + r^2 - 2ar \cos \alpha$$

Note that the z axis is not necessarily in the plane of the triangle POQ and that, therefore, in general, $\alpha + \theta \neq \pi/2$.   However, since $r \cos \alpha$ is the projection of r onto the radial line OQ, we can express it easily by projecting it first onto the y axis and subsequently onto OQ.   Thus we obtain

$$r \cos \alpha = r \sin \theta \sin \varphi$$

and
$$R = (a^2 + r^2 - 2ar \sin \theta \sin \varphi)^{\frac{1}{2}}$$

If $r \gg a$,

$$R \approx r\left(1 - 2\frac{a}{r}\sin \theta \sin \varphi\right)^{\frac{1}{2}} \approx r\left(1 - \frac{a}{r}\sin \theta \sin \varphi + \text{higher-order terms}\right)$$

where the last approximation follows from a binomial expansion.   To a first approximation, then,

$$\frac{1}{R} = \left(\frac{1}{r} + \frac{a}{r^2}\sin \theta \sin \varphi\right)$$

and
$$A_\varphi = \frac{\mu_0 I a}{4\pi} \int_0^{2\pi} \sin \varphi \left(\frac{1}{r} + \frac{a}{r^2}\sin \theta \sin \varphi\right) d\varphi$$

Since r is independent of φ, the integral of $\sin \varphi/r$ around the loop is zero.   It follows that

$$A_\varphi = \frac{\mu_0 I a^2 \sin \theta}{4\pi r^2} \int_0^{2\pi} \sin^2 \varphi \, d\varphi = \frac{\mu_0 \pi a^2 I \sin \theta}{4\pi r^2} \tag{4-33}$$

Using $\mathbf{B} = \nabla \times \mathbf{A}$, we now find

$$\mathbf{B} = \frac{\mu_0 \pi a^2 I}{4\pi r^3}(2 \cos \theta \, \mathbf{a}_r + \sin \theta \, \mathbf{a}_\theta) \tag{4-34}$$

and since $\mathbf{B} = \mu_0 \mathbf{H}$,

$$\mathbf{H} = \frac{\pi a^2 I}{4\pi r^3}(2 \cos \theta \, \mathbf{a}_r + \sin \theta \, \mathbf{a}_\theta) \tag{4-35}$$

Apart from an obvious difference in multiplicative constants, Eqs. (3-42) and (4-35) are otherwise identical.   It is clear, then, that if we limit consideration to points distant from the sources, the magnetic field of a loop will have the same spatial distribution as the electric field of an electric dipole.   For this reason, a small loop is often described as a *magnetic dipole* having a moment equal to the product of its area and the current I, and a direction parallel to the axis of the loop.   Thus

$$\mathbf{m} = \pi a^2 I \mathbf{a}_z \tag{4-36}$$

At this point it seems appropriate to summarize the expressions for static electric and magnetic singularities in free space of order zero and one.

THREE-DIMENSIONAL

Electric	*Magnetic*
Monopole	Monopole

$\phi = \dfrac{q}{4\pi\epsilon_0 r}$	Eq. (3-34)	Does not exist
$\mathbf{E} = \dfrac{q}{4\pi\epsilon_0 r^2}\,\mathbf{a}_r$	Eq. (3-33)	

Dipole	Dipole

$\phi = \dfrac{q\,d\cos\theta}{4\pi\epsilon_0 r^2}$	Eq. (3-39)	$\mathbf{A} = \dfrac{\mu_0(\pi a^2)I\sin\theta}{4\pi r^2}\,\mathbf{a}_\varphi$	Eq. (4-33)
$\mathbf{E} = \dfrac{qd}{4\pi\epsilon_0 r^3}\,(2\cos\theta\mathbf{a}_r + \sin\theta\mathbf{a}_\theta)$		$\mathbf{H} = \dfrac{(\pi a^2)I}{4\pi r^3}\,(2\cos\theta\mathbf{a}_r + \sin\,\theta\mathbf{a}_\theta)$	
	Eq. (3-42)		Eq. (4-35)

4.4 The Biot-Savart Law.

We now observe that the integral in Eq. (4-20) superimposes the increments of magnetic potential resulting from infinitesimal current elements $\mathbf{J}\,dv$ and that the $\mathbf{B}$ field can therefore be built up directly from the same current elements. It is reasonable to conclude that if we can find the $\mathbf{B}$ field due to a current element, we can find it for all currents by superposition. Therefore, what we need now is a differential formula that gives the magnitude and direction of the contribution to the $\mathbf{B}$ field due to a current flowing in an incremental length of wire.

Consider the wire shown in Fig. 4-5. We wish to relate the field $d\mathbf{B}$ at the point P due to the current I flowing in the length of wire dl. We begin

TWO-DIMENSIONAL

Electric	Magnetic
Monopole	Monopole

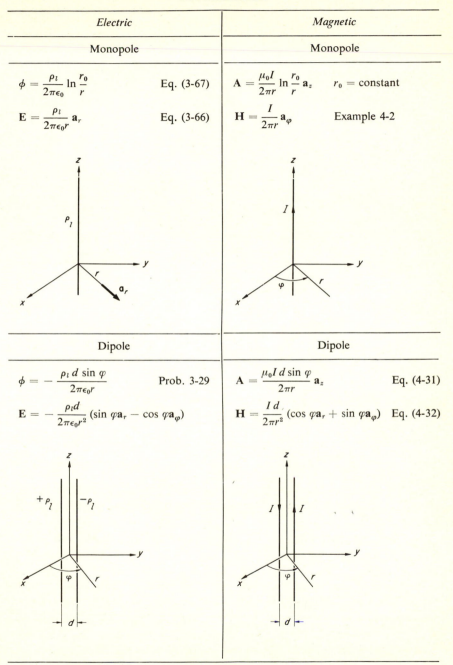

$$\phi = \frac{\rho_l}{2\pi\epsilon_0} \ln \frac{r_0}{r} \qquad \text{Eq. (3-67)} \qquad\qquad A = \frac{\mu_0 I}{2\pi r} \ln \frac{r_0}{r} \, a_z \qquad r_0 = \text{constant}$$

$$E = \frac{\rho_l}{2\pi\epsilon_0 r} \, a_r \qquad \text{Eq. (3-66)} \qquad\qquad H = \frac{I}{2\pi r} \, a_\varphi \qquad \text{Example 4-2}$$

Dipole	Dipole

$$\phi = -\frac{\rho_l \, d \sin \varphi}{2\pi\epsilon_0 r} \qquad \text{Prob. 3-29} \qquad A = \frac{\mu_0 I \, d \sin \varphi}{2\pi r} \, a_z \qquad \text{Eq. (4-31)}$$

$$E = -\frac{\rho_l d}{2\pi\epsilon_0 r^2} (\sin \varphi a_r - \cos \varphi a_\varphi) \qquad H = \frac{I \, d}{2\pi r^2} (\cos \varphi a_r + \sin \varphi a_\varphi) \quad \text{Eq. (4-32)}$$

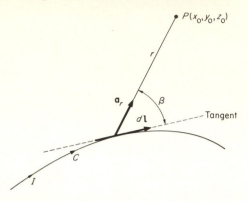

FIGURE 4-5. The geometry pertinent to the Biot-Savart law.

with Eq. (4-20), and suppose that a current I flows in a conductor whose cross section da is so small compared with r that, to a good approximation,

$$\mathbf{J}\,dv = J\,da\,dl = I\,d\mathbf{l} \tag{4-37}$$

where $d\mathbf{l}$ is a vector element of length measured along the tangent to the wire. In other words, we assume that the distribution of the current I is uniform over the cross-sectional area da, and that $\mathbf{J}$ is directed along the filament. Since a stationary current distribution is always solenoidal ($\nabla \cdot \mathbf{J} = 0$), the current density $\mathbf{J}$ can always be resolved into current filaments, all of which close upon themselves. Therefore the magnetic potential of an arbitrary current distribution may be obtained by summation over all the current filaments. It follows, then, that the contribution from a segment of one filament is

$$\mathbf{A}(x_0,y_0,z_0) = \frac{\mu}{4\pi} \int_V \frac{\mathbf{J}(x,y,z)}{r}\,dv = \frac{\mu I}{4\pi} \int_C \frac{d\mathbf{l}}{r} \tag{4-38}$$

where C specifies the filament contour, and r is the distance from the filament to the point (x_0,y_0,z_0). The magnetic vector $\mathbf{B}$ can be obtained directly from Eq. (4-38), using the familiar relation $\mathbf{B} = \nabla_0 \times \mathbf{A}$, where the zero in the operator ∇_0 denotes differentiation with respect to the subscripted (field) coordinates. We have

$$\mathbf{B}(x_0,y_0,z_0) = \frac{\mu I}{4\pi} \int_C \nabla_0 \times \frac{d\mathbf{l}}{r} \tag{4-39}$$

The integration and the curl operation in this equation can be interchanged because they involve different sets of variables.

The vector element of length $d\mathbf{l}$ is a function of the coordinate triplet (x,y,z), and $r = [(x - x_0)^2 + (y - y_0)^2 + (z - z_0)^2]^{1/2}$. The integrand in

Eq. (4-39) may be expanded according to the identity

$$\nabla \times \psi\mathbf{a} = \nabla\psi \times \mathbf{a} + \psi\nabla \times \mathbf{a}$$

where ψ is a scalar and $\mathbf{a}$ is a vector. In the notation of Fig. 4-5, this gives

$$\nabla_0 \times \left(\frac{1}{r}\,d\mathbf{l}\right) = \nabla_0\left(\frac{1}{r}\right) \times d\mathbf{l} = \frac{d\mathbf{l} \times \mathbf{a}_r}{r^2} \tag{4-40}$$

where $\mathbf{a}_r$ is a unit vector. Since $d\mathbf{l}$ is independent of the zero-subscripted variables, the second term of the expansion $(\nabla_0 \times d\mathbf{l})/r = 0$. The relation

$$\mathbf{B}(x_0, y_0, z_0) = \frac{\mu I}{4\pi}\int_C \frac{d\mathbf{l} \times \mathbf{a}_r}{r^2} \tag{4-41}$$

expresses the magnetic flux density vector as a superposition of elementary fields,

$$d\mathbf{B} = \frac{\mu I\, d\mathbf{l} \times \mathbf{a}_r}{4\pi r^2} \tag{4-42}$$

produced by each linear current element $I\,d\mathbf{l}$ of the wire. This is the desired form of the result, which is known as the *Biot-Savart law*. The direction of $d\mathbf{B}$ is perpendicular to the plane defined by $d\mathbf{l}$ and P, and is positive in the direction given by the right-hand rule.

The magnitude of $d\mathbf{B}$ is

$$dB = \frac{\mu I\, dl \sin\beta}{4\pi r^2} \tag{4-43}$$

where β is the angle defined in Fig. 4-5.

Example 4-5 Straight-line Segment. Let us use the Biot-Savart law to find the answer for Example 4-2. From Eq. (4-42), we have

$$dB_z = \frac{\mu_0 I\, dl \sin\beta}{4\pi r^2} = \frac{\mu_0 I y_0\, dx}{4\pi r^3} = \frac{\mu_0 I y_0\, dx}{4\pi[(x_0 - x)^2 + y_0^2]^{3/2}}$$

where β is the angle shown in Fig. 4-5. Therefore

$$B_z = \frac{\mu_0 I y_0}{4\pi}\int_a^b \frac{dx}{[(x_0 - x)^2 + y_0^2]^{3/2}} = \frac{\mu_0 I y_0}{4\pi}\left[\frac{1}{y_0^2}\frac{-(x_0 - x)}{\sqrt{(x_0 - x)^2 + y_0^2}}\right]_a^b$$

$$= \frac{\mu_0 I}{4\pi y_0}\left[\frac{x_0 - a}{\sqrt{(x_0 - a)^2 + y_0^2}} - \frac{x_0 - b}{\sqrt{(x_0 - b)^2 + y_0^2}}\right] = \frac{\mu_0 I}{4\pi y_0}(\sin\alpha - \sin\beta)$$

which fully agrees with Eq. (4-24).

Summarizing, an alternative method of finding the magnetic field of a known current distribution has been discussed. The two methods have a common analytical origin, and, as in electrostatics, the choice is one of expediency: we can integrate first and then differentiate (preceding section) or we can proceed directly to the evaluation of a vector integral (present section). Both methods are equally rigorous. The Biot-Savart law is frequently easier to use.

4.5 Magnetostatic Force. Although a more general study of electromagnetic forces is planned for Chap. 6, it is instructive at this point to undertake at least a cursory examination of the forces exerted between current-carrying wires.

Consider a conductor in a magnetic field **B**, and let I be the current which flows in the conductor. Experiments show that the presence of a magnetic field **B** causes a force to be exerted upon every element of length of the conductor. The magnitude and direction of this force are given by the following special form of the Lorentz force density vector:

$$\mathbf{f} = \mathbf{J} \times \mathbf{B} \tag{4-44}$$

The direction of the force, **f**, is implicit in the vector symbolism of Eq. (4-44). Thus **f** is normal to the plane formed by the local directions of the vectors **J** and **B**, and **B** is the local field due to *all* sources. In other words, the effect of the particular conductor under observation is always included.

First, let us examine the force on a current-carrying conductor arising from the field due to the current in the conductor. We have seen that for a very long conductor, the magnetic field displays cylindrical symmetry, and its direction is everywhere peripheral. If the conductor is completely isolated, the density of the current is uniform. By an elementary application of Ampère's circuital law, the field at any point within the wire (Fig. 4-6a) is found to be

$$\mathbf{H} = \frac{I}{2\pi a^2}\, r\mathbf{a}_\varphi = \frac{J}{2}\, r\mathbf{a}_\varphi \tag{4-45}$$

The complete functional dependence of the magnetic field intensity on radial distance is displayed graphically in Fig. 4-6b.

Assuming that the permeability of the wire is that of free space,

$$\mathbf{B} = \mu_0 \frac{J}{2}\, r\mathbf{a}_\varphi \tag{4-46}$$

inside the wire. Since $\mathbf{J} = J\mathbf{a}_z$,

$$\mathbf{f} = J\mathbf{a}_z \times \left(\frac{J}{2}\, r\mu_0\mathbf{a}_\varphi\right) = -\frac{J^2}{2}\, r\mu_0\mathbf{a}_r \tag{4-47}$$

The force is everywhere radial and constant for a fixed value of radius r. On diametrically opposite elements of the conductor, the forces will be equal

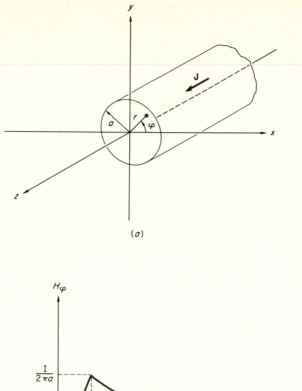

(a)

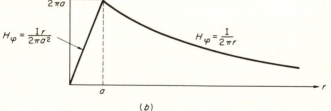

(b)

FIGURE 4-6. The field about a long wire. (a) Conductor carrying
a total current I; (b) field intensity versus radial distance in any plane
perpendicular to the z axis.

in magnitude and opposite in direction. Hence the net force per unit length
will be zero, and the wire will remain in static equilibrium.

This, however, is not the case if the wire is not straight. In general,
mutual forces do exist between elements of an arbitrarily shaped current-
carrying loop, and between elements of two infinitely long parallel wires, to be
examined next. The general conducting-loop problem presents formidable
analytical difficulties.

Example 4-6   Force between Parallel Wires.   Let us determine the force between two parallel conductors carrying equal and opposite currents of I A.   In particular, let us find the force per unit of length acting on the conductor on the right, shown in Fig. 4-3.

Let us assume that the conductors are far enough apart so that the presence of one in no way affects the distribution of the current in the other.   In other words, the distribution is assumed to be uniform in both conductors.   From our previous work we reason that the net force exerted on the left conductor due to the current in that conductor is zero; similarly, the right conductor experiences no self-produced net force.   From our previous work we also reason that the left conductor produces a magnetic field at the position of the right conductor, which is directed out of the plane of the paper (Fig. 4-3) and in the direction of the positive x axis.   At every point along the right conductor, this field is given by the expression

$$\mathbf{B} = \frac{\mu_0 I}{2\pi d}\mathbf{a}_x \tag{4-48}$$

On the other hand,

$$\mathbf{J} = \frac{I}{A}\mathbf{a}_z$$

where A is the wire cross-sectional area.   Therefore the force per unit length is

$$\mathbf{f} = \left(\frac{I}{A}\mathbf{a}_z\right) \times \left(\frac{\mu_0 I}{2\pi d}\mathbf{a}_x\right) = \frac{\mu_0 I^2}{2\pi A d}\mathbf{a}_y \tag{4-49}$$

This equation tells us that when currents flow in opposite directions, the two wires will repel each other.   It also tells us that the magnitude of $\mathbf{f}$ can be expressed as BI/A, and that therefore the total force acting on a length l of wire is

$$F = l\int \frac{BI}{A}\,da = BIl$$

to the approximation that B is uniform over the wire cross section.

Figure 4-7 shows a partial view of the field configuration around the right conductor in Fig. 4-3.   The net field due to both conductors is drawn in solid lines, and the field due to the conductor on the right is represented by

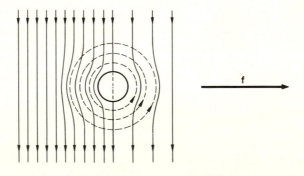

FIGURE 4-7.   Flux lines around one conductor of a two-wire line.

a dotted contour.   On the left, the two fields add; on the right, they subtract. Thus the field is stronger on the left and weaker on the right, and the force **f** is directed toward the weaker side of the field.

It is worth noting that by drawing a simple sketch of this sort it is always possible to predict the direction of the force acting on a current-carrying conductor.   The conductor will always tend to move toward the region where the conductor field and the externally applied field oppose each other.

It is also worth noting that the result, $F = BlI$, can be extended to include every case of a straight wire perpendicular to a uniform **B** field.   In that case

$$F = \int_V f \, dv = \int_C I \, |d\mathbf{l} \times \mathbf{B}| = BlI \tag{4-50}$$

4.6   Flux Linkages.   A concept developed from Maxwell's first equation,

$$\oint_C \mathbf{E} \cdot d\mathbf{l} = -\frac{d}{dt} \int_S \mathbf{B} \cdot \mathbf{n} \, da \tag{1}$$

is that of *flux linkages.*   It is developed as follows.

First replace the mathematical contour C with a conducting loop.   It will be found that a current is *induced* in the conducting loop.   Now open the loop at some point and measure the scalar potential difference between the ends of the loop.   Call this potential difference a voltage V.   It will be found that

$$V = -\frac{d}{dt} \int_S \mathbf{B} \cdot \mathbf{n} \, da \tag{4-51}$$

We say that the voltage V is *induced* in the conducting loop.

This result is known as *Faraday's law of induction.*   If we define the flux of **B** through S as Φ, the result can be written

$$V = -\frac{d\Phi}{dt}$$

For coils or other conductor configurations which are approximately a series of connected loops, it is convenient to modify our concepts and terminology slightly by observing that the flux Φ of **B** through the surface S is also the flux of **B** *encircled*, or *linked*, by the conductor and to speak of flux linkage by the conductor.   To emphasize this point of view, we change the notation and use the symbol λ instead of Φ when we speak of flux linkage by a conductor.   This proves advantageous when discussing multiple-loop (or turn) conductor configurations, as will be seen in the following discussion.

Consider first a single turn of a line conductor (Fig. 4-8a). For this single turn we have

$$\lambda_1 = \Phi_1 = \int_{S_1} \mathbf{B} \cdot \mathbf{n} \, da \tag{4-52}$$

where Φ_1 is the total flux linked by the wire contour, and S_1 is the surface spanned by the conductor. Similarly, if a second turn, bounding a surface S_2, is present, we have for this conductor

$$\lambda_2 = \Phi_2 = \int_{S_2} \mathbf{B} \cdot \mathbf{n} \, da \tag{4-53}$$

If these two loops are connected in such a way (Fig. 4-8b) that the flux is passing through them in the same direction, the total flux linkage of the two-turn loop (coil) is

$$\lambda = \lambda_1 + \lambda_2 = \Phi_1 + \Phi_2 \tag{4-54}$$

If the identical flux Φ passes through both loops, we should have, simply,

$$\lambda = 2\Phi \tag{4-55}$$

If there are N loops (turns in the coil), each with identical magnetic flux, we should obtain

$$\lambda = N\Phi \tag{4-56}$$

We should note that this last result would almost always be an idealization never achieved in practice since all N turns, because of their physical size, could never all be at exactly the same place, and hence will usually not enclose exactly the same flux. However, the idealization that all turns of a coil enclose identical flux is a good approximation for many problems, and leads to considerable simplification of the calculations.

When different amounts of flux pass through different parts of the same coil, the total flux linkage is equal to the sum of the linkages. For the situation of Fig. 4-8c,

$$\lambda = N_1\Phi_1 + N_2\Phi_2 + \cdots + N_n\Phi_n \tag{4-57}$$

In summary, flux linkage of a closed contour is the flux of the vector $\mathbf{B}$ through an area bounded by the contour.

4.7 The Inductance Concept. When a circuit (closed conducting path) is carrying a steady current I, it produces a magnetic field $\mathbf{B}$ in the medium surrounding the circuit. This magnetic field has definite flux linkage λ with the circuit itself. If the medium has a constant permeability, the flux linkage is proportional to the current, giving

$$\lambda = \sum_i N_i\Phi_i = LI \tag{4-58}$$

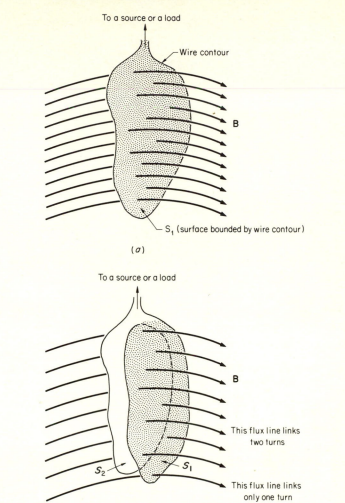

(a)

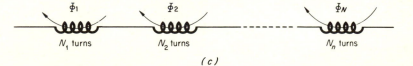

(b)

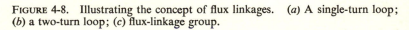

(c)

FIGURE 4-8. Illustrating the concept of flux linkages. (a) A single-turn loop; (b) a two-turn loop; (c) flux-linkage group.

where L is a constant of proportionality defined as the *self-inductance* of the circuit, or

$$\text{Inductance } L = \frac{\lambda}{I} \qquad (4\text{-}59)$$

When λ is given in weber-turns and I in amperes, Eq. (4-59) gives L in henrys. An *inductor* of 1 H is rather large; hence the units millihenry (10^{-3} H) and microhenry (10^{-6} H) are frequently employed.

Self-inductance specifies the magnitude of the magnetic coupling of a circuit with itself. The magnitude of the magnetic coupling of one circuit with a second circuit is called *mutual inductance*. The term inductance usually means self-inductance.

When the medium is linear, the inductance is determined entirely by the geometry and by the material of the inductor.

It is important to note that inductance is defined under *steady-state* conditions. In Chap. 11 we show that the behavior of circuit elements changes with frequency.

Example 4-7   Inductance of a Toroid.   As an example of the determination of inductance, let us examine the inductance of a toroid of square cross section, as shown in Fig. 4-9. Let the inside radius be R m and the total number of turns be N. If a current of I A flows through the turns, the flux density in the core of the toroid from Eq. (4-5) is

$$B_\varphi = \frac{\mu N I}{2\pi r}$$

All the flux which exits in one cross section of the core must link (or encircle) the turns on the inside of the toroid. Hence λ is $N\Phi$, where Φ is the flux over the crosssectional area of the toroid. We pick an element of area which is $a\,dr$, as shown. Note that **B** and **n** are in the same direction, so that the dot product becomes the product of the magnitudes. We have

$$\lambda = N \int_{R}^{R+a} \frac{\mu N I}{2\pi r} a\,dr = \frac{N^2 I \mu a}{2\pi} \ln \frac{R+a}{R} \qquad (4\text{-}60)$$

and the inductance is

$$L = \frac{\lambda}{I} = \frac{N^2 \mu a}{2\pi} \ln \frac{R+a}{R} \qquad (4\text{-}61)$$

Note that L is a function of μ, N, and inductor geometry only.

Example 4-8   The Solenoid.   In the preceding section we were able, with a few simplifying assumptions, to obtain an answer for the inductance of a toroid in closed form. Unfortunately, the analytical difficulties associated with an exact solution of the problem of the solenoid (helical coil) are overwhelming. Yet the most common form of an inductor used in practice is the solenoid. Let us therefore study such a coil.

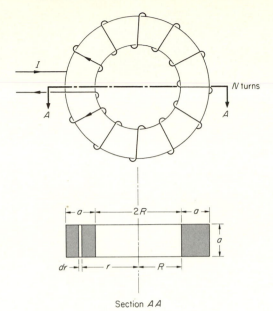

Section AA

FIGURE 4-9. A toroid of square cross section.

A solenoid is formed by winding a conductor as a helix, with the turns closely spaced at equal intervals along the length of the solenoid. Quite frequently the coil is wound on some kind of a paper tube, as indicated in Fig. 4-10a. In other cases, the winding may be completely unsupported, and may have the form of a spring. In still other instances the coil may have more than one layer. This is particularly true when the coil is wound on a permeable core (for example, output transformers in radios, power transformers in substations and on utility poles, and power-supply transformers in television sets).

The analysis problem of a solenoid is complicated by flux leakage near the ends of the solenoid (Fig. 4-10b). Thus some of the flux established by the coil links only a fractional part of the total number of turns. However, this leakage is mostly confined near the ends. Therefore, if the coil is long, a reasonable approximation would be to assume that the field is constant along the entire length of the solenoid and equal to its value at the center.

To begin the analysis of the solenoid, let us find the magnetic flux density vector on the axis of a loop of wire carrying a current I. To this end, let $\theta = 0$ in Fig. 4-4. From Eq. (4-43), we then have

$$dB = \frac{\mu I\, dl \sin \beta}{4\pi(z^2 + a^2)}$$

Note that β is $90°$ in Fig. 4-10c for every dl on the loop of wire; hence $\sin \beta = 1$. Since $d\mathbf{B}$ is perpendicular to the plane formed by $d\mathbf{l}$ and the point $P(0,0,z)$ on the z axis, $d\mathbf{B}$ has a z component but, because of symmetry, no radial or tangential components. It can be seen that

$$dB_z = dB \cos \alpha = \frac{\mu I\, dl}{4\pi(z^2 + a^2)} \frac{a}{\sqrt{z^2 + a^2}} \qquad (4\text{-}62)$$

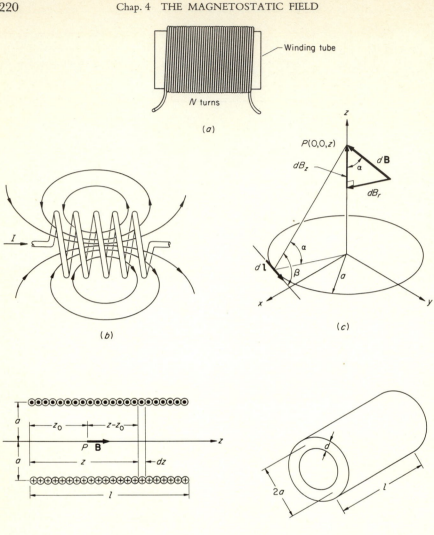

FIGURE 4-10. Cylindrical solenoids. (a) Single-layer cylindrical solenoid; (b) field distribution; (c) field on the axis of a single loop; (d) cross section of a long solenoid; (e) multilayer coil dimensions.

We obtain B_z at $P(0,0,z)$ by integrating over all dl's of the loop.

$$B_z = \frac{\mu I a}{4\pi(z^2 + a^2)^{3/2}} \oint_C dl = \frac{\mu I a^2}{2(z^2 + a^2)^{3/2}} \qquad (4\text{-}63)$$

At the center of the loop $(z = 0)$ this reduces to

$$B = \frac{\mu I}{2a} \qquad (4\text{-}64)$$

Let us now consider the field on the axis of a long straight solenoid as shown in Fig. 4-10d. We require that the turns on the solenoid be uniformly distributed, that there be a total of N turns, and that they be closely spaced so that the distance between turns will be negligible. We assume that the current can be treated as a sheet of current so that the current in an infinitesimal slice of dz of the solenoid will be $NI\,dz/l$. We want the flux density at P, which is a point on the axis and at a distance z_0 from one end of the solenoid.

At the point P the flux density due to the current in the slice dz is, from Eq. (4-63),

$$dB_z(z_0) = \frac{\mu a^2(NI\,dz/l)}{2[(z - z_0)^2 + a^2]^{3/2}} \tag{4-65}$$

The total B_z is obtained by integrating from one end of the coil to the other. Since B_z has the same direction for every element of current, Eq. (4-65) may be integrated to give the field at P due to the entire solenoid. Thus

$$B_z(z_0) = \frac{\mu a^2 NI}{2l} \int_0^l \frac{dz}{[(z - z_0)^2 + a^2]^{3/2}}$$

$$= \frac{\mu NI}{2l} \left[\frac{l - z_0}{\sqrt{(l - z_0)^2 + a^2}} + \frac{z_0}{\sqrt{z_0^2 + a^2}} \right] \tag{4-66}$$

This is the general result. If the point P is located at the center of the solenoid ($z_0 = l/2$), the flux density becomes

$$B_z\left(\frac{l}{2}\right) = \frac{\mu NI}{\sqrt{l^2 + 4a^2}} \tag{4-67}$$

On the other hand, when $z_0 = 0$, or when $z_0 = l$,

$$B_z(0) = B_z(l) = \frac{\mu NI}{2\sqrt{l^2 + a^2}} \tag{4-68}$$

It is seen that the flux density at the end of a long solenoid is less than at the center. This is caused by flux leakage near the ends of the solenoid.

If the length of the solenoid is large relative to the radius ($l \gg a$), the flux density at the center becomes

$$B_z\left(\frac{l}{2}\right) = \frac{\mu NI}{l} \tag{4-69}$$

while at the ends

$$B_z(0) = B_z(l) \approx \frac{\mu NI}{2l} \tag{4-70}$$

Thus, because of leakage, the flux density at the ends drops to about one-half of its value at the center.

We are now ready to obtain an expression for the inductance. If $l \gg a$, end effects may be neglected, and the magnetic flux density may be taken constant over the entire interior of the solenoid both on and off the central axis. From Eq. (4-69),

$$B_z = \frac{\mu NI}{l} \tag{4-71}$$

everywhere within the coil. As a result, the flux linkage of every individual turn of the coil is

$$\Delta \lambda = (\pi a^2)B_z = \pi a^2 \frac{\mu NI}{l} \tag{4-72}$$

Since there are N turns, the total flux linkage of the current in the solenoid with its own magnetic field is

$$\lambda = N \, \Delta\lambda = N\pi a^2 \frac{\mu NI}{l} \qquad (4\text{-}73)$$

The approximate inductance of a long solenoid of radius a and length l is therefore given by

$$L = \frac{\lambda}{I} = \frac{\mu \pi N^2 a^2}{l} \qquad (4\text{-}74)$$

To get a feeling for the order of magnitudes involved, let us calculate the inductance of a solenoid of 200 turns wound tightly on a cylindrical tube 6 cm in diameter. The length of the tube is 60 cm, and the medium is air. From Eq. (4-74) we have

$$L = \frac{4\pi \times 10^{-7} \times \pi \times (200)^2 \times (3 \times 10^{-2})^2}{60 \times 10^{-2}} = 0.237 \text{ mH}$$

Various approximate formulas which are more accurate than Eq. (4-74) have been developed for determining the inductance of solenoids. Most of these formulas are empirical or semiempirical. For example, the inductance of a single-layer helical coil is, approximately,

$$L = \frac{2.54N^2a^2}{9a + 10l} \times 10^{-8} \text{ H} \qquad (4\text{-}75)$$

where a is the mean radius of the coil (meters), and l is the axial length of the coil (also expressed in meters). Of course, N is the number of turns and is a dimensionless number. Equation (4-75) is accurate to within ± 1 percent when $l > 0.8a$.

The inductance of multilayer air-core coils (Fig. 4-10e) may be calculated from the formula

$$L = \frac{2.03N^2a^2}{6a + 9l + 10d} \times 10^{-8} \text{ H} \qquad (4\text{-}76)$$

where all dimensions are in meters.

Example 4-9 Inductance per Unit Length of a Long Two-wire Line. Let us calculate the self-inductance per unit length of a parallel pair of wires, each having radius a. The wires are infinitely long, and their axes are separated by a distance d.

For steady currents, the current distribution in the cross section of each cylindrical conducting wire is essentially uniform, particularly if the distance d is appreciably larger than the radius a. To a good approximation, then, the magnetostatic field outside the cylindrical conductors is identical with the field produced by a pair of *line* conductors, as depicted in Fig. 4-3. More precisely, the field in the region of the yz plane that extends from $y = -d/2 + a$ to $y = d/2 - a$ in Fig. 4-11 is approximately equal to the field of two infinitely long parallel lines which coincide with the axes of the cylinders. From Eqs. (4-28) and (4-29), then, we obtain, by setting $x = 0$,

$$H_x = \frac{I}{2\pi} \left(\frac{1}{y + d/2} - \frac{1}{y - d/2} \right) \qquad (4\text{-}77)$$

$$H_y = 0 \qquad (4\text{-}78)$$

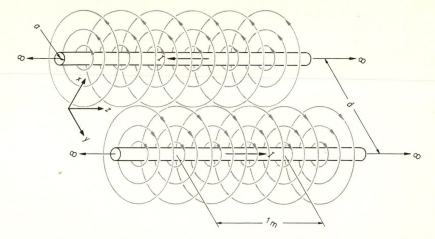

FIGURE 4-11. Magnetic field pattern between a parallel pair of cylindrical conductors carrying equal but opposite currents.

Now consider a 1-m length of line. The total magnetic flux passing between the conductors per meter is

$$\Phi = \int_{-(d/2-a)}^{(d/2-a)} \mu H_x(1)\, dy = \frac{\mu I}{2\pi}\left[\ln\left(y+\frac{d}{2}\right) - \ln\left(y-\frac{d}{2}\right)\right]_{-(d/2-a)}^{(d/2-a)} \qquad (4\text{-}79)$$

Therefore the flux linkage between the current I and its own magnetic field per unit of length of line is

$$\lambda = \Phi = \frac{\mu I}{\pi}\ln\frac{d-a}{a} \qquad (4\text{-}80)$$

If $d \gg a$,

$$\lambda \approx \frac{\mu I}{\pi}\ln\frac{d}{a} \qquad (4\text{-}81)$$

and in this case (which is the practical case), the inductance, in henrys per meter, is

$$L = \frac{\lambda}{I} \approx \frac{\mu}{\pi}\ln\frac{d}{a} \qquad (4\text{-}82)$$

It can be shown that, to remove the restriction $d \gg a$, we must use

$$L = \frac{\mu}{\pi}\cosh^{-1}\frac{d}{2a} \qquad (4\text{-}83)$$

Example 4-10 Inductance per Unit Length of a Coaxial Cable.† In the preceding example we neglected to take into account the partial flux linkages *inside* the wires,

† Some aspects of this example are difficult for all but the most capable student. Though the derivation is complex, the final results are useful from a practical point of view.

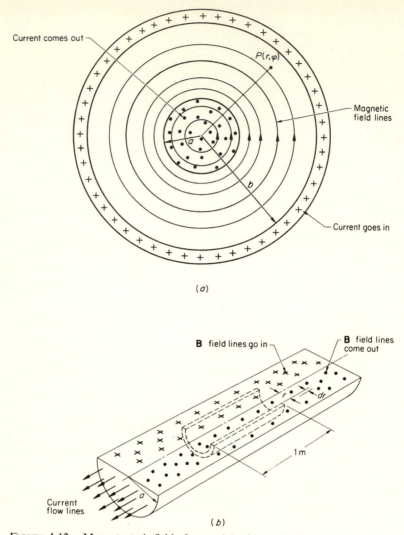

FIGURE 4-12.   Magnetostatic field of a coaxial cable.   (*a*) Geometrical arrangement and field lines; (*b*) half slice of inner conductor.

purposely, of course, to avoid the accompanying analytical difficulties. We shall
demonstrate the effect of partial flux linkages by considering a coaxial line made up
of a solid inner conductor and a thin-walled outer conductor (Fig. 4-12*a*).

Let us suppose that the inner conductor has a radius *a* and carries a current *I*
which flows out of the plane of the paper toward the observer. Let us also suppose
that the inside radius of the (thin) outer conductor is *b*, and that it carries an equal
and opposite current *I*. If the current *I* is distributed uniformly over the cross section
of the inner conductor, the current flowing in the portion of the inner conductor from

zero to any radius $r \leq a$ is $I(r^2/a^2)$. The field $\mathbf{H}$ is in the φ direction only, and, according to Ampère's circuital law, it is

$$H_\varphi = \frac{I}{2\pi a^2} r \qquad 0 \leq r \leq a \tag{4-84}$$

The permeability of good conductors is approximately that of free space. So

$$B_\varphi = \frac{\mu_0 I}{2\pi a^2} r \qquad 0 \leq r \leq a \tag{4-85}$$

Figure 4-12b shows a partial view of the current and field pattern within the inner conductor. Consider an annular ring from r to $r + dr$, 1 m long. The current flowing in this ring is $[I(2\pi r\, dr)]/\pi a^2$, and the flux in the ring has a magnitude of

$$B_\varphi(dr)(1) = \frac{\mu_0 I}{2\pi a^2} r\, dr \tag{4-86}$$

This flux is evidently a function of r, and links only a fraction of the total current flowing along the inner cylinder. In fact, the current in the annular ring is linked by that portion of the flux within the inner conductor which exists outside the annular ring, and also by the total flux which exists in the region between the inner and outer cylinders. In that region, of course,

$$B_\varphi = \mu_0 \frac{I}{2\pi r} \tag{4-87}$$

So the total magnetic flux linking the current flowing within the annular ring is, from Eqs. (4-86) and (4-87),

$$d\Phi = \int_r^a \frac{\mu_0 I}{2\pi a^2} r\, dr + \int_a^b \frac{\mu_0 I}{2\pi r}\, dr = \frac{\mu_0 I}{4\pi a^2}(a^2 - r^2) + \frac{\mu_0 I}{2\pi} \ln \frac{b}{a} \tag{4-88}$$

Since this flux links only a fraction $2\pi r\, dr/\pi a^2$ of the total current I, we must adjust the contribution of $d\Phi$ to the total flux linkages by precisely the same factor. In other words, for the purpose of calculating inductance per unit length of a coaxial line, we must take

$$d\lambda = \frac{2\pi r\, dr}{\pi a^2}\, d\Phi \tag{4-89}$$

and then integrate this expression from $r = 0$ to $r = a$ to get the net contribution of the entire current distribution to the flux linkages associated with this structure. We have

$$\lambda = \int_0^a d\lambda = \frac{\mu_0 I}{2\pi a^4} \int_0^a (a^2 - r^2)r\, dr + \frac{\mu_0 I}{\pi a^2} \ln \frac{b}{a} \int_0^a r\, dr$$

$$= \frac{\mu_0 I}{8\pi} + \frac{\mu_0 I}{2\pi} \ln \frac{a}{b} \tag{4-90}$$

Hence the inductance per unit length is

$$L = \frac{\lambda}{I} = \frac{\mu_0}{8\pi} + \frac{\mu_0}{2\pi} \ln \frac{b}{a} \tag{4-91}$$

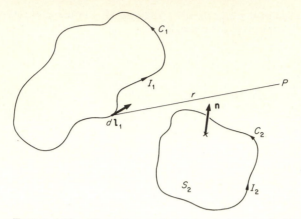

FIGURE 4-13.   Magnetic coupling between two conducting loops.

Obviously, the first term on the right arises from the flux linkages internal to the inner conductor. For this reason, it is known as the *internal inductance* of the inner conductor per unit of length of line. The second term arises from the flux linkages between the total current I and the flux which exists in the interelectrode space. It is sometimes referred to as the *external inductance* of the coaxial line per unit of its length.

Let us next consider the mutual inductance between two circuits.

Consider the conducting loops C_1 and C_2 shown in Fig. 4-13. A current I_1 flowing through the first loop will produce a magnetic field such that a certain amount of flux Φ_{12} will be linked by the second loop. The *mutual inductance* between the two loops is defined as

$$L_{12} = \frac{\Phi_{12}}{I_1} \tag{4-92}$$

Likewise, a current I_2 flowing in loop C_2 produces a magnetic flux Φ_{21} that will be linked by loop C_1. By definition,

$$L_{21} = \frac{\Phi_{21}}{I_2} \tag{4-93}$$

We shall now show that if the surrounding medium is linear, homogeneous, and isotropic,

$$L_{12} = L_{21} \tag{4-94}$$

To prove this statement, let us suppose that the wires forming the conducting loops are so thin that the flow of current is virtually confined to line contours. Then the vector potential caused by current I_1 is, according to Eq. (4-20),

$$\mathbf{A} = \frac{\mu I_1}{4\pi} \oint_{C_1} \frac{d\mathbf{l}_1}{r} \tag{4-95}$$

We recall that this formula is valid only when the permeability is a scalar constant. The magnetic flux density at P is obtained from

$$\mathbf{B} = \nabla \times \mathbf{A} \tag{4-96}$$

while the total flux linking circuit C_2 is

$$\Phi_{12} = \int_{S_2} \mathbf{B} \cdot \mathbf{n} \, da = \int_{S_2} (\nabla \times \mathbf{A}) \cdot \mathbf{n} \, da \tag{4-97}$$

By Stokes' theorem, the surface integral on the right of this expression may be converted to a contour integral:

$$\Phi_{12} = \int_{S_2} (\nabla \times \mathbf{A}) \cdot \mathbf{n} \, da = \oint_{C_2} \mathbf{A} \cdot d\mathbf{l}_2 \tag{4-98}$$

Hence

$$\Phi_{12} = \oint_{C_2} \left(\frac{\mu I_1}{4\pi} \oint_{C_1} \frac{d\mathbf{l}_1}{r} \right) \cdot d\mathbf{l}_2 \tag{4-99}$$

from which it follows that

$$L_{12} = \frac{\Phi_{12}}{I_1} = \frac{\mu}{4\pi} \oint_{C_1}\oint_{C_2} \frac{d\mathbf{l}_1 \cdot d\mathbf{l}_2}{r} \tag{4-100}$$

In this equation r is the distance from a point on C_1 to a point on C_2.

It is obvious that, by repeating the same development for a current I_2 flowing in circuit C_2, we should obtain

$$L_{21} = \frac{\Phi_{21}}{I_2} = \frac{\mu}{4\pi} \oint_{C_2}\oint_{C_1} \frac{d\mathbf{l}_2 \cdot d\mathbf{l}_1}{r} \tag{4-101}$$

Since the dot product is commutative, the right members in Eqs. (4-100) and (4-101) are identical. This proves the validity of Eq. (4-94). Equation (4-101) is known as the *Neumann formula*.

Example 4-11 Mutual Inductance of Two Coaxial Loops. Figure 4-14 shows two very thin coaxial loops in free space. The mutual inductance between the two loops is desired, assuming that the planes of the loops are parallel to one another and are spaced at a distance z much larger than the radii a and b.

From Eq. (4-33), the vector potential at P, caused by current I_1 flowing in the bottom loop, is

$$\mathbf{A} = \frac{\mu_0 a^2 I_1 \sin\theta}{4r^2} \mathbf{a}_\varphi = \frac{\mu_0 a^2 I_1 b}{4(b^2 + z^2)^{3/2}} \mathbf{a}_\varphi \approx \frac{\mu_0 a^2 I_1 b}{4z^3} \mathbf{a}_\varphi \tag{4-102}$$

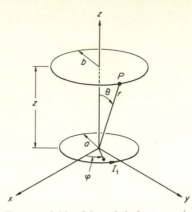

FIGURE 4-14. Mutual inductance between two coaxial circular loops.

when $z \gg a, b$. From Eq. (4-98),

$$\Phi_{12} = \oint_{C_2} \mathbf{A} \cdot d\mathbf{l}_2 = \frac{\mu_0 a^2 I_1 b}{4z^3} \oint_{C_2} dl_2 = \frac{\mu_0 \pi a^2 b^2 I_1}{2z^3} \tag{4-103}$$

Thus

$$L_{12} = \frac{\Phi_{12}}{I_1} = \frac{\mu_0 \pi a^2 b^2}{2z^3} \qquad z \gg a, b \tag{4-104}$$

In summary, this section has dealt with the subject of inductance. Inductance of a circuit specifies the degree of magnetic coupling of a circuit either with itself (self-inductance) or with a neighboring circuit (mutual inductance).

4.8 The Scalar Magnetic Potential.   In many instances, up to this point, the treatment of magnetostatics was facilitated considerably by the introduction of the vector magnetic potential function **A**.   The availability of this function as an artifice in analysis is always assured by the solenoidal character $(\mathbf{\nabla} \cdot \mathbf{B} = 0)$ of the field vector **B**.   But **A** is a vector.   Life would be simpler if, as in electrostatics, we could deal instead with a scalar potential.   We shall now discuss the possibility of such a scalar magnetic potential, which we shall call ϕ^* (phi star).

Consider a current-free region V.   Figure 4-15 shows a plane section through such a region, which, as indicated, has the form of a doughnut, and hence it is multiply connected.   The region V', which complements V, contains the sources of the magnetostatic field under consideration.   To fix ideas, we suppose that the field is produced by a coil wound around the region V, as shown in the figure.   The coil has N turns and carries a current I.

At every point in the current-free region V, Eq. (4-2) reduces to

$$\mathbf{\nabla} \times \mathbf{H} = 0 \tag{4-105}$$

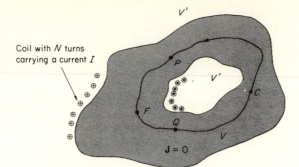

FIGURE 4-15. A multiply connected region that does not contain currents.

Hence, within V, $\mathbf{H}$ may be expressed as the negative gradient of a scalar function ϕ^*.

$$\mathbf{H} = -\boldsymbol{\nabla}\phi^* \tag{4-106}$$

The function ϕ^* is the *scalar magnetic potential*, and is measured in ampere-turns (At).   By virtue of this definition, it follows that

$$\int_{C_1} \mathbf{H} \cdot d\mathbf{l} = \int_{C_1} (-\boldsymbol{\nabla}\phi^*) \cdot d\mathbf{l} = \phi^*(P) - \phi^*(Q) \tag{4-107}$$

where C_1 denotes the portion *PFQ* of the closed contour C in Fig. 4-15. The difference in values that the potential function ϕ^* assumes at the extreme points P and Q of C_1 has been traditionally known as the *magnetomotive force* (mmf, for short) between these points.   This notion is analogous to that of potential difference, or voltage, normally associated with the electrostatic field.   The corresponding term, *electromotive force* (emf), is normally associated with the quasistatic field (Chap. 11).

We saw in Chap. 3 that the electrostatic scalar potential is a single-valued function of the coordinates.   In direct contrast, the magnetostatic scalar potential ϕ^* is generally a multivalued function of position except when the region is simply connected and current-free.   The validity of this statement can be demonstrated as follows.

From Ampère's circuital law we have

$$\oint_C \mathbf{H} \cdot d\mathbf{l} = NI \tag{4-108}$$

This equation states that if the contour C begins and terminates at the same point P, the initial and final values of the magnetic potential ϕ^* will be different, and will differ by an amount exactly equal to NI.   In other words, Eq. (4-108) tells us that

$$\oint_C \mathbf{H} \cdot d\mathbf{l} = \phi^*(P)\bigg|_{\text{start}} - \phi^*(P)\bigg|_{\text{finish}} = NI \tag{4-109}$$

If the contour is traversed twice,

$$\phi^*(P)\Big|_{\text{start}} - \phi^*(P)\Big|_{\text{finish}} = 2NI \qquad (4\text{-}110)$$

and if thrice,

$$\phi^*(P)\Big|_{\text{start}} - \phi^*(P)\Big|_{\text{finish}} = 3NI \qquad (4\text{-}111)$$

and so on.   Thus the discontinuity in ϕ^* is equal to the net current linked by the contour.   Only when the region is simply connected, or when the current density is everywhere zero, does the potential become a single-valued function of position.

To summarize the points we have made in this section, a magnetic potential ϕ^*, defined by Eq. (4-106), exists in a region V that does not contain currents.   When a closed contour C in the region V encircles a net current I, the potential function ϕ^* is discontinuous by an amount equal to I.

The difference in values of the scalar magnetic potential ϕ^* at any two points in V is called the magnetomotive force, or mmf.   In general, the value of ϕ^* is dependent on the particular path by which these points are connected.

4.9 Ferromagnetic Circuits.   The scalar magnetic potential is particularly useful in the analysis of ferromagnetic circuits of such practical devices as transformers, motors, generators, and certain microwave ferrite† devices (Fig. 4-16).   It is advisable then, to consider at this point the most pertinent aspects of ferromagnetism.

For most materials the permeability μ has a definite value which is very nearly equal to the value for air.   For a small but important group of materials, of which iron, its alloys, and ferrites are the most common, μ is not a constant; it has a value which is much larger than the value for air, it varies with the magnetic flux density, and it is not uniquely defined by **B**. These materials are called *ferromagnetic*.†

The most important properties of ferromagnetic substances are, from our present viewpoint, a nonlinear *B-H* characteristic (nonlinear constitutive relation) and *hysteresis*.

Let us investigate the *B-H* relationship for a toroid (Fig. 4-1) in which the core is a ferromagnetic material.   The flux density can be measured by any one of several available methods.   It is found that it depends not only on the value of the magnetic field intensity, but also on the past history.   A typical relationship is shown in Fig. 4-17a.   This figure shows the result that would be obtained if one starts with an unmagnetized iron core, increases the

† Ferrites are really ferrimagnetic materials; the distinction is unnecessary for our present purposes.

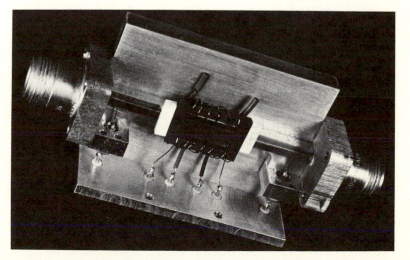

FIGURE 4-16. A latching ferrite phase shifter. (*a*) Core half slice; (*b*) phase-shifter assembly. (*Courtesy of Scientific-Atlanta, Inc.*)

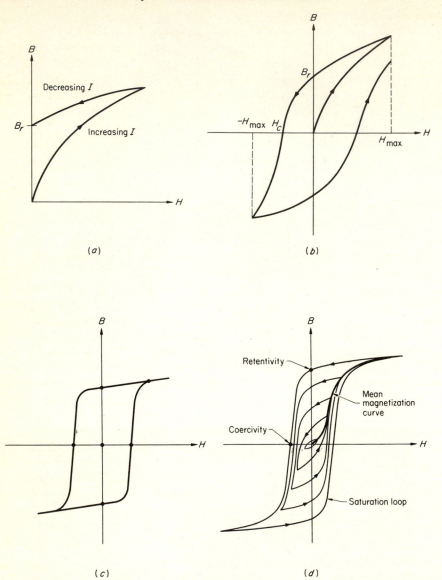

FIGURE 4-17.   The hysteresis effect in a ferromagnetic substance.   (*a*) The beginning of a hysteresis loop; (*b*) the first cycle of a hysteresis loop; (*c*) square hysteresis loop; (*d*) a family of hysteresis loops.

current (which increases H) from zero to I, and then reduces it to zero again. Since H and I are proportional, Fig. 4-17a is plotted against H.

It is shown in Fig. 4-17a that even when the current is decreased to zero, the material retains a certain amount of magnetic flux density B_r. This quantity is termed the *residual flux density*. It is a measure of the ability of the material to retain the magnetic flux. Accordingly, B_r is called the *retentivity* of the material.

If the current is reversed, we find that a certain negative value of current is necessary to give zero flux density. The (negative) field H_c necessary to overcome the retentivity is called the *coercive force*. If the current is increased below this to a value corresponding to $-H_{\max}$ and then returned to zero, we find a curve such as shown in Fig. 4-17b. If the current is taken back to the original value in the positive direction, a different curve is obtained. Hence we conclude that it is necessary to know the past history of a piece of iron in order to predict its performance.

If the current is alternated between definite extreme values, the curve settles into a definite closed loop called a *hysteresis loop*. A particularly important type of hysteresis loop is the *square* hysteresis loop shown in Fig. 4-17c. This type of hysteresis loop is characteristic of the ferromagnetic ceramic material used to construct the basic storage device in a magnetic-core memory of a digital computer. It is typical also of the material used to construct microwave ferrite devices.

In order to make the best estimate of μ, we utilize a mean magnetization curve, which can be obtained by starting with an unmagnetized piece of iron and increasing the alternating current gradually. Thus a series of hysteresis loops are obtained (Fig. 4-17d). A curve is then drawn through the peaks of these curves. Examples are shown in Fig. 4-18. We should point out that the ratio B/H along the mean magnetization curve is the permeability of the iron and that it varies with B and current.

Let us now consider the determination of the total flux in the simple configuration shown in Fig. 4-19a. Magnetic flux, like current in an electric circuit, seeks the "path of least resistance" (high permeability). If all the flux is in the iron, we need to set the line integral of **H** around a closed path in the iron equal to NI. We know, of course, that the result of the integration is NI for all paths around the iron loop, but we wish to determine **B**. The line integral of **H** around the two paths shown in the figure, that is, L_1 and L_2, will be the same, although **B** and **H** are different along the two paths. We wonder what path will yield the value of **B** which will give the total flux when multiplied by the area, that is, the average flux density. For most configurations a rigorous solution is almost impossible. For those cases where only engineering accuracy is desired, it turns out that the path along the middle of the iron gives the desired answer.

Under this assumption, the line integral of **H** around the closed magnetic

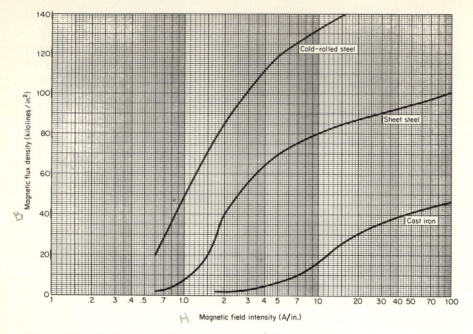

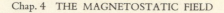

FIGURE 4-18.   Typical magnetization curves.

circuit becomes a summation of the average value of H times the mean path length; that is,

$$\oint_C \mathbf{H} \cdot d\mathbf{l} = \sum H_{av} \, l_{mean} = NI \tag{4-112}$$

For Fig. 4-19a, $l_{mean} = 2\pi R$, where R denotes the mean radius, so that, by Ampère's circuital law,

$$H_{av} = \frac{NI}{2\pi R} \tag{4-113}$$

For Fig. 4-19b, $l_{mean} = L_1 + L_2 + L_3 + L_4$; that is, we take a path from corner to corner, going down the center of each leg to obtain the mean length.   Hence

$$H_{av} = \frac{NI}{L_1 + L_2 + L_3 + L_4} \tag{4-114}$$

is the average magnetic field intensity along the mean path.   Since the magnetization curve for the metal of the core is also a plot of B_{av} versus H, we can obtain B_{av} from the magnetization curve.

The products $H_{av}L_1, H_{av}L_2, \ldots, H_{av}L_n$, where L_n is the mean path of the magnetic circuit's nth segment, are simply the differences in the values

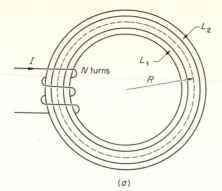

(a)

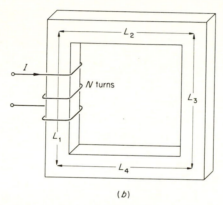

(b)

FIGURE 4-19.   Simple magnetic circuit configurations.   (a) Plane section of a toroid;
(b) a rectangular core.

assumed by the scalar potential function ϕ^* at the extremities of each path.
They are usually designated by the script letter $\mathscr{F}$ and, as noted earlier, are
called *magnetomotive forces*, or mmfs, for short.   In terms of this notation,
then, Eq. (4-112) reads

$$\mathscr{F}_1 + \mathscr{F}_2 + \mathscr{F}_3 + \mathscr{F}_4 = NI \tag{4-115}$$

or in general,

$$\sum_n \mathscr{F}_n = NI \tag{4-116}$$

This equation expresses a general law in magnetic circuit analysis, which is
analogous to Kirchhoff's voltage law and which effectively states that around
any closed path in the magnetic circuit, the algebraic sum of all the mmfs
acting on the circuit is equal to the net current enclosed by the path.

The second law of the magnetic circuit, which is analogous to Kirchhoff's current law, states that the net flux entering a junction in a magnetic

circuit equals the net flux leaving the junction; that is,

$$\sum_{n} \Phi_n = 0 \tag{4-117}$$

where the summation is performed algebraically. This law is a direct consequence of the continuity property ($\mathbf{\nabla} \cdot \mathbf{B} = 0$) of the magnetic flux density vector.

A useful concept is that of magnetic *reluctance*. The reluctance of any path in a magnetic circuit is defined

$$R = \frac{\mathscr{F}}{\Phi} \tag{4-118}$$

where R = reluctance, A/Wb
$\quad \mathscr{F}$ = mmf drop along the path, At
$\quad \Phi$ = total flux, Wb

If the path is l m long, has an area of A m², and a permeability μ H/m, then $B = \Phi/A$, $H = B/\mu$, and

$$\mathscr{F} = Hl = \frac{B}{\mu} l = \frac{\Phi}{A\mu} l \tag{4-119}$$

Therefore reluctance can be expressed algebraically in a form analogous to resistance (Chap. 5) as

$$R = \frac{l}{\mu A} \tag{4-120}$$

Of course, μ generally depends upon H, and hence the reluctance R is not a constant.

It is sometimes helpful to draw a comparison between magnetic circuit quantities and electric circuit quantities. This comparison can be summarized as follows:

Magnetic Circuit	DC Circuit
mmf $\mathscr{F}$	Voltage, emf, V
Flux Φ	Current I
Reluctance $R = \dfrac{\mathscr{F}}{\Phi}$	Resistance $R = \dfrac{V}{I}$
Flux density $B = \dfrac{\Phi}{A}$	Current density $J = \dfrac{I}{A}$
Permeability	Conductivity
Σ mmfs $= NI$	Kirchhoff's voltage law
$\Sigma \Phi$'s $= 0$	Kirchhoff's current law

Magnetic-circuit computations in English-speaking countries are usually carried out in the so-called *mixed English system* of units. These units, as well as the corresponding units in the rationalized mksc system, are tabulated below.

Quantity	Mixed English	Rationalized mksc
$\mathscr{F}$	ampere-turn	ampere-turn
Φ	line, or kiloline	weber ($= 10^8$ lines)
B	line per square inch, or kiloline per square inch	weber per meter squared ($= 64.5$ kilolines/in.^{2})
H	ampere-turn per inch	ampere-turn per meter ($= 0.0254$ At/in.)
μ_0	3.19×10^{-3} kiloline per ampere-inch	$4\pi \times 10^{-7}$ henry per meter

A useful relation to remember is that, in free space,

$$H = 313B \qquad \text{At/in.} \qquad (4\text{-}121)$$

where B is measured in kilolines per square inch.

Before proceeding with the analysis of a few typical magnetic circuits, it should be pointed out that many magnetic circuits include one or more air elements, called *air gaps*, as integral parts of the circuit. A typical example of such a magnetic circuit is shown in Fig. 4-20a. The flux will tend to spread out in the air gap much in the manner portrayed by Fig. 4-20b. Correction for such fringing may be made by basing the calculation of effective area on the formula

$$A = (b + t)(a + t) \qquad (4\text{-}122)$$

In other words, the gap length t is added to each of the two transverse dimensions of the air gap.

It should be noted that magnetic cores of physically large electrical devices are usually made from *laminations* which are punched out of thin sheets of ferromagnetic material. The reason for this type of construction is to be found in the electrical properties of matter. When flux varies with time, voltages are induced in the cores, in accordance with Faraday's law, which in turn cause the circulation of *eddy* currents in the cores; power is then lost in the form of heat. However, when the laminations are employed, they are always insulated from each other by thin layers of nonconducting material. The resistance offered to the flow of eddy currents is increased, thus decreasing the power lost. As a result of the space occupied by the insulating layers, the cross-sectional area of the actual magnetic path is effectively reduced. To account for this reduction, a quantity known as the *stacking factor* of the laminations is multiplied by the actual area of the core. That is, if the physical dimensions of the core are a and b in., the magnetic cross-sectional area is set equal to ab times the stacking factor (for example, $0.9ab$).

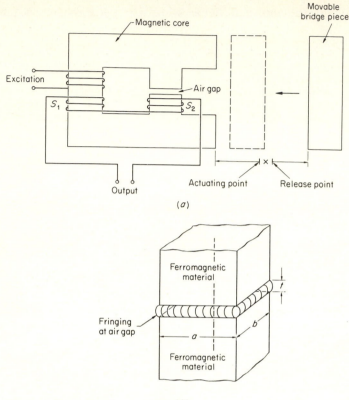

(a)

(b)

FIGURE 4-20.   Magnetic circuit with air gap.   (a) A variable-reluctance switch consisting of a wire-wound magnetic core and movable bridge piece that alters the core flux pattern to produce an electric output useful for switching purposes. (*Courtesy of the National Aeronautics and Space Administration.*)   (b) Fringing of magnetic flux at an air gap.

The most satisfactory way of demonstrating the application of the foregoing concepts is by means of a few illustrative examples.

Example 4-12   A Series Magnetic Circuit.   In the magnetic circuit of Fig. 4-19b, $L_1 = L_3 = 7$ in. and $L_2 = L_4 = 8$ in. The core is made of cast iron (stacking factor $= 1$), and has a square cross section of 2×2 in. If the flux in the core is 120,000 lines, find the ampere-turns of the coil.

We proceed as follows:

$$\text{Core area } A = 2 \times 2 = 4 \text{ in.}^2$$

$$\text{Mean core length } l = 2(7 + 8) = 30 \text{ in.}$$

$$\text{Flux density } B = 120,000/4 = 30,000 \text{ lines/in.}^2$$

Entering the graph (Fig. 4-18) at this value of B, we read off the cast-iron curve $H = 20$ At/in. Therefore the mmf in the core is $Hl = 20 \times 30 = 600$ At. And according to Ampère's circuital law, $NI = 600$ At.

The inverse problem is that of finding the flux in the core for a given number of ampere-turns. Suppose, for example, the coil has 1200 At. Then $H = 1200/30 = 40$ At/in., and from the magnetization curve, $B = 40$ kilolines/in.^{2} Therefore $\Phi = 40 \times 4 = 160$ kilolines.

It is instructive at this point to insert a few observations:

1. Since the *B-H* relationship is nonlinear, the problem of magnetic circuit analysis is a problem in *nonlinear analysis*. So the techniques which will be demonstrated here are quite general, and may be carried over to other types of nonlinear problems.
2. In the preceding example, the flux was assumed to be constant everywhere in the *series* magnetic circuit. This is a definite constraint on the problem. We shall find that different constraints are imposed on different types of magnetic circuits.
3. If the cross section of the core is not uniform throughout its length, the flux density B, and hence H, will be different at different portions of the series magnetic circuit; however, the flux will still have the same value everywhere in the core.
4. The relation of flux direction to that of current flow can often be established by the right-hand rule if, as in Example 4-12, the winding directions are known. However, in more complex arrangements, it may be impossible to establish by inspection the direction of flux in a given leg of the magnetic circuit, even though winding and current directions may both be known. In such instances, flux directions in various parts of the circuit are assigned arbitrarily. Then, if at the conclusion of the solution the algebraic sign of the flux in a particular leg is positive, the initially assumed direction is correct; but if the algebraic sign of that flux is negative, the actual direction is opposite to what was initially assumed.
5. Note that the abbreviation At for ampere-turn is used in this section and hereafter.

Example 4-13 Series Circuit with Air Gap. Figure 4-21 shows a magnetic circuit constructed of sheet steel. The cross section of the core is uniform and measures 2×2 in. Compute the number of ampere-turns required to produce a flux of 252 kilolines in the circuit. Assume a stacking factor of 0.9 for the laminations.

This is a series circuit in which the total mmf is the sum of the mmf in the iron plus the mmf in the air gap. The details of the calculations are

Core:
$$\text{Area} = 2 \times 2 \times 0.9 = 3.6 \text{ in.}^2$$
$$\text{Length} = 30 \text{ in.}$$
$$B = 252/3.6 = 70 \text{ kilolines/in.}^2$$

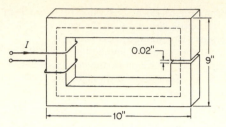

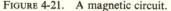

FIGURE 4-21.   A magnetic circuit.

From the graph:
$$H = 5 \text{ At/in.}$$
$$Hl = 5 \times 30 = 150 \text{ At}$$

Air gap:
$$\text{Area} = (2 + 0.02)^2 = 4.08 \text{ in.}^2$$
$$\text{Length} = 0.02 \text{ in.}$$
$$B = 252/4.08 = 61.8 \text{ kilolines/in.}^2$$
$$H = 313 \times 61.8 = 19,343 \text{ At/in.}$$
$$Hl = 19,343 \times 0.02 = 387 \text{ At}$$

Therefore the total mmf drop in the magnetic circuit is $150 + 387 = 537$ At. And in accordance with the mmf law, $NI = 537$ At.

The inverse problem of finding the flux resulting from, say, 270 At is considerably more difficult.   Because of the nonlinearity of the magnetization curve, it is impossible to determine the flux by linear-analysis methods.   The solution of the problem can be effected, however, by trial and error; that is, the flux in the core is assigned arbitrarily, and the mmf required to produce this flux is found by the previous method. If this mmf agrees with the specified ampere-turns, the solution of the problem is terminated.   Otherwise the magnetic flux is assigned different values successively until one is found which requires the specified excitation.   Accuracy to within 5 percent is acceptable for our purposes.

A second method of solution is graphical, and will be illustrated later.

The first trial is such that the assumed flux falls within the range of the magnetization curves.   In the problem at hand, a more accurate estimate of flux can be made with the aid of two extreme assumptions:

1.   The entire mmf is used in the air gap.
2.   The entire mmf is used in the core.

Either of these assumptions leads to a value of flux which is higher than the actual one.   Therefore a value lower than either of these should be selected as a first trial.

If the specified 270-At mmf is used in its entirety in the air gap, then
$$H = 270/0.02 = 13,500 \text{ At/in.}$$
$$B = 13,500/313 = 43.1 \text{ kilolines/in.}^2$$
$$\Phi = 4.08 \times 43.1 = 176 \text{ kilolines}$$

If it is used in the core,
$$H = 270/30 = 90 \text{ At/in.}$$
$$B = 100 \text{ kilolines/in.}^2$$
$$\Phi = 100 \times 3.6 = 360 \text{ kilolines}$$

As a first trial we take $\Phi = 170$ kilolines.   Next we compute the total mmf.

Core:
$$B = 170/3.6 = 47.2 \text{ kilolines/in.}^2$$
$$H = 2.3 \text{ At/in.}$$
$$Hl = 2.3 \times 30 = 69 \text{ At}$$

Air gap:
$$B = 170/4.08 = 41.6 \text{ kilolines/in.}^2$$
$$H = 313 \times 41.6 = 13{,}000 \text{ At/in.}$$
$$Hl = 13{,}000 \times 0.02 = 260 \text{ At}$$
$$\text{Total mmf} = 69 + 260 = 329 \text{ At}$$

Since the difference, $329 - 270 = 59$, is higher than 5 percent of 270, a second, lower value of Φ must be assumed.   We try $\Phi = 140$ kilolines.   Then, as before,

Core:
$$B = 140/3.6 = 38.9 \text{ kilolines/in.}^2$$
$$H = 2 \text{ At/in.}$$
$$Hl = 2 \times 30 = 60 \text{ At}$$

Air gap:
$$B = 140/4.08 = 34.3 \text{ kilolines/in.}^2$$
$$H = 313 \times 34.3 = 10{,}700 \text{ At/in.}$$
$$Hl = 10{,}700 \times 0.02 = 214 \text{ At}$$
$$\text{Total mmf} = 214 + 60 = 274 \text{ At}$$

This value of the total mmf is, for our purposes, sufficiently close to the given excitation.   So the solution of our problem is now complete.

As previously stated, a direct graphical analysis requiring no guessing is also possible.   Let us refer once again to Fig. 4-21.   Ampère's circuital law requires that, around the magnetic circuit, the algebraic sum of the mmf drops be equal to zero, or

$$\text{mmf}_{\text{core}} + \text{mmf}_{\text{air gap}} = NI$$

From Eq. (4-121),

$$\text{mmf}_{\text{air gap}} = H_{\text{air gap}} l_{\text{air gap}} = (313 B_{\text{air gap}})(0.02)$$

The same flux exists in both the core and in the air gap.   However, because of fringing, the flux density in the air is smaller than in the core.   More precisely,

$$B_{\text{air gap}} = \frac{2 \times 2 \times 0.9}{(2 + 0.02)(2 + 0.02)} B_{\text{core}} = \frac{3.6}{4.08} B_{\text{core}}$$

So
$$\text{mmf}_{\text{air gap}} = 5.52 B_{\text{core}}$$

On the other hand,
$$\text{mmf}_{\text{core}} = H_{\text{core}} l_{\text{core}} = 30 H_{\text{core}}$$

Therefore the equation of equilibrium becomes

$$30 H_{\text{core}} + 5.52 B_{\text{core}} = 270$$

This is a constraint imposed by the magnetic circuit.   A second constraint is imposed by the material of the core in the form of a nonlinear relationship, shown in Fig. 4-18.

The variables of the problem must satisfy both conditions simultaneously. Hence the point of intersection of the proper magnetization curve with the plot of the last equation marks the values of H_{core} and B_{core}, which satisfy both constraints at the same time. If the magnetization curves were plotted on linear scales, the plot of the equation of equilibrium would be a straight line and, hence, easy to draw. In this case, the plot must be made point by point. The student can verify for himself that the two curves cross at $B = 38.5$ kilolines/in.^{2}, $H = 1.95$ At/in.

Example 4-14 Incremental Inductance. The topic of incremental inductance actually belongs under time-varying fields. For present purposes, however, let us assume that the field quantities vary so slowly with time that, to a good approximation, the magnetostatic field equations apply.

Consider a magnetic circuit with two coils, as in Fig. 4-22a. Let I_2 be a fixed current and i_1 be a slowly varying current whose amplitude is at all times very small compared with I_2. Then it is reasonable to assume that the effect of coil 2 is to establish a constant value of incremental permeability (Sec. 2.7) for the core. This value of μ_{inc} is the slope $\Delta B/\Delta H$, at the value of H fixed by coil 2, neglecting coil 1.

Now any problem concerning coil 1 can be worked by assuming μ_{inc} to have the constant value established by coil 2. Assuming that $N_1 = 1000$ turns, $N_2 = 2000$ turns, core length $l = 10$ in., core area $= 1$ in.^{2}, and the core is made of sheet steel, determine the value of I_2 required to give coil 1 an incremental inductance of 3.5 H.

By definition,

$$L_{inc} = N_1 \frac{\Delta \Phi}{i_1}$$

where $\Delta \Phi$ denotes the increment in total flux resulting from the flow of the current i_1 in coil 1. Since $B = \Delta \Phi/1$,

$$L_{inc} = N_1 \frac{\Delta B}{i_1}$$

Multiplying numerator and denominator of this expression by the factor N_1/l gives

$$L_{inc} = N_1 \frac{\Delta B(N_1/l)}{i_1(N_1/l)} = \frac{N_1^2}{l} \frac{\Delta B}{\Delta H}$$

Now, we need

$$L_{inc} = 3.5 \text{ H} = 3.5 \text{ Wb/A} = 3.5 \times 10^5 \text{ kilolines/A}$$

Hence

$$3.5 \times 10^5 = \frac{(1000)^2}{10} \frac{\Delta B}{\Delta H}$$

and

$$\frac{\Delta B}{\Delta H} = 3.5 \text{ kilolines/At} \cdot \text{in.}$$

From Fig. 4-18 it is seen that the slope of the sheet-steel magnetization curve has this value in the neighborhood of $H = 6$ At/in. Therefore

$$\frac{N_2 I_2}{l} = \frac{2000 I_2}{10} = 6$$

from which it follows that $I_2 = 0.03$ A.

This example is actually a simplification of the true problem. Ferromagnetic substances exhibit dynamic nonlinearities of the type illustrated in Fig. 4-22c. The minor hysteresis loop shown superimposed on a relatively large hysteresis loop is the

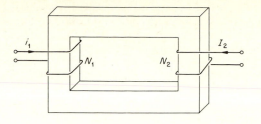

(a)

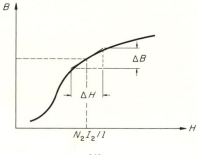

(b)

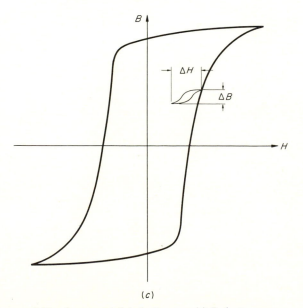

(c)

FIGURE 4-22. Incremental inductance. (a) Inductor configuration; (b) illustrating the slope $\Delta B/\Delta H$; (c) minor
hysteresis loop.

result of a small time-varying current superimposed on a relatively large direct current. Exact analysis of the problem requires information about the dynamic behavior of the material. However, this information is not readily available, and consequently the approximate value of μ_{inc} obtained as the slope of the magnetization curve is frequently used. The true incremental permeability is usually less than the slope of the mean magnetization curve.

To summarize, this section has been devoted to a development of analytical techniques applicable to ferromagnetic circuits. Because of the many simplifying assumptions, the results obtained by such techniques are only approximate.

4.10   The Boundary-value Problem in Magnetostatics.

Let us consider once again the scalar magnetic potential, discussed earlier in Sec. 4.8. It was then shown that in a current-free region the magnetic field intensity vector may be derived from the gradient of the scalar magnetic potential:

$$\mathbf{H} = -\nabla\phi^* \tag{4-123}$$

If the region we are considering happens to be linear, homogeneous, and isotropic, the permeability μ is a scalar constant. Under these conditions, the pair

$$\mathbf{B} = \mu\mathbf{H} \qquad \nabla \cdot \mathbf{B} = 0 \tag{4-124}$$

implies that

$$\nabla \cdot \mathbf{H} = 0 \tag{4-125}$$

and in conjunction with Eq. (4-123),

$$\nabla^2\phi^* = 0 \tag{4-126}$$

In every current-free region whose magnetic properties are independent of position and of level and direction of excitation, the magnetic potential ϕ^* is thus seen to satisfy Laplace's equation, and the methods of Chap. 3 can be applied directly.

Example 4-15   Sphere in a Uniform Field.   A sphere of radius a and permeability μ is in a magnetic field which, far from the sphere, is uniform. The magnetic field inside and outside the sphere is desired.

Since the entire region is current-free and simply connected, the magnetic field may be derived from a single-valued scalar potential which satisfies Laplace's equation. Therefore, if the center of the sphere is made to coincide with the origin of a spherical system of coordinates (Fig. 4-23), the problem is that of finding a solution to the boundary-value problem in which:

$\nabla^2\phi^* = 0$ everywhere.

ϕ^* is continuous at $r = a$.

The normal component of the magnetic flux density is continuous at $r = a$.

$\phi^* = H_0 r \cos\theta$, $r \gg a$, where H_0 is a constant.

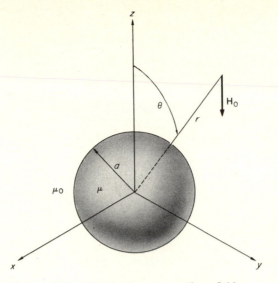

FIGURE 4-23.   Metal sphere in a uniform field.

The solution to a similar problem (sphere in a uniform electric field) was obtained in Example 3-9.   Following the same general procedure, we find that admissible solutions [Eq. (3-173)] for the magnetic potential in the sphere and in the free region outside are, respectively,

$$\phi^* = A + \frac{B}{r} + Cr\cos\theta + \frac{D}{r^2}\cos\theta$$

$$\phi_0^* = A_0 + \frac{B_0}{r} + C_0 r\cos\theta + \frac{D_0}{r^2}\cos\theta$$

From the boundary condition at $r \gg a$, we obtain $A_0 = 0$, $C_0 = H_0$.   To avoid infinite potentials at the origin ($r = 0$), we must take $B = D = 0$.   So

$$\phi^* = A + Cr\cos\theta$$

$$\phi_0^* = \frac{B_0}{r} + H_0 r\cos\theta + \frac{D_0}{r^2}\cos\theta$$

Now, continuity of the normal component of the magnetic flux density vector at the boundary $r = a$ requires that

$$\mu\frac{\partial\phi^*}{\partial r}\bigg|_{r=a} = \mu_0\frac{\partial\phi_0^*}{\partial r}\bigg|_{r=a}$$

that is,

$$\mu(C\cos\theta) = \mu_0\left(-\frac{B_0}{a^2} + H_0\cos\theta - \frac{2D_0}{a^3}\cos\theta\right)$$

Hence $B_0 = 0$, and

$$\mu C = \mu_0\left(H_0 - \frac{2D_0}{a^3}\right) \tag{4-127}$$

On the other hand, continuity of ϕ^* itself at $r = a$ requires that

$$A + Ca \cos \theta = H_0 a \cos \theta + \frac{D_0}{a^2} \cos \theta$$

Hence $A = 0$, and

$$Ca = H_0 a + \frac{D_0}{a^2} \qquad (4\text{-}128)$$

Simultaneous solution of Eqs. (4-127) and (4-128) gives

$$C = \frac{3\mu_0}{2\mu_0 + \mu} H_0 \qquad D_0 = a^3 \frac{\mu_0 - \mu}{2\mu_0 + \mu} H_0$$

so that

$$\phi^* = H_0 \frac{3\mu_0}{2\mu_0 + \mu} r \cos \theta \qquad (4\text{-}129)$$

$$\phi_0^* = H_0 \left[1 + \left(\frac{a}{r}\right)^3 \frac{\mu_0 - \mu}{2\mu_0 + \mu} \right] r \cos \theta \qquad (4\text{-}130)$$

This pair of equations is of the same form as the answers to Prob. 3-14.   Uniqueness of the solution is assured by the theorems proved in Sec. 3.11.

The magnetic field intensity within the sphere is

$$\mathbf{H} = -\nabla\phi^* = H_0 \frac{3\mu_0}{2\mu_0 + \mu} (-\cos \theta \, \mathbf{a}_r + \sin \theta \, \mathbf{a}_\theta)$$

$$= -H_0 \frac{3\mu_0}{2\mu_0 + \mu} \mathbf{a}_z \qquad r \leq a \qquad (4\text{-}131)$$

from which it is clear that the sphere is uniformly magnetized in one direction. Outside the sphere,

$$\mathbf{H}_0 = -\nabla\phi_0^* = -H_0 \left[1 - 2\left(\frac{a}{r}\right)^3 \frac{\mu_0 - \mu}{2\mu_0 + \mu} \right] \cos \theta \, \mathbf{a}_r$$

$$+ H_0 \left[1 + \left(\frac{a}{r}\right)^3 \frac{\mu_0 - \mu}{2\mu_0 + \mu} \right] \sin \theta \, \mathbf{a}_\theta \qquad r > a \qquad (4\text{-}132)$$

which, for $r \gg a$, reduces to a constant value

$$\mathbf{H}_0 \approx H_0(-\cos \theta \, \mathbf{a}_r + \sin \theta \, \mathbf{a}_\theta) = -H_0 \mathbf{a}_z \qquad (4\text{-}133)$$

as required.

For $\mu \gg \mu_0$, the field becomes

$$\phi^* \approx 0 \qquad \mathbf{H} \approx 0$$

$$\phi_0^* \approx H_0 \left[1 - \left(\frac{a}{r}\right)^3 \right] r \cos \theta \qquad (4\text{-}134)$$

$$\mathbf{H}_0 \approx -H_0 \left[1 + 2\left(\frac{a}{r}\right)^3 \right] \cos \theta \, \mathbf{a}_r + H_0 \left[1 - \left(\frac{a}{r}\right)^3 \right] \sin \theta \, \mathbf{a}_\theta$$

These expressions show a clear similarity to those obtained in Example 3-9 for a metal sphere in a uniform electric field.   Inside the sphere the magnetic field intensity is essentially zero when the permeability is high.   However, the magnetic flux density in the sphere remains nonzero and finite because

$$\lim_{\mu \to \infty} B = \lim_{\mu \to \infty} \mu H = \lim_{\mu \to \infty} H_0 \frac{-3\mu\mu_0}{2\mu_0 + \mu} \mathbf{a}_z = -3\mu_0 H_0 \mathbf{a}_z \qquad (4\text{-}135)$$

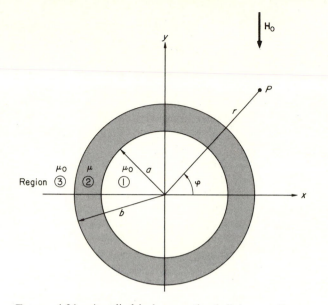

FIGURE 4-24. A cylindrical magnetic shell in a uniform field.

Example 4-16 Magnetic Shield. As a second example, let us compute the magnetic shielding in air afforded by a cylindrical shell of radii a and b and permeability μ.

A $-y$-directed uniform magnetic field H_0 (that is, $\phi^* = H_0 r \sin \varphi$) exists in an extended region of space. Into this uniform field is introduced a long cylindrical magnetic shell of permeability μ. We wish to determine the field within the shell, assuming that the axis of the cylinder is coincident with the z axis of a cylindrical coordinate system (Fig. 4-24).

The entire region is current-free. Hence a magnetic potential exists everywhere and satisfies Laplace's equation. Boundary conditions are

$$\phi_3^* = H_0 r \sin \varphi \qquad r \gg b$$

$$\phi_3^* = \phi_2^* \quad \text{and} \quad \mu_0 \frac{\partial \phi_3^*}{\partial r} = \mu \frac{\partial \phi_2^*}{\partial r} \qquad \text{at} \qquad r = b$$

$$\phi_2^* = \phi_1^* \quad \text{and} \quad \mu \frac{\partial \phi_2^*}{\partial r} = \mu_0 \frac{\partial \phi_1^*}{\partial r} \qquad \text{at} \qquad r = a$$

Following the techniques developed in Chap. 3 (see Prob. 4-19), we obtain for the field in region 1

$$\phi_1^* = \frac{4\mu_0 \mu H_0}{(\mu + \mu_0)^2 - (a^2/b^2)(\mu - \mu_0)^2} r \sin \varphi \qquad (4\text{-}136)$$

$$\mathbf{H}_1 = -\nabla \phi_1^* = -\frac{4\mu_0 \mu H_0}{(\mu + \mu_0)^2 - (a^2/b^2)(\mu - \mu_0)^2} \mathbf{a}_y \qquad (4\text{-}137)$$

For $\mu \gg \mu_0$, the field intensity becomes

$$\mathbf{H}_1 \approx -\frac{4(\mu_0/\mu)H_0}{1 - (a^2/b^2)}\, \mathbf{a}_y \tag{4-138}$$

and is seen to vanish with increasing μ.   As in Example 4-15, $\lim\limits_{\mu \to \infty} B_1 \neq 0$.

The foregoing development, together with the subsequent examples, clearly shows that within a current-free ($\mathbf{J} = 0$) region, the magnetostatic boundary-value problem is mathematically equivalent to an electrostatic boundary-value problem.   This equivalence is reflected by the identical forms of the equations which describe the two types of fields.   Therefore the same methods discussed in Chap. 3 for solving Laplace's equations apply.

In case the region is not current-free, the magnetostatic boundary-value problem is not amenable to solution through the scalar potential ϕ^*, and recourse must be taken to the magnetic vector potential $\mathbf{A}$ or to the field vectors themselves.

Example 4-17   Magnetic Field within a Current-carrying Wire.   As an example of the formal solution using the magnetic vector potential, let us take a long wire, of radius a, carrying a longitudinal current along the z axis of a cylindrical coordinate system (Fig. 4-6a).   We suppose that $\mathbf{J} = J\mathbf{a}_z$.   In this case, the relation $\nabla^2 A = -\mu\mathbf{J}$, written in cylindrical coordinates, becomes

$$\frac{d^2 A_z}{dr^2} + \frac{1}{r}\frac{dA_z}{dr} = -\mu J \tag{4-139}$$

Because of cylindrical symmetry, the current density vector $\mathbf{J}$ is at most r-dependent. In fact, experience suggests that J might be constant throughout the wire cross section. Therefore, assuming that J is independent of r, Eq. (4-139) gives

$$A_z = C_1 + C_2 \ln r - \frac{\mu J}{4} r^2$$

The boundary conditions are that the potential must be finite at $r = 0$; in fact, we can arbitrarily set $A_z = 0$ at $r = 0$, and thus obtain $C_1 = C_2 = 0$, and

$$\mathbf{A} = -\frac{\mu J}{4} r^2 \mathbf{a}_z$$

From this, we have

$$\mathbf{B} = \nabla \times \mathbf{A} = \frac{\mu J}{2} r\mathbf{a}_\varphi$$

If the wire is carrying a total current I, then $J = I/\pi a^2$.   Therefore, in terms of I,

$$\mathbf{B} = \frac{\mu I r}{2\pi a^2}\, \mathbf{a}_\varphi$$

and

$$\mathbf{H} = \frac{I r}{2\pi a^2}\, \mathbf{a}_\varphi \tag{4-140}$$

Now, as shown in Example 2-2, the field outside the conductor is

$$\mathbf{H} = \frac{I}{2\pi r}\,\mathbf{a}_\varphi \qquad (4\text{-}141)$$

so long as the current distribution is symmetric about the axis of the conductor. At $r = a$, Eqs. (4-140) and (4-141) reduce to the same result, namely,

$$\mathbf{H} = \frac{I}{2\pi a}\,\mathbf{a}_\varphi \qquad (4\text{-}142)$$

thus guaranteeing the continuity of tangential **H**. The uniqueness of the solution of Poisson's equation, proved in Sec. 3.11, assures the validity of these results, and confirms that our initial *guess* of a uniform current distribution was indeed correct.

To summarize, within a current-free ($\mathbf{J} = 0$) region, the magnetostatic field may be derived from a scalar potential ϕ^*, which is finite and continuous everywhere and satisfies Laplace's equation. For such regions the magneto-static boundary-value problem is mathematically equivalent to an electro-static boundary-value problem, and is amenable to solution by the same analytical techniques. In case the region is not current-free, recourse is taken to the vector potential **A**, which is finite and continuous everywhere and subject to constraints imposed by the equations $\nabla^2\mathbf{A} = -\mu\mathbf{J}$ and $\nabla \cdot \mathbf{A} = 0$. In certain cases it is found advantageous to work directly with the field vectors themselves, rather than through the vector potential **A** (Prob. 4-23).

4.11 Synthesis of Magnetostatic Fields. The synthesis problem in magnetostatics is the problem of finding volume ($\mathbf{J} = \nabla \times \mathbf{H}$) and surface ($\mathbf{K} = \mathbf{n} \times \mathbf{H}$) distributions of current which will give rise to a desired magnetic field. That is to say, the synthesis problem is really the design problem. Only in a few instances can an exact solution be obtained. Moreover, the design problem is always complicated by extraneous considerations, such as cost, weight, size, and time limitations, and by such esoteric considerations as experience, intuition, and divine inspiration, which usually compromise the purely technical aspects of the problem and make the exact solution in-appropriate. The extreme complexity of the practical design problem created by the combination of extraneous, esoteric, and technical consider-ations defies a short presentation of real-life design procedures, but poses no serious difficulty for the experienced individual who is competent in analysis techniques.

4.12 Summary. This chapter has dealt with magnetostatic fields. Every magnetostatic field is a solenoidal field; this means that the divergence of the magnetic flux density vector is always zero.

We showed that a vector potential, which is finite and continuous every-where, including at the boundaries, can be associated with every magneto-static field, and that this potential satisfies a vector Poisson equation in the general case, or a vector Laplace equation in the restricted, current-free case. Both equations follow directly from Maxwell's equations. The relation of this vector potential to its sources (direct currents) is of the same form as the relation of the electrostatic potential to its sources (stationary charges).

The magnetic flux density vector may be determined indirectly from the curl of the vector potential, or directly from the Biot-Savart law.

The magnetostatic force density expression is a special case of the general Lorentz force density expression. In this connection, we stressed the point that **B** is the *total* field due to *all* currents.

Then we defined inductance as flux linkages divided by current. Self-inductance specifies the magnitude of the magnetic coupling of a circuit (closed conducting loop) with itself. Mutual inductance specifies the magnitude of the magnetic coupling of one circuit with a second circuit. Inductance of a particular structure can be calculated from a knowledge of the magnetostatic field associated with the structure, or by means of the Neumann formula.

We showed that a scalar magnetic potential exists in every current-free ($\mathbf{J} = 0$) region. This potential is discontinuous in the region if a net non-zero current is enclosed by at least one closed contour in the region. Hence, in general, the value of this potential at a point is not unique, and the magnetic potential difference at two points, that is, the mmf, is therefore dependent on the particular path by which the points are connected. (In contrast, the value of the electrostatic scalar potential is always unique except at surface dipole layers, and potential difference in the electric case is always independent of path.) The scalar magnetic potential is useful in the solution of ferromagnetic circuit problems. It is also useful as a tool in the analysis of magnetostatic boundary-value problems when the current density vector is everywhere zero. This type of boundary-value problem is mathematically equivalent to an electrostatic boundary-value problem because the same types of equations apply. When the region is not free of current, the magnetostatic boundary-value problem must be formulated in terms of either the vector potential or the field vectors themselves.

Problems

4-1   Field of a Rectangular Loop.   Find the expression for the magnetic flux density **B** at a point a distance h above the center of a rectangular loop of wire d m on one side and a m on the other side. The loop carries a current of 1 A.

4-2   Field of a Solenoid.   Derive Eq. (4-69) using Ampère's circuital law.

4-3   Single-layer Coil.   In order to acquire an appreciation for the magnitude of the currents needed to obtain a strong magnetic field, calculate the current which will produce a field of 1 gauss ($1\ \mathrm{G} = 10^{-4}\ \mathrm{Wb/m^2}$) at the center of a long single-layer solenoid suspended in air. Assume that the outside diameter of the wire used is approximately 50 mils

(1 mil $= 10^{-3}$ in.) and that the separation between adjacent turns is essentially zero. Does your answer appear to be practical?

4-4 Magnetic Field of a Current Distribution. A circular disk of radius b and thickness t carries a current in a circular direction about the center. A cylindrical coordinate system has the origin at the center of the disk and has the z axis coinciding with the axis of symmetry of the disk. The current density in the disk is $\mathbf{J} = kr\mathbf{a}_\varphi$ A/m², where k is a real constant greater than zero. Derive a formula for the magnetic intensity $\mathbf{H}$ at the center of the disk, assuming that the thickness t is very small compared with b.

4-5 Magnetic Field of a Current Distribution. Two identical *pancake* coils share a common axis and carry equal currents in the same direction. Assuming that the radius a is much smaller than the distance of separation d, determine the magnetic field in the plane of symmetry, $z = d/2$.

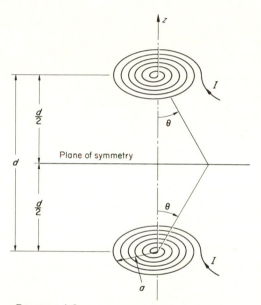

PROBLEM 4-5

4-6 Superposition. The figure shows a long straight wire carrying a current of 10 A. Specify a configuration of wire, or wires, and the current in this wire, or wires, so that the **B** field at the point P due to the long straight wire and the added wire(s) is zero.

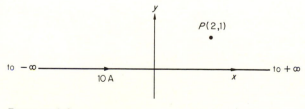

PROBLEM 4-6

4-7   Superposition.   A straight circular conductor of radius a contains an eccentric hole of radius $b(b < a/4)$.   The axes of the hole and conductor are parallel and a distance D apart.   A direct current of uniform density J_0 flows along the conductor parallel to its axis.   Determine the magnetic field intensity at any point in the hole.

4-8   Convergence.   It was shown in Example 4-2 that **B** is finite even when $\alpha = -\beta = \pi/2$ [Eq. (4-24)].   Yet the vector potential is undefined for an infinite line.   How do you explain this apparent discrepancy?

4-9   Magnetostatic Force.   A very long straight conductor, carrying a current I, is contained in the plane of a triangular conductor, which is carrying a current I', so that one side of the triangle is parallel to the straight conductor, as shown in the figure.   Determine the mutual force between the two conductors.

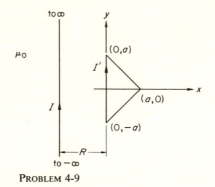

PROBLEM 4-9

4-10   Flux Linkages.   A current I flows in a long straight wire as shown.   Calculate the flux linked by the coplanar conducting loop.

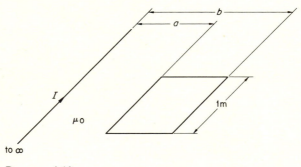

PROBLEM 4-10

4-11   Induced Voltage.   A single long straight conductor carries a current which may be expressed as $i = 100 \sin 400t$ A, where t is time.   Determine the voltage induced in a unit length of a two-conductor telephone line located as shown in the sketch.   Assume that the theory of static magnetic fields gives reasonably accurate results at $400/2\pi$ Hz (Chap. 11).

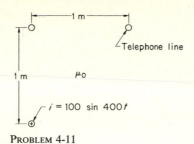

PROBLEM 4-11

4-12 Magnetic Force. An initially unmagnetized sphere of permeability μ and mass m is introduced in the center of the air gap of a permanent magnet. The radius R of the sphere is much smaller than the gap length L, so that, far from the sphere, the magnetic field is uniform and equal to $-H_0\mathbf{a}_z$. If the z axis coincides with the local vertical, in which direction will the sphere move, if at all? Explain.

4-13 Inductance. The first term on the right-hand side of Eq. (4-91) gives the internal inductance of the inner conductor per unit length of coaxial line. Since a single-wire line has an internal field structure which is identical with that of the center conductor of a coaxial line, it follows that the internal inductance of an infinitely long single wire of circular cross section is $\mu_0/8\pi$ H per unit length.

Show that the internal inductance of an arbitrary closed loop of circular wire is approximately equal to $\mu_0 L/8\pi$, where L is the mean length of the loop. Assume that the radius of curvature is much greater than the wire radius at every point along the perimeter of the loop. In particular, show that the internal inductance of a circular loop of wire of mean radius R is simply $\mu_0 R/4$.

4-14 Incremental Inductance. Given a rectangular cast-iron yoke as shown. The cross section is square, 1/2 in. on each side. The coil has 900 turns and carries a current of 1/2 A. (*a*) Find the incremental inductance of the coil. (*b*) What is the incremental inductance of the coil when it carries 1 A?

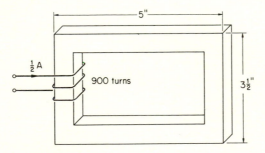

PROBLEM 4-14

4-15 Parallel Magnetic Circuit with Air Gap. The magnetic circuit shown is made of sheet steel. Assuming that

$$N = 40 \text{ turns}$$

$$I = 5 \text{ A}$$

Cross section $= 3 \times 4$ in. everywhere

Stacking factor $= 0.9$

Air-gap length $= 0.03$ in.

$l_{bad} = 30$ in., $l_{bd} = 10$ in., $l_{bcd} = 30$ in.

find the flux Φ_c through the path bcd.

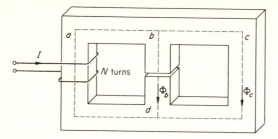

PROBLEM 4-15

4-16 Multiple Excitation. Find the current in the 50-turn coil shown if $\Phi = 90$ kilo-lines. The cross section of the sheet-steel magnetic circuit measures 1 by 1 in. everywhere.

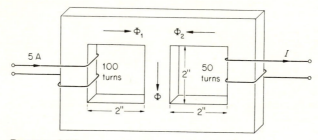

PROBLEM 4-16

4-17 Magnetic Circuit with Nonuniform Core. Given the sheet-steel core with windings as shown. Find the flux in the core. The thickness is 1 in. everywhere.

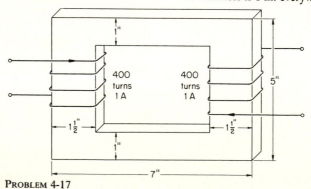

PROBLEM 4-17

4-18 Permanent Magnetization. A long magnetized cylindrical rod of radius a has a constant magnetization $\mathbf{M}_0$ perpendicular to the rod axis. Neglecting end effects, prove that the magnetic field in the free space surrounding the rod is identical with that of a two-dimensional magnetic dipole.

4-19 Formal Solution. Derive Eq. (4-136) of Example 4-16.

4-20 Magnetic Shield. For the cylindrical magnetic shield of Example 4-16, let $\mu/\mu_0 = 1000$, $a = 5$ cm, and $b = 5.1$ cm. Calculate and sketch the magnetic field intensity and the flux density both inside and outside the cylinder.

4-21 Perturbation Method. Repeat the solution of the magnetic-shield problem considered in Example 4-16, assuming initially that the permeability of the cylinder is infinitely large. Determine $\mathbf{B}$ and $\mathbf{H}$ in the cylindrical shell for an infinite μ.

Next obtain a solution for the case when μ is finite but still large compared with μ_0, assuming that, outside the cylinder and in the shell, the magnetic flux density vector of the new solution is essentially the same as before. Based on this assumption, determine the magnetic field intensity in the shell and inside the cavity (this is known as the *perturbation method* of solution).

4-22 Magnetic Shield. A uniform magnetic field B_0 exists in an extended region of space. Into this field is introduced a spherical shell of radii a (inner) and b (outer) and permeability μ. Determine the shielding properties of the shell, and compare its effectiveness with the equivalent electrostatic case.

4-23 Formal Solution. Show that in a linear, homogeneous, and isotropic medium and for a static magnetic field, Maxwell's equations give

$$\nabla^2 \mathbf{H} = -\nabla \times \mathbf{J}$$

As an example of the use of this equation obtain the solution to the problem treated in Example 4-17.

4-24 Formal Solution. A sinusoidal surface current $\mathbf{K} = K_0[\sin(\pi/a)x]\mathbf{a}_z$ exists at every point in the plane $y = b/2$. An equal and opposite surface current, $\mathbf{K} = -K_0[\sin(\pi/a)x]\mathbf{a}_z$, exists at every point in the plane $y = -b/2$. Determine the magnetic field set up in the space between the two sheets. (*Hint:* Find ϕ^* as a solution of Laplace's equation.)

4-25 Formal Solution. A magnetic rod of infinite length and constant permeability μ is immersed in a magnetic field which is normal to the axis of the rod and which, far from the rod, is uniform. Find the magnetic field inside and outside the rod, and compare your answers with the results of Example 4-16. Also, determine the magnetization vector $\mathbf{M}$.

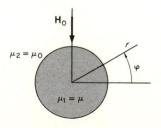

PROBLEM 4-25

4-26 "Frozen" Magnetic Field. Prove that a static magnetic field in a perfect conductor, once established, is "frozen" inside the conductor, never to collapse.

4-27 Formal Solution. A three-phase induction motor is made, basically, of a *stator* and a *rotor*. One type of rotor, known as a *squirrel-cage rotor*, consists of straight conducting bars, one in each *slot* of the rotor, and short-circuited at each end by conducting *rings*. These bars carry three-phase currents which are induced as the rotor moves through the three-phase magnetic field produced by the stator windings.

Let one phase of the induced currents be approximated by a surface current $\mathbf{K} = K_0 \sin \varphi \mathbf{a}_z$, where K_0 is a scalar constant and φ and z are the usual cylindrical coordinates. At 60 Hz the theory of static fields provides reasonably accurate predictions of the actual fields. Using the theory of magnetostatics, find the magnetic field in the air gap and in the rotor, assuming that the permeability of the stator is infinitely large and that of the rotor is μ; that the stator thickness is essentially infinite; and that variations of the field in the z direction are negligible.

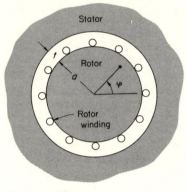

PROBLEM 4-27

4-28 Synthesis. Current flows along a very long circular cylindrical shell of radius R. Determine the distribution of the current on the surface of the shell which will produce a constant field inside the shell and a field outside such that $\phi^* = (\sin \varphi)/r$. The shell is in free space.

4-29 Field around a Toroid with an Air Gap. A winding with N turns is wound around a highly permeable ($\mu \gg \mu_0$) toroid with a short air gap. For a current I flowing through the winding, and neglecting fringing in the air gap, the flux density is

$$B_\varphi = \frac{NI\mu\mu_0}{2\pi r \mu_0 + (\mu - \mu_0)t}$$

Now, in the permeable substance, $H_{\varphi_i} = B_\varphi/\mu$, while in the air gap, $H_{\varphi_g} = B_\varphi/\mu_0$. If the toroid substance is ferromagnetic and is initially unmagnetized, then, when the coil current is dropped to zero, the relation between H_{φ_i} and H_{φ_g} becomes

$$\frac{H_{\varphi_i}}{H_{\varphi_g}} = -\frac{t}{2\pi r - t}$$

as required by Ampère's circuital law.   In other words, H_{φ_i} and H_{φ_g} are oppositely directed.   In the air gap, B_{φ_g} and H_{φ_g} are in the same direction.   However, in the toroid, B_{φ_i} and H_{φ_i} are oppositely directed, and μ is negative.   Is this a paradox?   Or can you offer a logical explanation?

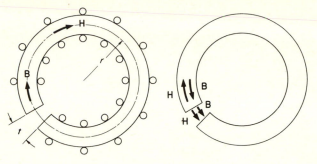

PROBLEM 4-29

CHAPTER 5

THE ELECTROMAGNETOSTATIC FIELD

5.1 Introduction. An *electromagnetostatic field* is a field which consists of a static electric field and a static magnetic field coupled together in a current-carrying region. We saw in Chaps. 3 and 4 that an electrostatic field arises from a static distribution of charges and that a magnetostatic field arises from a static distribution of currents. An electromagnetic field arises from a static distribution of charges *and* currents in the same region. The most common media in which an electromagnetostatic field can be maintained are conducting media, as, for example, practical resistors. We know that in such media a relation exists between the current density vector **J** and the magnetic field intensity vector **H**. This is evident by inspection of the second Maxwell equation (Ampère's circuital law). On the other hand, a relation always exists in a conducting medium between **J** and the electric field intensity vector **E**. This, in general, takes on the familiar form of the constitutive relation $\mathbf{J} = J(\mathbf{E})$, or in the more specialized case of a linear and isotropic medium, $\mathbf{J} = \sigma\mathbf{E}$. Therefore, when σ is finite, the vector field **J** can exist only in the presence of **E**, and conversely. Current can exist in the absence of a finite, nonzero **E** if the medium is superconducting ($\sigma \to \infty$). Superconductivity occurs only under very special conditions; however, it is frequently useful to consider good conductors to have approximately infinite conductivity.

It is important to note that *an electrostatic field is associated with a pure capacitor, a magnetostatic field with a pure inductor, and an electromagnetostatic field with a pure resistor.*

5.2 Electromagnetostatic Field Equations. For time-invariant fields, the field laws are as shown on page 259.

It is evident from Eq. (5-1) that the static electric field **E** is conservative. On the other hand, Eq. (3) shows that, as in all cases, the magnetic field **B** is solenoidal.

In order to determine analytically the field vectors of an electromagnetostatic field, we must first determine the electric field by finding the solution of

Integral form	Differential form
$\displaystyle\oint_C \mathbf{E} \cdot d\mathbf{l} = 0$ (5-1)	$\mathbf{\nabla} \times \mathbf{E} = 0$ (5-3)
$\displaystyle\oint_C \mathbf{H} \cdot d\mathbf{l} = \int_S \mathbf{J} \cdot \mathbf{n}\, da = I$ (5-2)	$\mathbf{\nabla} \times \mathbf{H} = \mathbf{J}$ (5-4)
$\displaystyle\oint_\Sigma \mathbf{B} \cdot \mathbf{n}\, da = 0$ (3)	$\mathbf{\nabla} \cdot \mathbf{B} = 0$ (III)
$\displaystyle\oint_\Sigma \mathbf{D} \cdot \mathbf{n}\, da = \int_V \rho\, dv$ (4)	$\mathbf{\nabla} \cdot \mathbf{D} = \rho$ (IV)
$\displaystyle\mathbf{F} = \int_V \rho(\mathbf{E} + \mathbf{v} \times \mathbf{B})\, dv$ (5)	$\mathbf{f} = \rho(\mathbf{E} + \mathbf{v} \times \mathbf{B})$ (V)

Eqs. (5-3) and (IV) which fits the prescribed conditions at the boundaries. Next, we can determine $\mathbf{J}$ from the constitutive relation $\mathbf{J} = J(\mathbf{E})$. Finally, we can find the magnetic field using Eqs. (5-4) and (III).

Let us now be a bit more specific and consider a conducting region which is also linear (or at least piecewise linear), homogeneous, and isotropic. We showed in Chap. 2 that in such a medium there can be no permanent distribution of free charge. Therefore $\rho = 0$, and Eq. (IV) then becomes

$$\mathbf{\nabla} \cdot \mathbf{E} = 0 \tag{5-5}$$

By virtue of Eq. (5-3),

$$\mathbf{E} = -\mathbf{\nabla}\phi \tag{5-6}$$

where ϕ is a scalar potential. When this expression is substituted in Eq. (5-5), the result is

$$\mathbf{\nabla}^2\phi = 0 \tag{5-7}$$

In other words, when a conducting medium is linear, homogeneous, and isotropic, the electric field can be derived from a scalar potential which satisfies Laplace's equation. In this respect, then, the electromagnetostatic field is mathematically analogous to an electrostatic field in a charge-free region. The methods developed in Chap. 3 for the solution of Laplace's equation apply here as well. What is more, the value of ϕ at a point is unique. This means that in a conducting medium of the type we are considering, potential difference can be defined uniquely.

Example 5-1   Leaky Capacitor. As an example, consider the case of a parallel-plate capacitor with an imperfect dielectric. We suppose that the parallel plates forming the capacitor are separated by a dielectric substance which is linear, homogeneous,

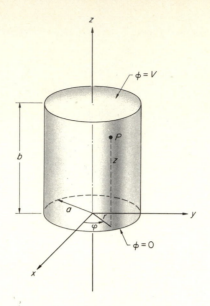

FIGURE 5-1.   A leaky capacitor.

and isotropic, but which possesses a nonvanishing conductivity, and therefore amounts to an imperfect insulator. Let ϵ, μ_0, and σ be the parameters of the leaky dielectric. To simplify the analysis, we assume that the plates are circular in shape, of radius a, and separated by a distance b (Fig. 5-1). The upper plate is held at a potential $\phi = V$, while the lower plate is grounded ($\phi = 0$). A point anywhere in the dielectric is located by the usual cylindrical coordinates r, φ, and z.

The field within the dielectric is electromagnetostatic. The electric field is a solution of the following boundary-value problem:

$$\nabla^2 \phi = 0$$

$$\phi = V \qquad \text{at } z = b$$

$$\phi = 0 \qquad \text{at } z = 0 \tag{5-8}$$

$$\frac{\partial \phi}{\partial r} = 0 \qquad \text{at } r = a$$

The last condition follows from a consideration of the fact that the normal component of $\mathbf{J}$ at every point on the cylindrical boundary must be zero. In other words, no direct current can flow out into the surrounding free space. Since $\mathbf{J} = \sigma \mathbf{E} = -\sigma \nabla \phi$, it follows that the normal component of $\mathbf{J}$ is $-\sigma\, \partial \phi / \partial r$, and for $\sigma \neq 0$, the condition $\partial \phi / \partial r$ at $r = a$ follows at once.

Now a solution of Laplace's equation which satisfies all three boundary conditions is, fortunately, a simple one:

$$\phi = \frac{V}{b} z \tag{5-9}$$

Therefore
$$\mathbf{E} = -\nabla\phi = -\frac{V}{b}\,\mathbf{a}_z \qquad (5\text{-}10)$$

$$\mathbf{J} = \sigma\mathbf{E} = -\frac{\sigma V}{b}\,\mathbf{a}_z \qquad (5\text{-}11)$$

and this shows that the electric field intensity and the current density vector are both constant everywhere within the dielectric. Now, at all interior points of the dielectric,

$$\nabla \times \mathbf{H} = \mathbf{J} = -\frac{\sigma V}{b}\,\mathbf{a}_z \qquad (5\text{-}12)$$

$$\nabla \cdot \mathbf{H} = 0 \qquad (5\text{-}13)$$

Because of circular symmetry ($\partial/\partial\varphi = 0$), H_φ is the only admissible nonzero component of $\mathbf{H}$. Equation (5-12) then gives

$$\frac{1}{r}\frac{\partial}{\partial r}(rH_\varphi) = -\frac{\sigma V}{b} \qquad (5\text{-}14)$$

from which it follows that

$$\mathbf{H} = -\frac{\sigma V}{2b}\,r\mathbf{a}_\varphi \qquad (5\text{-}15)$$

The complete solution to the electromagnetostatic problem of the leaky capacitor is now

$$\mathbf{E} = -\frac{V}{b}\,\mathbf{a}_z \qquad \mathbf{D} = \epsilon\mathbf{E} \qquad \mathbf{H} = -\frac{\sigma V}{2b}\,r\mathbf{a}_\varphi \qquad \mathbf{B} = \mu_0\mathbf{H} \qquad \mathbf{J} = -\frac{\sigma V}{b}\,\mathbf{a}_z \quad (5\text{-}16)$$

In summary, the equations of an electromagnetostatic field are obtained by placing the time derivatives in the field equations equal to zero. The solution must be accomplished in the following sequence. First, solve the mathematically equivalent electrostatic problem; second, solve the mathematically equivalent magnetostatic problem.

5.3 The Resistance Concept.

A concept intimately allied to an electromagnetostatic field is that of resistance.

Consider two equipotential surfaces within or on the boundary of a finitely conducting region. Let ϕ_1 be the constant potential of one surface, and ϕ_2 that of the other; then $V = \phi_1 - \phi_2$ is the potential difference between these two surfaces. Every point whose potential is ϕ_1 lies on one surface, and every point whose potential is ϕ_2 lies on the other. From previous work we know that the electric field intensity vector, $\mathbf{E} = -\nabla\phi$, is normal to the equipotential surfaces. We know, also, that the current density vector $\mathbf{J} = \sigma\mathbf{E}$ is always divergenceless:

$$\nabla \cdot \mathbf{J} = 0 \qquad (5\text{-}17)$$

In other words, a stationary flow of electricity is always solenoidal, and this means that all streamlines, or current filaments, close upon themselves.

Additionally, this means that any particular streamline will penetrate both equipotential surfaces and that the total current which crosses one surface will also cross the other. Let I be that current. Then the *resistance* between the equipotential surfaces is *defined* as the ratio

$$R = \frac{V}{I} = \frac{\phi_1 - \phi_2}{I} \tag{5-18}$$

This is the famous *Ohm's law*. As an experimental law, it defines resistance, and forms part of the basic theory incorporated into Maxwell's equations. We have started with Maxwell's equations and deduced Ohm's law. This deductive approach is typical of the point of view stressed in this text.

Resistance is measured in volts per ampere. The unit of resistance is the ohm (Ω). When the unit is inconveniently small, units such as the kilohm ($10^3\ \Omega$) or the megohm ($10^6\ \Omega$) are employed. Resistance is dependent only on the conductivity of the medium and on the particular geometry of the conducting body.

A simple expression can be derived for the resistance of a conducting body which is uniform in cross section and carries a current whose density is uniform in all planes normal to its flow. For such a body, we have

$$\int_C \mathbf{E} \cdot d\mathbf{l} = \int_C \frac{\mathbf{J}}{\sigma} \cdot d\mathbf{l} = \int_C \frac{I}{\sigma A}\, dl = I \frac{l}{\sigma A} \tag{5-19}$$

where I is the total current carried by the conductor, σ its conductivity, A its cross-sectional area, and l its length. The letter C stands for a continuous path, connecting a point a on one end of the conductor to a point b on the other along a particular current streamline. In terms of the scalar potential, then,

$$\int_C \mathbf{E} \cdot d\mathbf{l} = -\int_C \nabla\phi \cdot d\mathbf{l} = -[\phi(b) - \phi(a)] = \phi(a) - \phi(b) \tag{5-20}$$

If the difference in potentials $[\phi(a) - \phi(b)]$ is denoted by V, then

$$V = I \frac{l}{\sigma A}$$

and consequently

$$R = \frac{l}{\sigma A} = \rho \frac{l}{A} \tag{5-21}$$

where $\rho = 1/\sigma$ is the *resistivity* of the medium, and is measured in ohm-meters ($\Omega \cdot m$).

Equation (5-21) may be applied directly to the leaky capacitor of Example 5-1. The resistance between the upper and lower plates is clearly given by $R = b/\pi a^2 \sigma$. However, this formula does not apply in the general case. To show this, let us consider the following illustrative example.

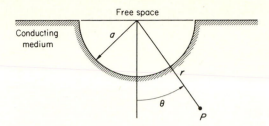

FIGURE 5-2. A grounded electrode.

Example 5-2 A Grounded Electrode. Figure 5-2 shows a perfectly conducting hemi-spherical electrode sunk into an extended region of space, the conductivity of which is nonzero but finite. Let us determine the resistance between this electrode and a concentric, hemispherical, perfectly conducting surface of infinite radius.

Let us establish a spherical system of coordinates (upside down) where φ (not shown) measures angles on the horizontal plane separating the conductor from the free space above. Since the spherical electrode is a perfect conductor, the potential will be constant at every point within and on the surface of this electrode. The potential of the concentric infinite hemisphere will also be constant, and may be arbitrarily set equal to zero. Because of symmetry about a vertical axis, all field quantities will be independent of the variable φ. Accordingly, $\phi = A/r$, where A is a constant, is an admissible general solution of Laplace's equation (Sec. 3.13) that satisfies all the requirements.

The analysis is facilitated by assuming that the electrode is held at some arbitrary potential V. The actual value of V is immaterial since in this case we are interested only in the ratio of V to the total current crossing the outer surface of the electrode. Thus, at $r = a$, we have $\phi = V$, and from this it follows that $A = aV$. Then $\phi = Va/r$.

One other condition remains to be fulfilled. At every point on the interface between the conducting region and the free space above, the normal component of $\mathbf{J}$ must vanish. The vector

$$\mathbf{J} = \sigma\mathbf{E} = \sigma(-\nabla\phi) = -\sigma\left(\frac{\partial\phi}{\partial r}\,\mathbf{a}_r\right) = \frac{\sigma Va}{r^2}\,\mathbf{a}_r \qquad (5\text{-}22)$$

satisfies this condition. Moreover, the total current which crosses any concentric hemisphere of radius r is

$$I = (\text{current density})(\text{area}) = \frac{\sigma Va}{r^2}\,\frac{4\pi r^2}{2} = 2\pi\sigma aV \qquad (5\text{-}23)$$

Therefore the required expression for resistance is

$$R = \frac{V}{I} = \frac{V}{2\pi\sigma aV} = \frac{1}{2\pi\sigma a} \qquad (5\text{-}24)$$

and is obviously different from the result that would be obtained through a careless application of Eq. (5-21). *The resistance of any configuration depends upon its geometry.*

A pure resistor does not exist physically. Some inductance and some capacitance are always associated with every practical resistor. Conversely, some resistance and some capacitance are always associated with every practical inductor, while some resistance and some inductance are always associated with every practical capacitor.

5.4 Duality. Two physical systems, or phenomena, are called *dual* if they are described by equations of the same mathematical form.

In linear, homogeneous, isotropic, and source-free media, the static field vectors $\mathbf{D}$ and $\mathbf{J}$ satisfy equations and boundary conditions of the same mathematical form, as shown in the following comparison, and thus are duals.

Field of $\mathbf{D}$	*Field of* $\mathbf{J}$
$\nabla \times \mathbf{D} = 0$	$\nabla \times \mathbf{J} = 0$
$\nabla \cdot \mathbf{D} = 0$	$\nabla \cdot \mathbf{J} = 0$
$\mathbf{D} = \epsilon \mathbf{E}$	$\mathbf{J} = \sigma \mathbf{E}$
$\mathbf{n} \cdot (\mathbf{D}_2 - \mathbf{D}_1) = 0$	$\mathbf{n} \cdot (\mathbf{J}_2 - \mathbf{J}_1) = 0$
$\mathbf{n} \times \left(\dfrac{\mathbf{D}_2}{\epsilon_2} - \dfrac{\mathbf{D}_1}{\epsilon_1} \right) = 0$	$\mathbf{n} \times \left(\dfrac{\mathbf{J}_2}{\sigma_2} - \dfrac{\mathbf{J}_1}{\sigma_1} \right) = 0$

It is important to note that either set may be obtained from the other by a mere interchange of $\mathbf{D}$ with $\mathbf{J}$ and ϵ with σ. This fact is important in itself, but assumes added significance when one notes that a solution of one set of equations, with only minor changes in notation, is also a solution of the second set of equations.

When duality exists, there is a relation between resistance and capacitance. This relation will now be established.

Consider Fig. 5-3. Two perfectly conducting electrodes are embedded in a conducting medium of infinite extent. Let a total charge Q be distributed on the surface of each perfectly conducting electrode by virtue of a potential difference set up by a battery suitably connected between electrodes. The wire current I in Fig. 5-3 is equal to the current crossing an imaginary closed surface Σ, completely surrounding one electrode:

$$I = \oint_{\Sigma} \mathbf{J} \cdot \mathbf{n} \, da = \frac{\sigma}{\epsilon} \oint_{\Sigma} \mathbf{D} \cdot \mathbf{n} \, da = \frac{\sigma}{\epsilon} Q \qquad (5\text{-}25)$$

The last step follows directly from Gauss' law. Since $Q = CV$, where C is the capacitance of the electrodes, it follows that

$$I = \frac{\sigma}{\epsilon} CV$$

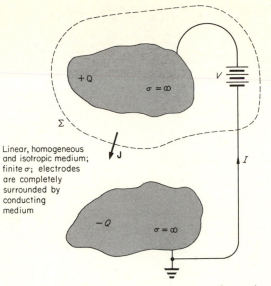

FIGURE 5-3. Duality between resistance and capacitance.

or that resistance

$$R = \frac{V}{I} = \frac{\epsilon}{\sigma} \frac{1}{C}$$

Therefore

$$RC = \frac{\epsilon}{\sigma} \qquad\qquad (5\text{-}26)$$

is the relation between resistance and capacitance. It should be noted that Eq. (5-26) is valid only if the perfectly conducting electrodes are completely immersed in the conducting medium. It cannot be applied to the leaky-capacitor problem of Example 5-1 without violating this condition. The principal difference is that although the electrostatic field can fringe into the free space surrounding the leaky capacitor, the conduction field is totally confined to the region of the leaky dielectric. Thus Eq. (5-26) can be applied to the leaky-capacitor problem only if the assumption of negligible fringing is introduced in the formulation of the problem, in which case $C = \epsilon \pi a^2/b$. Since $R = b/\sigma \pi a^2$, it follows that $RC = \epsilon/\sigma$, in full agreement with Eq. (5-26).

Duality allows the solution of electrostatic problems by means of conduction analogs. We saw earlier in this chapter that the scalar potential associated with an electromagnetostatic field in a linear, homogeneous, and isotropic medium satisfies Laplace's equation. Therefore, under the same (but otherwise completely arbitrary) boundary conditions, a conduction field problem may be employed as an analog of an electrostatic field problem.

FIGURE 5-4.   An electrolytic tank.   (*Courtesy of D. L. Amort, Oregon State University.*)

A convenient method of creating a practical conducting analog is by means of an electrolytic tank† (Fig. 5-4).   The physical dimensions of the tank are normally very large compared with the dimensions of the structure under investigation.   The tank is filled with an electrolytic solution (water, for example), and the lowered electrodes are constructed with proper shapes and relative orientation.   A source is connected between the electrodes, and the conduction field is probed by means of a high-input-impedance voltmeter connected to a probe.   Equipotential surfaces are established as the locus of all points in the liquid for which the voltmeter maintains a constant reading. The flow lines (**J** lines, or **E**-field lines) are later drawn by hand normal to the equipotential surfaces.

A practical method of implementing conduction analogs of two-dimensional fields is by means of resistive-coated paper called *teledeltos* paper.‡   In this case, electrodes are painted on the paper with silver paint (Fig. 5-5).   It is possible to record constant-potential lines automatically.

† D. L. Amort, The Electrolytic Tank Analog, *Electro-Technol.* (*New York*), vol. 70, pp. 86–92, July, 1962.
‡ D. Vitkovitch, "Field Analysis," D. Van Nostrand Company, Inc., Princeton, N.J., 1966.

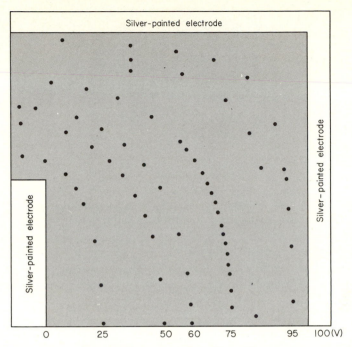

FIGURE 5-5.   Partial field map of one-quarter of a coaxial (rectangular) conductor, showing measured points on equipotential surfaces.   (*Courtesy of D. C. Ray, Georgia Institute of Technology.*)

5.5   The Boundary-value Problem in Electromagnetostatics.

The fundamental approach to the boundary-value problem in electromagnetostatics is no different than it is in electrostatics or magnetostatics.   If the conducting medium is linear, homogeneous, and isotropic, the electromagnetostatic field, as shown earlier, is derivable from a scalar potential which satisfies Laplace's equation.   The theory developed in Chap. 3 for the solution of Laplace's equation applies here as well.   In particular, the numerical method may be used most effectively to find the resistance of conducting configurations with geometrically irregular boundaries.   For example, it would be extremely difficult, if not impossible, to formulate mathematically the boundary-value problem for the arbitrary resistor configuration, shown in Fig. 5-6, even if the problem of the actual resistor were to be idealized to a two-dimensional geometry.   In direct contrast, the numerical method always allows us to obtain a solution.

The problem is set up in exactly the same way as in Sec. 3.15.   As before, the basis is Property 3 of Laplace's equation.   The only difference is that an electromagnetostatic problem will normally involve mixed boundary conditions.   For example, in Fig. 5-6, values for the scalar potential would be

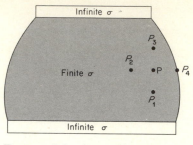

FIGURE 5-6. An arbitrary resistor configuration.

specified on the sides adjacent to the infinitely conducting regions. Then a solution to Laplace's equation would be sought which fits the specified potential on these boundaries; on the remaining portion of the boundary, the normal derivative of this solution would have to vanish. With reference to Fig. 5-6, $\phi(P_4)$ would be set equal to $\phi(P)$. Then, applying the average-value theorem at P, we should have (assuming a two-dimensional configuration)

$$\phi(P) = \frac{\phi(P_1) + \phi(P_2) + \phi(P_3) + \phi(P_4)}{4} \tag{5-27}$$

or by virtue of the equality $\phi(P_4) = \phi(P)$,

$$\phi(P) = \frac{\phi(P_1) + \phi(P_2) + \phi(P_3)}{3} \tag{5-28}$$

The normal averaging implied by Eq. (5-27) would, naturally, apply to all interior points (such as P_2) of the conductor configuration. Aside from this small difference, the computer program for this problem would be identical with that discussed in Example 3-11.

5.6 Summary. In the general time-varying case, a coupling exists between the vectors **D** and **E**, which describe the electric field, and the vectors **B** and **H**, which describe the magnetic field; therefore, in general, neither pair can be determined independently of the other. Coupling between the electric and magnetic fields disappears if and only if $\partial/\partial t = 0$ and $\mathbf{J} = 0$. It is preserved when $\mathbf{J} \neq 0$, even though the field itself may not vary with time. In that case the field is a coupled time-invariant electric and magnetic field, called an *electromagnetostatic* field.

An electromagnetostatic field is encountered in conducting regions for which σ is finite. For such regions, Ohm's law states that resistance equals voltage divided by current. Strictly speaking, resistance can be defined only if $\partial/\partial t = 0$. A *resistor* is a conductor configuration, of arbitrary shape, which impedes the flow of current. The field within a resistor is electro-magnetostatic.

In a linear, homogeneous, and isotropic medium, an electromagneto-static field can be derived from a scalar potential which satisfies Laplace's equation. As a result, the static fields of the vectors $\mathbf{D}$ and $\mathbf{J}$ are duals. This property establishes a relation ($RC = \epsilon/\sigma$) between resistance and capaci-tance, and also provides the theoretical basis for the solution of electrostatic field problems by conduction analogs. In practice, these analogs normally take the form of an electrolytic tank (three-dimensional fields) or resistance paper (two-dimensional fields).

The boundary-value problem in electromagnetostatics, from the analysis point of view, is basically the same as the boundary-value problem in electro-statics or in magnetostatics.

Problems

5-1 Leaky Capacitor. Using the integral laws, obtain the solution of the leaky-capacitor problem of Example 5-1.

5-2 Boundary Conditions. A plane surface S separates two linear, homogeneous, and isotropic conducting media. Show that, if a time-invariant current crosses from one medium into the other, a surface charge will appear on S whose density is

$$\rho_s = \left(\frac{\epsilon_2}{\sigma_2} - \frac{\epsilon_1}{\sigma_1} \right) \mathbf{J}_1 \cdot \mathbf{n}$$

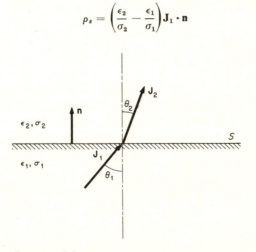

PROBLEM 5-2

5-3 Boundary Conditions. Determine the surface distribution of charge on the top and bottom plates of the capacitor shown in Fig. 5-1.

5-4 Relaxation. The parallel-plate capacitor of Fig. 5-1 is disconnected from a battery of V volts. Calculate the time after which the potential across the condenser has fallen to $V/2$. Use $\epsilon_r = 3.6$ and $\sigma = 10^{-12}$ ℧/m.

5-5 Duality. A battery is connected to a pair of probe electrodes the tips of which make electric contact with two points on a conducting plane (thin conducting sheet) of infinite extent. Show that the current flows from one electrode to the other along arcs of circles. [*Hint:* Use Eqs. (3-189) and (3-200).]

5-6 Inhomogeneous Conducting Medium. A spherical electrode of radius a is surrounded by a concentric spherical shell of radius b. The intervening space is filled with a medium whose conductivity varies proportionally ($\sigma/r = $ constant) with distance from the common center of the spheres. If a potential difference of V volts is maintained between spheres, with the inner electrode grounded, what is the potential at a distance r ($a < r < b$) from the center?

5-7 Noninductive Resistor. Wire-wound resistors are constructed by winding the wire on a cylindrical tube. How would you wind such a resistor so as to minimize the inductance of the configuration?

5-8 Formal Solution. A large conducting plane carries a direct current which, far from the center, is uniform, $\mathbf{J} = \mathbf{J}_0 \mathbf{a}_y$. Obtain an expression for $\mathbf{J}$ after a hole of radius a has been cut out of the metal sheet at the point $x = 0$, $y = 0$. Find the maximum magnitude of J and the point(s) on the sheet where this maximum occurs.

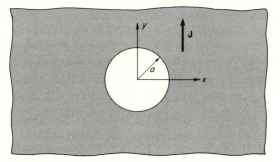

PROBLEM 5-8

5-9 Resistance. Find the surface resistance between the silver-painted equipotentials. Assume that the *surface resistivity* of the sheet is σ_s $\mho$ per square.

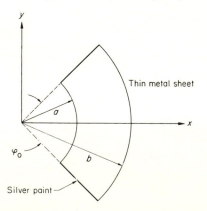

PROBLEM 5-9

5-10 Inhomogeneous Conducting Medium. Let σ and ϵ be functions of position. Then show that a time-invariant current flowing through a conducting medium will establish a volume distribution of charge whose density is

$$\rho = -\frac{1}{\sigma}\,(\sigma\mathbf{\nabla}\epsilon - \epsilon\mathbf{\nabla}\sigma)\cdot\mathbf{\nabla}\phi$$

Note that the electromagnetostatic field that arises from the flow of current can still be derived from a scalar potential, but the potential itself no longer satisfies Laplace's equation, because $\rho \neq 0$.

5-11 Formal Solution. A current I flows through diametrically opposite arcs, of very small width $2\Delta\varphi$, on the perimeter of a coin, such as a copper penny. The conductivity of the coin is uniform, and the thickness is so small that all field quantities within the coin are virtually independent of depth. Assuming that in the intervals $\pi/2 - \Delta\varphi < \varphi < \pi/2 + \Delta\varphi$ and $3\pi/2 - \Delta\varphi < \varphi < 3\pi/2 + \Delta\varphi$, on the perimeter of the coin, the current density vector is radial and independent of φ, calculate the distribution of the potential in the (two-dimensional) region of the coin.

PROBLEM 5-11

5-12 Formal Solution. The vertical, perfectly conducting wire shown in the figure carries a current I into a semi-infinite medium of uniform conductivity σ. Determine the field in both media.

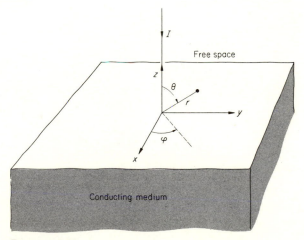

PROBLEM 5-12

5-13 Electrolytic Battery. The open-circuit voltage of a cylindrical electrolytic battery is $V_{oc} = V_1 + V_2$, where V_1 is the potential difference between the two sides of the dipole layer built up on the surface of the inner electrode, and V_2 is the potential difference of the outer electrode.   Show that the internal resistance of the battery is given by the relation $R = \ln (b/a)/2\pi\sigma L$.

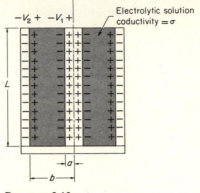

PROBLEM 5-13

5-14 Numerical Analysis. Write a computer program to calculate the surface resistance between a point $P(x_0, y_0)$ and the silver-painted electrode at $x = a$, in terms of σ_s, the surface resistivity of the conducting sheet.

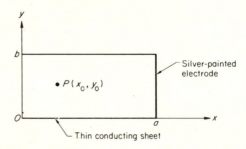

PROBLEM 5-14

CHAPTER 6

ELECTROMAGNETIC ENERGY, POWER, STRESS, AND MOMENTUM

6.1 Introduction. Although our ideas about forces stem largely from everyday experiences, the precise definition of the concept of force presents a somewhat difficult problem.   It is true that force is the time rate of change of momentum.   However, this definition is deceptively simple, since it is based on the assumption that we have a fairly clear understanding of what we mean by momentum, which in turn implies that we have a clear understanding of mass, length, and time.

Closely related to the idea of force are the ideas of energy and power. The importance of these ideas in electromagnetic theory cannot be overemphasized.   *Electromagnetic energy is the only form of energy which can be transmitted through vacuum under controlled conditions, and by means of which intelligence can be processed and transferred effectively even over extremely long distances.*

In electromechanics we have a wedding of electrical and mechanical phenomena.   For this reason, we find ourselves using concepts and definitions such as force, energy, power, and momentum in contexts and situations which are foreign to our basic (and sometimes intuitive) mechanical definitions of the terms.   In particular, we find it necessary to extend our definitions of energy and power to include situations which are wholly electrical.   The resulting electrical definitions are seemingly divorced from their mechanical origin.   With this warning of a possible conceptual difficulty associated with the need for the use of mechanical and electrical concepts simultaneously, we state that those situations where both of these aspects of a phenomenon or device must be considered simultaneously are the subject of this chapter.

6.2 General Energy Relations. An excellent point of departure in a discussion of energy relations in the electromagnetic field is furnished by

Poynting's theorem.   This was first taken up in Sec. 2.11.   We recall that

$$-\nabla \cdot (\mathbf{E} \times \mathbf{H}) = \mathbf{E} \cdot \mathbf{J} + \left(\mathbf{E} \cdot \frac{\partial \mathbf{D}}{\partial t} + \mathbf{H} \cdot \frac{\partial \mathbf{B}}{\partial t} \right) \tag{6-1}$$

is the differential form of Poynting's theorem, and that

$$-\oint_{\Sigma} (\mathbf{E} \times \mathbf{H}) \cdot \mathbf{n} \, da = \int_{V} \mathbf{E} \cdot \mathbf{J} \, dv + \int_{V} \left(\mathbf{E} \cdot \frac{\partial \mathbf{D}}{\partial t} + \mathbf{H} \cdot \frac{\partial \mathbf{B}}{\partial t} \right) dv \tag{6-2}$$

is the integral form.   We also recall that either of these expressions is simply a restatement of the law of conservation of energy, and that each term in the integral form has the dimensions of power, and hence can be written as the time rate of change of an energy.   Accordingly, let W be the energy associated with the last integral on the right of Eq. (6-2).   Then

$$\frac{\partial W}{\partial t} = \int_{V} \left(\mathbf{E} \cdot \frac{\partial \mathbf{D}}{\partial t} + \mathbf{H} \cdot \frac{\partial \mathbf{B}}{\partial t} \right) dv \tag{6-3}$$

If the substitutions

$$\mathbf{E} = -\nabla \phi - \frac{\partial \mathbf{A}}{\partial t} \qquad \mathbf{B} = \nabla \times \mathbf{A}$$

are introduced in the integrand of Eq. (6-3), we obtain

$$\mathbf{E} \cdot \frac{\partial \mathbf{D}}{\partial t} + \mathbf{H} \cdot \frac{\partial \mathbf{B}}{\partial t} = -\nabla \phi \cdot \frac{\partial \mathbf{D}}{\partial t} - \frac{\partial \mathbf{A}}{\partial t} \cdot \frac{\partial \mathbf{D}}{\partial t} + \mathbf{H} \cdot \frac{\partial (\nabla \times \mathbf{A})}{\partial t}$$

Using the vector identities of Chap. 1, we find

$$\nabla \phi \cdot \frac{\partial \mathbf{D}}{\partial t} = \nabla \cdot \left(\phi \frac{\partial \mathbf{D}}{\partial t} \right) - \phi \frac{\partial}{\partial t} (\nabla \cdot \mathbf{D})$$

and

$$\mathbf{H} \cdot \frac{\partial (\nabla \times \mathbf{A})}{\partial t} = \nabla \cdot \left(\frac{\partial \mathbf{A}}{\partial t} \times \mathbf{H} \right) + \frac{\partial \mathbf{A}}{\partial t} \cdot (\nabla \times \mathbf{H})$$

It follows that

$$\frac{\partial W}{\partial t} =$$

$$\int_{V} \left[\phi \frac{\partial}{\partial t} (\nabla \cdot \mathbf{D}) + \left(\nabla \times \mathbf{H} - \frac{\partial \mathbf{D}}{\partial t} \right) \cdot \frac{\partial \mathbf{A}}{\partial t} - \nabla \cdot \left(\phi \frac{\partial \mathbf{D}}{\partial t} \right) - \nabla \cdot \left(\mathbf{H} \times \frac{\partial \mathbf{A}}{\partial t} \right) \right] dv$$

$$= \int_{V} \left(\phi \frac{\partial \rho}{\partial t} + \mathbf{J} \cdot \frac{\partial \mathbf{A}}{\partial t} \right) dv - \oint_{\Sigma} \left(\phi \frac{\partial \mathbf{D}}{\partial t} + \mathbf{H} \times \frac{\partial \mathbf{A}}{\partial t} \right) \cdot \mathbf{n} \, da \tag{6-4}$$

Equations **(II)** and **(IV)** and the divergence theorem were used to arrive at Eq. (6-4).

It was shown earlier that when all the sources are within a finite distance from the origin, and when the time variations are very slow, the field vectors vary as $1/r^2$, where r is the distance from the origin to an arbitrary point of observation. Under the same conditions, the potentials vary as $1/r$. If Σ denotes the surface of a sphere centered about the origin, then, as the radius of this sphere is allowed to increase without limit, the integrand under the surface integral will vanish as $1/r$, and the integral itself will become zero. In this case,†

$$\frac{\partial W}{\partial t} = \int_V \left(\phi \frac{\partial \rho}{\partial t} + \mathbf{J} \cdot \frac{\partial \mathbf{A}}{\partial t} \right) dv \tag{6-5}$$

and

$$W = \int_0^t \frac{\partial W}{\partial t} \, dt = \int_0^t \left[\int_V \left(\phi \frac{\partial \rho}{\partial t} + \mathbf{J} \cdot \frac{\partial \mathbf{A}}{\partial t} \right) dv \right] dt$$

Interchanging the order of integration allows us to make a change of variables by noting the equality

$$\left(\phi \frac{\partial \rho}{\partial t} + \mathbf{J} \cdot \frac{\partial \mathbf{A}}{\partial t} \right) dt = \phi \, \delta\rho + \mathbf{J} \cdot \delta\mathbf{A}$$

where δ denotes a differential change at a point. Thus we obtain

$$W = \int_V \left(\int_0^\rho \phi \, \delta\rho + \int_0^\mathbf{A} \mathbf{J} \cdot \delta\mathbf{A} \right) dv \tag{6-6}$$

Equation (6-6) represents the sum of the work required to set up a continuous charge density ρ, that is,

$$W_e = \int_V \left(\int_0^\rho \phi \, \delta\rho \right) dv \tag{6-7}$$

plus the work required to set up a continuous current distribution $\mathbf{J}$, that is,

$$W_m = \int_V \left(\int_0^\mathbf{A} \mathbf{J} \cdot \delta\mathbf{A} \right) dv \tag{6-8}$$

In Eq. (6-7), electric energy is referred to the zero state at $\rho = 0$, and in Eq. (6-8) magnetic energy is referred to the zero state at $J = 0$.

When the medium is linear, multiplication of the charge density at all points by a scalar constant k multiplies the potential at all points by the same scalar constant. In the integral

$$\int_0^\rho \phi \, \delta\rho$$

of Eq. (6-7), the charge buildup from zero to ρ is actually a continuous change during which the constant k varies from 0 to 1. From this point of view, the intermediate values of potential and charge are $k\phi$ and $k\rho$, so that

† The same result is true for arbitrary time variations since only the transverse components of a *radiated* field (Chap. 10) vary as $1/r$. All others go down faster than $1/r$.

$d\rho = \rho dk$, and the integral itself is

$$\int_0^1 \phi \rho k \, dk$$

where ϕ and ρ are now the final values corresponding to $k = 1$. By simple integration this integral gives $\frac{1}{2}\phi\rho$, and Eq. (6-7) then becomes

$$W_e = \int_V \tfrac{1}{2}\phi\rho \, dv \qquad (6\text{-}9)$$

In a similar manner, we can show that for linear media Eq. (6-8) leads to

$$W_m = \int_V \tfrac{1}{2}\mathbf{J} \cdot \mathbf{A} \, dv \qquad (6\text{-}10)$$

The extension of the preceding energy expressions to the case in which the source distributions are other than continuous is an elementary task. For example, if charge is distributed in thin layers over the surfaces of conductors, Eq. (6-7) reduces to

$$W_e = \int_S \left(\int_0^{\rho_s} \phi \, \delta\rho_s \right) da \qquad (6\text{-}11)$$

Also, from Eq. (6-9) it follows that the energy of a group of N point charges arbitrarily placed in a linear medium is

$$W_e = \frac{1}{2} \sum_{i=1}^N \phi_i q_i \qquad (6\text{-}12)$$

where ϕ_i is the potential at q_i due to the remaining $N - 1$ charges of the group. This represents the amount of work done when the charges are brought from infinity to specified points in the medium when *no other sources are present*.

6.3 Energy Relations in the Electrostatic Field. Equation (6-7) shows that the work ΔW_e required to increase the charge density at every point in V by an amount $\delta\rho$ is

$$\Delta W_e = \int_V \phi \, \delta\rho \, dv \qquad (6\text{-}13)$$

where ϕ is the potential due to the initial distribution ρ. Based on this expression we shall now derive an alternative expression for electrostatic energy as a function of field intensity.

From the field equation $\nabla \cdot \mathbf{D} = \rho$, we have†

$$\delta\rho = \delta(\nabla \cdot \mathbf{D}) = \nabla \cdot \delta\mathbf{D}$$

---

† This proof assumes that the volume V contains only dielectrics. If conductors are also present, the proof is more complicated, but the result, Eq. (6-15), is still the same as shown in J. A. Stratton, "Electromagnetic Theory," pp. 107–111, McGraw-Hill Book Company, New York, 1941.

and using $\mathbf{E} = -\nabla\phi$,

$$\phi\,\delta\rho = \phi\nabla\cdot\delta\mathbf{D} = \nabla\cdot\phi\,\delta\mathbf{D} - \delta\mathbf{D}\cdot\nabla\phi = \nabla\cdot\phi\,\delta\mathbf{D} + \mathbf{E}\cdot\delta\mathbf{D}$$

where the second equality follows identically. Thus

$$\Delta W_e = \int_V \nabla\cdot\phi\,\delta\mathbf{D}\,dv + \int_V \mathbf{E}\cdot\delta\mathbf{D}\,dv = \oint_\Sigma \phi\,\delta\mathbf{D}\cdot\mathbf{n}\,da + \int_V \mathbf{E}\cdot\delta\mathbf{D}\,dv$$

If the surface Σ is allowed to expand into a sphere of infinite radius about some arbitrary origin, then, at every point on this surface, ϕ will vary as $1/r$ and $\mathbf{D}$ as $1/r^2$. Hence the surface integral itself will vary as $1/r$, and will therefore vanish in the limit as $r \to \infty$. For an electrostatic field, then,

$$\Delta W_e = \int_V \mathbf{E}\cdot\delta\mathbf{D}\,dv \qquad (6\text{-}14)$$

To find the total energy stored in such a field, the increment ΔW_e must be integrated between the limits zero and $\mathbf{D}$.

$$W_e = \int_V \left(\int_0^{\mathbf{D}} \mathbf{E}\cdot\delta\mathbf{D}\right) dv \qquad (6\text{-}15)$$

Equation (6-15) is an expression for the electrostatic energy as a function of field intensity. For fields in the stationary state, this expression is completely equivalent to, and always interchangeable with, Eq. (6-7). If the medium is linear and isotropic, it becomes

$$W_e = \int_V \tfrac{1}{2}\mathbf{E}\cdot\mathbf{D}\,dv = \int_V \tfrac{1}{2}\epsilon E^2\,dv \qquad (6\text{-}16)$$

in full agreement with Eq. (2-71).

Actually, in view of its origin, the result expressed by Eq. (6-15) is not very surprising. We started out by integrating the right side of Eq. (6-3) over a time interval $[0,t]$, during which the field quantities were generally expected to change. At that point of the development, and through a change of independent variables, we could have written at once

$$W_e = \int_0^t \left(\int_V \mathbf{E}\cdot\frac{\partial\mathbf{D}}{\partial t}\,dv\right) dt = \int_V \left(\int_0^{\mathbf{D}} \mathbf{E}\cdot\delta\mathbf{D}\right) dv$$

Though longer, the former route is more instructive in some aspects.

A most useful relation is easily derived from Eq. (6-11). Consider a capacitor with charge Q and potential difference $V = \phi_+ - \phi_-$, as shown in Fig. 6-1. If the conductors and dielectrics which make up this capacitor are linear, an analysis similar to that which led to Eq. (6-9) will show that

$$W_e = \oint_\Sigma \tfrac{1}{2}\phi\rho_s\,da \qquad (6\text{-}17)$$

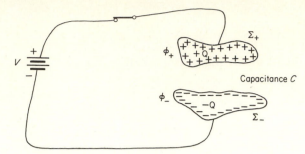

FIGURE 6-1.   A charged capacitor.

where the integration extends over the surfaces of both conductors.   Noting that each conductor surface is an equipotential, we obtain

$$W_e = \tfrac{1}{2}\phi_+ \int_{\Sigma_+} \rho_s \, da + \tfrac{1}{2}\phi_- \int_{\Sigma_-} \rho_s \, da = \tfrac{1}{2}\phi_+ Q - \tfrac{1}{2}\phi_- Q = \tfrac{1}{2}VQ \qquad (6\text{-}18)$$

or making use of the defining relation $Q = CV$,

$$W_e = \tfrac{1}{2}VQ = \frac{1}{2}\frac{Q^2}{C} = \tfrac{1}{2}CV^2 \qquad (6\text{-}19)$$

Equation (6-19) is an expression for the amount of energy stored in a capacitor C, the conductors of which are held at a fixed potential difference V.

It is important to note that the preceding expressions apply only when the system is complete, or closed.   Assume that some external system of charges gives rise to a static field **E**.   Then the force exerted on a point charge q is $q\mathbf{E}$, and the work done *by* the forces of the field, in a displacement of q from a point a to a second point b along a contour C (Fig. 6-2), is given by

$$W = q \int_a^b \mathbf{E} \cdot d\mathbf{l} = q[\phi(a) - \phi(b)] \qquad (6\text{-}20)$$

When a field extends over all space, it is customary to set $\phi(\infty) = 0$.   Therefore, if we let the point b recede to infinity, the work done against the forces

FIGURE 6-2.   An arbitrary contour C.

of the field in bringing q from infinity to the point a is

$$W = q\phi(a) \tag{6-21}$$

In view of this, the scalar potential $\phi(a)$ may be interpreted as the work done against the forces of the field when a unit positive charge is brought from infinity to the point a.

$$\phi(a) = \int_a^\infty \mathbf{E} \cdot d\mathbf{l} = -\int_\infty^a \mathbf{E} \cdot d\mathbf{l} \tag{6-22}$$

This, by the way, is the classical definition of potential.

On the basis of the foregoing discussion we conclude that the energy of N discrete charges in an external field ϕ_0 is obtained by combining Eqs. (6-12) and (6-21).

$$W = \frac{1}{2} \sum_{i=1}^N \phi_i q_i + \sum_{i=1}^N \phi_0 q_i \tag{6-23}$$

To summarize, an expression for the electrostatic energy of a closed system as a function of field intensity was derived. This is given by Eq. (6-15). An alternative expression is Eq. (6-7). The energy stored in a capacitor, with linear elements, is specified by Eq. (6-19). The effects of an external field are taken into account by Eq. (6-23).

6.4 Thomson's Theorem. Thomson's theorem expresses a principle which is the electrostatic analog of the familiar phenomenon that water always flows downhill.

If we place a system of charges on the surfaces of conductors which are embedded in a linear and isotropic dielectric, Thomson's theorem asserts that, when the charges come to a complete rest, *the energy of the resulting electrostatic field is always a minimum.*

We could simply accept Thomson's theorem as a postulate and proceed to use it. However, a formal proof is a good exercise in some typical electromagnetic field manipulations, and is given below.

To prove Thomson's theorem, we suppose to the contrary that there is another possible charge distribution and its associated electrostatic field, the energy of which is less than that of the actual field. Both the actual field and the second field satisfy Maxwell's equations. In particular, we have for the actual field

$$\mathbf{\nabla} \times \mathbf{E} = 0 \qquad \mathbf{\nabla} \cdot \mathbf{D} = \rho \qquad \mathbf{E} = -\mathbf{\nabla}\phi \tag{6-24}$$

and for the second field,

$$\mathbf{\nabla} \times \mathbf{E}' = 0 \qquad \mathbf{\nabla} \cdot \mathbf{D}' = \rho \qquad \mathbf{E}' = -\mathbf{\nabla}\phi' \tag{6-25}$$

The volume charge density ρ is presumed to be bound in the dielectric, and hence stays the same. Although the total charge on each conductor surface, Σ_i, remains constant, the surface charge density will be different in the two cases, and as a result the potential ϕ_i' on each conductor surface will not necessarily remain constant. Thus

$$\nabla \cdot (\mathbf{D}' - \mathbf{D}) = 0 \qquad (6\text{-}26)$$

$$\oint_{\Sigma_i} (\mathbf{D}' - \mathbf{D}) \cdot \mathbf{n} \, da = 0 \qquad (6\text{-}27)$$

According to Eq. (6-16), the difference in the electrostatic energies of the two fields is

$$W_e' - W_e = \int_V \tfrac{1}{2}(\mathbf{E}' \cdot \mathbf{D}' - \mathbf{E} \cdot \mathbf{D}) \, dv$$

The fields at all interior points in the conductors are zero. Hence, in this equation, the integration covers the region occupied by the dielectric, which is bounded by the surface at infinity and the surfaces Σ_i. Since $\mathbf{D} = \epsilon\mathbf{E}$ and $\mathbf{D}' = \epsilon\mathbf{E}'$, it follows that

$$\mathbf{E} \cdot \mathbf{D}' = \mathbf{E} \cdot \epsilon\mathbf{E}' = \epsilon\mathbf{E} \cdot \mathbf{E}' = \mathbf{D} \cdot \mathbf{E}'$$

and that, consequently,

$$W_e' - W_e = \int_V \tfrac{1}{2}(\mathbf{E}' - \mathbf{E}) \cdot (\mathbf{D}' - \mathbf{D}) \, dv + \int_V \mathbf{E} \cdot (\mathbf{D}' - \mathbf{D}) \, dv \quad (6\text{-}28)$$

With the aid of the identity

$$\nabla \cdot \psi\mathbf{A} = \mathbf{A} \cdot \nabla\psi + \psi\nabla \cdot \mathbf{A}$$

the second integrand in Eq. (6-28) can be written

$$\mathbf{E} \cdot (\mathbf{D}' - \mathbf{D}) = -\nabla\phi \cdot (\mathbf{D}' - \mathbf{D})$$
$$= -\nabla \cdot [\phi(\mathbf{D}' - \mathbf{D})] + \phi\nabla \cdot (\mathbf{D}' - \mathbf{D}) = -\nabla \cdot [\phi(\mathbf{D}' - \mathbf{D})]$$

where the last equality follows directly from Eq. (6-26). Thus

$$\int_V \mathbf{E} \cdot (\mathbf{D}' - \mathbf{D}) \, dv = -\int_V \nabla \cdot [\phi(\mathbf{D}' - \mathbf{D})] \, dv = -\oint_\Sigma \phi(\mathbf{D}' - \mathbf{D}) \cdot \mathbf{n} \, da$$

On the conductor surfaces $\phi_i = $ constant; hence it may be pulled out from under the surface integral, and by virtue of (6-27), the integral itself then vanishes. Moreover, on the surface at infinity, the scalar potential vanishes as $1/r$, while the vectors $\mathbf{D}$ and $\mathbf{D}'$ vanish as $1/r^2$; hence the surface integral itself also vanishes, and Eq. (6-28) reduces to

$$W_e' - W_e = \int_V \tfrac{1}{2}(\mathbf{E}' - \mathbf{E}) \cdot (\mathbf{D}' - \mathbf{D}) \, dv = \int_V \tfrac{1}{2}\epsilon(E' - E)^2 \, dv \quad (6\text{-}29)$$

Since the integrand on the right of this expression is an essentially positive quantity, the difference $W_e' - W_e$ is always positive, and this means that W_e' is greater than W_e.

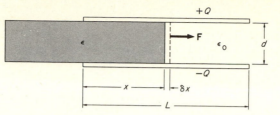

FIGURE 6-3. Parallel-plate capacitor with partially inserted dielectric.

Example 6-1. Parallel-plate Capacitor. A capacitor is formed by two rectangular parallel conducting plates of area A as shown in Fig. 6-3. Each plate carries an electric charge of magnitude Q. A slab of dielectric is inserted between the plates so that only a portion of the slab actually lies between the plates. In which direction will the slab tend to move?

We consider two extreme cases. In one case, the space between the plates of the capacitor is completely empty, and in the second case the space between the plates is completely filled by the slab. Ignoring end effects (fringing), the energies of the resulting electrostatic fields in the two cases are, respectively,

$$W_0 = \int_V \frac{1}{2}\, \epsilon_0 E^2 \, dv = \int_V \frac{1}{2}\, \epsilon_0 \left(\frac{D}{\epsilon_0}\right)^2 dv = \int_V \frac{Q^2}{2\epsilon_0 A^2}\, dv$$

and

$$W = \int_V \frac{1}{2}\, \epsilon E^2 \, dv = \int_V \frac{1}{2}\, \epsilon \left(\frac{D}{\epsilon}\right)^2 dv = \int_V \frac{Q^2}{2\epsilon A^2}\, dv$$

Since $\epsilon > \epsilon_0$, it follows that $W < W_0$. By Thomson's theorem, this system must seek the state of lowest energy. Hence the force acting on the dielectric will tend to pull it into the interelectrode region.

6.5 Energy Relations in the Magnetostatic Field.

According to Eq. (6-8),

$$\Delta W_m = \int_V \mathbf{J} \cdot \delta \mathbf{A} \, dv \qquad (6\text{-}30)$$

is the differential amount of work required to increase the vector potential at every point in the field by an amount $\delta \mathbf{A}$. Based on this expression, we shall now derive an alternative expression for magnetostatic energy as a function of field intensity.

We know that, in the stationary state, $\nabla \times \mathbf{H} = \mathbf{J}$. In terms of this relation the integrand in Eq. (6-30) becomes

$$\mathbf{J} \cdot \delta \mathbf{A} = (\nabla \times \mathbf{H}) \cdot \delta \mathbf{A} = \nabla \cdot (\mathbf{H} \times \delta \mathbf{A}) + \mathbf{H} \cdot \nabla \times \delta \mathbf{A}$$

where the last equality is a vector identity. Furthermore, using $\mathbf{B} = \nabla \times \mathbf{A}$,

$$\mathbf{J} \cdot \delta \mathbf{A} = \nabla \cdot (\mathbf{H} \times \delta \mathbf{A}) + \mathbf{H} \cdot \delta \mathbf{B}$$

Introducing this result in Eq. (6-30) and applying the divergence theorem gives

$$\Delta W_m = \int_V \mathbf{H} \cdot \delta\mathbf{B} \, dv + \oint_\Sigma (\mathbf{H} \times \delta\mathbf{A}) \cdot \mathbf{n} \, da$$

If the surface Σ is allowed to expand into a sphere of infinite radius about some arbitrary origin, the contribution of the surface integral vanishes. This is so because $\mathbf{A}$ and $\mathbf{B}$ vanish as $1/r$ and $1/r^2$, respectively, and the surface increases only as r^2. Thus, if the integration extends over all space, the expression

$$\Delta W_m = \int_V \mathbf{H} \cdot \delta\mathbf{B} \, dv \tag{6-31}$$

specifies the incremental energy required to establish a stationary current distribution *in the absence of a fixed external field*, such as might be produced by one or more permanent magnets.

To find the total energy stored in such a field, the increment ΔW_m must be integrated from zero to $\mathbf{B}$. Interchanging the order of integration, we obtain

$$W_m = \int_V \left(\int_0^\mathbf{B} \mathbf{H} \cdot \delta\mathbf{B} \right) dv \tag{6-32}$$

as an expression for magnetostatic energy in terms of field intensity. For fields in the stationary state, this expression is completely equivalent to, and always interchangeable with, Eq. (6-8).

If the medium is linear and isotropic, Eq. (6-32) becomes

$$W_m = \int_V \tfrac{1}{2}\mathbf{H} \cdot \mathbf{B} \, dv = \int_V \tfrac{1}{2}\mu H^2 \, dv \tag{6-33}$$

which is obviously in accord with Eq. (2-71). If, on the contrary, the medium is nonlinear, the energy required to set up a magnetostatic field must be found by integration, as indicated by the shaded area in Fig. 6-4.

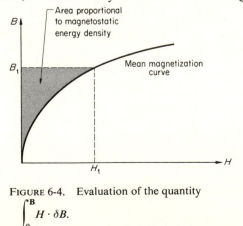

FIGURE 6-4. Evaluation of the quantity

$$\int_0^\mathbf{B} H \cdot \delta B.$$

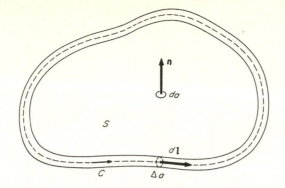

FIGURE 6-5. A current filament. The contour C is represented by the dotted line.

Let us apply the preceding results to a current filament. Figure 6-5 shows a filament C which bounds an open surface S. The area of the filament cross section is denoted by Δa, and an element of area on S is represented by da. The direction of the unit vector $\mathbf{n}$, drawn normal to S, is fixed relative to C by the conventional right-hand-screw rule.

We wish to determine the amount of energy required to force a steady current I through the filament C. According to Eq. (6-8),

$$W_m = \int_\Gamma \left(\int_0^A \mathbf{J} \cdot \delta\mathbf{A} \right) dv = \oint_C \left(\int_0^A \mathbf{J} \cdot \delta\mathbf{A} \right) \Delta a \, dl$$

Since $\mathbf{J} \, \Delta a \, d\mathbf{l} = I \, d\mathbf{l}$,

$$W_m = \int_0^A \oint_C I \, \delta\mathbf{A} \cdot d\mathbf{l}$$

By Stokes' theorem,

$$W_m = \int_0^A \int_S I(\nabla \times \delta\mathbf{A}) \cdot \mathbf{n} \, da = \int_0^B \int_S I \, \delta\mathbf{B} \cdot \mathbf{n} \, da$$

Denoting the incremental magnetic flux threading the contour C by $\delta\Phi$, we obtain

$$W_m = \int_0^\Phi I \, \delta\Phi \tag{6-34}$$

Equation (6-34) is an expression for the total energy stored in the magnetic field of a single filament carrying a current I. When the relation between Φ and I is linear,

$$W_m = \tfrac{1}{2}\Phi I = \tfrac{1}{2}LI^2 \tag{6-35}$$

where $L = \Phi/I$ is the self-inductance of the loop. For a configuration of N distinct filaments, all of which carry the same current I and also thread the same flux Φ (as in ideal inductors), the magnetic energy is

$$W_m = \sum_{i=1}^{N} \tfrac{1}{2}\Phi_i I_i = \tfrac{1}{2}N\Phi I = \tfrac{1}{2}LI^2 \qquad (6\text{-}36)$$

where $L = N\Phi/I$ is the self-inductance of the new configuration.

In summary, an expression for the magnetostatic energy of a closed system as a function of field intensity was derived. This is given by Eq. (6-32). An alternative expression is Eq. (6-8). The energy stored in a linear inductor is specified by Eq. (6-36).

6.6 Energy Considerations in Physical Media. Let us now consider some of the aspects of power transfer through a closed surface Σ, within the boundaries of which are present not only *free* charges and currents, but also *bound* charges and currents, accounting for the presence of polarizable and magnetizable matter within the enclosed volume V.

Let us first determine the amount of work required to introduce a dielectric substance into the field of fixed distribution of static charges.

Consider Fig. 6-6a. Let V_1 be a linear and isotropic dielectric of permittivity ϵ_1, and V_2 be the volume displaced by a linear and isotropic dielectric body of permittivity ϵ_2. The total charge on each conductor surface is assumed to be fixed. With $\partial/\partial t = 0$, the energy change resulting from the introduction of the dielectric body is

$$\Delta W_e = \int_V \tfrac{1}{2}(\mathbf{E}' \cdot \mathbf{D}' - \mathbf{E} \cdot \mathbf{D}) \, dv$$

$$= \int_V \tfrac{1}{2}(\mathbf{E}' + \mathbf{E}) \cdot (\mathbf{D}' - \mathbf{D}) \, dv - \int_V \tfrac{1}{2}(\mathbf{E} \cdot \mathbf{D}' - \mathbf{E}' \cdot \mathbf{D}) \, dv \qquad (6\text{-}37)$$

where $V = V_1 + V_2$; $\mathbf{E} = -\boldsymbol{\nabla}\phi$ and $\mathbf{D}$ represent the initial field; and $\mathbf{E}' = -\boldsymbol{\nabla}\phi'$ and $\mathbf{D}'$ represent the final field. With the aid of the identity

$$\boldsymbol{\nabla} \cdot \psi\mathbf{A} = \mathbf{A} \cdot \boldsymbol{\nabla}\psi + \psi\boldsymbol{\nabla} \cdot \mathbf{A}$$

and the divergence theorem, the first integral on the right of Eq. (6-37) can be written

$$-\int_V \tfrac{1}{2}(\boldsymbol{\nabla}\phi' + \boldsymbol{\nabla}\phi) \cdot (\mathbf{D}' - \mathbf{D}) \, dv = -\oint_\Sigma \tfrac{1}{2}(\phi' + \phi)(\mathbf{D}' - \mathbf{D}) \cdot \mathbf{n} \, da$$

$$+ \int_V \tfrac{1}{2}(\phi' + \phi)\boldsymbol{\nabla} \cdot (\mathbf{D}' - \mathbf{D}) \, dv$$

Now the right side of this expression is equal to zero because, by assumption, the source charges are fixed. Therefore $\boldsymbol{\nabla} \cdot (\mathbf{D}' - \mathbf{D}) = 0$ throughout V, and the volume integral vanishes. Over the conductor surfaces, which form

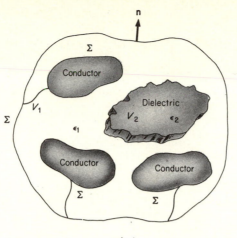

(a)

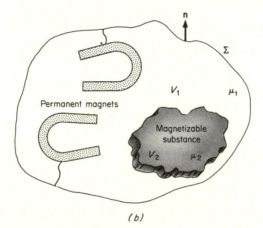

(b)

FIGURE 6-6. Pertaining to discussion on energy
in physical media. (a) Polarizable media; (b)
magnetizable media ($J = 0$).

the inner boundary of Σ, the potential is constant, and the total surface
charge is fixed; hence that portion of the surface integral vanishes identically.
Each of the finlike surfaces from the interior conductors to the outer surface
are charge-free and contribute nothing to the integral.   Finally, the outer
boundary of Σ may be allowed to expand to infinity so as to include all space.
Under these conditions the corresponding portion of the surface integral
goes to zero, since at infinity the integrand vanishes faster than the surface
area.   Hence

$$\int_V \tfrac{1}{2}(\mathbf{E}' + \mathbf{E}) \cdot (\mathbf{D}' - \mathbf{D})\, dv = 0$$

The second integral on the right of Eq. (6-37) can be written as the sum

$$-\int_{V_1} \tfrac{1}{2}(\mathbf{E} \cdot \mathbf{D}' - \mathbf{E}' \cdot \mathbf{D})\, dv - \int_{V_2} \tfrac{1}{2}(\mathbf{E} \cdot \mathbf{D}' - \mathbf{E}' \cdot \mathbf{D})\, dv$$

$$= -\int_{V_2} \tfrac{1}{2}(\mathbf{E} \cdot \mathbf{D}' - \mathbf{E}' \cdot \mathbf{D})\, dv$$

since in V_1

$$\mathbf{E} \cdot \mathbf{D}' = \mathbf{E} \cdot \epsilon_1\mathbf{E}' = \epsilon_1\mathbf{E} \cdot \mathbf{E}' = \mathbf{D} \cdot \mathbf{E}'$$

which makes the integral over V_1 identically zero. In V_2, $\mathbf{D}' = \epsilon_2\mathbf{E}'$ and $\mathbf{D} = \epsilon_1\mathbf{E}$, so that

$$-\int_{V_2} \tfrac{1}{2}(\mathbf{E} \cdot \mathbf{D}' - \mathbf{E}' \cdot \mathbf{D})\, dv = -\int_{V_2} \tfrac{1}{2}(\epsilon_2 - \epsilon_1)\mathbf{E} \cdot \mathbf{E}'\, dv$$

and thus, finally,

$$\Delta W_e = -\int_{V_2} \tfrac{1}{2}(\epsilon_2 - \epsilon_1)\mathbf{E} \cdot \mathbf{E}'\, dv \qquad (6\text{-}38)$$

This is the required result. If $\epsilon_1 = \epsilon_0$, then in V_2

$$\mathbf{D}' = \epsilon_2\mathbf{E}' = \epsilon_0\mathbf{E}' + \mathbf{P}'$$

so that

$$\mathbf{P}' = (\epsilon_2 - \epsilon_0)\mathbf{E}'$$

and Eq. (6-38) becomes

$$\Delta W_e = -\int_{V_2} \tfrac{1}{2}\mathbf{P}' \cdot \mathbf{E}\, dv \qquad (6\text{-}39)$$

Equation (6-39) expresses, in terms of the polarization vector $\mathbf{P}'$, the energy required to introduce a dielectric substance into a fixed-source static electric field in free space.

The development for magnetostatics follows a similar course. Figure 6-6b shows a system of permanent magnets and a configuration of linear and isotropic substances in which $\mathbf{J} = 0$. The energy change resulting from the introduction of the magnetizable body is

$$\Delta W_m = \int_V \tfrac{1}{2}(\mathbf{H}' \cdot \mathbf{B}' - \mathbf{H} \cdot \mathbf{B})\, dv$$

$$= \int_V \tfrac{1}{2}(\mathbf{H}' + \mathbf{H}) \cdot (\mathbf{B}' - \mathbf{B})\, dv - \int_V \tfrac{1}{2}(\mathbf{H} \cdot \mathbf{B}' - \mathbf{H}' \cdot \mathbf{B})\, dv \qquad (6\text{-}40)$$

Noting that $\nabla \cdot \mathbf{B} = 0$, that $(\mathbf{B}' - \mathbf{B}) \cdot \mathbf{n} = 0$ on the boundaries of the permanent magnets, and that in a current-free region $\mathbf{H} = -\nabla\phi^*$, we can trace, step by step, the development leading to Eq. (6-38), and obtain the corresponding result

$$\Delta W_m = -\int_{V_1} \tfrac{1}{2}(\mu_2 - \mu_1)\mathbf{H} \cdot \mathbf{H}'\, dv \qquad (6\text{-}41)$$

If $\mu_1 = \mu_0$, then in V_2

$$\mathbf{M}' = \frac{1}{\mu_0}(\mu_2 - \mu_0)\mathbf{H}'$$

and Eq. (6-41) becomes

$$\Delta W_m = -\int_{V_2} \tfrac{1}{2}\mu_0 \mathbf{M}' \cdot \mathbf{H} \, dv \tag{6-42}$$

Example 6-2 Energy of a Magnetized Sphere. As an example, let us determine the work required to introduce an initially unmagnetized sphere of radius R and permeability μ into the air gap of a permanent magnet. We suppose that the radius of the sphere is sufficiently small compared with the dimensions of the gap so that, to a good approximation, the field around the entire boundary of the air gap is essentially uniform, has a magnitude H_0, and is in the negative z direction.

If the center of the sphere coincides with the origin of a spherical system of coordinates, the field at every interior point of the sphere is, according to Eq. (4-131),

$$\mathbf{H} = \frac{3\mu_0}{\mu + 2\mu_0} H_0(-\cos\theta\,\mathbf{a}_r + \sin\theta\,\mathbf{a}_\theta) = -\frac{3\mu_0}{\mu + 2\mu_0} H_0 \mathbf{a}_z$$

Introducing this result into Eq. (6-41), we obtain

$$W_m = -\int_V \tfrac{1}{2}(\mu - \mu_0)\frac{3\mu_0}{\mu + 2\mu_0} H_0^2 \, dv = -2\pi R^3 \frac{\mu_0(\mu - \mu_0)}{\mu + 2\mu_0} H_0^2 \tag{6-43}$$

If $\mu \gg \mu_0$, then

$$\Delta W_m = -2\pi\mu_0 R^3 H_0^2$$

6.7 Circuit Concept of Power. To preface the development of this concept, let us discuss a question arising in network theory. Let V denote the difference $[\phi(a) - \phi(b)]$ in Eq. (6-20). Then $W = qV$ is the amount of energy transferred in a displacement of q through a potential difference V. Accepting from mechanics the definition of power, we have

$$\text{Power} = \frac{dW}{dt} = V\frac{dq}{dt} + q\frac{dV}{dt} = VI + q\frac{dV}{dt} \tag{6-44}$$

in clear violation, however, of the well-known principle that circuit power is always equal to the product of voltage times current. How are we then to account for the second term on the right?

Fortunately, we need not think very hard to detect the fallacy. We recall that Eq. (6-20) is valid only for fields in the stationary state $(\partial/\partial t = 0)$. Hence we are not entitled to take derivatives with respect to time, expecting to arrive at valid results. Instead, the answers we are seeking must be sought elsewhere, at the very origin of Eq. (6-20), which, we recall, is Poynting's theorem.

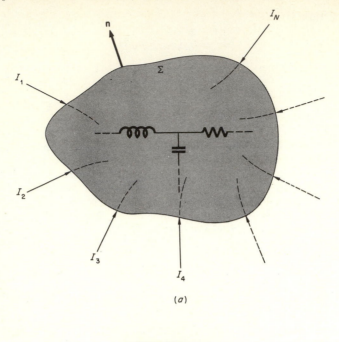

(a)

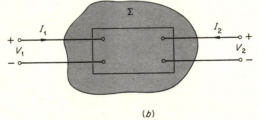

(b)

FIGURE 6-7.   The power aspects of a circuit.   (a) A general circuit; (b) a two-port circuit.

Let us apply this theorem to the geometry shown in Fig. 6-7a.   We suppose that the closed surface Σ encloses a circuit, the elements of which are fed by a group of N filamentary wires carrying currents $I_1, I_2, I_3, I_4, \ldots, I_N$. The circuit itself may include active as well as passive elements.   The power entering the volume enclosed by Σ is given by

$$P = -\oint_\Sigma (\mathbf{E} \times \mathbf{H}) \cdot \mathbf{n}\, da = -\int_V \mathbf{\nabla} \cdot (\mathbf{E} \times \mathbf{H})\, dv \qquad (6\text{-}45)$$

and according to Eq. (6-1),

$$-\mathbf{\nabla} \cdot (\mathbf{E} \times \mathbf{H}) = \mathbf{E} \cdot \mathbf{J} + \left(\mathbf{E} \cdot \frac{\partial \mathbf{D}}{\partial t} + \mathbf{H} \cdot \frac{\partial \mathbf{B}}{\partial t} \right)$$

The first term on the right side of this equation can be transformed with the aid of $\mathbf{E} = -\nabla\phi - \partial\mathbf{A}/\partial t$ and the identity

$$\nabla \cdot \psi\mathbf{G} = \mathbf{G} \cdot \nabla\psi + \psi\nabla \cdot \mathbf{G} \qquad (6\text{-}46)$$

We have

$$\mathbf{E} \cdot \mathbf{J} = \left(-\nabla\phi - \frac{\partial\mathbf{A}}{\partial t}\right) \cdot \mathbf{J} = -\nabla\phi \cdot \mathbf{J} - \mathbf{J} \cdot \frac{\partial\mathbf{A}}{\partial t}$$

$$= -\nabla \cdot \phi\mathbf{J} + \phi\nabla \cdot \mathbf{J} - \mathbf{J} \cdot \frac{\partial\mathbf{A}}{\partial t}$$

$$= -\nabla \cdot \phi\mathbf{J} - \phi\frac{\partial\rho}{\partial t} - \mathbf{J} \cdot \frac{\partial\mathbf{A}}{\partial t} \qquad (6\text{-}47)$$

Consequently,

$$-\nabla \cdot (\mathbf{E} \times \mathbf{H}) = -\nabla \cdot \phi\mathbf{J} - \left(\phi\frac{\partial\rho}{\partial t} + \mathbf{J} \cdot \frac{\partial\mathbf{A}}{\partial t}\right) + \left(\mathbf{E} \cdot \frac{\partial\mathbf{D}}{\partial t} + \mathbf{H} \cdot \frac{\partial\mathbf{B}}{\partial t}\right)$$

and

$$P = -\int_V \nabla \cdot \phi\mathbf{J}\,dv - \int_V \left(\phi\frac{\partial\rho}{\partial t} + \mathbf{J} \cdot \frac{\partial\mathbf{A}}{\partial t}\right)dv + \int_V \left(\mathbf{E} \cdot \frac{\partial\mathbf{D}}{\partial t} + \mathbf{H} \cdot \frac{\partial\mathbf{B}}{\partial t}\right)dv$$

$$(6\text{-}48)$$

If time variations are sufficiently slow, the last two terms may be neglected, and power is then given by

$$P = -\int_V \nabla \cdot \phi\mathbf{J}\,dv \qquad (6\text{-}49)$$

Finally, the divergence theorem gives

$$P = -\oint_\Sigma \phi\mathbf{J} \cdot \mathbf{n}\,da = \oint_\Sigma \phi\mathbf{J} \cdot (-\mathbf{n})\,da = \sum_{k=1}^{N} \phi_k I_k \qquad (6\text{-}50)$$

which is the familiar relation for power. Specifically, for the two-port shown in Fig. 6-7b, this relation gives

$$P = V_1 I_1 + V_2 I_2$$

6.8 Electromagnetic Stress and Momentum. In this section we discuss an expression for the total force acting on a system of charges and/or currents distributed with arbitrary densities over an extended region of space. This development is confined to the case of a linear, homogeneous, and isotropic medium.

Starting with the field equations and following a rather long, involved series of manipulations,† it can be shown that the total force is given by

$$\mathbf{F} = \int_V (\rho \mathbf{E} + \mathbf{J} \times \mathbf{B})\, dv$$

$$= \oint_\Sigma \left[\epsilon \mathbf{E}(\mathbf{E} \cdot \mathbf{n}) - \frac{\epsilon}{2} E^2 \mathbf{n} \right] da$$

$$+ \oint_\Sigma \left[\mu \mathbf{H}(\mathbf{H} \cdot \mathbf{n}) - \frac{\mu}{2} H^2 \mathbf{n} \right] da - \epsilon \int_V \frac{\partial}{\partial t} (\mathbf{E} \times \mathbf{B})\, dv \quad (6\text{-}51)$$

In this expression the field vectors represent the *total* field, produced in part by charges and currents within the region V, and in part by sources exterior to it.   If we suppose that the sources within were purposely introduced to measure the intensity of the field, it is obvious that the field is altered by the probe sources themselves, and because of this distortion it is never possible to attain a true measure of the undistorted fields.

We saw in Sec. 2.4 that an electromagnetic field may be defined as the transmitter of force interactions between charges at rest and/or in motion. Accordingly, Eq. (6-51) is an expression for the force transmitted from one region to another, much like the force transmitted from one point of a stretched rubber band to another.   Indeed, if we think of lines of force as miniature rubber bands, which conceptually cross every imaginary surface, the region itself may be viewed as an imaginary elastic medium under a fictitious state of stress.   We must be careful, however, not to carry this analogy too far because electromagnetic forces can be transmitted through space even in the absence of material substances.

For fields in the stationary state, the volume integral vanishes, and the force is given by the surface integrals only.   It is then expressible in terms of stresses, frequently called *Maxwell stresses*, which are transmitted through the elements of the surface Σ.   In particular, purely electrostatic forces can be correctly evaluated from

$$\mathbf{F}_e = \int_V \rho \mathbf{E}\, dv = \oint_\Sigma \left[\epsilon \mathbf{E}(\mathbf{E} \cdot \mathbf{n}) - \frac{\epsilon}{2} E^2 \mathbf{n} \right] da \quad (6\text{-}52)$$

while purely magnetostatic forces can be calculated from

$$\mathbf{F}_m = \int_V (\mathbf{J} \times \mathbf{B})\, dv = \oint_\Sigma \left[\mu \mathbf{H}(\mathbf{H} \cdot \mathbf{n}) - \frac{\mu}{2} H^2 \mathbf{n} \right] da \quad (6\text{-}53)$$

† The proof is rather involved, and is omitted. It may be found in J. A. Stratton, "Electromagnetic Theory," chap. 2, McGraw-Hill Book Company, New York, 1941. Properly modified, this entire derivation can be duplicated step by step for arbitrary media, provided $\mathbf{J}$ is replaced by $\mathbf{J} + \nabla \times \mathbf{M} + (\partial \mathbf{P}/\partial t)$, and ρ is replaced by $\rho - \nabla \cdot \mathbf{P}$ (Prob. 2-26).

Obviously, the net force transmitted through Σ is zero when the region bounded by this closed surface contains no sources, and $\partial/\partial t = 0$. If, on the contrary, $\partial/\partial t \neq 0$, then the force $F \neq 0$, even though there are no sources in the region bounded by Σ. The classical explanation of this apparent paradox is based on the mechanical definition of momentum. In newtonian mechanics, force is defined as the time derivative of momentum. Using this notation as a basis, we conclude from the last term in Eq. (6-51) that, with every time-varying electromagnetic field, we can associate a momentum whose density is

$$\mathbf{g} = \epsilon(\mathbf{E} \times \mathbf{B}) \tag{6-54}$$

at every point in V.

In closing we should note that the force expressed by Eq. (6-51) is purely electromagnetic in nature and that equilibrium can be established and maintained only by means of mechanical forces.

Example 6-3   Attracting Force of Magnets.   A coil of 60 turns is wound around an iron ring ($\mu_r = 1000$) having a 20-cm diameter and a 10-cm² cross section. As indicated in Fig. 6-8, the ring contains an air gap of length $t = 0.1$ mm, which is small enough so that fringing in the air gap may be neglected.   We wish to determine the force acting between the pole pieces of the iron ring when the current $I = 1$ A.

This problem involves a direct application of Eq. (6-53). The field may be determined by the methods developed in Chap. 4.   Thus the magnetic flux density in the ring is

$$B = \frac{\mu_0 \mu_r N I}{\pi D + (\mu_r - 1)t} = \frac{4\pi \times 10^{-7} \times 1000 \times 60 \times 1}{0.2\pi + (1000 - 1) \times 10^{-4}} = 0.111 \qquad \text{Wb/m}^2$$

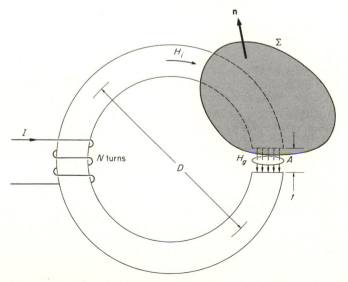

FIGURE 6-8.   The attracting force of magnets.

The magnetic field intensities in the iron and in the air gap are $H_i = B/\mu_0\mu_r$ and $H_g = B/\mu_0$, respectively. It is clear that $H_i \ll H_g$, and that the contribution of H_i to the force expression may therefore be neglected. Accordingly,

$$\mathbf{F}_m = \int_S \left[\mu_0 \mathbf{H}_g (\mathbf{H}_g \cdot \mathbf{n}) - \frac{\mu_0}{2} H_g{}^2 \mathbf{n} \right] da = \int_S \left(\frac{\mu_0 H_g{}^2}{2} \right) \mathbf{n} \, da$$

The magnitude of this force is

$$F_m = \frac{\mu_0 H_g{}^2 A}{2} = \frac{B^2 A}{2\mu_0} = \frac{(0.111)^2 \times 10 \times 10^{-4}}{2 \times 4\pi \times 10^{-7}} = 5 \quad N$$

and the direction along the normal vector $\mathbf{n}$. This means that the pole pieces tend to attract each other.

6.9   The Principle of Virtual Work.

In its simplest form, this principle is merely a restatement of the principle of conservation of energy. If a system were given a small displacement, the mechanical work performed in the process, when added to the increments in electromagnetic energy storage and losses, must equal the sum of energy increments from all sources. Put mathematically, this statement is

$$\delta W_s = \delta W_{\text{mech}} + \delta W_{\text{em}} + \delta W_{\text{loss}} \tag{6-55}$$

All energy terms affected by the virtual displacement must be included in this energy balance. Subsequently, this allows us to determine the mechanical force associated with such a displacement, because by definition

$$\delta W_{\text{mech}} = F_{\text{mech}} \, \delta x \tag{6-56}$$

where δx is the mechanical displacement. Upon introducing the relation (6-56) in Eq. (6-55) and transposing terms, we obtain

$$F_{\text{mech}} = \frac{\delta W_s}{\delta x} - \frac{\delta W_{\text{em}}}{\delta x} - \frac{\delta W_{\text{loss}}}{\delta x} \tag{6-57}$$

or in the limit as $\delta x \to 0$,

$$F_{\text{mech}} = \frac{\partial W_s}{\partial x} - \frac{\partial W_{\text{em}}}{\partial x} - \frac{\partial W_{\text{loss}}}{\partial x} \tag{6-58}$$

Although Eq. (6-58) is correct, it is difficult to include loss terms because dissipative effects generally depend on the rate at which the system is perturbed. Hence the principle of virtual work is usually applied only to quasistatic reversible processes.

Example 6-4   Force on a Dielectric Slab. As an example, let us determine the force tending to draw the dielectric into the parallel-plate capacitor shown in Fig. 6-3. We shall consider that the depth of the capacitor is b, although for our purposes the depth is immaterial so long as fringing is neglected. The electrostatic energy stored in the system, when the slab edge is located at x, is given by

$$W_e = \tfrac{1}{2}\epsilon_0\left(\frac{V}{d}\right)^2(L - x)bd + \tfrac{1}{2}\epsilon\left(\frac{V}{d}\right)^2(x)bd$$

where V is the potential difference between the plates, considered in this case to be held constant by an externally applied source (not shown in Fig. 6-3). If a virtual displacement δx is postulated, the energy storage will change by an amount δW_e such that the total energy is

$$W_e + \delta W_e = \tfrac{1}{2}\epsilon_0\left(\frac{V}{d}\right)^2(L - x - \delta x)bd + \tfrac{1}{2}\epsilon\left(\frac{V}{d}\right)^2(x + \delta x)bd$$

The change δW_e may be found by subtracting the last two expressions. We obtain

$$\delta W_e = \tfrac{1}{2}\epsilon_0\left(\frac{V}{d}\right)^2(-\delta x)bd + \tfrac{1}{2}\epsilon\left(\frac{V}{d}\right)^2(\delta x)bd = \frac{1}{2}\frac{V^2 b}{d}(\epsilon - \epsilon_0)\,\delta x$$

To hold the potential difference V at a fixed value, the external source must furnish a charge δQ to the plates such that

$$\delta Q = \left(\epsilon\frac{V}{d}\right)b\,\delta x - \left(\epsilon_0\frac{V}{d}\right)b\,\delta x = \frac{Vb}{d}(\epsilon - \epsilon_0)\,\delta x$$

where $\epsilon_0 V/d$ and $\epsilon V/d$ represent surface charge densities on the infinitesimal strip δx of the plates, before and after the virtual displacement. In so doing, the external source furnishes an amount of energy

$$\delta W_s = V\,\delta Q = \frac{V^2 b}{d}(\epsilon - \epsilon_0)\,\delta x$$

If the system is lossless, $\delta W_{\text{loss}} = 0$ and Eq. (6-57) then gives

$$F_{\text{mech}} = \frac{\delta W_s}{\delta x} - \frac{\delta W_e}{\delta x} = \frac{V^2 b}{d}(\epsilon - \epsilon_0) - \frac{1}{2}\frac{V^2 b}{d}(\epsilon - \epsilon_0) = \frac{1}{2}\frac{V^2 b}{d}(\epsilon - \epsilon_0)$$

Since $\epsilon > \epsilon_0$, the direction of F_{mech} is such as to draw the dielectric farther into the parallel-plate capacitor.

A little consideration will show that the principle of virtual work is a direct consequence of Poynting's theorem:

$$-\oint_{\Sigma}(\mathbf{E} \times \mathbf{H}) \cdot \mathbf{n}\,da = \int_V \mathbf{E} \cdot \mathbf{J}\,dv + \int_V \frac{\partial}{\partial t}(\tfrac{1}{2}\epsilon E^2 + \tfrac{1}{2}\mu H^2)\,dv \quad (6\text{-}59)$$

According to classical interpretations, the left side of this equation represents power entering a region V through the bounding surface Σ. On the other hand, the first term on the right side represents the power transformed irreversibly in the volume V, while the second integral on the right represents the time rate of change of energy which is stored in the electromagnetic field and is always recoverable. Upon integrating both sides of Eq. (6-59) with respect to time, we obtain an energy balance equation, which may be stated thus:

$$\text{Total energy input} = \frac{\text{energy transformed}}{\text{irreversibly}} + \frac{\text{electromagnetic}}{\text{energy storage}}$$

The first term on the right may include losses due to heat, as well as to transformation of electric energy into mechanical energy. In mathematical notation we have, then,

$$W_s = W_{\text{loss}} + W_{\text{mech}} + W_{\text{em}} \tag{6-60}$$

which is just another form of Eq. (6-55).

Summarizing, the principle of virtual work provides a simple method of calculating forces in any system having electrical as well as mechanical aspects. The principle states simply that energy is conserved.

6.10   Summary.   This chapter has been devoted to a discussion of energy exchanges and forces in electromagnetic systems.   Starting with Poynting's theorem, we derived, first, an expression for the total work required to set up continuous source densities, ρ and $\mathbf{J}$, at every point in the field.   The formula

$$W = \int_V \left(\int_0^{\rho} \phi \, \delta\rho + \int_0^{\mathbf{A}} \mathbf{J} \cdot \delta\mathbf{A} \right) dv \tag{6-6}$$

represents total energy storage when referred to a zero state $\rho = 0$, $\mathbf{J} = 0$. Based on this expression, special forms for electrostatic and magnetostatic energy were derived in terms of field intensities:

$$W_e = \int_V \left(\int_0^{\mathbf{D}} \mathbf{E} \cdot \delta\mathbf{D} \right) dv \tag{6-15}$$

$$W_m = \int_V \left(\int_0^{\mathbf{B}} \mathbf{H} \cdot \delta\mathbf{B} \right) dv \tag{6-32}$$

The same general expression for W was also used to derive the familiar expressions $CV^2/2$ and $LI^2/2$ for energy stored in a capacitor C and in an inductor L, respectively.

A theorem on energy minimization, Thomson's theorem, was discussed. The essence of this theorem is that the arrangement of charges under static conditions is such as to minimize the *total* energy stored in the resulting electrostatic field.

Next we showed that the familiar expression for circuit power (power = VI) follows from Poynting's theorem, which in turn follows from Maxwell's equations.

We then displayed the relation

$$\mathbf{F} = \oint_{\Sigma} \left[\epsilon \mathbf{E}(\mathbf{E} \cdot \mathbf{n}) - \frac{\epsilon}{2} E^2 \mathbf{n} \right] da + \oint_{\Sigma} \left[\mu \mathbf{H}(\mathbf{H} \cdot \mathbf{n}) - \frac{\mu}{2} H^2 \mathbf{n} \right] da$$

$$- \epsilon \int_V \frac{\partial}{\partial t} (\mathbf{E} \times \mathbf{B}) \, dv \quad (6\text{-}51)$$

for the total force acting on a system of sources contained within a region V. This formula defines force in terms of the *total* field present in the region, that due to the sources within V as well as exterior to it. The surface integrals are usually interpreted in terms of stresses in the medium, while the volume integral is interpreted in terms of electromagnetic momentum.

If a system is given in infinitesimal displacement about a position of equilibrium, then conservation of energy requires that

$$\delta W_s = \delta W_{\text{mech}} + \delta W_{\text{em}} + \delta W_{\text{loss}} \quad (6\text{-}55)$$

This is the statement of the principle of virtual work.

Problems

6-1 Electrostatic Energy. Find the total energy stored in a concentric spherical capacitor of radii a and b, assuming that equal and opposite amounts of charge Q are distributed uniformly on both electrode surfaces.

6-2 Magnetostatic Pressure. A long straight wire of radius a carries a steady current I. Find the net force per unit length which at any radius r tends to reduce the current cross section.

6-3 Magnetostatic Energy. Prove that the total energy stored in a system of two conducting loops is given by

$$W_m = \tfrac{1}{2}I_1{}^2 L_{11} + \tfrac{1}{2}I_2{}^2 L_{22} + I_1 I_2 L_{12}$$

where I_1 and I_2 = currents in the two loops

L_{11} and L_{22} = self-inductances

L_{12} = mutual inductance between the loops

6-4 Energy Minimization. Prove that a static distribution of currents is such that the heat generated is minimum. Assume that the medium is linear and isotropic.

6-5 Electrostatic Deflection. This problem is concerned with the motion of charged particles (mass m, charge e) in a uniform electrostatic field, as it occurs in certain types of cathode-ray tubes (see figure).

In a cathode-ray tube, a beam of electrons enters the region between deflecting plates with a velocity imparted to all elements of the beam by a high accelerating potential V_0. The beam is deflected by a transverse electric field set up between the plates by applying a

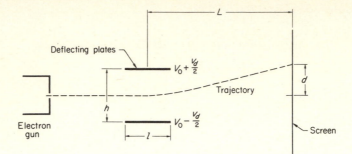

PROBLEM 6-5

potential difference V_d between the plates. The beam so deflected then strikes the fluorescent screen at a distance

$$d = \frac{lLV_d}{2hV_0}$$

Derive this expression for d.

6-6 Magnetostatic Deflection. Assume that the transverse electric field of Prob. 6-5 has been replaced by a uniform static magnetic field B_0 directed out of the plane of the page. Derive the new expression for d valid for small deflections.

6-7 Force in Terms of Energy Exchanges (Vacuum Diode). Charge is distributed with density ρ throughout a region of space bounded by two parallel plates. The plates are separated by a distance which is small relative to their lateral dimensions, and are maintained at a potential difference V_0, with one plate grounded. The mass of the charged particles is m, and their charge is e. Distance from the grounded plate is measured by the coordinate x.

Let ϕ denote the potential at any point in the region bounded by the plates, and show that Poisson's equation and the principle of conservation of energy lead to

$$\left(\frac{d\phi}{dx}\right)^2 = -4 \frac{J}{\epsilon_0} \left(\frac{m\phi}{2e}\right)^{1/2} + \text{constant}$$

which is the determinantal equation for ϕ. In this expression J denotes convection current density (charge density times velocity).

Assuming that $d\phi/dx = 0$ at the grounded plate, show that the solution of the above differential equation is

$$J = KV_0^{3/2}$$

where K is a constant. This is known as the *Child-Langmuir law*.

6-8 Lorentz Force. Find the force between a conducting sphere of radius R carrying a charge Q and a point charge q of the same sign located a distance d ($d > R$) from the center of the sphere. (*Hint:* See Prob. 3-17.)

Prove that under certain conditions the charges may attract each other, and calculate the point at which the force changes direction. (Is this attraction physically possible? Explain.)

6-9 Magnetostatic Force. An infinitely long conductor in free space, carrying a current I, is parallel to an infinite plane surface marking the boundary of a semi-infinite slab of iron of permeability μ. Calculate the force acting on the conductor per unit of its length.

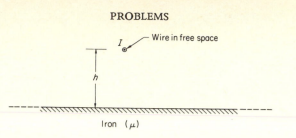

PROBLEM 6-9

6-10 Lorentz Force. A conducting bar is placed on a line which is fed by a time-independent current source. Will the bar move away or toward the source?

PROBLEM 6-10

6-11 Electrostatic Stress. Determine the mutual force acting between the plates of a parallel-plate capacitor. Assume that $\epsilon = \epsilon_0$.

6-12 Electrostatic Stress. Two equal and opposite point charges q are separated by a distance $2h$ in free space. Find the force acting on the negative charge by integrating directly the surface integral in Eq. (6-52) over an infinite plane surface bisecting perpendicularly the distance between the two like charges.

6-13 Principle of Virtual Work. Solve the problem of Example 6-4 for the case in which the voltage across the capacitor plates is allowed to vary but the total charge Q is held constant.

6-14 The Hall Effect. The Hall effect is the appearance of an electric field mutually perpendicular to a longitudinal electric field and a transverse magnetic field. Thus, if an electric current is flowing in the x direction, an electric field E_H will appear in the y direction when a magnetic field B is applied in the z direction. Prove that in metals, $E_H = vB$, where v is the velocity of electrons in the x direction.

6-15 Principle of Virtual Work. Two parallel metal plates separated by a distance of 1 cm are charged by a battery of 10 V. The battery is then disconnected, and the plates separated to a distance of 2 cm. If the area of the plates is 400 cm², find the work done in separating the plates. Neglect fringing, and assume that the medium between the plates is free space.

6-16 Principle of Virtual Work. An electrostatic voltmeter is shown in the accompanying sketch. Two semicircular fixed plates are connected to the potential to be measured. A movable conducting plate is free to rotate into the region between the plates, but is constrained from doing so by a torsional spring. Develop an expression for the angle θ in terms of V. Neglect edge effects. Introduce necessary parameters for the development. State clearly what each parameter signifies.

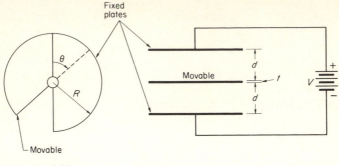

PROBLEM 6-16

6-17   Electromechanical Force.   The operation of rotating electric machinery is based on the interaction between magnetic fields and current-carrying conductors.   The rotor conductors are normally insulated, and are placed in slots around the periphery of the rotor.   It would seem logical to surmise that, since forces act on the current-carrying conductors, the insulation would eventually deteriorate as a result of the conductors' pressing against the rotor slot walls.   This, however, is not what happens ordinarily. How do you explain the phenomenon?

6-18   Force on Polarized Matter.   A dielectric sphere of radius a is in an electric field that, far from the sphere, is uniform.   If the sphere is cut in half by a plane perpendicular to the field, find the mutual force between the hemispheres.   Assume that the dielectric constant of the sphere is ϵ_r, that the magnitude of the external field is E_0, and that the surrounding medium is free space.

6-19   Dipole Alignment.   Prove that the torque exerted on a dipole in free space by an external field is

$$\mathbf{T} = \mathbf{p} \times \mathbf{E}$$

where $\mathbf{p}$ is the vector dipole moment.

Also prove that the magnetic counterpart is

$$\mathbf{T} = \mathbf{m} \times \mu_0\mathbf{H}$$

6-20   Dipole Energy.   Prove that the magnitude of the energy of a dipole in an external field $\mathbf{E}$ is equal to the dot product of $\mathbf{E}$ and the dipole moment $\mathbf{p}$.

CHAPTER 7

TRANSVERSE ELECTROMAGNETIC (TEM) WAVES

7.1 Introduction. Maxwell's equations imply the possibility of energy propagation by means of electromagnetic waves. These waves can be extremely complex in form or, in special cases, rather simple.   A particularly useful special case, which also serves as an introduction to the more general problem, is the subject of this chapter.

A general statement of an electromagnetic wave problem is as follows. Given a set of conditions and restrictions, what nonzero, nonstatic, electromagnetic field configurations *can exist*, and what is the characteristic behavior?   If a sufficient number of conditions and restrictions are assumed, only one field configuration will be allowed.   However, most electromagnetic wave problems *allow* the *independent coexistence* of several field configurations. It is profitable to adopt the waveguide terminology and call an allowed field configuration a *mode*.   It should be emphasized that the fact that a mode *can* exist does not mean that it *does* exist.   In fact, if the modes *are* independent, we can arbitrarily assume that all of them except one, or more, do not exist (zero amplitude), and then examine the remaining field configurations.   This, of course, is equivalent to imposing further restrictions on the original problem, but it avoids the necessity of repeating steps which are common to a class of problems.

The main objective of this chapter is a study of transverse electromagnetic (TEM, for short) waves.   *If at every point in space the vectors of a time-varying field are contained in a local plane the space orientation of which is independent of time, then the field is said to be a transverse electromagnetic wave.*   It is to be understood that, in general, the orientation of the local planes is different for different points in space.   An exception is the case of *plane waves*, discussed below.

7.2 The Wave Equation. In this section we wish to derive the so-called *wave equation* for linear, homogeneous, isotropic media, to express this wave equation in rectangular coordinates, and then to restrict the problem further to its simplest meaningful form.   We shall see that these conditions allow a particular set of wave propagation modes, called *uniform plane waves*, which have their **E** and **H** field components transverse to the direction of propagation. Hence our solutions will be one type of *transverse electromagnetic waves*.

For a linear, homogeneous, isotropic, and source-free medium, Maxwell's equations can be written

$$\nabla \times \mathbf{E} = -\mu \frac{\partial \mathbf{H}}{\partial t} \qquad (7\text{-}1)$$

$$\nabla \times \mathbf{H} = \sigma \mathbf{E} + \epsilon \frac{\partial \mathbf{E}}{\partial t} \qquad (7\text{-}2)$$

$$\nabla \cdot \mathbf{H} = 0 \qquad (7\text{-}3)$$

$$\nabla \cdot \mathbf{E} = 0 \qquad (7\text{-}4)$$

Taking the curl of Eq. (7-2), we obtain

$$\nabla \times \nabla \times \mathbf{H} = \sigma(\nabla \times \mathbf{E}) + \epsilon \frac{\partial}{\partial t}(\nabla \times \mathbf{E})$$

Substituting for $\nabla \times \mathbf{E}$ from Eq. (7-1) and transposing, we find

$$\nabla \times \nabla \times \mathbf{H} + \mu\epsilon \frac{\partial^2 \mathbf{H}}{\partial t^2} + \mu\sigma \frac{\partial \mathbf{H}}{\partial t} = 0 \qquad (7\text{-}5)$$

A similar exercise shows that

$$\nabla \times \nabla \times \mathbf{E} + \mu\epsilon \frac{\partial^2 \mathbf{E}}{\partial t^2} + \mu\sigma \frac{\partial \mathbf{E}}{\partial t} = 0 \qquad (7\text{-}6)$$

Making use of the vector identity

$$\nabla \times \nabla \times \mathbf{A} = \nabla(\nabla \cdot \mathbf{A}) - \nabla^2 \mathbf{A}$$

and the third and fourth of Maxwell's equations allows us to write Eqs. (7-5) and (7-6), respectively, as

$$-\nabla^2 \mathbf{H} + \mu\epsilon \frac{\partial^2 \mathbf{H}}{\partial t^2} + \mu\sigma \frac{\partial \mathbf{H}}{\partial t} = 0 \qquad (7\text{-}7)$$

and

$$-\nabla^2 \mathbf{E} + \mu\epsilon \frac{\partial^2 \mathbf{E}}{\partial t^2} + \mu\sigma \frac{\partial \mathbf{E}}{\partial t} = 0 \qquad (7\text{-}8)$$

Equations (7-7) and (7-8) are the general forms of the so-called wave equations obeyed by the field vectors **E** and **H**. They are in fact too general for our purposes. We shall specialize our problem in order to reduce its complexity sufficiently so that we can solve it.

Our first simplification is to require that the source-free medium be nonconducting; that is, we require that $\sigma = 0$ everywhere. Under these conditions Eqs. (7-7) and (7-8) reduce to

$$\nabla^2 \mathbf{H} - \mu\epsilon \frac{\partial^2 \mathbf{H}}{\partial t^2} = 0 \tag{7-9}$$

$$\nabla^2 \mathbf{E} - \mu\epsilon \frac{\partial^2 \mathbf{E}}{\partial t^2} = 0 \tag{7-10}$$

The mathematical symmetry of Eqs. (7-9) and (7-10) implies that the solutions of both will have the same mathematical form. Hence we can solve either and infer the solution to the other. Let us choose to deal first with the wave equation in E, Eq. (7-10).

We recall that in rectangular coordinates

$$\nabla^2 \mathbf{A} = \nabla^2 A_x \mathbf{a}_x + \nabla^2 A_y \, \mathbf{a}_y + \nabla^2 A_z \, \mathbf{a}_z \tag{7-11}$$

Hence Eq. (7-10) may be presented as three scalar equations, namely,

$$\nabla^2 E_x - \mu\epsilon \frac{\partial^2 E_x}{\partial t^2} = 0$$

$$\nabla^2 E_y - \mu\epsilon \frac{\partial^2 E_y}{\partial t^2} = 0 \tag{7-12}$$

$$\nabla^2 E_z - \mu\epsilon \frac{\partial^2 E_z}{\partial t^2} = 0$$

It is seen that expressing Eq. (7-10) in rectangular coordinates results in a further simplification, that of mathematical symmetry in the equations governing the individual field components.

One final specialization is required in order to reduce the problem to its simplest meaningful form. This specialization will be stated in the form of the following questions.

For linear, homogeneous, isotropic, source-free, nonconducting, unbounded media, are there solutions to the wave equations such that there exists only a single component of the electric field which varies only with time and with the direction perpendicular to the field component? What is the significance of such a solution? To be specific, is there a field such that

$$E_x(z,t) \neq 0 \qquad \frac{\partial E_x}{\partial x} = 0 \qquad \frac{\partial E_x}{\partial y} = 0 \tag{7-13}$$

$$E_y = 0 \qquad E_z = 0$$

If, indeed, there is, then the set (7-12) reduces to the single equation

$$\frac{\partial^2 E_x}{\partial z^2} - \mu\epsilon \frac{\partial^2 E_x}{\partial t^2} = 0 \tag{7-14}$$

This is the desired result. Its implications are examined in this and the following chapters.

7.3 Solution of the Simplified Wave Equation.

Equation (7-14) is a second-order partial differential equation. As such, it has two linearly independent solutions, in addition to a non-time-dependent solution. Specifically,

$$E_x(z,t) = E_x^+(z - vt) + E_x^-(z + vt) \tag{7-15}$$

where $v = 1/\sqrt{\mu\epsilon}$ and E_x^+ and E_x^- denote arbitrary functions of their arguments. We can verify this result by direct substitution into Eq. (7-14). Thus, treating the arguments $z - vt$ and $z + vt$ as composite variables, we obtain

$$\frac{\partial^2 E_x}{\partial z^2} = E_x^{+''} + E_x^{-''} \tag{7-16}$$

and

$$\frac{\partial^2 E_x}{\partial t^2} = v^2 E_x^{+''} + v^2 E_x^{-''} \tag{7-17}$$

where $E_x^{+''}$ and $E_x^{-''}$ indicate differentiations with respect to the composite variables. Substitution into Eq. (7-14) yields

$$E_x^{+''} + E_x^{-''} - \mu\epsilon v^2 E_x^{+''} - \mu\epsilon v^2 E_x^{-''} = 0 \tag{7-18}$$

or

$$(1 - \mu\epsilon v^2)E_x^{+''} + (1 - \mu\epsilon v^2)E_x^{-''} = 0 \tag{7-19}$$

which is an identity if $v = 1/\sqrt{\mu\epsilon}$.

7.4 Interpretation of the Solutions.

Let us now focus our attention on Eq. (7-15). Let us examine $E_x^+(z - vt)$ first.

In order to visualize a function of the two variables, it is convenient to choose two or more fixed values of one variable and plot the dependence of the function on the other variable. Let us assume that $E_x^+(z - vt)$ has the form shown in Fig. 7-1 at the time $t = t_0$.† Careful examination of the functional form of E_x^+ will show that, at a later time $t_1 > t_0$, E_x^+ will have moved, unchanged in shape, a distance $z_1 - z_0 = v(t_1 - t_0)$ in the $+z$ direction with a velocity v. Hence we see the significance of the $+$ sign on E_x^+. Recalling that we demand that $\partial E_x/\partial x = \partial E_x/\partial y = 0$, we can summarize our interpretation of E_x^+ as follows. The function $E_x^+(z - vt)$ represents a *uniform*

---

† Maxwell's equations and our solution of the resulting wave equation do not specify E_x^+ except to require that it be a function of $z - vt$ and that it have continuous first and second derivatives. The exact form will depend on the initial (time) and boundary (space) conditions.

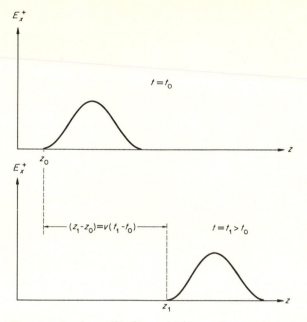

FIGURE 7-1. A possible form of $E_x^+(z - vt)$ at two instants of time.

plane wave traveling in the $+z$ direction with a velocity $v = 1/\sqrt{\mu\epsilon}$. It is a uniform plane wave because at every point in space, the **E** vector is contained in a plane (the xy plane) perpendicular to the direction of propagation $(+z)$, and because the magnitude of the field vectors are independent of the *transverse* (x and y) coordinates.

In general, if the projections of the vectors, defining a time-varying field along some straight line, all vanish identically at every instant of time and for all points in space, then the time-varying field is called a *plane wave*.

A similar analysis shows that $E_x^-(z + vt)$ is a uniform plane wave traveling in the $-z$ direction. Note that in each case the **E**-field direction is perpendicular to the direction of propagation. Hence the waves are *transverse* waves.

A similar analysis can be carried out for the **H**-field vector. The results will be of the same form, interpretable as transverse, uniform plane waves. However, as the following analysis will show, there is an intimate connection between the **E** and **H** fields.

For concreteness let us look at waves traveling in the $+z$ direction. Then we have

$$E_z = H_z = 0 \qquad \frac{\partial}{\partial x} = \frac{\partial}{\partial y} = 0 \qquad (7\text{-}20)$$

For the linear, homogeneous, isotropic, nonconducting, and source-free regions which we are considering, Maxwell's equations (**I**) and (**II**) reduce to

$$\nabla \times \mathbf{E} = -\mu \frac{\partial \mathbf{H}}{\partial t} \qquad (7\text{-}21)$$

$$\nabla \times \mathbf{H} = \epsilon \frac{\partial \mathbf{E}}{\partial t} \qquad (7\text{-}22)$$

As the student can verify, Eq. (7-20) substituted into Eqs. (7-21) and (7-22) will yield

use 2 if $\bar{E} = E_x \bar{a}_x$

use 4 if $\bar{E} = E_x \bar{a}_x + E_y \bar{a}_y$

$$\frac{\partial E_x}{\partial z} = -\mu \frac{\partial H_y}{\partial t} \qquad \frac{\partial E_y}{\partial z} = \mu \frac{\partial H_x}{\partial t}$$

$$\frac{\partial H_y}{\partial z} = -\epsilon \frac{\partial E_x}{\partial t} \qquad \frac{\partial H_x}{\partial z} = \epsilon \frac{\partial E_y}{\partial t} \qquad (7\text{-}23)$$

These equations reveal a connection between E_x and H_y and between E_y and H_x. Direct substitution of the solutions for E_x, H_y, E_y, and H_x, which are all of the form of Eq. (7-15), followed by integration† yields the results

$$E_x^+ = \eta H_y^+ \qquad E_x^- = -\eta H_y^- \qquad (7\text{-}24)$$

$$E_y^+ = -\eta H_x^+ \qquad E_y^- = \eta H_x^- \qquad (7\text{-}25)$$

where

$$\eta = \sqrt{\frac{\mu}{\epsilon}} \qquad (7\text{-}26)$$

and is called the *characteristic impedance of the medium*. The quantity η has the dimensions

$$\left(\frac{\text{henrys/meter}}{\text{farads/meter}}\right)^{1/2} = \left(\frac{\text{henrys} \times \text{seconds}}{\text{farads} \times \text{seconds}}\right)^{1/2} = \left(\frac{\text{ohms}}{\text{mhos}}\right)^{1/2} = \text{ohms}$$

This intimate relationship between the electric and magnetic field vectors is the motivation for the term *electromagnetic waves*.

7.5 The Sinusoidal Steady State.

Before proceeding with our discussion of uniform plane waves, we need to digress slightly in order to develop some concepts and notations which facilitate the discussion.

Consider an electric field vector $\mathscr{E}$ which is a function of position and a

† Example: $\partial E_x^+/\partial z = E_x^{+\prime}$ and $\partial H_y^+/\partial t = -vH_y^{+\prime}$. Therefore the first of Eqs. (7-23) becomes $E_x^{+\prime} = \mu v H_y^\prime$, which, upon integration, yields $E_x^+ = \mu(1/\sqrt{\mu\epsilon})H_y^+$, or $E_x^+ = \sqrt{\mu/\epsilon}\, H_y^+$. The constant of integration is disregarded since we are interested only in time-dependent fields.

sinusoidal function of time. In rectangular component form we should write

$$\mathscr{E} = \mathscr{E}_x \mathbf{a}_x + \mathscr{E}_y \mathbf{a}_y + \mathscr{E}_z \mathbf{a}_z$$

where $\mathscr{E}_x$, $\mathscr{E}_y$, and $\mathscr{E}_z$ are real functions of x, y, z and cosinusoidal functions of time. That is,

$$\mathscr{E}_x = E_{x0} \cos (\omega t + \phi_x)$$
$$\mathscr{E}_y = E_{y0} \cos (\omega t + \phi_y)$$
$$\mathscr{E}_z = E_{z0} \cos (\omega t + \phi_z)$$

where E_{x0}, E_{y0}, E_{z0} are real functions of position (x,y,z), and ϕ_x, ϕ_y, ϕ_z are phase angles with respect to some arbitrary time reference.

The advantages of complex phasor notation in circuit theory are well known. A complex vector notation has similar advantages in electromagnetic field theory. In this notation we have

$$\mathscr{E} = \mathrm{Re}\, (E_{x0}e^{j(\omega t+\phi_x)}\mathbf{a}_x + E_{y0}e^{j(\omega t+\phi_y)}\mathbf{a}_y + E_{z0}e^{j(\omega t+\phi_z)}\mathbf{a}_z)$$
$$= \mathrm{Re}\, [(E_{x0}e^{j\phi_x}\mathbf{a}_x + E_{y0}e^{j\phi_y}\mathbf{a}_y + E_{z0}e^{j\phi_z}\mathbf{a}_z)e^{j\omega t}]$$
$$= \mathrm{Re}\, [(E_x\mathbf{a}_x + E_y\mathbf{a}_y + E_z\mathbf{a}_z)e^{j\omega t}] \tag{7-27}$$

where $E_x = E_{x0}e^{j\phi_x}$, etc., and Re ( ) means *real part of*. The quantities E_x, E_y, and E_z are now *complex* components of a vector **E**, and

$$\mathscr{E} = \mathrm{Re}\, (\mathbf{E}e^{j\omega t}) \tag{7-28}$$

It is sometimes advantageous to note that

$$\mathrm{Re}\, (\mathbf{E}e^{j\omega t}) = \tfrac{1}{2}(\mathbf{E}e^{j\omega t} + \mathbf{E}^*e^{-j\omega t}) \tag{7-29}$$

where **E*** is the complex conjugate of **E**.

The other field quantities may be written in a similar form. However, unless the constitutive relations

$$\mathbf{D} = D(\mathbf{E}) \qquad \mathbf{B} = B(\mathbf{H}) \qquad \mathbf{J} = J(\mathbf{E})$$

are linear, then cross-product terms result in generation of harmonics of the fundamental frequency ω, and we cannot have the sinusoidal steady state.

We next wish to write Maxwell's equations in terms of the complex field vectors, assuming the medium is linear, homogeneous, and isotropic.

In terms of our current notation, Maxwell's equations are

$$\nabla \times \mathscr{E} = -\frac{\partial \mathscr{B}}{\partial t} \tag{I}$$

$$\nabla \times \mathscr{H} = \mathscr{J} + \frac{\partial \mathscr{D}}{\partial t} \tag{II}$$

$$\nabla \cdot \mathscr{B} = 0 \tag{III}$$

$$\nabla \cdot \mathscr{D} = \rho \tag{IV}$$

Substituting Eq. (7-28) into these equations and noting that

$$\nabla \times \text{Re} (\mathcal{A}) = \text{Re} (\nabla \times \mathcal{A})$$

$$\frac{\partial}{\partial t} [\text{Re} (\mathcal{A})] = \text{Re} \left(\frac{\partial \mathcal{A}}{\partial t}\right)$$

and after canceling the common $e^{j\omega t}$ term which appears on both sides of each equation, we obtain the results

$$\nabla \times \mathbf{E} = -j\omega\mu\mathbf{H} \tag{7-30}$$

$$\nabla \times \mathbf{H} = (\sigma + j\omega\epsilon)\mathbf{E} + \mathbf{J}_s \tag{7-31}$$

$$\nabla \cdot \mathbf{H} = 0 \tag{7-32}$$

$$\nabla \cdot \mathbf{E} = \rho/\epsilon \tag{7-33}$$

as Maxwell's equations for the complex vector fields in the sinusoidal steady state.† Use of these complex vector fields results in the same sort of simplification in mathematical manipulation that the use of complex scalar quantities provides in circuit theory. In Eq. (7-31), $\mathbf{J}_s$ is a source current.

Taking the curl of Eq. (7-30) and substituting Eq. (7-31) into the result yields

$$\nabla \times \nabla \times \mathbf{E} + j\omega\mu(\sigma + j\omega\epsilon)\mathbf{E} = -j\omega\mu\mathbf{J}_s \tag{7-34}$$

and, similarly, taking the curl of Eq. (7-31) and substituting Eq. (7-30) into the result yields

$$\nabla \times \nabla \times \mathbf{H} + j\omega\mu(\sigma + j\omega\epsilon)\mathbf{H} = \nabla \times \mathbf{J}_s \tag{7-35}$$

Alternative forms of Eqs. (7-34) and (7-35) may be displayed by using the expansion, Eq. (1-84), and defining

$$\gamma^2 = j\omega\mu(\sigma + j\omega\epsilon) = -\omega^2\mu\epsilon + j\omega\mu\sigma \tag{7-36}$$

The results are

$$\nabla(\nabla \cdot \mathbf{E}) - \nabla^2\mathbf{E} + \gamma^2\mathbf{E} = -j\omega\mu\mathbf{J}_s$$
$$\nabla(\nabla \cdot \mathbf{H}) - \nabla^2\mathbf{H} + \gamma^2\mathbf{H} = \nabla \times \mathbf{J}_s \tag{7-37}$$

In the case of linear, homogeneous, and isotropic regions of space which contain no free sources, this pair of equations reduces to

$$\nabla^2\mathbf{E} - \gamma^2\mathbf{E} = 0$$
$$\nabla^2\mathbf{H} - \gamma^2\mathbf{H} = 0 \tag{7-38}$$

† Note that ρ is of the form $\rho = \rho(x,y,z) \cos(\omega t + \phi_\rho) = \text{Re} (\bar{\rho}e^{j\omega t})$, where $\bar{\rho}$ is a complex scalar. It simplifies the notation if we just remember this, rather than using $\bar{\rho}$.

A second, and very important, point is to note that writing Eq. (7-31) in the form $\nabla \times \mathbf{H} = (\epsilon - j\sigma/\omega)j\omega\mathbf{E} + \mathbf{J}_s$ allows us to define *complex permittivity* as $\epsilon = j\sigma/\omega$, and *complex dielectric constant* as $\epsilon_r - j\sigma/\omega\epsilon_0$, both quantities being widely used in practice, principally in specifying the properties of low-loss dielectrics.

β = phase propagation constant

Equations (7-38) are vector Helmholtz equations. Their solutions and the properties of their solutions have been extensively studied and are well known. A large part of the rest of this text is devoted to examination of electromagnetic fields in situations where the vector Helmholtz equations apply.

7.6 Uniform Plane Waves in Unbounded, Lossless, Source-free Media.

In Secs. 7.2 to 7.4 we observed that Maxwell's equations, applied to unbounded, linear, homogeneous, isotropic, nonconducting, charge-free regions, allowed solutions which were interpreted as uniform plane waves. It is the purpose of this section to examine these solutions for the sinusoidal steady state in rectangular coordinates. We shall use the complex vector-field notation.

For concreteness, we once again assume that the wave is traveling in the z direction and that it is uniform over the xy plane. Mathematically, this is stated by

$$E_z = H_z = 0 \qquad \frac{\partial}{\partial x} = \frac{\partial}{\partial y} = 0 \qquad (7\text{-}20)$$

Substitution of these conditions into the vector Helmholtz equations (7-38) results in the set of scalar equations

$$\frac{d^2 E_x}{dz^2} - \gamma^2 E_x = 0$$

$$\frac{d^2 E_y}{dz^2} - \gamma^2 E_y = 0$$

$$\frac{d^2 H_x}{dz^2} - \gamma^2 H_x = 0 \qquad (7\text{-}39)$$

$$\frac{d^2 H_y}{dz^2} - \gamma^2 H_y = 0$$

Since E_x, E_y, H_x, and H_y are now functions only of z, these are ordinary differential equations, and as may be readily verified by direct substitution, each of them has a solution of a form typified by

$$E_x = E_x^+ e^{-\gamma z} + E_x^- e^{+\gamma z} \qquad (7\text{-}40)$$

It is advantageous to note that, for our current discussion, $\sigma = 0$, and hence

$$\gamma^2 = -\omega^2 \mu\epsilon = (\pm j\beta)^2$$

where

$$\beta = \omega\sqrt{\mu\epsilon} \qquad (7\text{-}41)$$

$$E = E_x^+ \cos\left[\omega\left(t - \frac{1}{v}z\right)\right] = E_x^+ \cos\left(\omega t - \frac{\omega}{v}z\right) \qquad v = \frac{1}{\sqrt{\mu\epsilon}}$$

but now $\beta = \omega\sqrt{\mu\epsilon}$ so $E = E_x^+ \cos(\omega t - \beta z) = E_x^+ e^{-j\beta z}$

is the *phase propagation constant* (or *wave number*). Thus we have as solutions to our problem

$$E_x = E_x^+ e^{-j\beta z} + E_x^- e^{+j\beta z}$$
$$E_y = E_y^+ e^{-j\beta z} + E_y^- e^{+j\beta z}$$
$$H_x = H_x^+ e^{-j\beta z} + H_x^- e^{+j\beta z}$$
$$H_y = H_y^+ e^{-j\beta z} + H_y^- e^{+j\beta z}$$

(7-42)

where the factor $e^{j\omega t}$ is understood.

Once again, we need to look for the connection between the **E**- and **H**-field components.

By direct substitution into Eqs. (7-30) and (7-31), we obtain the two sets of equations

$$\frac{\partial E_x}{\partial z} = -j\omega\mu H_y \qquad \frac{\partial E_y}{\partial z} = j\omega\mu H_x$$

$$\frac{\partial H_y}{\partial z} = -j\omega\epsilon E_x \qquad \frac{\partial H_x}{\partial z} = j\omega\epsilon E_y$$

(7-43)

Even more clearly than in Eqs. (7-23), we see a connection between E_x and H_y and between E_y and H_x. Direct substitution of Eqs. (7-42) into Eqs. (7-43) shows that the relationships are

$$E_x^+ = \eta H_y^+ \qquad E_x^- = -\eta H_y^-$$
$$E_y^+ = -\eta H_x^+ \qquad E_y^- = \eta H_x^-$$

(7-44)

where $\eta = \sqrt{\mu/\epsilon}$ as before.

From Eqs. (7-44) we see that each electric field component is in phase (+ sign) or 180° out of phase (− sign) with its magnetic field counterpart. In order to interpret further the significance of our results, let us once again focus our attention on a particular part of the solution. Specifically, assume that[†]

$$E_x^+ \neq 0 \qquad E_x^- = 0$$
$$E_y^+ = 0 \qquad E_y^- = 0$$

Then we have

$$\mathbf{E} = E_x^+ e^{-j\beta z}\mathbf{a}_x$$

$$\mathbf{H} = \frac{E_x^+}{\eta} e^{-j\beta z}\mathbf{a}_y$$

or

$$\mathscr{E} = \mathrm{Re}\,(E_{x0}^+ e^{j(\omega t - \beta z + \phi_x)})\mathbf{a}_x$$

$$\mathscr{H} = \mathrm{Re}\left(\frac{E_{x0}^+}{\eta} e^{j(\omega t - \beta z + \phi_x)}\right)\mathbf{a}_y$$

(7-45)

[†] Since E_x is independent of E_y, and E_x^+ and E_x^- are linearly independent, this is a possible situation, and will lead to a nontrivial result.

Phase velocity v_p occurs when $(\omega t - \beta z) = const$

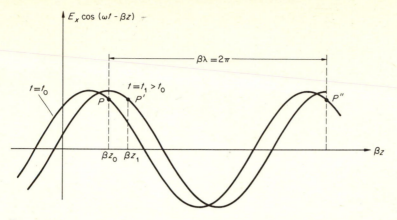

FIGURE 7-2. A spatially harmonic wave plotted at two instants of time.

It is apparent that $\mathscr{E}$ and $\mathscr{H}$ are orthogonal in space and in time phase. Since the relative phase ϕ_x is with respect to an arbitrary time and space reference, we can usually omit it. It can be reinserted when relative phase is important.

If we rewrite Eqs. (7-45), taking the real part and dropping the plus superscript as redundant, we have

$$\mathscr{E} = E_x \cos(\omega t - \beta z)\mathbf{a}_x$$
$$\mathscr{H} = \frac{E_x}{\eta} \cos(\omega t - \beta z)\mathbf{a}_y \qquad (7\text{-}46)$$

Aside from direction and magnitude, the space and time behavior of both the $\mathscr{E}$ and $\mathscr{H}$ fields of Eqs. (7-46) are the same. Let us look at the $\mathscr{E}$ field. A plot of this field at two instants of time, $t = t_0$ and $t = t_1 > t_0$, is shown in Fig. 7-2, from which it is obvious that the wave moves in the $+z$ direction. If we focus our attention on the points P and P', where the amplitude and slope of the two curves are the same, we see that

$$\cos(\omega t_0 - \beta z_0) = \cos(\omega t_1 - \beta z_1) = \text{constant}$$

Thus points P, P', and similar points for other times are *points of constant phase*. Hence, in general, points of constant phase are defined by

$$\omega t - \beta z = \text{constant}$$

By differentiation we obtain the velocity with which a constant phase point travels, and obtain

$$v_p = \frac{dz}{dt}\bigg|_{\omega t - \beta z \,=\, \text{constant}} = \frac{\omega}{\beta} \qquad (7\text{-}47)$$

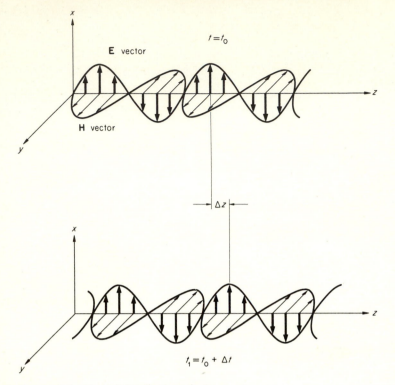

FIGURE 7-3.   Spatial distribution of a harmonic wave.

This velocity is known as the *phase velocity* of the wave, and is implied in Fig. 7-3, which shows the spatial distribution of both **E** and **H** at two instants of time.

A second descriptive parameter of the wave can be determined from examination of Fig. 7-2.   In this figure the points P and P'' are separated in phase by $\beta \, \Delta z = 2\pi$ and in space by $\Delta z = \lambda$.   This space change, required to change phase by 2π (at any fixed time), is called the *wavelength* of the wave. Thus

$$\lambda = \frac{2\pi}{\beta} \tag{7-48}$$

is a quantity which is constant for uniform plane waves.   Notice that wavelength, phase velocity, phase propagation constant, and frequency are interrelated.   In fact, we find easily from Eqs. (7-47) and (7-48) that

$$v_p = \lambda f \tag{7-49}$$

where $f = \omega/2\pi$ is the frequency of the wave.

It should be stressed that *the distance λ, in which the phase changes by 2π, is always measured along the direction of propagation.*

Example 7-1   Wave Parameters. A 10-GHz plane wave traveling in free space has an amplitude $E_x = 1$ V/m.

    (*a*)   Find the phase velocity, the wavelength, and the propagation constant.
    (*b*)   Determine the characteristic impedance of the medium.
    (*c*)   Find the amplitude and direction of the magnetic field intensity.
    (*d*)   Repeat part (*a*) if the wave is traveling in a lossless, unbounded medium, having a permeability the same as free space, but permittivity four times that of free space.

SOLUTIONS

    (*a*)   For free space

$$v_p = c = \frac{1}{\sqrt{\mu_0 \epsilon_0}} = \frac{1}{\sqrt{(4\pi \times 10^{-7})(8.854 \times 10^{-12})}} \approx 3 \times 10^8 \text{ m/s}$$

so that

$$\lambda = \frac{c}{f} = \frac{3 \times 10^8}{10 \times 10^9} = 3 \times 10^{-2} \text{ m} = 3 \text{ cm}$$

and

$$\beta = \frac{2\pi}{\lambda} = \frac{2\pi}{3} = 2.093 \text{ rad/cm}$$

    (*b*)   The characteristic (intrinsic) impedance of free space is

$$\eta = \sqrt{\frac{\mu_0}{\epsilon_0}} = \sqrt{\frac{4\pi \times 10^{-7}}{8.854 \times 10^{-12}}} = 377 \ \Omega$$

    (*c*)   From Eq. (7-44)

$$H_y^+ = \frac{E_x^+}{\eta} = \frac{1}{377} = 2.65 \times 10^{-3} \text{ A/m}$$

if the wave is traveling in the $+z$ direction, or

$$H_y^- = -\frac{E_x^+}{\eta} = -\frac{1}{377} = -2.65 \times 10^{-3} \text{ A/m}$$

if the wave is traveling in the $-z$ direction. Here the minus sign indicates a reversal in space direction.

    (*d*)   For a dielectric medium with $\epsilon_r = 4$,

$$v_p = \frac{1}{\sqrt{\mu_0 \epsilon_0 \epsilon_r}} = \frac{1}{2\sqrt{\mu_0 \epsilon_0}} = 1.5 \times 10^8 \text{ m/s}$$

so that

$$\lambda = \frac{v_p}{f} = \frac{1.5 \times 10^8}{10 \times 10^9} = 1.5 \text{ cm}$$

and

$$\beta = \frac{2\pi}{\lambda} = \frac{2\pi}{1.5} = 4.186 \text{ rad/cm}$$

A similar analysis of the other possibilities contained in our original solutions, Eqs. (7-42), shows that identical phase velocity and wavelength will be obtained for each of the $e^{-j\beta z}$ forms, and that the $e^{+j\beta z}$ form will yield a phase velocity of ω/β in the negative z direction. Thus each of the field components of Eqs. (7-42) is to be interpreted as the sum of two independent waves, one traveling in the $+z$ direction and one traveling in the $-z$ direction. It is to be remembered that our choice of the z direction as the direction of propagation was arbitrary. Our results show that uniform plane waves in an unbounded medium can travel in any direction independently of each other. The form of the results for any other direction of propagation can be inferred from Eqs. (7-42) by noting that the possible field components are transverse to the direction of propagation (Sec. 7.10). The connection between the E-field component and its H-field counterpart may be inferred from the field equations.

7.7   The Complex Poynting's Theorem.

We have shown in Chap. 6 that Poynting's vector

$$\mathcal{P} = \mathcal{E} \times \mathcal{H}$$

is the instantaneous power flow density vector. We wish to obtain the form of Poynting's vector which is appropriate for use with our complex vector representation of the sinusoidal steady state.

From Eq. (7-29) we have

$$\mathcal{E} = \text{Re} (\mathbf{E}e^{j\omega t}) = \tfrac{1}{2}(\mathbf{E}e^{j\omega t} + \mathbf{E}^*e^{-j\omega t})$$

$$\mathcal{H} = \text{Re} (\mathbf{H}e^{j\omega t}) = \tfrac{1}{2}(\mathbf{H}e^{j\omega t} + \mathbf{H}^*e^{-j\omega t})$$

as the instantaneous values of $\mathcal{E}$ and $\mathcal{H}$. In the same way we can write

$$\mathcal{P} = \tfrac{1}{2}(\mathbf{E}e^{j\omega t} + \mathbf{E}^*e^{-j\omega t}) \times \tfrac{1}{2}(\mathbf{H}e^{j\omega t} + \mathbf{H}^*e^{-j\omega t})$$

$$= \tfrac{1}{4}(\mathbf{E} \times \mathbf{H}^* + \mathbf{E}^* \times \mathbf{H}) + \tfrac{1}{4}(\mathbf{E} \times \mathbf{H}e^{j2\omega t} + \mathbf{E}^* \times \mathbf{H}^*e^{-j2\omega t})$$

Using Eq. (7-29) in reverse and noting that

$$(\mathbf{E} \times \mathbf{H}^*)^* = \mathbf{E}^* \times \mathbf{H} \qquad (\mathbf{E} \times \mathbf{H})^* = \mathbf{E}^* \times \mathbf{H}^*$$

we obtain

$$\mathcal{P} = \tfrac{1}{2} \text{Re} [\mathbf{E} \times \mathbf{H}^*] + \tfrac{1}{2} \text{Re} [\mathbf{E} \times \mathbf{H}e^{j2\omega t}] \qquad (7\text{-}50)$$

Notice that the first term is independent of time and that the second term has a magnitude

$$\tfrac{1}{2} |\mathbf{E} \times \mathbf{H}| \cos (2\omega t + \phi)$$

The two terms, in general, have unequal amplitudes. The second term has a time dependence of twice the frequency of the wave and a time-average value which is zero. A plot of the instantaneous value of $\mathcal{P}_x$ is shown in Fig. 7-4.

Instantaneous - $\bar{P} = \bar{E} \times \bar{H}$

Sinusoidal ave - $P_{ave} = \frac{1}{2} \mathrm{Re}[\bar{E} \times \bar{H}^*]$

In Fields we assume given values are Peak values

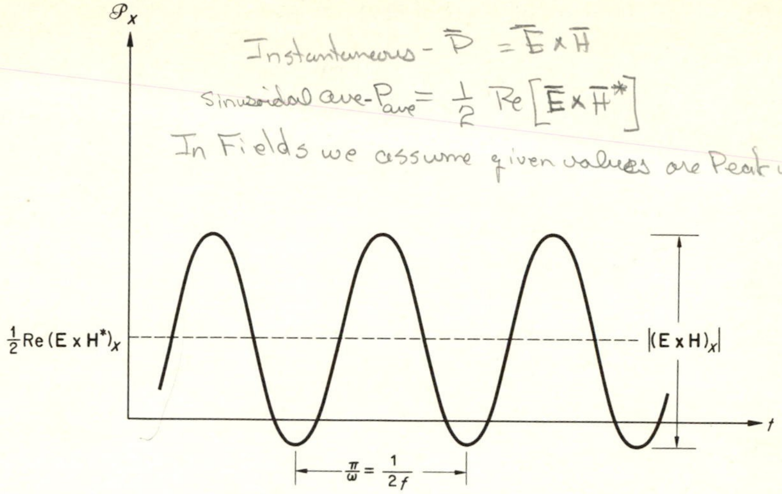

FIGURE 7-4. The time dependence of $\mathscr{P}_x$.

In many problems, the time-average power flow density is the only part of power flow which is of interest. This suggests that we should define a *complex Poynting's vector*

$$\mathbf{S}_c = \tfrac{1}{2}(\mathbf{E} \times \mathbf{H}^*) \tag{7-51}$$

which we can use to calculate the time-average power flow density directly from the complex field vectors without returning to their instantaneous forms. Actually, the motivation for defining a complex Poynting's vector, rather than just noting that the time-average power flow density equals the real part of $\mathbf{S}_c$, is to obtain and to interpret a complex Poynting's theorem. This we can do from a consideration of the complex form of Maxwell's equations,

$$\nabla \times \mathbf{E} = -j\omega\mathbf{B}$$
$$\nabla \times \mathbf{H} = \mathbf{J} + j\omega\mathbf{D}$$

First, we dot-multiply the first equation by $\mathbf{H}^*$ and the conjugate of the second equation by $\mathbf{E}$. We have

$$\mathbf{H}^* \cdot (\nabla \times \mathbf{E}) = -j\omega\mathbf{H}^* \cdot \mathbf{B}$$
$$\mathbf{E} \cdot (\nabla \times \mathbf{H}^*) = \mathbf{E} \cdot \mathbf{J}^* - j\omega\mathbf{E} \cdot \mathbf{D}^*$$

Next, we subtract the first equation from the second, and find

$$-\mathbf{H}^* \cdot (\nabla \times \mathbf{E}) + \mathbf{E} \cdot (\nabla \times \mathbf{H}^*) = \mathbf{E} \cdot \mathbf{J}^* + j\omega(\mathbf{B} \cdot \mathbf{H}^* - \mathbf{E} \cdot \mathbf{D}^*)$$

Finally, we transform the left side of this equation using the vector identity (1-82) and multiply both sides by the factor ½. We obtain

$$-\nabla \cdot \tfrac{1}{2}(\mathbf{E} \times \mathbf{H}^*) = \tfrac{1}{2}\mathbf{E} \cdot \mathbf{J}^* + j\omega(\tfrac{1}{2}\mathbf{B} \cdot \mathbf{H}^* - \tfrac{1}{2}\mathbf{E} \cdot \mathbf{D}^*)$$

$P_{z\,ave} = \frac{1}{2} \frac{E_{x0}^2}{\eta}$ (Peak values)

ave Power through an area S ⊥ to Propagation is

$P_{z\,ave} = \frac{1}{2} \frac{E_{x0}^2}{\eta} S$

The left side is obviously equal to $-\nabla \cdot \mathbf{S}_c$. So

$$-\nabla \cdot \mathbf{S}_c = \tfrac{1}{2}\mathbf{E} \cdot \mathbf{J}^* + j\omega(\tfrac{1}{2}\mathbf{B} \cdot \mathbf{H}^* - \tfrac{1}{2}\mathbf{E} \cdot \mathbf{D}^*) \qquad (7\text{-}52)$$

This is the differential form of the complex Poynting's theorem. The corresponding integral form follows at once.

$$-\oint_\Sigma \mathbf{S}_c \cdot \mathbf{n}\, da = \int_V \tfrac{1}{2}\mathbf{E} \cdot \mathbf{J}^*\, dv + j\omega\int_V (\tfrac{1}{2}\mathbf{B} \cdot \mathbf{H}^* - \tfrac{1}{2}\mathbf{E} \cdot \mathbf{D}^*)\, dv \quad (7\text{-}53)$$

Inspection shows that the second term on the right of Eq. (7-53) is purely imaginary. The first term, however, could have a real and an imaginary part. In general, the current density vector $\mathbf{J}$ consists of a conduction component and a source component

$$\mathbf{J} = \mathbf{J}_c + \mathbf{J}_s = \sigma\mathbf{E} + \mathbf{J}_s \qquad (7\text{-}54)$$

so that

$$\tfrac{1}{2}\mathbf{E} \cdot \mathbf{J}^* = \tfrac{1}{2}\sigma E^2 + \tfrac{1}{2}\mathbf{E} \cdot \mathbf{J}_s^* \qquad (7\text{-}55)$$

Although the first term on the right of Eq. (7-55) is purely real, the second term is generally complex, because of the arbitrary phase relationship between the vector $\mathbf{E}$ and the source component of $\mathbf{J}$.

In view of this discussion, it is seen that

$$\langle P_d \rangle = \int_V \tfrac{1}{2}\mathbf{E} \cdot \mathbf{J}_c^*\, dv = \int_V \tfrac{1}{2}\sigma E^2\, dv \qquad (7\text{-}56)$$

represents the power dissipation within V *averaged* over a complete cycle of the wave.   (The symbol $\langle\ \rangle$ is used to denote the time average of a quantity.) Similarly,

$$\langle P_s \rangle = \mathrm{Re}\left(\int_V \tfrac{1}{2}\mathbf{E} \cdot \mathbf{J}_s^*\, dv\right) \qquad (7\text{-}57)$$

and

$$j\langle Q_s \rangle = j\,\mathrm{Im}\left(\int_V \tfrac{1}{2}\mathbf{E} \cdot \mathbf{J}_s^*\, dv\right) \qquad (7\text{-}58)$$

denote the real and imaginary components of the average power exchanged between the field and the sources in V.

Now, using the same notation, we see that

energy
Densities

$$\langle W_m \rangle = \int_V \tfrac{1}{4}\mathbf{B} \cdot \mathbf{H}^*\, dv \qquad (7\text{-}59)$$

$$\langle W_e \rangle = \int_V \tfrac{1}{4}\mathbf{E} \cdot \mathbf{D}^*\, dv \qquad (7\text{-}60)$$

are the magnetic and electric energy storage terms averaged over a complete cycle of the wave.   Therefore the total complex power flow across the surface Σ may now be written

$$-\oint_\Sigma \mathbf{S}_c \cdot \mathbf{n}\, da = (\langle P_d \rangle + \langle P_s \rangle) + j[2\omega(\langle W_m \rangle - \langle W_e \rangle) + \langle Q_s \rangle] \quad (7\text{-}61)$$

for Lossy dielectric

$$P_{z,ave} = \frac{1}{2}\frac{E_{xo}^2}{\eta_m}\, e^{-2\alpha z}\cos\theta_\eta \qquad \text{where } \eta = \eta_m \angle\theta_\eta$$

thus displaying a clear identification of the complex Poynting's vector with the real and imaginary parts of the power transmitted through the surface Σ. A very useful application of Eq. (7-61) is made in Sec. 11.5, where impedance is defined in terms of complex power.

In conclusion, let us note that the relations $\mathbf{D} = \epsilon\mathbf{E}$ and $\mathbf{B} = \mu\mathbf{H}$ allow us to write Eq. (7-53) in the equivalent form

$$-\oint_{\Sigma} \tfrac{1}{2}(\mathbf{E} \times \mathbf{H}^*) \cdot \mathbf{n}\, da = \int_V \tfrac{1}{2}\mathbf{E} \cdot \mathbf{J}^*\, dv + j\omega \int_V (\tfrac{1}{2}\mu H^2 - \tfrac{1}{2}\epsilon E^2)\, dv \quad (7\text{-}62)$$

Wave Polarized in x direction means E is in x direction

7.8 Polarization. Polarization of a uniform plane wave is a term which describes the behavior of the instantaneous electric field vector. ($\mathscr{H}$ is perpendicular to $\mathscr{E}$ and need not be described separately.) It is best defined in terms of specific examples.

Consider the uniform plane wave whose total description is given by

$$\mathscr{E} = \mathrm{Re}\,[(E_x^+ e^{-j\beta z})e^{j\omega t}]\mathbf{a}_x = E_{x0}\cos\,(\omega t - \beta z + \phi_x)\mathbf{a}_x$$

$$\mathscr{H} = \mathrm{Re}\left[\left(\frac{E_x^+}{\eta}\,e^{-j\beta z}\right)e^{j\omega t}\right]\mathbf{a}_y = \frac{E_{x0}}{\eta}\cos\,(\omega t - \beta z + \phi_x)\mathbf{a}_y \quad (7\text{-}63)$$

It is seen that the instantaneous electric field vector is always in the x direction. Thus this wave is said to be *linearly polarized in the x direction*, or alternatively, the x plane is the *plane of polarization*, or the wave is *x-polarized*. Notice that these descriptions are incomplete in that the direction of propagation is not stated. Notice also that writing out the instantaneous $\mathscr{E}$ and $\mathscr{H}$ fields has made it easier to visualize the behavior of the fields. The complex vector form

$$\mathbf{E} = E_x^+ e^{-j\beta z}\mathbf{a}_x$$

$$\mathbf{H} = \frac{E_x^+}{\eta}\,e^{-j\beta z}\mathbf{a}_y \quad (7\text{-}64)$$

is actually a sufficient description of the wave.

As a second example consider the field given by

$$\mathbf{E} = E_x^- e^{j\beta z}\mathbf{a}_x$$

$$\mathbf{H} = \frac{E_x^-}{\eta}\,e^{j\beta z}\mathbf{a}_y \quad (7\text{-}65)$$

This is also linearly polarized in the x direction. Clearly, the polarization of a wave does not specify its direction of propagation. For instance, the field given by

$$\mathbf{E} = (E_x^+ e^{-j\beta z} + E_x^- e^{j\beta z})\mathbf{a}_x$$

$$\mathbf{H} = \left(\frac{E_x^+}{\eta}\,e^{-j\beta z} - \frac{E_x^-}{\eta}\,e^{j\beta z}\right)\mathbf{a}_y \quad (7\text{-}66)$$

In time phase

$$E = E_x^+ e^{-j\beta z} + E_y^+ e^{-j\beta z}$$

$$|E| = \sqrt{E_x^2 + E_y^2}$$

for linear Polarization $\quad \theta = \tan\dfrac{E_y^+}{E_x^+}$

$$\theta = \sin^{-1}\frac{A}{B}$$

when 0

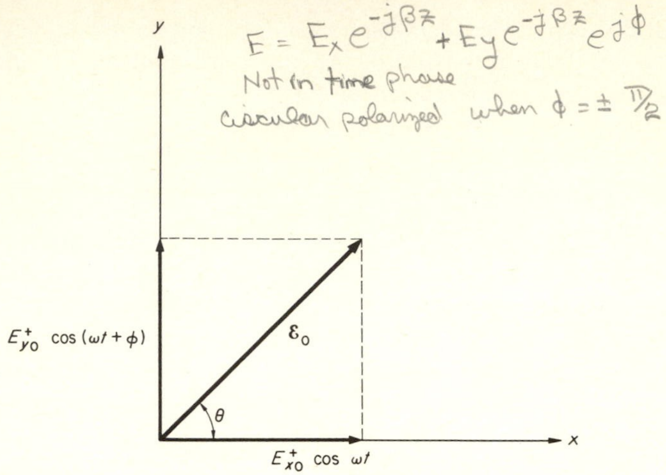
$E = E_x e^{-j\beta z} + E_y e^{-j\beta z} e^{j\phi}$
Not in time phase
circular polarized when $\phi = \pm \pi/2$
y
x
E^+_{y0} cos ($\omega t + \phi$)
$\mathcal{E}_0$
θ
E^+_{x0} cos ωt

FIGURE 7-5. The instantaneous value of $\mathcal{E}_0$.

obviously consists of an x-polarized wave traveling in the $+z$ direction plus an x-polarized wave traveling in the $-z$ direction, both waves being linearly polarized in the x direction.

Consider now the case of a uniform plane wave traveling in the $+z$ direction. Let

$$\mathbf{E} = E_x^+ e^{-j\beta z}\mathbf{a}_x + E_y^+ e^{-j\beta z + j\phi}\mathbf{a}_y$$
$$\mathbf{H} = \frac{E_x^+}{\eta} e^{-j\beta z}\mathbf{a}_y - \frac{E_y^+}{\eta} e^{-j\beta z + j\phi}\mathbf{a}_x \tag{7-67}$$

where E_x^+ and E_y^+ are arbitrary complex amplitudes. Obviously, the field is the sum of the orthogonal, linearly polarized uniform plane waves of some as yet unspecified magnitudes and relative phase. Both are traveling in the $+z$ direction. Visualization of the resultant field is not simple. If we write out the instantaneous $\mathcal{E}$ field, we obtain

$$\mathcal{E} = E_{x0}^+ \cos(\omega t - \beta z)\mathbf{a}_x + E_{y0}^+ \cos(\omega t - \beta z + \phi)\mathbf{a}_y \tag{7-68}$$

where ϕ is the relative phase of the y component of $\mathcal{E}$ with respect to the x component of $\mathcal{E}$.

Let us examine $\mathcal{E}$ at $z = 0$.† We have

$$\mathcal{E}_0 = E_{x0}^+ \cos \omega t\, \mathbf{a}_x + E_{y0}^+ \cos(\omega t + \phi)\mathbf{a}_y \tag{7-69}$$

A plot of this instantaneous $\mathcal{E}_0$ field in the xy plane is shown in Fig. 7-5.

† The discussion applies to any other plane $z = $ constant, since the term βz is common and changes only the phase of the x and y components of $\mathcal{E}$. Furthermore, if the reversal in the direction of wave travel is properly taken into account, the same analysis applies to a wave traveling in the negative z direction.

Analysis of the figure shows that

$$|\mathscr{E}_0|^2 = (E_{x0}^+)^2 \cos^2 \omega t + (E_{y0}^+)^2 \cos^2 (\omega t + \phi) \qquad (7\text{-}70)$$

and that $\mathscr{E}_0$ is at an angle θ with respect to the x axis, given by

$$\tan \theta = \frac{E_{y0}^+ \cos (\omega t + \phi)}{E_{x0}^+ \cos \omega t} \qquad (7\text{-}71)$$

The implications of these equations will be more apparent if we now give a precise definition of polarization.

The polarization of an electromagnetic wave is specified by the plane curve, whose parametric equation is defined by the instantaneous values of the electric field components at a fixed point in space. A second definition states that the polarization is the curve traced out by the end point of the arrow representing the $\mathscr{E}_0$ field in Fig. 7-5. The two definitions are identical. This is particularly evident if we use E_x and E_y as the rectangular coordinates of the planar curve. From this point of view, we have from Eq. (7-69)

$$E_x = E_{x0}^+ \cos \omega t \qquad (7\text{-}72)$$

$$E_y = E_{y0}^+ \cos (\omega t + \phi) \qquad (7\text{-}73)$$

which generally represent an ellipse. Therefore the wave is said to be *elliptically polarized*, and the curve is called the *polarization ellipse*.

Some special cases of polarization are of interest, as follows.

Case 1

$$\phi = 0$$

$$E_x = E_{x0}^+ \cos \omega t$$

$$E_y = E_{y0}^+ \cos \omega t \qquad (7\text{-}74)$$

$$\tan \theta = \frac{E_y}{E_x} = \frac{E_{y0}^+}{E_{x0}^+}$$

The locus is a straight line. The field is *linearly polarized* in the θ direction (Fig. 7-6a).

Case 2

$$E_{x0}^+ = E_{y0}^+ = E \qquad \phi = -\frac{\pi}{2}$$

$$E_x = E \cos \omega t \qquad (7\text{-}75)$$

$$E_y = E \cos \left(\omega t - \frac{\pi}{2}\right) = E \sin \omega t$$

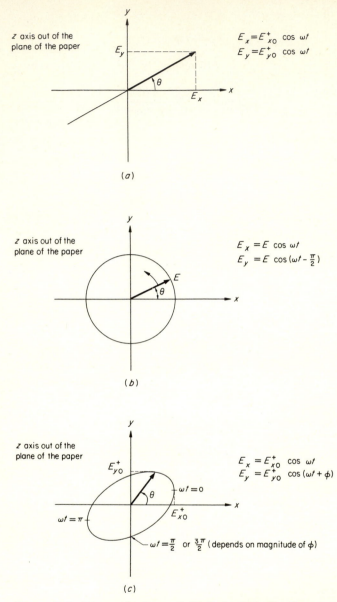

FIGURE 7-6.   Linear, circular, and elliptical polarization for waves traveling in the positive z direction.   (a) Linear polarization; (b) circular polarization (clockwise rotation); (c) elliptical polarization.

The locus is a circle (Fig. 7-6b) of radius E, and

$$\tan \theta = \frac{E \sin \omega t}{E \cos \omega t} = \tan \omega t \tag{7-76}$$

$$\theta = \omega t$$

The electric field is constant in magnitude, and it rotates clockwise with an angular frequency ω. The field is said to be *circularly polarized* in the *clockwise* (right-hand) direction, looking along (not against) the direction $+z$ of propagation.

Case 3

$$E_{x0}^+ = E_{y0}^+ = E \qquad \phi = \frac{\pi}{2} \tag{7-77}$$

The student will verify that this field is *circularly polarized* in the *counterclockwise* (left-hand) direction.

Case 4

$$E_{x0}^+ \neq E_{y0}^+ \qquad \phi = -\frac{\pi}{2}$$

$$E_x = E_{x0}^+ \cos \omega t$$

$$E_y = E_{y0}^+ \sin \omega t \tag{7-78}$$

$$\tan \theta = \frac{E_{y0}^+}{E_{x0}^+} \tan \omega t$$

This is an ellipse whose major and minor axes are along the x and y axes, and θ is not equal to ωt. The wave is *elliptically polarized*. The electric field rotates in the clockwise direction. The *period* of its rotational frequency is given by ω, but its angular velocity of rotation is *not* constant (Prob. 7-18).

Case 5

$$E_{x0}^+ \neq E_{y0}^+ \qquad \phi \neq \pm\frac{\pi}{2} \tag{7-79}$$

This is the general case of elliptical polarization (Fig. 7-6c). Construction of the locus and an examination of it will show that the major axis of the ellipse is inclined at an angle with respect to the x axis; this angle will be seen to depend upon E_{y0}^+/E_{x0}^+ and upon ϕ. In fact, the locus traced by the tip of the vector $\mathbf{E}$ is

$$\left(\frac{E_x}{E_{x0}^+}\right)^2 - 2\left(\frac{E_x E_y}{E_{x0}^+ E_{y0}^+}\right)\cos\phi + \left(\frac{E_y}{E_{y0}^+}\right)^2 = \sin^2\phi \tag{7-80}$$

an expression obtained by elimination of the variable component ωt (Prob. 7-5). The various cases considered earlier follow easily from a consideration of this equation under specialized conditions. For example, for Case 3, Eq. (7-80) reduces to

$$E_x^2 + E_y^2 = E^2$$

which, obviously, is the equation of a circle.

Example 7-2   Polarization. The wave represented by

$$E_x = 2 \cos (\omega t - \beta z)$$

is linearly polarized in the x direction. Similarly, the wave represented by

$$E_y = 4 \cos (\omega t - \beta z)$$

is linearly polarized in the y direction. On the other hand, the wave denoted by

$$E_x = 2 \cos (\omega t - \beta z)$$
$$E_y = 4 \cos (\omega t - \beta z)$$

is linearly polarized also, but in a direction determined by $\tan \theta = 2$ (Fig. 7-6a).
    The wave described by

$$E_x = 2 \cos (\omega t - \beta z)$$
$$E_y = 2 \cos (\omega t - \beta z - 90°)$$

→ clockwise

is circularly polarized with clockwise rotation, whereas

→ counterclockwise

$$E_x = 2 \cos (\omega t - \beta z)$$
$$E_y = 2 \cos (\omega t - \beta z + 90°)$$

is circularly polarized also, but with counterclockwise rotation. The last statement applies also in the case of the wave represented by

$$E_x = 2 \cos (\omega t + \beta z)$$
$$E_y = 2 \cos (\omega t + \beta z - 90°)$$

Finally,

$$E_x = 2 \cos (\omega t - \beta z)$$
$$E_y = 4 \cos (\omega t - \beta z + 60°)$$

is an elliptically polarized wave rotating counterclockwise. The equation of the ellipse is

$$\left(\frac{E_x}{2}\right)^2 - \frac{1}{8} E_x E_y + \left(\frac{E_y}{4}\right)^2 = \frac{3}{4}$$

From the foregoing discussion it is apparent that an arbitrarily polarized, uniform plane wave can be considered to result from the superposition of two linearly polarized waves. Moreover, every linearly polarized wave can be resolved into two circularly polarized waves rotating in opposite directions with the same angular speed (Prob. 7-4). Consequently, an arbitrarily polarized wave can be considered as the sum of two circularly polarized waves.

Rate at which $\rho \downarrow$ in a material $\rho(t) = \rho_0 e^{-\frac{\sigma}{\epsilon} t}$

Relaxation time = time it takes for charge density ρ to reduce to $\frac{\rho_0}{e}$

$\tau = $ Relax time $= \frac{\epsilon}{\sigma}$ sec

Summarizing, this section has dealt with elliptical polarization and its two special limiting cases of circular and linear polarization. Linear polarization requires either one of two transverse field components, or if both are present, no phase difference between them. Circular polarization requires two equal field components in space and phase quadrature.

7.9 Uniform Plane Waves in Conducting Media.

In this section we examine the behavior of uniform plane waves in unbounded, linear, homogeneous, isotropic, charge-free regions whose conductivity is finite but not zero.

We recall that the governing equations are Eqs. (7-36) and (7-38), with solutions of the form

$$E_x = E_x^+ e^{-\gamma z} + E_x^- e^{+\gamma z} \tag{7-40}$$

and similarly for E_y, H_x, H_y. However, we must now note that γ is complex. The definition of γ was through Eq. (7-36). It follows, then, that

$$\gamma = \sqrt{j\omega\mu(\sigma + j\omega\epsilon)} \tag{7-81}$$

The radical has two roots whose choice will be governed by the requirement that the imaginary part of γ shall always be positive. However, for the sake of discussion, we consider momentarily four distinct possibilities which result from taking both roots while at the same time allowing σ to be negative as well as positive. These are

$$\gamma = \begin{cases} \alpha + j\beta \\ -\alpha - j\beta \end{cases} \tag{7-82}$$

$$\gamma = \begin{cases} \alpha - j\beta \\ -\alpha + j\beta \end{cases} \tag{7-83}$$

We shall call α the *attenuation constant* and β, as previously, the phase-shift constant. The significance of the two forms, and of α and β, can be seen from a substitution in Eq. (7-40).

We look first at $\gamma = \alpha + j\beta$. In this case, Eq. (7-40) becomes

$$E_x = E_x^+ e^{-\alpha z} e^{-j\beta z} + E_x^- e^{+\alpha z} e^{+j\beta z} \tag{7-84}$$

and similarly for E_y, H_x, H_y. From Eq. (7-84) it is apparent that the resultant wave consists of two components, a component traveling in the $+z$ direction with an amplitude $E_x^+ e^{-\alpha z}$ and a component traveling in the $-z$ direction with an amplitude $E_x^- e^{+\alpha z}$. Both components are exponentially attenuated in the direction of propagation. This is to be expected. It states that the ohmic losses in the conducting medium result in an exponential attenuation of the wave amplitude.

The definition $\gamma = -(\alpha + j\beta)$ leads to identical results, the only difference being the interchange of E_x^+ and E_x^- in Eq. (7-84).

Now, the definition $\gamma = \alpha - j\beta$ gives

$$E_x = E_x^+ e^{+\alpha z} e^{-j\beta z} + E_x^- e^{-\alpha z} e^{+j\beta z} \tag{7-85}$$

Interpretation of this equation shows that both components are exponentially growing in the direction of propagation. Although this solution is mathematically valid, it is physically unreal for passive media since it implies that energy is being extracted from the medium. Use of exponentially growing waves is useful in analysis of the behavior of certain active devices, such as traveling-wave tubes, but is ignored in the balance of this book.

Similar comments apply to $\gamma = -(\alpha - j\beta)$.

Having found the solutions for E_x, E_y, H_x, H_y, we need to establish the connection between the field components and to obtain the values of α and β.

First, we write the complex vector form of Maxwell's equations for our special case, which is expressed by the set (7-20). With a little algebra, we obtain equations similar to Eqs. (7-43), namely,

$$\frac{\partial E_x}{\partial z} = -j\omega\mu H_y \qquad \frac{\partial E_y}{\partial z} = j\omega\mu H_x$$

$$\frac{\partial H_y}{\partial z} = -(\sigma + j\omega\epsilon)E_x \qquad \frac{\partial H_x}{\partial z} = (\sigma + j\omega\epsilon)E_y \tag{7-86}$$

As in the lossless case, we see that the equations separate into two independent sets, and once again a connection exists between E_x and H_y and between E_y and H_x. To obtain the actual relationships, we substitute our solution into these equations, noting that, for E_x in the form of Eq. (7-40),

$$\frac{\partial E_x}{\partial z} = -\gamma E_x^+ e^{-\gamma z} + \gamma E_x^- e^{+\gamma z} \tag{7-87}$$

and similarly for $\partial E_y/\partial z$, $\partial H_x/\partial z$, $\partial H_y/\partial z$. We obtain for the pair on the left of Eqs. (7-86)

$$-\gamma E_x^+ e^{-\gamma z} + \gamma E_x^- e^{+\gamma z} = -j\omega\mu H_y^+ e^{-\gamma z} - j\omega\mu H_y^- e^{+\gamma z}$$

$$-\gamma H_y^+ e^{-\gamma z} + \gamma H_y^- e^{+\gamma z} = -(\sigma + j\omega\epsilon)E_x^+ e^{-\gamma z} - (\sigma + j\omega\epsilon)E_x^- e^{+\gamma z} \tag{7-88}$$

Since the wave represented by $e^{-\gamma z}$ is really independent of the wave represented by $e^{\gamma z}$, each of the equations above implies two equalities. Specifically,

$$E_x^+ = \frac{j\omega\mu}{\gamma} H_y^+ = \sqrt{\frac{j\omega\mu}{\sigma + j\omega\epsilon}} H_y^+ \tag{7-89}$$

$$E_x^- = -\frac{j\omega\mu}{\gamma} H_y^- = -\sqrt{\frac{j\omega\mu}{\sigma + j\omega\epsilon}} H_y^- \tag{7-90}$$

A similar analysis yields relations between E_y^+ and H_x^+ and E_y^- and H_x^-. The complete results tabulated for reference are

$$E_x^+ = \eta H_y^+ \qquad E_y^+ = -\eta H_x^+$$
$$E_x^- = -\eta H_y^- \qquad E_y^- = \eta H_x^-$$

(7-91)

where

$$\eta = \sqrt{\frac{j\omega\mu}{\sigma + j\omega\epsilon}}$$

(7-92)

These relations are formally the same as those obtained for the lossless case, the only difference being in the definition of the intrinsic impedance. It will be noted that, for the lossless case ($\sigma = 0$), Eq. (7-92) reduces to the previously obtained value, $\eta = \sqrt{\mu/\epsilon}$ [Eq. (7-26)].

The complex nature of the intrinsic impedance implies that, although the electric and magnetic field vectors of a uniform plane wave in conducting media are still orthogonal to each other and are still transverse to the direction of propagation, they are no longer in time phase with each other.[†]

We still need to obtain explicit expressions for α and β. Setting

$$\gamma^2 = (\alpha + j\beta)^2 = j\omega\mu(\sigma + j\omega\epsilon)$$

then expanding,

$$\alpha^2 - \beta^2 + 2j\alpha\beta = j\omega\mu\sigma - \omega^2\mu\epsilon$$

and equating the real and the imaginary parts yields

$$\alpha^2 - \beta^2 = -\omega^2\mu\epsilon$$

$$2\alpha\beta = \omega\mu\sigma$$

This pair can be solved simultaneously to give

$$\alpha = \omega\sqrt{\mu\epsilon}\left\{\frac{1}{2}\left[\sqrt{1 + \left(\frac{\sigma}{\omega\epsilon}\right)^2} - 1\right]\right\}^{1/2}$$

(7-93)

$$\beta = \omega\sqrt{\mu\epsilon}\left\{\frac{1}{2}\left[\sqrt{1 + \left(\frac{\sigma}{\omega\epsilon}\right)^2} + 1\right]\right\}^{1/2}$$

(7-94)

Although this pair gives the exact results, not much insight into their significance can be obtained in this form. As a practical matter, most materials fall into one of two widely separated classes, which we shall call *good conductors* and *good dielectrics*. Their definition and behavior are described below.

[†] The student may wish to compare the significance of $E_x^+ = \eta H_y^+$, where η is complex, with $V = ZI$ of circuit theory, where Z is complex.

Class 1. Good conductors By convention we shall define a good conductor as the case where

$$\left(\frac{\sigma}{\omega\epsilon}\right)^2 \gg 1$$

This definition depends upon frequency, as well as conductivity. Most metallic conductors are good conductors at frequencies below 10^{10} Hz. So, for good conductors, it follows from Eqs. (7-93) and (7-94) that

$$\alpha \approx \beta \approx \sqrt{\frac{\omega\mu\sigma}{2}} \gg \omega\sqrt{\mu\epsilon}$$

$$\eta \approx \sqrt{\frac{j\omega\mu}{\sigma}} = \sqrt{\frac{\omega\mu}{\sigma}} \; \underline{|45°}$$

(7-95)

This implies that the wave is attenuated very rapidly, that its phase velocity is very low compared with phase velocity in lossless media, and that the **E**-field vector leads its associated **H**-field vector by 45°.

For good conductors, a distance called the *skin depth*, or *depth of penetration*, δ, is defined as the distance in which the wave amplitude is attenuated to $1/e$ of its initial value. From a consideration of the first term on the right of Eq. (7-84) we see that

$$\delta = \frac{1}{\alpha} = \sqrt{\frac{2}{\omega\mu\sigma}}$$

(7-96)

The skin depth δ is very small for good conductors at high frequencies. Typically, for copper, $\delta = 0.85$ cm at 60 Hz and $\delta = 0.007$ cm at 10^6 Hz.

Class 2. Good dielectrics We shall define a *good dielectric* as the case where†

$$\left(\frac{\sigma}{\omega\epsilon}\right)^2 \ll 1$$

In this case, one needs to use the binomial expansion of the exact expressions for α and β. Retaining only the leading terms yields

$$\alpha \approx \frac{\sigma}{2}\sqrt{\frac{\mu}{\epsilon}}$$

$$\beta \approx \omega\sqrt{\mu\epsilon}\left[1 + \frac{1}{8}\left(\frac{\sigma}{\omega\epsilon}\right)^2 + \cdots\right] \approx \omega\sqrt{\mu\epsilon}$$

(7-97)

Also

$$\eta \approx \sqrt{\frac{\mu}{\epsilon}}$$

(7-98)

† It is common practice to refer to the ratio $\sigma/\omega\epsilon$ by the name *loss tangent* and to denote it as $\tan\delta = \sigma/\omega\epsilon$. This δ, however, is different from the skin depth δ.

It is seen that, for good dielectrics, the wave is attenuated very slowly, that the phase velocity is nearly the same as though the medium were lossless, and that the **E**-field vector and its associated **H**-field vector are nearly in phase.

In summarizing propagation phenomena in conducting media, we should point out that, analytically at least, this case is no different from the case of propagation in nonconducting media. As always, the point of departure is the set of Maxwell's equations. And, as in lossless media, a TEM wave can propagate through a conducting medium. However, there is a very significant difference in the structural characteristics of the field. In lossy media the wave is attenuated at a rate determined by the amplitude of the attenuation constant α. This in turn is determined by the conductivity of the medium. The larger the conductivity, the larger the attenuation constant. In the limit as $\sigma \rightarrow \infty$, the attenuation constant becomes infinitely large, and propagation stops. In other words, no wave can go through a truly perfect conductor.

onaxit

7.10 Propagation of Uniform Plane Waves in Arbitrary Space Directions. We shall now consider the general problem of uniform-plane-wave propagation along a specified axis, orientated in a completely arbitrary manner relative to a rectangular system of coordinates.† To fix ideas, we suppose that a TEM wave is propagated in a direction fixed by the unit vector **n** (Fig. 7-7), so that, at every instant, the vectors **E** and **H** are constant in both magnitude and direction over planes defined by

$$\mathbf{r} \cdot \mathbf{n} = \text{constant} \qquad (7\text{-}99)$$

where
$$\mathbf{r} = x\mathbf{a}_x + y\mathbf{a}_y + z\mathbf{a}_z \qquad (7\text{-}100)$$

That this is the equation of the plane follows from a consideration of the fact that, for every point on the plane, the projection of the radius vector **r** on the line fixed by the unit vector **n** is just equal to the distance from the origin to the point of intersection of the plane with the z' axis.

It can be shown (Prob. 7-17) that an *admissible* solution to Maxwell's equations in a linear, homogeneous, isotropic, and charge-free medium is

$$\mathbf{E} = (\mathbf{E}_1 e^{-\gamma \mathbf{n} \cdot \mathbf{r}} + \mathbf{E}_2 e^{\gamma \mathbf{n} \cdot \mathbf{r}})$$

$$\mathbf{H} = \frac{-j\gamma}{\omega\mu} (\mathbf{n} \times \mathbf{E}_1 e^{-\gamma \mathbf{n} \cdot \mathbf{r}} - \mathbf{n} \times \mathbf{E}_2 e^{\gamma \mathbf{n} \cdot \mathbf{r}}) \qquad (7\text{-}101)$$

where the vectors $\mathbf{E}_1$ and $\mathbf{E}_2$ are contained in planes normal to **n** and are

† A simpler, but less general, development is given in Sec. 8.2.

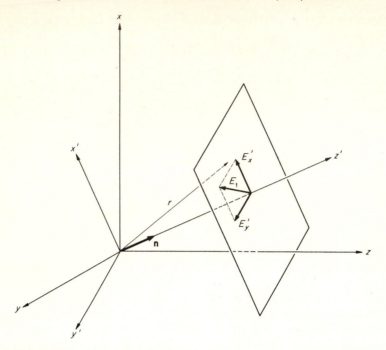

FIGURE 7-7.   The space orientation of a TEM wave traveling in the z' direction.

independent of the coordinates x', y', and z'.   In addition,

$$\gamma^2 = -\omega^2\mu\epsilon + j\omega\mu\sigma \qquad (7\text{-}102)$$

is the complex propagation constant previously defined in Sec. 7.5.

Evidently, the terms associated with $\mathbf{E_1}$ represent harmonic waves traveling in the positive z' direction, with a phase velocity determined by the imaginary part of $\gamma = \alpha + j\beta$.   Thus, setting the time derivative of the constant-phase expression

$$-\beta\mathbf{n} \cdot \mathbf{r} + \omega t = \text{constant} \qquad (7\text{-}103)$$

equal to zero, we find

$$v_p = \mathbf{n} \cdot \frac{d\mathbf{r}}{dt} = \frac{\omega}{\beta} \qquad (7\text{-}104)$$

Similarly, the terms associated with $\mathbf{E_2}$ represent harmonic waves traveling in the negative z' direction, with the same phase velocity $v_p = \omega/\beta$.

It will be noted, further, that the negative sign introduces a sign reversal in Poynting's vector for the two component waves, and thus correctly predicts energy travel in opposite directions. For simplicity, assume $\gamma = j\beta$. Then

$$\mathbf{S_1} = \tfrac{1}{2}\mathbf{E_1} \times \mathbf{H_1^*} = \tfrac{1}{2}\mathbf{E_1} \times \left(\frac{\beta}{\omega\mu}\,\mathbf{n} \times \mathbf{E_1^*}\right)$$

$$= \frac{1}{2}\frac{\beta}{\omega\mu}[(\mathbf{E_1} \cdot \mathbf{E_1^*})\mathbf{n} - (\mathbf{E_1} \cdot \mathbf{n})\mathbf{E_1^*}] = \frac{1}{2}\frac{\beta}{\omega\mu}\,E_1{}^2\mathbf{n} \qquad (7\text{-}105)$$

and $$\mathbf{S_2} = \tfrac{1}{2}\mathbf{E_2} \times \mathbf{H_2^*} = \tfrac{1}{2}\mathbf{E_2} \times \left(-\frac{\beta}{\omega\beta}\,\mathbf{n} \times \mathbf{E_2^*}\right)$$

$$= \frac{1}{2}\frac{\beta}{\omega\mu}[-(\mathbf{E_2} \cdot \mathbf{E_2^*})\mathbf{n} + (\mathbf{E_2} \cdot \mathbf{n})\mathbf{E_2^*}] = \frac{1}{2}\frac{\beta}{\omega\mu}\,E_2{}^2(-\mathbf{n}) \qquad (7\text{-}106)$$

Now, we stated earlier that the vectors $\mathbf{E_1}$ and $\mathbf{E_2}$ are contained in planes normal to the unit vector $\mathbf{n}$. Therefore $\mathbf{n} \cdot \mathbf{E_1} = 0$ and $\mathbf{n} \cdot \mathbf{E_2} = 0$. Since $\mathbf{n} \times \mathbf{E}$ is a vector perpendicular to both $\mathbf{n}$ and $\mathbf{E}$, it follows that both component vectors of $\mathbf{H}$ are normal to both $\mathbf{n}$ and $\mathbf{E}$ and therefore are coplanar with $\mathbf{E_1}$ and $\mathbf{E_2}$. We conclude that the electric and magnetic field vectors associated with either the forward $(+z')$ or the backward $(-z')$ traveling waves are orthogonal to each other and to the direction of propagation.

A final property of the wave represented by Eqs. (7-101) is revealed through examination of the phase factor

$$\phi = \pm\beta\mathbf{n} \cdot \mathbf{r} + \omega t \qquad (7\text{-}107)$$

At a fixed instant of time, the second term is a constant, which we can denote as k, and, without loss in generality, we can focus attention on the phase factor which carries the plus sign. Then we can write

$$\phi = \beta\mathbf{n} \cdot \mathbf{r} + k \qquad (7\text{-}108)$$

If we let

$$\mathbf{n} = n_x\mathbf{a}_x + n_y\mathbf{a}_y + n_z\mathbf{a}_z \qquad (7\text{-}109)$$

where n_x, n_y, n_z are the fixed direction cosines of $\mathbf{n}$, the phase becomes

$$\phi = \beta(xn_x + yn_y + zn_z) + k \qquad (7\text{-}110)$$

so that $$\nabla\phi = \beta(n_x\mathbf{a}_x + n_y\mathbf{a}_y + n_z\mathbf{a}_z) = \beta\mathbf{n} \qquad (7\text{-}111)$$

We conclude that the phase of the field changes most rapidly in the direction of $\mathbf{n}$, which, we note, is the unit normal to the planes of constant phase; moreover, the maximum rate of change of ϕ per unit distance is exactly equal to β, the phase constant of the medium. It is worth noting that in any direction, other than $\mathbf{n}$, the phase change per unit distance is smaller, and the phase velocity correspondingly larger. In fact, as this direction approaches a normal to $\mathbf{n}$, *the phase velocity tends to infinity.*

Example 7-3 Characteristic Features of a General TEM Wave. For the uniform plane wave defined by

$$\mathbf{E} = [\mathbf{a}_x + E_y\mathbf{a}_y + (2 + j5)\mathbf{a}_z]e^{-j2.3(-0.6x+0.8y)+j\omega t}$$

$$\mathbf{H} = (H_x\mathbf{a}_x + H_y\mathbf{a}_y + H_z\mathbf{a}_z)e^{-j2.3(-0.6x+0.8y)+j\omega t}$$

where H_x, H_y, H_z are all independent of x, y, and z, determine
- (a) The components E_y, H_x, H_y, H_z, assuming that $\mu = \mu_0$ and $\epsilon = \epsilon_0$
- (b) The frequency and corresponding wavelength
- (c) The equation of the surface of constant phase
- (d) The state of polarization of the wave

SOLUTIONS

(a) By comparison with the first exponential in the set (7-101) it is clear that

$$\gamma\mathbf{n} \cdot \mathbf{r} = j2.3(-0.6x + 0.8y) = j2.3\mathbf{n} \cdot \mathbf{r}$$

where

$$\mathbf{n} = -0.6\mathbf{a}_x + 0.8\mathbf{a}_y$$

and since $\gamma = \alpha + j\beta$, $\alpha = 0$, and $\beta = 2.3$.   Since $\mathbf{n} \cdot \mathbf{E} = 0$, it follows that

$$-0.6 + 0.8E_y = 0$$

and

$$E_y = 0.75$$

Now $-j\gamma = \beta = \omega\sqrt{\mu_0\epsilon_0}$; hence

$$\mathbf{H} = \frac{\beta}{\omega\mu_0}\,\mathbf{n} \times \mathbf{E} = \frac{1}{\eta}\,\mathbf{n} \times \mathbf{E} = \frac{1}{377}\begin{vmatrix} \mathbf{a}_x & \mathbf{a}_y & \mathbf{a}_z \\ -0.6 & 0.8 & 0 \\ 1 & 0.75 & 2+j5 \end{vmatrix}$$

so that

$$H_x = \frac{0.8(2 + j5)}{377} = (4.24 + j10.6) \times 10^{-3}$$

$$H_y = \frac{0.6(2 + j5)}{377} = (3.18 + j7.95) \times 10^{-3}$$

$$H_z = -\frac{0.6 \times 0.75 + 0.8}{377} = 3.31 \times 10^{-3}$$

(b) The wavelength is determined as

$$\lambda = \frac{2\pi}{\beta} = 2.73 \text{ m}$$

and the frequency as

$$f = \frac{c}{\lambda} = \frac{3 \times 10^8}{2.73} = 1.1 \times 10^8 \text{ Hz}$$

(c) The equation of the surface of constant phase

$$\mathbf{n} \cdot \mathbf{r} = -0.6x + 0.8y = \text{constant}$$

obviously defines a plane which is parallel to the z axis.

(d) Separating the real and imaginary parts of the magnitude of **E**, we find

$$\mathbf{E}_r = \mathbf{a}_x + 0.75\mathbf{a}_y + 2\mathbf{a}_z$$
$$\mathbf{E}_i = 5\mathbf{a}_z$$

Clearly, $\mathbf{E}_r$ and $\mathbf{E}_i$ are neither collinear nor mutually orthogonal in space. Therefore the polarization of the wave cannot possibly be linear nor circular; it must be elliptical.

To summarize, we have stated that a field of the type specified by Eqs. (7-101) is a uniform plane wave, characterized by a three-way orthogonality property between the vectors **E**, **H**, and **n**. Additionally, we have shown that **n** defines the direction of maximum rate of change of the phase of the wave with distance.

7.11 Group Velocity. Let us consider plane wave propagation in a medium in which waves of different frequencies travel at different phase velocities. Since all information is transmitted in the form of signals, which are made up of different frequency components, it is clear that as a signal travels through such a medium, a relative phase displacement of the various signal components will occur, and the signal, reconstructed at some distant point, will not be a true replica of the transmitted waveform. We say then that *distortion* results because of the *dispersive* nature of the medium.

An important question that should be answered is just what is meant by wave velocity when dispersion occurs, which is the rule rather than the exception. Strictly speaking, the concept of phase velocity applies to an infinite wave of a single frequency. Therefore, when we speak of the velocity of a wave containing many frequency components and traveling in a dispersive medium, we must be thinking of the speed of some as yet undefined "shape." But if the shape changes somewhat with distance, we lose the point of reference. What is then meant by *velocity*?

To answer this very important question, we need to introduce the concept of *group velocity*.

We saw earlier in this chapter that the x or y component of any electromagnetic vector associated with a wave traveling in the z direction satisfies an equation of the form

$$\frac{\partial^2 \psi}{\partial z^2} - \gamma^2 \psi = 0 \tag{7-112}$$

where

$$\gamma^2 = (\alpha + j\beta)^2 = -\omega^2 \mu \epsilon + j\omega\mu\sigma \tag{7-113}$$

In a nonconducting ($\sigma = 0$) medium, Eq. (7-112) reduces to

$$\frac{\partial^2 \psi}{\partial z^2} + \beta^2 \psi = 0 \tag{7-114}$$

where

$$\beta = \omega\sqrt{\mu\epsilon} \tag{7-115}$$

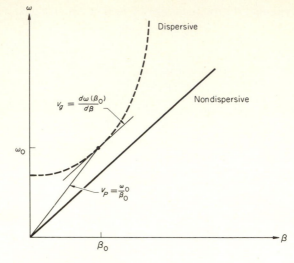

FIGURE 7-8. Typical ω-β diagrams for dispersive (dotted line) and nondispersive (solid line) media.

and has a particular solution of the form

$$\psi(z,t) = Ae^{j(\omega t - \beta z)} + Be^{j(\omega t + \beta z)} \qquad (7\text{-}116)$$

Here both A and B are either scalar or, at most, complex functions of frequency. For purposes of this discussion let us consider only the forward traveling wave. Then

$$\psi(z,t) = A(\omega)e^{j(\omega t - \beta z)} \qquad (7\text{-}117)$$

Since Eq. (7-112) is linear, the general solution can be constructed as a linear superposition of solutions of the form of Eq. (7-117) over a range of the angular frequency ω. For truly periodic waves, the summation will extend over a discrete set of frequencies. For aperiodic waves or for periodic waves of finite duration, the summation will extend continuously over the entire frequency spectrum. Thus an aperiodic wave will be represented by the integral

$$\psi(z,t) = \int_{-\infty}^{\infty} A(\omega)e^{j(\omega t - \beta z)} \, d\omega \qquad (7\text{-}118)$$

Now, Eq. (7-115) shows an explicit relationship between β and ω. In a nondispersive medium this relationship is purely linear. However, in a dispersive medium the relationship between β and ω is generally nonlinear, as typified by the typical (dotted) curve shown in Fig. 7-8. To allow for the possibility of dispersion, we shall consider ω as an arbitrary function of β, $\omega = \omega(\beta)$. But whatever the actual dependence, either ω or β can be taken as the independent variable when summations of particular solutions of the form of Eq. (7-117) are considered. Thus, in place of Eq. (7-118), we can write

$$\psi(z,t) = \int_{-\infty}^{\infty} A(\beta)e^{j(\omega t - \beta z)}\, d\beta \tag{7-119}$$

where, obviously, $A(\beta)$ is a different function from $A(\omega)$ in Eq. (7-118).

Next, we suppose that the wave is very nearly periodic, so that the signal is *band-limited;* that is, $A(\beta)$ is virtually zero everywhere except over a small interval $\beta_1 \leq \beta \leq \beta_2$. In this case, the limits of integration can be changed so that

$$\psi(z,t) = \int_{\beta_1}^{\beta_2} A(\beta)e^{j(\omega t - \beta z)}\, d\beta \tag{7-120}$$

while the angular frequency ω can be represented by the first two terms of the Taylor series expansion of $\omega(\beta)$ about β_0, where $\beta_1 \leq \beta_0 \leq \beta_2$. Thus

$$\omega(\beta) \approx \omega(\beta_0) + (\beta - \beta_0)\frac{d\omega(\beta_0)}{d\beta} \tag{7-121}$$

and

$$\omega t - \beta z \approx \left[\omega(\beta_0) + (\beta - \beta_0)\frac{d\omega(\beta_0)}{d\beta}\right]t - \beta z \tag{7-122}$$

If we set $\omega_0 = \omega(\beta_0)$, and if we add and subtract the term $\beta_0 z$ from the right-hand side of Eq. (7-122), we obtain

$$\omega t - \beta z \approx (\omega_0 t - \beta_0 z) + (\beta - \beta_0)\left[\frac{d\omega(\beta_0)}{d\beta}t - z\right] \tag{7-123}$$

Equation (7-120) then becomes

$$\psi(z,t) = \psi_0(z,t)e^{j(\omega_0 t - \beta_0 z)} \tag{7-124}$$

where the amplitude

$$\psi_0(z,t) = \int_{\beta_1}^{\beta_2} A(\beta)\exp\left\{j(\beta - \beta_0)\left[\frac{d\omega(\beta_0)}{d\beta}t - z\right]\right\} d\beta \tag{7-125}$$

is only a function of z and t, once the integration with respect to β is performed. Clearly, the magnitude of ψ_0 is constant over surfaces defined by

$$\frac{d\omega(\beta_0)}{d\beta}t - z = \text{constant} \tag{7-126}$$

from which we conclude that the "shape" we spoke of earlier, which is known as the *wave packet*, or the *group*, is propagated with the *group velocity*

$$v_g = \frac{d\omega(\beta_0)}{d\beta} \tag{7-127}$$

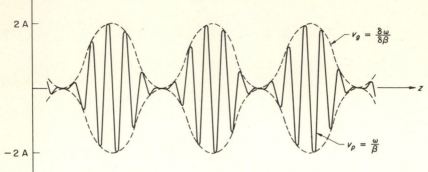

FIGURE 7-9. A simple wave packet.

As shown in Fig. 7-8, v_g is the slope of the ω-β diagram evaluated at β_0. In dispersive media, this slope differs from the ratio ω_0/β_0, which is the phase velocity for a particular frequency.   In nondispersive media, however, the group velocity and the phase velocity are one and the same thing, since the slope of a straight line [Eq. (7-115)] is just the ratio of the ordinate to the abscissa of any point along the line.

Example 7-4 Group Velocity of a Simple Wave Packet. As an illustration let us consider the case of a signal made up of two harmonic waves which differ only slightly in ω and β.   We suppose that the components of the signal are

$$\psi_1 = A \cos \left[(\omega + \delta\omega)t - (\beta + \delta\beta)z \right]$$

$$\psi_2 = A \cos \left[(\omega - \delta\omega)t - (\beta - \delta\beta)z \right]$$

where A is a scalar constant.   Then the signal in question, obtained by superposition of ψ_1 and ψ_2, is

$$\psi = \psi_1 + \psi_2 = 2A \cos (t\,\delta\omega - z\,\delta\beta) \cos (\omega t - \beta z)$$

and has an amplitude

$$2A \cos (t\,\delta\omega - z\,\delta\beta)$$

which is a function of both distance and time, and therefore represents a wave whose frequency is very small compared with ω.   At a given instant of time, a plot of ψ versus z appears as a series of periodically repeated groups, as shown in Fig. 7-9, which travel with the group velocity

$$v_g = \frac{\delta\omega}{\delta\beta}$$

On the other hand, the wave represented by

$$\cos (\omega t - \beta z)$$

$$\cos(A+B) + \cos(A-B) = 2\cos A \cos B$$

varies more rapidly along the time axis, and moves along the space axis with the phase velocity

$$v_p = \frac{\omega}{\beta}$$

thus appearing to slip through the moving envelope (dotted) of the groups.

In summary, the concept of group velocity is extremely important in the study of wave propagation through a dispersive medium. It applies only to narrow-band signals and loses its physical significance when $\delta\beta$ is large. The transport of information, or, loosely speaking, of energy, occurs at the group velocity.

omit

7.12 Wave Propagation in Linear Anisotropic Media. Crystalline materials frequently have different mechanical and electrical properties when measured along different crystalline axes. These differences are due to the internal atomic structure of the material. In some cases the crystalline axes are not at right angles to each other, and the analysis is very complicated. Analysis of a comparatively simple case will illustrate the effects of anisotropy.

Consider a calcite crystal. Chemically, calcite is $CaCO_3$. Structurally, the calcium and the oxygen atoms are displaced about the carbon atom in such a manner that they occupy the normal regular tetrahedral bonding positions of the carbon atom in an arrangement such that, on an alternating basis, there are parallel planes containing oxygen atoms, carbon atoms, calcium atoms, and repeating, except that in the second oxygen plane the three oxygen atoms are rotated 60° with respect to the first plane. The so-called *optic axis* of calcite is perpendicular to these planes. Mechanically weak planes exist such that calcite may be readily cleaved. The cleavage planes are along planes defined by the calcium atom and two of the oxygen atoms. Thus the cleavage planes are skewed with respect to the optic axis. Cleavage faces are parallelograms with interior angles of 71 and 108°. The optic axis trisects the corner where three of the 108° cleavage parallelograms meet to form an obtuse apex angle. But we are interested in calcite's electrical properties.

The disposition of the atoms in calcite gives rise to an electric polarizability which has one value parallel to the optic axis and a different value perpendicular to the optic axis. This means that the constitutive relations between **E** and **D** will depend upon the direction of **E**. For convenience let us take the optic axis as the z direction. Then the constitutive relations will be of the form

$$D_x = \epsilon_\perp E_x \qquad D_y = \epsilon_\perp E_y \qquad D_z = \epsilon_\| E_z \qquad (7\text{-}128)$$

where $\epsilon_\perp$ is the value of ϵ perpendicular to the optic axis, and $\epsilon_\|$ is its value parallel to the optic axis.

Now consider a uniform plane wave which is propagating in some specified direction in calcite and which, at least at some reference point, is linearly polarized. Obviously, the behavior of this wave will depend upon the direction of propagation and the direction of polarization at the reference point. Solution of the general problem of arbitrary direction of propagation and polarization is complicated, and tends to obscure the physical behavior of the wave. Since our presentation is introductory rather than exhaustive, it will be sufficient to examine two special cases from which some aspects of the general case can be inferred.

Case 1. Propagation along the optic axis This is the simpler case. We have $E_z = H_z = 0$, and hence $\epsilon_\parallel$ does not enter the problem. At any point,

$$\mathbf{E} = E_x^+ e^{-j\beta_\perp z}\mathbf{a}_x + E_y^+ e^{-j\beta_\perp z}\mathbf{a}_y \qquad (7\text{-}129)$$

where $\beta_\perp = \omega\sqrt{\mu\epsilon_\perp}$, and the phase velocity is

$$v_\perp = \frac{1}{\sqrt{\mu\epsilon_\perp}}$$

Also, the $\mathbf{E}$ field will be related to the $\mathbf{H}$ field by

$$E_x^+ = \eta_\perp H_y^+ \qquad E_y^+ = -\eta_\perp H_x^+$$

where $\eta_\perp = \sqrt{\mu/\epsilon_\perp}$. Thus the wave behaves like an ordinary uniform plane wave in an isotropic medium of permittivity $\epsilon_\perp$.

Case 2. Propagation perpendicular to the optic axis For definiteness, let the x direction be the direction of propagation. With $E_x = H_x = 0$ and $\partial/\partial y = \partial/\partial z = 0$, Maxwell's equations yield

$$\frac{\partial^2 E_y}{\partial x^2} + \omega^2\mu\epsilon_\perp E_y = 0 \qquad (7\text{-}130)$$

$$\frac{\partial^2 E_z}{\partial x^2} + \omega^2\mu\epsilon_\parallel E_z = 0 \qquad (7\text{-}131)$$

Therefore, for a wave traveling in the positive x direction,

$$\mathbf{E} = E_y^+ e^{-j\beta_\perp x}\mathbf{a}_y + E_z^+ e^{-j\beta_\parallel x}\mathbf{a}_z \qquad (7\text{-}132)$$

where $$\beta_\perp = \omega\sqrt{\mu\epsilon_\perp} \qquad \beta_\parallel = \omega\sqrt{\mu\epsilon_\parallel} \qquad (7\text{-}133)$$

resulting in two phase velocities,

$$v_\perp = \frac{1}{\sqrt{\mu\epsilon_\perp}} \qquad v_\parallel = \frac{1}{\sqrt{\mu\epsilon_\parallel}} \qquad (7\text{-}134)$$

The **E** and **H** field vectors are related by

$$E_y^+ = \eta_\perp H_z^+ \qquad E_z^+ = -\eta_\| H_y^+$$

where
$$\eta_\perp = \sqrt{\frac{\mu}{\epsilon_\perp}} \qquad \eta_\| = \sqrt{\frac{\mu}{\epsilon_\|}} \tag{7-135}$$

The significant feature of these results is that the two component waves have different phase propagation constants and phase velocities. If the two component waves are in time phase at a position $x_0 = 0$, E_y^+ and E_z^+ are real. If we define

$$\Delta\beta = \beta_\perp - \beta_\|$$

we can rewrite Eq. (7-132) as

$$\mathbf{E} = E_y^+ e^{-j\beta_\perp x}\mathbf{a}_y + E_z^+ e^{j(\Delta\beta)x}e^{-j\beta_\perp x}\mathbf{a}_z \tag{7-136}$$

Comparison with the discussion in Sec. 7.8 will show that this is an elliptically polarized wave. The relative phase between the two components of **E** is a linear function of x, and is given by

$$\phi = (\Delta\beta)x \tag{7-137}$$

Hence, at positions such that

$$\phi = \pm n\pi \qquad n = 0, 1, 2, 3, \ldots$$

the wave is linearly polarized and makes an angle θ with respect to the y axis, given by (Fig. 7-10)

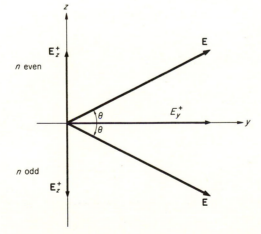

FIGURE 7-10. The plane of polarization for $\phi = \pm n\pi$.

$$\theta = \begin{cases} \tan^{-1} \dfrac{E_z^+}{E_y^+} & \text{for } n \text{ even} \\[2em] -\tan^{-1} \dfrac{E_z^+}{E_y^+} & \text{for } n \text{ odd} \end{cases}$$

We see now that the planes of linear polarization for odd values of n are rotated by an angle 2θ with respect to the even values of n.

For the special case $E_y^+ = E_x^+ = E$, $\theta = 45°$, the alternating planes of linear polarization are perpendicular to each other, and are inclined $45°$ with respect to the optic axis.

Example 7-5 Propagation in an Ionized Atmosphere. In further illustration, let us consider propagation of TEM waves through the earth's ionosphere. A reasonably accurate description of the ionosphere can be given on the assumption that there are N particles of charge e and mass m at every point of an otherwise empty region. Actually, both electrons and positive ions are present in equal numbers, but because of their relatively large mass and attendant slow velocity, positive ions play rather insignificant roles in the propagation of waves through the ionosphere.

We suppose that propagation is in the z direction, parallel to the local direction of the earth's magnetic field, $\mathbf{B_0}$.

Under the influence of the passing wave, the charged particles will be forced to move with time.   If

$$\mathbf{u} = \dot{x}\mathbf{a}_x + \dot{y}\mathbf{a}_y + \dot{z}\mathbf{a}_z \tag{7-138}$$

denotes the instantaneous velocity of the particles, and

$$\dot{\mathbf{u}} = \ddot{x}\mathbf{a}_x + \ddot{y}\mathbf{a}_y + \ddot{z}\mathbf{a}_z$$

the instantaneous acceleration at a point (x,y,z), then

$$m\dot{\mathbf{u}} = e(\mathbf{E} + \mathbf{u} \times \mathbf{B}) \tag{7-139}$$

is the equation of dynamic equilibrium, which, when expanded, reads

$$\ddot{x} = \frac{e}{m} E_x + \frac{e}{m} (\dot{y}B_0 - \dot{z}B_y)$$

$$\ddot{y} = \frac{e}{m} E_y + \frac{e}{m} (\dot{z}B_x - \dot{x}B_0) \tag{7-140}$$

$$\ddot{z} = \frac{e}{m} (\dot{x}B_y - \dot{y}B_x)$$

and thus describes the motion of the particles.   The wave, of course, is assumed to be TEM, and $E_z = 0$. For plane waves in free space, $|\mathbf{E}/\mathbf{H}| = \sqrt{\mu_0/\epsilon_0}$.   Hence

$$E = \frac{B}{\mu_0} \sqrt{\frac{\mu_0}{\epsilon_0}} = \frac{B}{\sqrt{\mu_0 \epsilon_0}} = cB$$

and

$$|\mathbf{u} \times \mathbf{B}| = \frac{1}{c} |\mathbf{u} \times \mathbf{E}|$$

This means that in Eq. (7-139) the ratio of the second term to the first is equal in magnitude to the ratio of u to c, which would normally be extremely small.   Therefore all terms in Eqs. (7-140) which involve B_x and B_y may be neglected, so that

$$\ddot{x} = \frac{e}{m} E_x + \frac{e}{m} \dot{y} B_0$$

$$\ddot{y} = \frac{e}{m} E_y - \frac{e}{m} \dot{x} B_0 \tag{7-141}$$

$$\ddot{z} = 0$$

This shows, essentially, that the motion of the particles is largely confined to the transverse xy plane.

Let us neglect particle collisions and assume that the field vectors contain the time only as a factor $e^{j\omega t}$.   Then

$$j\omega \dot{x} = \frac{e}{m} E_x + \frac{e}{m} \dot{y} B_0$$

$$j\omega \dot{y} = \frac{e}{m} E_y - \frac{e}{m} \dot{x} B_0 \tag{7-142}$$

This is a pair of simultaneous equations in $\dot{x}$ and $\dot{y}$.   Routine solution gives

$$\dot{x} = \frac{j\omega(m/e)E_x + B_0 E_y}{B_0{}^2 - \omega^2(m/e)^2} \tag{7-143}$$

$$\dot{y} = \frac{j\omega(m/e)E_y - B_0 E_x}{B_0{}^2 - \omega^2(m/e)^2} \tag{7-144}$$

suggesting anisotropic behavior, because $\dot{x}$ is affected by E_y and $\dot{y}$ by E_x.   This will become even more apparent as the development is continued.

The denominator in both Eqs. (7-143) and (7-144) is a quantity which vanishes when $\omega = \omega_g = eB_0/m$.   For electrons ($e/m = 1.76 \times 10^{11}$ C/kg) in the ionosphere ($B_0 = 0.5 \times 10^{-4}$ Wb/m²), the so-called *gyrofrequency* $f_g = \omega_g/2\pi = 1400$ kHz. At this frequency (broadcast band), the charged electrons are set in violent motion, absorbing virtually all of the energy of the wave, and thus causing propagation to stop.   More favorable conditions, of course, prevail at frequencies different from the gyro-frequency of the medium.

Setting

$$\mathbf{J} = N e \mathbf{u} = N e (\dot{x} \mathbf{a}_x + \dot{y} \mathbf{a}_y) \tag{7-145}$$

in Maxwell's equations, we now obtain

$$\nabla \times \mathbf{E} = -j\omega \mathbf{B}$$

$$\nabla \times \mathbf{H} = j\omega [\epsilon_{ij}][\mathbf{E}] \tag{7-146}$$

where

$$[\mathbf{E}] = \begin{bmatrix} E_x \\ E_y \\ 0 \end{bmatrix} \tag{7-147}$$

and

$$[\epsilon_{ij}] = \begin{bmatrix} \epsilon' & -j\epsilon'' & 0 \\ j\epsilon'' & \epsilon' & 0 \\ 0 & 0 & 0 \end{bmatrix} \tag{7-148}$$

is the equivalent *tensor* permittivity of the medium.   In this expression

$$\epsilon' = \epsilon_0 \left(1 - \frac{\omega_0^2}{\omega^2 - \omega_g^2} \right)$$

$$\epsilon'' = -\epsilon_0 \frac{\omega_0^2(\omega_g/\omega)}{\omega^2 - \omega_g^2}$$

where $\omega_g = eB_0/m$ designates the previously defined angular gyrofrequency, and where $\omega_0 = e\sqrt{N/\epsilon_0 m}$ is the *plasma*, or the *critical angular frequency*.

The anisotropic behavior of the medium, clearly indicated by Eq. (7-148), is caused by the presence of the fixed magnetic field $\mathbf{B}_0$.   To show that this is so, we note that if $B_0 = 0$, then $\omega_g = 0$, $\epsilon'' = 0$, and $[\epsilon_{ij}] = \epsilon'$, where now $\epsilon' = \epsilon_0(1 - \omega_0^2/\omega^2)$.

In expanded form, Eqs. (7-146) read

$$\frac{\partial E_y}{\partial z} = j\omega\mu_0 H_x \tag{7-149}$$

$$\frac{\partial E_x}{\partial z} = -j\omega\mu_0 H_y \tag{7-150}$$

$$\frac{\partial H_y}{\partial z} = -j\omega(\epsilon' E_x - j\epsilon'' E_y) \tag{7-151}$$

$$\frac{\partial H_x}{\partial z} = j\omega(\epsilon' E_y + j\epsilon'' E_x) \tag{7-152}$$

If we eliminate H_x from Eqs. (7-149) and (7-152), and also eliminate H_y from Eqs. (7-150) and (7-151), we obtain

$$\frac{\partial^2 E_y}{\partial z^2} = \omega^2\mu_0(j\epsilon'' E_x - \epsilon' E_y)$$
$$\frac{\partial^2 E_x}{\partial z^2} = \omega^2\mu_0(-\epsilon' E_x + j\epsilon'' E_y) \tag{7-153}$$

Substituting an assumed solution of the form

$$E_x = Ae^{j\omega(t-z/v)} \qquad E_y = Be^{j\omega(t-z/v)}$$

where A and B are (generally complex) constants, in the set (7-153), and subsequently dividing out the common exponential factor, we find

$$\left(-\frac{1}{v^2} + \mu_0\epsilon' \right) B = -j\mu_0\epsilon'' A$$
$$\left(-\frac{1}{v^2} + \mu_0\epsilon' \right) A = j\mu_0\epsilon'' B \tag{7-154}$$

The condition $A^2 = -B^2$ reduces both equations in this set to identities, and therefore the solutions corresponding to the two possible cases $B = jA$ and $B = -jA$ are as follows.

SOLUTION 1. $B = jA$

$$v = v_1 = \frac{1}{\sqrt{\mu_0(\epsilon' + \epsilon'')}} = \frac{1/\sqrt{\mu_0\epsilon_0}}{\sqrt{1 - \omega_0^2/[\omega(\omega - \omega_g)]}} \tag{7-155}$$

$$E_x = A \cos \omega \left(t - \frac{z}{v_1} \right) \qquad E_y = -A \sin \omega \left(t - \frac{z}{v_1} \right) \tag{7-156}$$

SOLUTION 2. $B = -jA$

$$v = v_2 = \frac{1}{\sqrt{\mu_0(\epsilon' - \epsilon'')}} = \frac{1/\sqrt{\mu_0\epsilon_0}}{\sqrt{1 - \omega_0{}^2/[\omega(\omega + \omega_g)]}} \qquad (7\text{-}157)$$

$$E_x = A \cos \omega \left(t - \frac{z}{v_2} \right) \qquad E_y = A \sin \omega \left(t - \frac{z}{v_2} \right) \qquad (7\text{-}158)$$

The physical interpretation of these results can be realized by noting that a linearly polarized wave traveling in the (effectively) anisotropic medium is decomposed into two circularly polarized waves rotating in opposite directions. One of these waves (for example, with the clockwise rotation) is propagated with the velocity v_2, while the other is propagated with the velocity v_1. Therefore a wave impinging upon the ionosphere, though linearly polarized, may be reflected back to the earth with a circularly polarized component (which, incidentally, causes a fading in the returned signal).

This phenomenon may be explained with the aid of Eqs. (7-155) and (7-157). For very large ω, $v_1 \approx v_2$, and both waves go through the ionized medium under essentially identical conditions. For $\omega \approx \omega_g$, $v_1 \approx 0$, and absorption of one of the circularly polarized components occurs; the remaining component goes through virtually unaffected and accounts for the change in polarization previously noted. For small ω, both v_1 and v_2 are imaginary, and the wave is attenuated as it goes through the ionized medium.

Summarizing, the theory of uniform plane waves was extended to anisotropic media. It was found that propagation with two distinct velocities is generally to be expected.

7.13 Cylindrical and Spherical Waves.

We have seen that in linear, homogeneous, isotropic, and source-free media, the wave equation predicts the possible existence of propagating modes which are interpretable as uniform plane waves, with **E** and **H** fields transverse to the direction of propagation. We found these possible modes by (in essence) asking what the consequences would be of demanding that a particular field component (in rectangular coordinates) be a function only of time and one other coordinate [see the discussion following Eq. (7-10)]. A similar question could be asked for the wave equation expressed in cylindrical or in spherical coordinates. There are, of course, three choices for each coordinate system for the coordinate of variation. We shall display for each coordinate system the solutions for one choice.

Cylindrical waves If we require that E_z be a function only of radius r, we shall find that, in the limit of large r, the wave equation will allow a solution of the type

$$E_z = E^+ \frac{e^{-j\beta r}}{\sqrt{r}} + E^- \frac{e^{j\beta r}}{\sqrt{r}} \qquad (7\text{-}159)$$

with **H** perpendicular to **E** and to the radial direction and with a magnitude given by $|\mathbf{H}| = |\mathbf{E}|/\eta$ in a manner similar to that for uniform plane waves.

This solution is uniform over cylinders $r =$ constant. The wave propagates along the radial direction, and the field components are transverse to the direction of propagation. Hence it is called a *uniform cylindrical wave*.

Spherical waves If we require in spherical coordinates that E_θ be a function only of r and θ, we shall find that, off the $\theta = 0$ axis, the wave equation will allow solutions of the type

$$E_\theta = E^+ \frac{e^{-j\beta r}}{r \sin \theta} + E^- \frac{e^{j\beta r}}{r \sin \theta} \tag{7-160}$$

with **H** perpendicular to **E** and to the radial direction and with $|\mathbf{H}| = |\mathbf{E}|/\eta$.

The wave propagates along the radial direction, and the field components are transverse to the direction of propagation.

7.14 Nonuniform Waves. [Omit] Uniform waves are constant in amplitude and phase over surfaces normal to the direction of propagation. Although it is not at all obvious at this stage, it seems intuitively possible to have wave propagation for which the surfaces of constant amplitude and the surfaces of constant phase are distinct. This intuitive result turns out to be indeed valid. Such waves are called *nonuniform waves*.

As an example, we can show easily that, if $E_y = E_z = 0$ and $\partial/\partial x = 0$, then a solution to Maxwell's equations is

$$\mathbf{E} = E^+ e^{-\alpha y} e^{-j\beta z} \mathbf{a}_x$$

$$\mathbf{H} = \frac{\beta}{\omega\mu} E^+ e^{-\alpha y} e^{-j\beta z} \mathbf{a}_y + \frac{j\alpha}{\omega\mu} E^+ e^{-\alpha y} e^{-j\beta z} \mathbf{a}_z \tag{7-161}$$

with

$$\alpha^2 - \beta^2 = \omega^2 \mu \epsilon \tag{7-162}$$

A more general description of a nonuniform wave can be given by writing

$$\psi = \psi_0 e^{-\alpha \mathbf{n}_1 \cdot \mathbf{r}} e^{j(\omega t - \beta \mathbf{n}_2 \cdot \mathbf{r})} \tag{7-163}$$

where $\mathbf{n}_1$ and $\mathbf{n}_2$ are two distinct unit vectors, for either of the two transverse components of a plane wave in rectangular coordinates. This wave may then be readily shown (Prob. 7-13) to satisfy Maxwell's equations by expressing the spatial dependence of the exponent in the form

$$\gamma \mathbf{n} = \alpha' \mathbf{n}_1 + j\beta' \mathbf{n}_2 \tag{7-164}$$

where **n** defines a complex propagation direction, and $\alpha' \neq \alpha$, $\beta' \neq \beta$. Clearly, the planes of constant amplitude and the planes of constant phase, respectively, defined by

$$\mathbf{n}_1 \cdot \mathbf{r} = \text{constant} \qquad \mathbf{n}_2 \cdot \mathbf{r} = \text{constant} \qquad (7\text{-}165)$$

are distinct, and, in fact, at times orthogonal, to each other.

The physical implications of the vectors $\mathbf{n}$, $\mathbf{n}_1$, and $\mathbf{n}_2$ are examined in greater detail in Chap. 8.

Example 7-6 Characteristic Features of a Nonuniform Wave. Let us suppose that the electric field associated with the wave considered in Example 7-3 is slightly modified to read

$$\mathbf{E} = [\mathbf{a}_x + E_y\mathbf{a}_y + (2 + j5)\mathbf{a}_z]e^{-j2.3[-0.6x+(0.8-j0.6y)]+j\omega t}$$

If this is to represent a TEM wave, we should have $\gamma\, \mathbf{n} \cdot \mathbf{E} = 0$. This in turn requires that

$$-0.6 + (0.8 - j0.6)E_y = 0$$

$$E_y = 0.48 + j0.36$$

In order to better understand the nature of this field, we need to examine the character of the exponent. We have

$$-j2.3[-0.6x + (0.8 - j0.6)y] = 2.3[-0.6y + j(0.6x - 0.8y)]$$

$$= -1.38\mathbf{a}_y \cdot (x\mathbf{a}_x + y\mathbf{a}_y + z\mathbf{a}_z) + j2.3(0.6\mathbf{a}_x - 0.8\mathbf{a}_y) \cdot (x\mathbf{a}_x + y\mathbf{a}_y + z\mathbf{a}_z)$$

$$= -1.38\mathbf{n}_1 \cdot \mathbf{r} - j2.3\mathbf{n}_2 \cdot \mathbf{r}$$

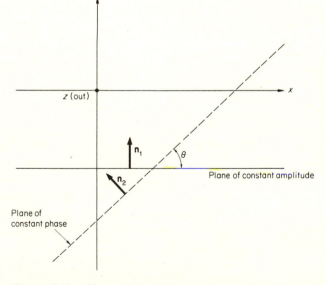

FIGURE 7-11. The spatial relationship between the planes of constant amplitude and planes of constant phase for a nonuniform wave.

where
$$n_1 = a_y$$

$$n_2 = -(0.6a_x - 0.8a_y)$$

This places in evidence the fact that the planes of constant amplitude ($y = $ constant) are not coincident with the planes of constant phase ($0.6x - 0.8y = $ constant). In fact, as shown in Fig. 7-11, the two planes are inclined relative to each other at an angle

$$\theta = \cos^{-1} \frac{n_1 \cdot n_2}{n_1 n_2} = \cos^{-1} 0.8 = 36.8°$$

On an equiphase front, the amplitude of **E** is therefore *not constant* for this wave.

We may summarize by defining a nonuniform TEM wave, in general, as a wave whose amplitude is dependent upon one, two, or even three space coordinates.

7.15 Summary. This chapter has given an introduction to wave propagation by considering the important subject of transverse electromagnetic waves, with primary emphasis on uniform plane waves. A brief introduction to uniform-plane-wave propagation in anisotropic media was presented, and the case of nonuniform plane waves was discussed preparatory to a presentation of some of their more specific aspects, in Chap. 8.

The importance of plane waves stems from a very basic and practical consideration. *At sufficiently large distances from their sources, all waves traveling in ordinary regions behave locally like uniform plane waves;* in other words, the vectors **E** and **H** are orthogonal to each other and to the direction of propagation. We shall see, in Chap. 10, that the wave emanating from a *Hertzian dipole* (time-varying *point* source) is locally transverse electromagnetic at points outside a sphere, centered about the dipole, with a minimum radius of about one-sixth of a wavelength.

The types of waves considered in this chapter were determined as *admissible* solutions of the field equations without regard to the nature of the sources which generate them. This is in keeping with the traditional approach to analysis, where simple solutions are determined first, from which more elaborate solutions are subsequently constructed by superposition. An ordinary example is furnished by the circular functions (sines and cosines), linear combinations of which comprise familiar Fourier series representations of arbitrary functions.

Strictly speaking, a uniform plane wave can be generated only by an unbounded system of sources spread over an infinite plane at infinity. Obviously, this arrangement is physically impossible. However, waves which are locally plane can be generated by systems of sources which are bounded and which are located at finite distances away from all points of observation.

For all types of harmonic ($e^{j\omega t}$) waves, the direction of energy flow is determined by the complex Poynting vector $\mathbf{S}_c$; the real part of this vector gives the rate of energy flow per unit averaged over a complete cycle of the wave. In nondispersive media, the transport of this energy occurs with the phase velocity; in dispersive media, it occurs with the group velocity, provided the group is band-limited. No meaningful statement can be made in this regard for signals with unlimited frequency spectra because of the extreme deformation of the original shape of the signal.

The most general polarization of a plane wave is elliptical. Special polarizations are the linear and the circular. The rotation (clockwise or counterclockwise) of a circularly or elliptically polarized wave is established by standing behind the wave, so to speak, looking in the direction of propagation. The distance between two successive peaks of a wave, measured along the axis of propagation, is the wavelength.

Problems

7-1 Uniform-plane-wave Field Configurations. Given a uniform plane wave whose complex vector **E** field is given by

$$\mathbf{E} = E_x e^{-j\beta z}\mathbf{a}_x$$

where E_x is real.

    (*a*) Sketch the actual **E** and **H** fields as functions of z for $t = 0$.
    (*b*) Sketch **E** and **H** as functions of t for $z = 0$.
(*Hint:* Write out the real fields and sketch them.)

7-2 Wave Parameters. For the wave of Example 7-6 determine the frequency and the wavelength if $\mu = \mu_0$ and $\epsilon = \epsilon_0$. Also, determine the polarization, and state the direction of rotation. (*Hint:* See results of Prob. 7-13.)

7-3 Polarization. Given a uniform plane wave whose complex vector **E** field is

$$\mathbf{E} = 3e^{-j\beta z}\mathbf{a}_x + 4e^{-j\beta z}e^{j\pi/4}\mathbf{a}_y$$

    (*a*) Write the parametric equations for the polarization ellipse and sketch the ellipse.
    (*b*) Sketch the direction and magnitude of the electric field vector at several values of time for a fixed position z.
    (*c*) Repeat part (*b*) for several values of z for a fixed time t.

7-4 Polarization. Prove that every linearly polarized wave can be resolved into two circularly polarized waves rotating in opposite directions with the same angular speed.

7-5 Polarization. Derive Eq. (7-80).

7-6 Wave Velocities. Prove that, for β real,

$$v_g = v_p + \beta\frac{dv_p}{d\beta} = v_p - \lambda\frac{dv_p}{d\lambda}$$

7-7   Wave Velocities.   Given the ω-β relation

$$\omega^2 \mu \epsilon - \beta^2 = \left(\frac{\pi}{a}\right)^2$$

Determine v_p and v_g, and prove that, in this case, $v_p v_g = 1/\mu\epsilon$.

7-8   Lossy Medium.   The conductivity σ of ordinary metals is in the range from 10^5 to 10^8 $\mho/m$; for ordinary dielectrics, the range is from 10^{-10} to 10^{-15} $\mho/m$.

Calculate the approximate values of the attenuation factor α, the propagation factor β, the phase velocity v_p, and the wavelength λ, for a frequency 10^8 Hz. Assume $\mu = \mu_0$.

(a)   For a material of conductivity $\sigma = 10^5$ $\mho/m$ and $\epsilon_r = 1$.

(b)   For a material of conductivity $\sigma = 10^{-10}$ $\mho/m$ and $\epsilon_r = 2.6$.

7-9   Energy.   Prove that, for a uniform plane wave, in a lossless medium, $\frac{1}{2}\epsilon E^2 = \frac{1}{2}\mu H^2$.

7-10   Poynting's Vector.   Consider the plane wave of Prob. 7-1 in free space, and let $E_x = 0.2$ V/m.   Find the time-average power propagated by this wave.

7-11   Energy and Power.   A 100-MHz wave is traveling in free space and has an amplitude $E_0 = 10^{-3}$ V/m.   Find

(a)   The parameters λ and β

(b)   The average energy densities $\langle w_e \rangle$ and $\langle w_m \rangle$ per unit volume

(c)   The complex Poynting's vector and its peak amplitude

7-12   Poynting's Vector and Nonuniform Spherical Waves.   Given a spherical wave with a complex E-field vector

$$\mathbf{E} = E_0 \frac{1}{r \sin \theta} e^{-j\beta r} \mathbf{a}_\theta$$

where E_0 is the complex constant $20e^{j\pi/2}$ V/m.   Calculate the total outward power flow for this wave.   (*Hint:* Integrate from $\theta = \theta_0$ to $\theta = \pi - \theta_0$.)

7-13   Nonuniform Waves.   According to Eqs. (7-163) and (7-164), a nonuniform wave may be represented as

$$\psi = \psi_0 e^{j\omega t - \gamma \mathbf{n} \cdot \mathbf{r}}$$

Prove that this type of wave reduces the field equations to

$$\gamma \mathbf{n} \times \mathbf{E} = j\omega \mathbf{B} \qquad\qquad \gamma \mathbf{n} \cdot \mathbf{B} = 0$$

$$-\gamma \mathbf{n} \times \mathbf{H} = \mathbf{J} + j\omega \mathbf{D} \qquad \gamma \mathbf{n} \cdot \mathbf{D} = 0$$

where the last equation on the right holds only if the medium is free of charge.

Also prove that

$$\gamma^2 \mathbf{n} \cdot \mathbf{n} = -\omega^2 \mu \epsilon + j\omega\mu\sigma$$

Finally, prove that the vector $\mathbf{E}$ is perpendicular to both $\mathbf{n}_1$ and $\mathbf{n}_2$ and that the plane of $\mathbf{n}_1$ and $\mathbf{n}_2$ contains the magnetic field intensity vector $\mathbf{H}$, *provided* the medium is linear and isotropic and the $\mathbf{E}$ field is linearly polarized.

7-14   Complex Notation.   Given two complex vectors $\mathbf{G}_1$ and $\mathbf{G}_2$ with arbitrary dependence on the space coordinates.   Determine all possible states of polarization for which space orthogonality in the frequency domain ($\mathbf{G}_1 \cdot \mathbf{G}_2 = 0$) implies space orthogonality in the time domain [Re $(\mathbf{G}_1) \cdot$ Re $(\mathbf{G}_2) = 0$], and conversely.

7-15   Propagation in an Ionized Medium.   Consider the problem of Example 7-5, and examine in full detail the case in which the static magnetic field is perpendicular to the direction of propagation. Neglect the effect of particle collisions, and assume two-dimensional particle motion.

7-16 Propagation in an Ionized Medium. The purpose of this exercise is to study the effects of particle collisions on plane-wave propagation through a plasma with zero static magnetic field.

When the gas-molecule density in the plasma is sufficiently high, energy stored in electron motion is lost through collisions, and the wave is attenuated. Since the collision process is random, it is necessary to write the force equation for an electron which, on the average, experiences one collision per mean-free-path length. The energy loss is expressed by a damping term in the equation of equilibrium:

$$\frac{d\mathbf{u}}{dt} + g\mathbf{u} = \frac{e}{m}\,\mathbf{E}$$

For certain special gases, g is independent of electron velocity and reduces to the collision frequency f_c (a fixed number). In these cases the preceding differential equation is easily solved:

$$u = \frac{e/m}{f_c + j\omega}\,E$$

Suppose that a plane wave is propagated in the direction of the positive z axis of a rectangular coordinate system. For the sake of example, we assume only that there are N particles per unit volume of charge e and mass m in otherwise empty space. A static magnetic field is *not* present in this case.

(*a*) Write the field equations for this problem.

(*b*) Does the medium behave anisotropically?

(*c*) Determine $\gamma\mathbf{n}\cdot\gamma\mathbf{n}$, where γ is the familiar propagation constant.

(*d*) Determine the equivalent (real) permittivity of the medium.

(*e*) Determine the equivalent (real) conductivity of the medium.

Express all your answers (where pertinent) in terms of the critical (plasma) frequency $\omega_0 = e\sqrt{N/\epsilon_0 m}$.

7-17 General TEM Wave Expressions. By direct substitution, prove that Eqs. (7-101) are solutions of

$$\nabla \times \mathbf{E} = -j\omega\mu\mathbf{H} \qquad \nabla \cdot \mathbf{H} = 0$$
$$\nabla \times \mathbf{H} = (\sigma + j\omega\epsilon)\mathbf{E} \qquad \nabla \cdot \mathbf{E} = 0$$

where ϵ, μ, and σ are scalar constants.

7-18 Polarization. Consider a uniform plane wave propagating in the positive z direction, and let

$$\mathbf{E}(z,t) = \mathbf{E}_1 \cos(\omega t - \beta z)\mathbf{a}_x + E_2 \cos(\omega t - \beta z + \phi)\mathbf{a}_y$$

If $\theta(t)$ denotes the angle between $\mathbf{E}$ and the positive x axis, show that the angular speed of rotation of the electric field vector is given by

$$\frac{d\theta(t)}{dt} = \frac{\mathbf{a}_z \cdot [\mathbf{E} \times (d\mathbf{E}/dt)]}{|\mathbf{E}|^2} = \frac{-\omega E_1 E_2 \sin\phi}{E_1^2 \cos^2(\omega t - \beta z) + E_2^2 \cos^2(\omega t - \beta z + \phi)}$$

CHAPTER 8

REFLECTION AND REFRACTION OF PLANE WAVES

for interface with medium 2 being perfect conductor σ = ∞, there can be no ∂/∂t fields in 2 so all energy must be reflected back

8.1 Introduction. In this chapter we consider the problem of uniform plane waves propagating in a region which has an abrupt change in the constitutive relations. The surface where this abrupt change takes place is called the *boundary* between the two regions. For simplicity, we consider only planar boundaries.† Thus we discuss situations such as a uniform plane wave in free space incident upon a plane conducting region, uniform plane waves incident upon a plane interface between two dielectrics, and uniform plane waves incident upon a plane dielectric slab. In each situation we implicitly or explicitly require that there be only one plane-wave source, located so remotely from our region of interest that interactions with the source can be ignored. This source will produce a plane *incident wave*. Our problem in this chapter is the effect of the boundary upon this incident wave.

8.2 General Direction of Propagation. In Sec. 7.10 we developed the proper description for uniform plane waves propagating in arbitrary directions. In this section we repeat a portion of that development from a slightly different point of view. Our motivation is that, in the general case, our incident wave will be traveling in some arbitrary direction with respect to the interface between our two regions. It will usually be found desirable to express the behavior of the waves in terms of a right-hand rectangular coordinate system referenced to the surface normal.

Let us consider, then, a linearly polarized uniform plane wave traveling in a specified x', y', z' coordinate system. Let the wave be x'-polarized, and let it be traveling in the $+z'$ direction. Then, from Chap. 7, the x', y', z'

† We should stress at the outset that the theory we develop is applicable even when the surface of discontinuity is not plane, provided *the wavelength is small compared with the radius of curvature of the surface.*

346

description of this wave is

$$\mathbf{E} = E_{x'}e^{-\gamma z'}\mathbf{a}_{x'} \tag{8-1}$$

$$\mathbf{H} = \frac{E_{x'}}{\eta}e^{-\gamma z'}\mathbf{a}_{y'} \tag{8-2}$$

where $e^{-\gamma z'}$ has been used to allow the possibility of conducting regions.

It should be apparent that the problem of describing this wave in the x, y, z coordinate system is a problem in transformation of coordinates. That is to say, we need to express z', $E_{x'}$, $\mathbf{a}_{x'}$, and $\mathbf{a}_{y'}$ in the x, y, z coordinate system. With little loss of generality and great reduction in complexity, we may demand that the origins of the two coordinate systems coincide. This means that the two systems are related by rotation only. Translational transformations are seldom necessary and, if required, can usually be made after rotation has been accounted for.

As will be recalled from analytic geometry, for rotational transformations in rectangular coordinates, the relations are

$$x' = l_{x'}x + m_{x'}y + n_{x'}z$$
$$y' = l_{y'}x + m_{y'}y + n_{y'}z$$
$$z' = l_{z'}x + m_{z'}y + n_{z'}z$$

where the l's, m's, and n's are the direction cosines of the primed axes with respect to the unprimed axes. Thus the z' factor of the exponential becomes

$$z' = l_{z'}x + m_{z'}y + n_{z'}z \tag{8-3}$$

It is frequently convenient to write $\gamma z'$ as the dot product of two vectors. To this end, let $\mathbf{r}$ be the radius vector from the common origin of the coordinate systems to some point P. Then we have

$$\mathbf{r} = x'\mathbf{a}_{x'} + y'\mathbf{a}_{y'} + z'\mathbf{a}_{z'} = x\mathbf{a}_x + y\mathbf{a}_y + z\mathbf{a}_z \tag{8-4}$$

Now define

$$\boldsymbol{\gamma} = \gamma\mathbf{a}_{z'} \tag{8-5}$$

Form the dot product of Eqs. (8-4) and (8-5), and obtain

$$\boldsymbol{\gamma} \cdot \mathbf{r} = \gamma z' = \gamma x\mathbf{a}_{z'} \cdot \mathbf{a}_x + \gamma y\mathbf{a}_{z'} \cdot \mathbf{a}_y + \gamma z\mathbf{a}_{z'} \cdot \mathbf{a}_z \tag{8-6}$$

But the dot products of the unit vectors in Eq. (8-6) are just the direction cosines $l_{z'}$, $m_{z'}$, $n_{z'}$. Therefore we can write a general form of the transformation for the exponential factor as

$$e^{-\gamma z'} = e^{-\boldsymbol{\gamma} \cdot \mathbf{r}} = e^{-\alpha(l_{z'}x + m_{z'}y + n_{z'}z)}e^{-j\beta(l_{z'}x + m_{z'}y + n_{z'}z)} = e^{-\boldsymbol{\alpha} \cdot \mathbf{r}}e^{-j\boldsymbol{\beta} \cdot \mathbf{r}} \tag{8-7}$$

where $\boldsymbol{\alpha} = \alpha(l_{z'}\mathbf{a}_x + m_{z'}\mathbf{a}_y + n_{z'}\mathbf{a}_z)$ and $\boldsymbol{\beta} = \beta(l_{z'}\mathbf{a}_x + m_{z'}\mathbf{a}_y + n_{z'}\mathbf{a}_z)$. It is worth noting that $\boldsymbol{\beta} \cdot \mathbf{r} = $ constant defines planes of constant z', and thus planes of constant phase of the wave. The normal to these planes is the $\boldsymbol{\beta}$ direction, which is the $+z'$ direction.

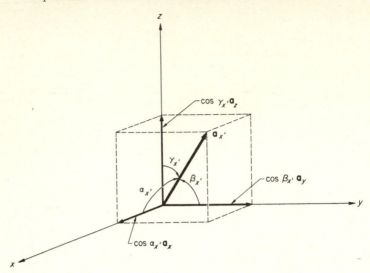

FIGURE 8-1. The geometrical relations between the unit vector $\mathbf{a}_{x'}$ and $\mathbf{a}_x$, $\mathbf{a}_y$, $\mathbf{a}_z$ of a rotated coordinate system. $\mathbf{a}_{x'} = \cos \alpha_{x'} \, \mathbf{a}_x + \cos \beta_{x'} \, \mathbf{a}_y + \cos \gamma_{x'} \mathbf{a}_z$.

A careful consideration of geometrical relations between $\mathbf{a}_{x'}$, $\mathbf{a}_{y'}$, $\mathbf{a}_{z'}$ and $\mathbf{a}_x$, $\mathbf{a}_y$, $\mathbf{a}_z$, as displayed in Fig. 8-1 for $\mathbf{a}_{x'}$, will show that the relations obtained by vector addition rules are

$$\mathbf{a}_{x'} = l_{x'}\mathbf{a}_x + m_{x'}\mathbf{a}_y + n_{x'}\mathbf{a}_z$$
$$\mathbf{a}_{y'} = l_{y'}\mathbf{a}_x + m_{b'}\mathbf{a}_y + n_{y'}\mathbf{a}_z \tag{8-8}$$
$$\mathbf{a}_{z'} = l_{z'}\mathbf{a}_x + m_{z'}\mathbf{a}_y + n_{z'}\mathbf{a}_z$$

From Eqs. (8-3) and (8-8) it is apparent that the complete transformation may be readily obtained if the direction cosines of the x', y', z' coordinate system with respect to the x, y, z coordinate system are known.   Returning to our original wave, we have

$$\mathbf{E} = E_{x'}(l_{x'}\mathbf{a}_x + m_{x'}\mathbf{a}_y + n_{x'}\mathbf{a}_z)e^{-\gamma \cdot \mathbf{r}}$$
$$\mathbf{H} = \frac{E_{x'}}{\eta}(l_{y'}\mathbf{a}_x + m_{y'}\mathbf{a}_y + n_{y'}\mathbf{a}_z)e^{-\gamma \cdot \mathbf{r}} \tag{8-9}$$

Similar expressions would be obtained for other polarizations and directions of propagation.   These expressions seem complicated, and they are.   Fortunately, in most problems it is possible to retain one coordinate axis common to both coordinate systems, thus using only a single angle of rotation about the common axis.   This reduces the nine direction angles to four, which take the simple values of 0, $\pi/2$, θ, and $\pi/2 - \theta$, where θ is the angle of rotation.

8.3 Snell's Laws. The general laws of reflection and refraction which describe the behavior of plane waves at plane interfaces between two linear, homogeneous, isotropic media are developed in this section. These laws, specialized for lossless media, are known as *Snell's laws*, which are the familiar laws of geometrical optics. They state that the angle of incidence equals the angle of reflection, and that the relation between the angle of incidence, θ_i, and the angle of refraction, θ_r, is given by

$$\frac{\sin \theta_i}{\sin \theta_r} = \frac{v_i}{v_r} \qquad \left[v_p = \frac{\omega}{\beta} = \frac{\omega \lambda}{2\pi} \right] \tag{8-10}$$

where v_i and v_r are the phase velocities of uniform plane waves in the medium of incidence and of refraction, respectively.

In geometrical optics, these laws are derived, usually, on a semi-intuitive physical reasoning basis. We propose to show that they follow from our previous results for TEM waves.

For notational and mathematical manipulative convenience, it is desirable to note that our expressions for a linearly polarized plane wave, Eqs. (8-9), can be rewritten in a more compact notation, which in fact is also more general. As shown in Sec. 7.10, the first of Eqs. (8-9) can be written

$$\mathbf{E} = \mathbf{E}_0 e^{-\gamma \mathbf{n}_0 \cdot \mathbf{r}} \tag{8-11}$$

where $\mathbf{E}_0$ is the complex vector magnitude of $\mathbf{E}$, $\gamma^2 = j\omega\mu(\sigma + j\omega\epsilon)$ is the propagation factor for the medium, and $\mathbf{n}_0$ is a unit vector in the direction of propagation.

Now, since we have a TEM wave, $\mathbf{H}$ is perpendicular to $\mathbf{E}$, and both $\mathbf{E}$ and $\mathbf{H}$ are perpendicular to $\mathbf{n}_0$, which is the direction of propagation. This allows us to write the expression for the $\mathbf{H}$ field as

$$\mathbf{H} = \frac{\mathbf{n}_0 \times \mathbf{E}}{\eta} e^{-\gamma \mathbf{n}_0 \cdot \mathbf{r}} \tag{8-12}$$

Note that, since Eqs. (8-11) and (8-12) hold for any component of a TEM wave, they also hold for the more general case of elliptical polarization.

Now consider the situation depicted in Fig. 8-2, where we show a two-dimensional projection of a plane wave propagating in the $\mathbf{n}_0$ direction in medium 1 and incident upon the plane surface S separating media 1 and 2; $\mathbf{n}$ is the unit normal to S taken positive into medium 1. This wave is partially reflected and partially refracted as two new TEM plane waves in the $\mathbf{n}_1$ and $\mathbf{n}_2$ directions.

In terms of the notation of the figure and Eqs. (8-11) and (8-12), we have an incident wave $\mathbf{E}_i$, $\mathbf{H}_i$, a reflected wave $\mathbf{E}_r$, $\mathbf{H}_r$, and a transmitted, or

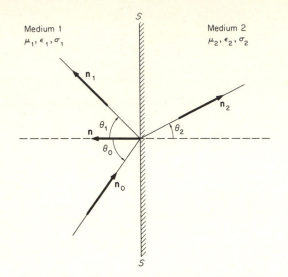

FIGURE 8-2.   The geometry and notation for reflection and refraction at a plane surface.

refracted, wave $\mathbf{E}_t$, $\mathbf{H}_t$, given by

$$\mathbf{E}_i = \mathbf{E}_0 e^{-\gamma_1 \mathbf{n}_0 \cdot \mathbf{r}} \qquad \mathbf{H}_i = \frac{1}{\eta_1} \mathbf{n}_0 \times \mathbf{E}_0 e^{-\gamma_1 \mathbf{n}_0 \cdot \mathbf{r}}$$

$$\mathbf{E}_r = \mathbf{E}_1 e^{-\gamma_1 \mathbf{n}_1 \cdot \mathbf{r}} \qquad \mathbf{H}_r = \frac{1}{\eta_1} \mathbf{n}_1 \times \mathbf{E}_1 e^{-\gamma_1 \mathbf{n}_1 \cdot \mathbf{r}} \qquad (8\text{-}13)$$

$$\mathbf{E}_t = \mathbf{E}_2 e^{-\gamma_2 \mathbf{n}_2 \cdot \mathbf{r}} \qquad \mathbf{H}_t = \frac{1}{\eta_2} \mathbf{n}_2 \times \mathbf{E}_2 e^{-\gamma_2 \mathbf{n}_2 \cdot \mathbf{r}}$$

where $\gamma_0 = \gamma_1$, $\eta_0 = \eta_1$ has been used in the first pair in Eqs. (8-13).

The derivation of Snell's laws from these relations involves the use of vector manipulations and of vector identities, and interpretation of the resulting equations.

First, we note that continuity of the tangential components of the total $\mathbf{E}$ and $\mathbf{H}$ fields everywhere on the boundary S requires that they be in phase with each other at the boundary, and that this requires that the exponential factors be equal at all points on the interface.   (Two complex numbers are equal if and only if they have equal real and imaginary parts.)

Second, we note that the interface plane S is given, mathematically, by $\mathbf{n} \cdot \mathbf{r} = $ constant, and that

$$\mathbf{n} \cdot \mathbf{r} = 0$$

places the origin of our coordinate system on the S plane.

Third, we use the vector identity

$$\mathbf{n} \times (\mathbf{n} \times \mathbf{r}) = (\mathbf{n} \cdot \mathbf{r})\mathbf{n} - (\mathbf{n} \cdot \mathbf{n})\mathbf{r}$$

or since $\mathbf{n} \cdot \mathbf{r} = 0$ on S,

$$\mathbf{r} = -\mathbf{n} \times (\mathbf{n} \times \mathbf{r}) \tag{8-14}$$

for a point on the interface. Now, equating the exponents in Eqs. (8-13) and using Eq. (8-14), so that r is on the surface, we obtain

$$\gamma_1\mathbf{n}_0 \cdot \mathbf{n} \times (\mathbf{n} \times \mathbf{r}) = \gamma_1\mathbf{n}_1 \cdot \mathbf{n} \times (\mathbf{n} \times \mathbf{r})$$

$$\gamma_1\mathbf{n}_0 \cdot \mathbf{n} \times (\mathbf{n} \times \mathbf{r}) = \gamma_2\mathbf{n}_2 \cdot \mathbf{n} \times (\mathbf{n} \times \mathbf{r})$$

Next, using the vector identity

$$\mathbf{n}_0 \cdot \mathbf{n} \times (\mathbf{n} \times \mathbf{r}) = (\mathbf{n}_0 \times \mathbf{n}) \cdot (\mathbf{n} \times \mathbf{r})$$

we find

$$(\mathbf{n}_0 \times \mathbf{n} - \mathbf{n}_1 \times \mathbf{n}) \cdot (\mathbf{n} \times \mathbf{r}) = 0$$
$$(\gamma_1\mathbf{n}_0 \times \mathbf{n} - \gamma_2\mathbf{n}_2 \times \mathbf{n}) \cdot (\mathbf{n} \times \mathbf{r}) = 0 \tag{8-15}$$

These equations look confusing, but actually, the geometrical interpretations of the cross product and the dot product make their interpretation fairly simple. Recall that the vector $\mathbf{n} \times \mathbf{r}$ is perpendicular to both $\mathbf{n}$ and $\mathbf{r}$. Also, recall that $\mathbf{A} \cdot \mathbf{B} = AB \cos \theta$, where θ is the angle between $\mathbf{A}$ and $\mathbf{B}$. Now, $\mathbf{n}_0 \times \mathbf{n}$, $\mathbf{n}_1 \times \mathbf{n}$, and $\mathbf{n}_2 \times \mathbf{n}$ are all perpendicular to $\mathbf{n}$, and hence lie in the plane of the interface. Note that $\mathbf{n} \times \mathbf{r}$ also lies in the plane of the interface and is perpendicular to $\mathbf{r}$ in this plane. Thus the direction of $\mathbf{n} \times \mathbf{r}$ in the interfacial plane is arbitrary, so that $\mathbf{n}_0 \times \mathbf{n}$, $\mathbf{n}_1 \times \mathbf{n}$, and $\mathbf{n}_2 \times \mathbf{n}$ are not in general perpendicular to $\mathbf{n} \times \mathbf{r}$. It is clear, then, that Eqs. (8-15) are of the form

$$|\mathbf{n}_0 \times \mathbf{n} - \mathbf{n}_1 \times \mathbf{n}|\, r \cos \delta = 0$$

$$|\gamma_1\mathbf{n}_0 \times \mathbf{n} - \gamma_2\mathbf{n}_1 \times \mathbf{n}|\, r \cos \delta = 0$$

where both r and δ have arbitrary values. This can be true if and only if

$$\mathbf{n}_0 \times \mathbf{n} - \mathbf{n}_1 \times \mathbf{n} = 0$$
$$\gamma_1\mathbf{n}_0 \times \mathbf{n} - \gamma_2\mathbf{n}_2 \times \mathbf{n} = 0 \tag{8-16}$$

which in turn can be true only if $\mathbf{n}_0 \times \mathbf{n}$, $\mathbf{n}_1 \times \mathbf{n}$, and $\mathbf{n}_2 \times \mathbf{n}$ are collinear. These cross products can be collinear only if the unit vectors $\mathbf{n}$, $\mathbf{n}_0$, $\mathbf{n}_1$, $\mathbf{n}_2$ are coplanar. Furthermore, the collinearity of the cross-product vectors and Eqs. (8-16) require that

$$|\mathbf{n}_0 \times \mathbf{n}| = |\mathbf{n}_1 \times \mathbf{n}|$$
$$|\gamma_1\mathbf{n}_0 \times \mathbf{n}| = |\gamma_2\mathbf{n}_2 \times \mathbf{n}| \tag{8-17}$$

In terms of the *angle of incidence* θ_0, the *angle of reflection* θ_1, and the *angle*

$\beta = \omega\sqrt{\mu\epsilon}$

of refraction θ_2, the set (8-17) yields the general form of the more familiar Snell's laws,

$\gamma^2 = -\omega^2\mu\epsilon + j\omega\mu\sigma$

$$\sin \theta_0 = \sin \theta_1 \qquad \text{useful} \qquad (8\text{-}18)$$

$$\frac{\sin \theta_0}{\sin \theta_2} = \frac{\gamma_2}{\gamma_1} \qquad (8\text{-}19)$$

It should be noted that the γ's of Eq. (8-19) are complex for conducting media, and thus imply that θ_0, θ_1, and θ_2 can be *complex angles* and that n_0, n_1, and n_2 can define complex propagation directions. We postpone further consideration of this general case until later, and note that for nonconducting media we have $\sigma_1 = \sigma_2 = 0$, and hence

j comes in due to $\sqrt{-1}$ in $\sqrt{-\omega^2\mu\epsilon + j(0)}$

$$\gamma_1 = j\omega\sqrt{\mu_1\epsilon_1} = j\beta_1$$
$$\gamma_2 = j\omega\sqrt{\mu_2\epsilon_2} = j\beta_2$$

and therefore

useful when medium 1 and 2 are not-conducting $(\sigma = 0)$

$$\frac{\sin \theta_0}{\sin \theta_2} = \frac{\beta_2}{\beta_1} = \frac{\sqrt{\mu_2\epsilon_2}}{\sqrt{\mu_1\epsilon_1}} = \frac{v_1}{v_2} \qquad (8\text{-}20)$$

which is the familiar form of Snell's law of refraction, valid for nonconducting media. If medium 1 is free space, the ratio on the right becomes

for medium 1 being free space →

$$\frac{c}{v_2} = \frac{\sqrt{\mu\epsilon}}{\sqrt{\mu_0\epsilon_0}} = \sqrt{\mu_r\epsilon_r} = n \qquad (8\text{-}21)$$

where the dimensionless number n is the *index of refraction* of medium 2. With the exception of ferrous substances, $\mu_r \approx 1$; hence $n \approx \sqrt{\epsilon_r}$. Typically, $n = 9$ for distilled water, and $n = 1.0003$ for air near the earth's surface.

8.4 Reflection and Refraction at Plane Dielectric Boundaries.

The solution to the general problem, reflection and refraction of uniform plane waves of arbitrary polarization incident at arbitrary angles, can be obtained by a rather straightforward application of the boundary conditions. The significance of the results is, however, somewhat obscured by their generality. In this text it seems more appropriate to solve special cases which cover most situations of practical interest.

Case 1. Perfect dielectric, normal incidence Consider an x- polarized uniform plane wave incident normally on the plane boundary between two media, as shown in Fig. 8-3. This wave is given by $\theta = 0$

$$\mathbf{E}_i = E_i e^{-j\beta_1 z}\mathbf{a}_x$$

$$\mathbf{H}_i = \frac{1}{\eta_1}\mathbf{a}_z \times \mathbf{E}_i = \frac{E_i}{\eta_1} e^{-j\beta_1 z}\mathbf{a}_y \qquad (8\text{-}22)$$

x-polarized means E field is in Ex direction

"going to right"
Incident wave has − exponent $-j\beta z$
reflected wave has + exponent

$\bar{E} \times \bar{H}$ gives direction of energy propagation

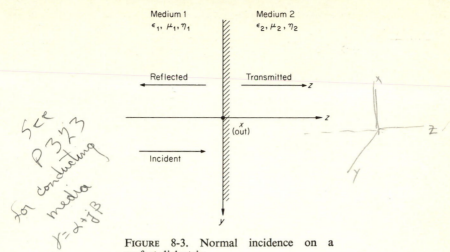

FIGURE 8-3. Normal incidence on a perfect dielectric.

The reflected wave will be given by

$$\mathbf{E}_r = E_r e^{j\beta_1 z}\mathbf{a}_x$$

$$\mathbf{H}_r = -\frac{E_r}{\eta_1} e^{j\beta_1 z}\mathbf{a}_y \qquad (8\text{-}23)$$

and the transmitted wave will be given by

$$\mathbf{E}_t = E_t e^{-j\beta_2 z}\mathbf{a}_x$$

$$\mathbf{H}_t = \frac{E_t}{\eta_2} e^{-j\beta_2 z}\mathbf{a}_y \qquad (8\text{-}24)$$

At the boundary ($z = 0$) we must have $\mathbf{E}_{\text{tan}}$ and $\mathbf{H}_{\text{tan}}$ continuous. Since each of the field components is tangential, this means that at $z = 0$

$$\mathbf{E}_i + \mathbf{E}_r = \mathbf{E}_t$$

$$\mathbf{H}_i + \mathbf{H}_r = \mathbf{H}_t$$

These conditions can be written as two scalar equations:

$$E_i + E_r = E_t$$

$$\frac{E_i}{\eta_1} - \frac{E_r}{\eta_1} = \frac{E_t}{\eta_2}$$

Algebraic solution of these equations yields

Reflection coefficient

$$\frac{E_r}{E_i} = \frac{\eta_2 - \eta_1}{\eta_2 + \eta_1} = \Gamma \tag{8-25}$$

Transmission Coefficient

$$\frac{E_t}{E_i} = \frac{2\eta_2}{\eta_2 + \eta_1} = T \tag{8-26}$$

Frequently, we define these ratios as the *reflection coefficient* Γ and as the *transmission coefficient* T, respectively.

The total fields in each region can be written in terms of the incident field and the reflection and transmission coefficients as

$$\mathbf{E}_1 = \mathbf{E}_i + \mathbf{E}_r = E_i(e^{-j\beta_1 z} + \Gamma e^{j\beta_1 z})\mathbf{a}_x$$

$$\mathbf{H}_1 = \mathbf{H}_i + \mathbf{H}_r = \frac{E_i}{\eta_1}(e^{-j\beta_1 z} - \Gamma e^{j\beta_1 z})\mathbf{a}_y$$

$$\mathbf{E}_2 = \mathbf{E}_t = E_i T e^{-j\beta_2 z}\mathbf{a}_x \tag{8-27}$$

$$\mathbf{H}_2 = \mathbf{H}_t = \frac{E_i T}{\eta_2} e^{-j\beta_2 z}\mathbf{a}_y$$

Example 8-1 Normal Incidence. Suppose, as an example, that medium 1 is free space, with $\mu_1 = \mu_0$, $\epsilon_1 = \epsilon_0$, and hence $\eta_1 = \eta_0 = \sqrt{\mu_0/\epsilon_0}$, and that medium 2 has $\mu_2 = \mu_0$ and a dielectric constant $\epsilon_r = 4$. Thus $\epsilon_2 = \epsilon_r\epsilon_0 = 4\epsilon_0$, and

$$\eta_2 = \sqrt{\frac{\mu_2}{\epsilon_2}} = \sqrt{\frac{\mu_0}{4\epsilon_0}} = \frac{1}{2}\eta_0$$

Substitution for η_1 and η_2 in Eqs. (8-25) and (8-26) gives

$$\frac{E_r}{E_i} = \Gamma = \frac{\tfrac{1}{2}\eta_0 - \eta_0}{\tfrac{1}{2}\eta_0 + \eta_0} = -\frac{1}{3}$$

$$\frac{E_t}{E_i} = T = \frac{2(\tfrac{1}{2}\eta_0)}{\tfrac{1}{2}\eta_0 + \eta_0} = \frac{2}{3}$$

The time-average power flow per square meter is obtained from the complex Poynting vector as

$$\mathbf{P}_i = \tfrac{1}{2}\,\mathrm{Re}\,(\mathbf{E}_i \times \mathbf{H}_i^*) = \frac{E_i^2}{2\eta_0}\,\mathbf{a}_z$$

$$\mathbf{P}_r = \tfrac{1}{2}\,\mathrm{Re}\,(\mathbf{E}_r \times \mathbf{H}_r^*) = -\frac{E_i^2}{18\eta_0}\,\mathbf{a}_z$$

$$\mathbf{P}_t = \tfrac{1}{2}\,\mathrm{Re}\,(\mathbf{E}_t \times \mathbf{H}_t^*) = \frac{4E_i^2}{18\eta_2}\,\mathbf{a}_z = \frac{8E_i^2}{18\eta_0}\,\mathbf{a}_z$$

It is seen that one-ninth of the incident power is reflected and that eight-ninths of it is transmitted into the second medium.

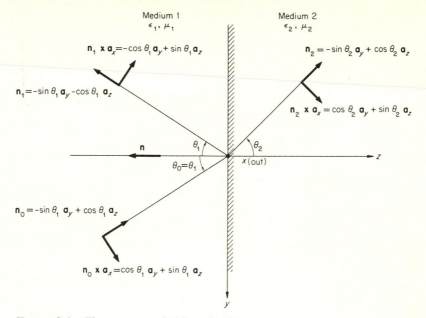

FIGURE 8-4. The geometry of oblique incidence when the xy plane is the plane of the interface.

Case 2. Oblique incidence A general vector formulation of oblique incidence based upon the geometry of Fig. 8-2 was given in Eq. (8-13). Now, let the z axis be perpendicular to the interface, and let the common plane of the unit vectors $\mathbf{n}_0$, $\mathbf{n}_1$, and $\mathbf{n}_2$, which specify the directions of propagation, be the yz plane, as shown in Fig. 8-4. The formulation is simplified somewhat since now $\mathbf{n}_0$, $\mathbf{n}_1$, and $\mathbf{n}_2$ each have a zero x component and the direction cosines involved can assume only the values 0, ± 1, $\pm \sin \theta_1$, $\pm \cos \theta_1$, $\pm \sin \theta_2$, $\cos \theta_2$. This geometry in particular allows us to write any polarization of the incident wave as the sum of two components, one with $\mathbf{E}_i$ parallel to the interface (that is, in the x direction), and a second component with $\mathbf{H}_i$ parallel to the interface. We shall solve for the behavior of these two components as two special cases.

Case 2a. Oblique incidence: $\underline{\mathbf{E}_i\ \textbf{\textit{parallel to the interface}}}\,(\mathbf{n} \cdot \mathbf{E}_i = 0)$ Referring to Fig. 8-4, which shows the geometry of oblique incidence when the interfacial plane is taken as the xy plane, <u>we see that $\mathbf{E}_i$ parallel to the interface means that $\mathbf{E}_i$, $\mathbf{E}_r$, and $\mathbf{E}_t$ have only an x component, and that $\mathbf{H}_i$, $\mathbf{H}_r$, and $\mathbf{H}_t$ all lie in the yz</u> plane (known as the *plane of incidence*) and are in the directions specified by $\mathbf{n}_0 \times \mathbf{a}_x$, $\mathbf{n}_1 \times \mathbf{a}_x$, and $\mathbf{n}_2 \times \mathbf{a}_x$, respectively. Thus we see that it is easier to obtain the description of the three fields from the geometry of Fig. 8-4, rather than from formal use of the vector and dot products involved.   Relative space orientations of the field vectors are shown in Fig. 8-5a.

The three fields are

$$\mathbf{E}_i = E_i e^{j\beta_1 y \sin \theta_1 - j\beta_1 z \cos \theta_1} \mathbf{a}_x$$

$$\mathbf{H}_i = \left(\frac{E_i}{\eta_1} \cos \theta_1 \, \mathbf{a}_y + \frac{E_i}{\eta_1} \sin \theta_1 \, \mathbf{a}_z \right) e^{j\beta_1 y \sin \theta_1 - j\beta_1 z \cos \theta_1}$$

$$\mathbf{E}_r = E_r e^{j\beta_1 y \sin \theta_1 + j\beta_1 z \cos \theta_1} \mathbf{a}_x$$

$$\mathbf{H}_r = \left(-\frac{E_r}{\eta_1} \cos \theta_1 \, \mathbf{a}_y + \frac{E_r}{\eta_1} \sin \theta_1 \, \mathbf{a}_z \right) e^{j\beta_1 y \sin \theta_1 + j\beta_1 z \cos \theta_1} \qquad (8\text{-}28)$$

$$\mathbf{E}_t = E_t e^{j\beta_2 y \sin \theta_2 - j\beta_2 z \cos \theta_2} \mathbf{a}_x$$

$$\mathbf{H}_t = \left(\frac{E_t}{\eta_2} \cos \theta_2 \, \mathbf{a}_y + \frac{E_t}{\eta_2} \sin \theta_2 \, \mathbf{a}_z \right) e^{j\beta_2 y \sin \theta_2 - j\beta_2 z \cos \theta_2}$$

Now the boundary conditions are that $\mathbf{E}_{\text{tan}}$ and $\mathbf{H}_{\text{tan}}$ are continuous in magnitude and phase at the interface $z = 0$.   Continuity of phase requires that

$$e^{j\beta_1 y \sin \theta_1} = e^{j\beta_2 y \sin \theta_2}$$

or simply,

$$\frac{\sin \theta_1}{\sin \theta_2} = \frac{\beta_2}{\beta_1} \qquad (8\text{-}29)$$

This is just Snell's law of refraction.

Continuity of magnitude of the tangential components of $\mathbf{E}$ and $\mathbf{H}$ requires that

$$E_i + E_r = E_t$$

$$\frac{E_i}{\eta_1} \cos \theta_1 - \frac{E_r}{\eta_1} \cos \theta_1 = \frac{E_t}{\eta_2} \cos \theta_2 \qquad (8\text{-}30)$$

These equations can be solved simultaneously to eliminate either E_t or E_r. The results are the *Fresnel equations*

$$\Gamma_\perp = \frac{E_r}{E_i} = \frac{\eta_2 \cos \theta_1 - \eta_1 \cos \theta_2}{\eta_2 \cos \theta_1 + \eta_1 \cos \theta_2} \qquad (8\text{-}31)$$

$$T_\perp = \frac{E_t}{E_i} = \frac{2\eta_2 \cos \theta_1}{\eta_2 \cos \theta_1 + \eta_1 \cos \theta_2} \qquad (8\text{-}32)$$

Clearly, the behavior depends both on the electromagnetic properties of the two media and upon the angle of incidence.

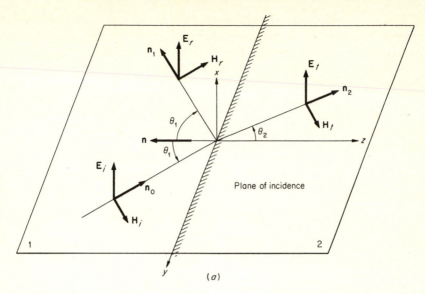

(a)

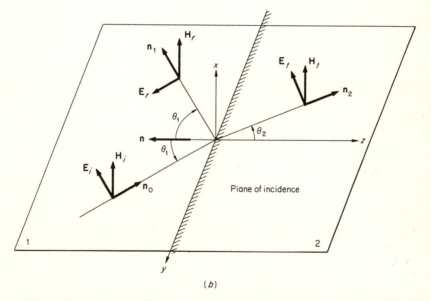

(b)

FIGURE 8-5. Oblique incidence. (a) E_i parallel to interface; (b) H_i parallel to interface.

Case 2b.   Oblique incidence: $\mathbf{H}_i$ *parallel to the interface* $(\mathbf{n} \cdot \mathbf{H}_i = 0)$ With some care, we can use the geometry of Fig. 8-4 for this case also, in conjunction with Fig. 8-5b.

We note, first, that the exponential factor in the wave's description depends only on the direction of propagation, and hence is unchanged. Second, we need to recall that $\mathbf{E}$ and $\mathbf{H}$ are both perpendicular to each other and to the direction of propagation in such a manner that $\mathbf{E} \times \mathbf{H}$ is in the positive direction of propagation.   Thus, for $\mathbf{H}_i = H_i \mathbf{a}_x = (E_i/\eta_1)\mathbf{a}_x$, $\mathbf{E}_i$ is in the $-\mathbf{n}_0 \times \mathbf{a}_x$ direction.   The new expressions for the field intensities are then

$$\mathbf{H}_i = \frac{E_i}{\eta_1} e^{j\beta_1 y \sin \theta_1 - j\beta_1 z \cos \theta_1} \mathbf{a}_x$$

$$\mathbf{E}_i = (-E_i \cos \theta_1 \, \mathbf{a}_y - E_i \sin \theta_1 \, \mathbf{a}_z)e^{j\beta_1 y \sin \theta_1 - j\beta_1 z \cos \theta_1}$$

$$\mathbf{H}_r = \frac{E_r}{\eta_1} e^{j\beta_1 y \sin \theta_1 + j\beta_1 z \cos \theta_1} \mathbf{a}_x \qquad\qquad (8\text{-}33)$$

$$\mathbf{E}_r = (E_r \cos \theta_1 \, \mathbf{a}_y - E_r \sin \theta_1 \, \mathbf{a}_z)e^{j\beta_1 y \sin \theta_1 + j\beta_1 z \cos \theta_1}$$

$$\mathbf{H}_t = \frac{E_t}{\eta_2} e^{j\beta_2 y \sin \theta_2 - j\beta_2 z \cos \theta_2} \mathbf{a}_x$$

$$\mathbf{E}_t = (-E_t \cos \theta_2 \, \mathbf{a}_y - E_t \sin \theta_2 \, \mathbf{a}_z)e^{j\beta_2 y \sin \theta_2 - j\beta_2 z \cos \theta_2}$$

The boundary conditions are once again continuity of the tangential components of $\mathbf{E}$ and $\mathbf{H}$.   The exponential factor yields Snell's law as in Case 2a, and will not be repeated.

Continuity of the magnitudes of the tangential components of $\mathbf{H}$ and $\mathbf{E}$ requires that

$$\frac{E_i}{\eta_1} + \frac{E_r}{\eta_1} = \frac{E_t}{\eta_2} \qquad\qquad (8\text{-}34)$$

$$-E_i \cos \theta_1 + E_r \cos \theta_1 = -E_t \cos \theta_2$$

Simultaneous solution of these equations to eliminate E_t or E_r yields the Fresnel equations for this case.

reflection coeff. for when H_i is $\parallel$ to plane boundary

$$\Gamma_\parallel = \frac{E_r}{E_i} = \frac{\eta_1 \cos \theta_1 - \eta_2 \cos \theta_2}{\eta_1 \cos \theta_1 + \eta_2 \cos \theta_2} \qquad\qquad (8\text{-}35)$$

E is parallel to plane of incidence.

$$T_\parallel = \frac{E_t}{E_i} = \frac{2\eta_2 \cos \theta_1}{\eta_1 \cos \theta_1 + \eta_2 \cos \theta_2} \qquad\qquad (8\text{-}36)$$

The results are different from the previous case.   This difference is frequently used to obtain a separation of the two linear polarizations.

It should be noted that when $\theta_0 = 0$, which is the normal incidence, there is no distinction between Cases 2a and 2b and that $T_\perp = T_\parallel$ while $|\Gamma_\perp| = |\Gamma_\parallel|$.   The reflection coefficients differ in sign at normal incidence because of the vector symbolism adopted in Fig. 8-5. Thus, whereas the vectors $\mathbf{E}_i$ and $\mathbf{E}_r$ are parallel in Fig. 8-5a, their tangential components have opposite directions in Fig. 8-5b.

Example 8-2 Oblique Incidence. This example will illustrate the different behavior of the two polarizations of the incident wave for oblique incidence.

Given that medium 1 is free space, which means $\epsilon_1 = \epsilon_0,\ \mu_1 = \mu_0,\ \eta_1 = \eta_0$, and that medium 2 has $\epsilon_2 = 4\epsilon_0,\ \mu_2 = \mu_0$, giving $\eta_2 = \sqrt{\mu_2/\epsilon_2} = \sqrt{\mu_0/4\epsilon_0} = \frac{1}{2}\eta_0$, then for Case 2$a$ ($\mathbf{E}_i$ parallel to interface), we have

$$\Gamma_\perp = \frac{\frac{1}{2}\eta_0 \cos\theta_1 - \eta_0 \cos\theta_2}{\frac{1}{2}\eta_0 \cos\theta_1 + \eta_0 \cos\theta_2} = \frac{\cos\theta_1 - 2\cos\theta_2}{\cos\theta_1 + 2\cos\theta_2}$$

and for Case 2b ($\mathbf{H}_i$ parallel to interface), we have

$$\Gamma_\parallel = \frac{\eta_0 \cos\theta_1 - \frac{1}{2}\eta_0 \cos\theta_2}{\eta_0 \cos\theta_1 + \frac{1}{2}\eta_0 \cos\theta_2} = \frac{2\cos\theta_1 - \cos\theta_2}{2\cos\theta_1 + \cos\theta_2}$$

The angles θ_1 and θ_2 are related by Snell's law, which is

$$\frac{\sin\theta_1}{\sin\theta_2} = \frac{\beta_2}{\beta_1}$$

and since $\beta = \omega\sqrt{\mu\epsilon}$,

$$\frac{\beta_2}{\beta_1} = \frac{\omega\sqrt{\mu_0(4\epsilon_0)}}{\omega\sqrt{\mu_0\epsilon_0}} = 2$$

giving $\sin\theta_2 = \frac{1}{2}\sin\theta_1$. For $\theta_1 = 30°$, $\Gamma_\perp = -0.381$ and $\Gamma_\parallel = 0.283$. These results show that there is a significant difference in the reflection for the two cases. Typical plots of $|\Gamma_\perp|^2$ and $|\Gamma_\parallel|^2$ are given in Fig. 8-6.

In summary, the general case of oblique incidence is best undertaken in two parts. In the first part, the electric field vector is considered to be parallel to the plane surface separating two dissimilar media, and in the second part, the magnetic field intensity is parallel to the reflecting plane. The two cases are distinct and are sometimes referred to as *perpendicular* (to the plane of incidence), or *horizontal, polarization*, and *parallel*, or *vertical, polarization*, respectively.

In the most general case, the incident wave will have a component of the E field parallel to the interface and a component parallel to the plane of incidence (Fig. 8-7). This case can best be treated by applying the developed theory separately to the two components of the field.

8.5 Brewster Angle (Polarizing Angle). In the preceding section we derived equations which show the magnitude of the reflected and transmitted waves relative to the incident wave magnitude for two special cases of linear

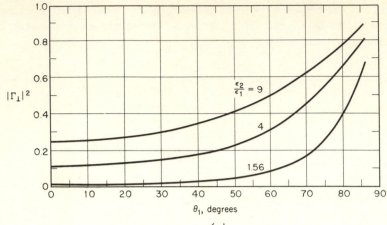

θ_1, degrees

(a)

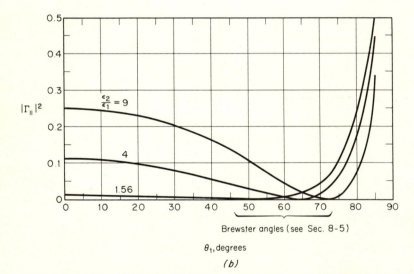

Brewster angles (see Sec. 8-5)

θ_1, degrees

(b)

FIGURE 8-6.   Plots of power reflection coefficients from a dielectric-dielectric interface.   *(a)* $\mathbf{E}_i$ parallel to the interface; *(b)* $\mathbf{H}_i$ parallel to the interface.

polarization.   We also noted that the equations showed a different dependence upon angle of incidence and refraction.

In this section we ask whether there is any incident angle such that either (or both) polarization(s) is totally transmitted into the second medium. By conservation of energy, this is equivalent to asking for the angle at which the reflected amplitude is zero.   We shall treat the two cases separately.

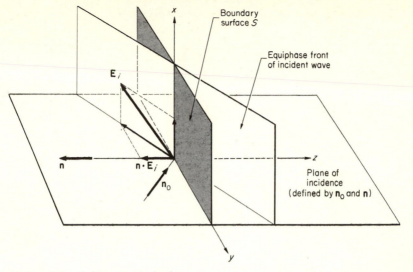

FIGURE 8-7. Arbitrary polarization.

For Case 2a, $\mathbf{E}_i$ parallel to the interface, we had

$$\Gamma_\perp = \frac{E_r}{E_i} = \frac{\eta_2 \cos \theta_1 - \eta_1 \cos \theta_2}{\eta_2 \cos \theta_1 + \eta_1 \cos \theta_2} \qquad (8\text{-}31)$$

For $E_r = 0$, this becomes

$$\eta_2 \cos \theta_1 = \eta_1 \cos \theta_2$$

Some algebraic and trigonometric manipulations plus application of Snell's law are required to see the significance of this result.

First, let us square both sides and make use of $\cos^2 \theta = 1 - \sin^2 \theta$. We obtain

$$\eta_2{}^2(1 - \sin^2 \theta_1) = \eta_1{}^2(1 - \sin^2 \theta_2)$$

Snell's law gives

$$\sin^2 \theta_2 = \frac{\beta_1{}^2}{\beta_2{}^2} \sin^2 \theta_1 \qquad (8\text{-}37)$$

and transforms the preceding equation to *for $\perp$ polarization*

$$\sin^2 \theta_1 = \frac{\mu_1 \epsilon_2 / \mu_2 \epsilon_1 - 1}{(\mu_1/\mu_2)^2 - 1} \qquad (8\text{-}38)$$

As a practical matter, the permeability of dielectric materials is approximately that of free space; hence we have $\mu_1 \approx \mu_2 \approx \mu_0$ and $\sin^2 \theta_1 \to \infty$, unless $\epsilon_1 = \epsilon_2$. This means that no angle exists for which there is zero reflection when $\mathbf{E}_i$ is parallel to the interface except when $\epsilon_1 = \epsilon_2$, which means that there is no interface.

For Case 2b, with $\mathbf{H}_i$ parallel to the interface,

$$\Gamma_\| = \frac{E_r}{E_i} = \frac{\eta_1 \cos \theta_1 - \eta_2 \cos \theta_2}{\eta_1 \cos \theta_1 + \eta_2 \cos \theta_2} \tag{8-35}$$

So for $E_r = 0$,

$$\eta_1 \cos \theta_1 = \eta_2 \cos \theta_2$$

Proceeding as before, we obtain

$$\eta_1{}^2(1 - \sin^2 \theta_1) = \eta_2{}^2(1 - \sin^2 \theta_2)$$

and making use of Eq. (8-37),

$$\sin^2 \theta_1 = \frac{\mu_2 \epsilon_1 / \mu_1 \epsilon_2 - 1}{(\epsilon_1 / \epsilon_2)^2 - 1} \tag{8-39}$$

Under the usual condition that $\mu_1 \approx \mu_2 \approx \mu_0$, this becomes

for // Polarization

$$\boxed{\sin^2 \theta_1 = \frac{\epsilon_2}{\epsilon_1 + \epsilon_2} \leq 1} \tag{8-40}$$

which always has a solution.

For certain purposes, it is desirable to express this angle, $\theta_1 = \theta_B$, known as the *Brewster angle*, in terms of its tangent. If we substitute $\sin^2 \theta_B = 1 - \cos^2 \theta_B$ into Eq. (8-40) and solve for $\cos^2 \theta_B$, we obtain

$$\cos^2 \theta_B = \frac{\epsilon_1}{\epsilon_1 + \epsilon_2} \tag{8-41}$$

for // Polarization in order to have $E_r = 0$

The formula

Brewster Angle Defined

$$\boxed{\tan \theta_B = \sqrt{\frac{\epsilon_2}{\epsilon_1}}} \tag{8-42}$$

is an alternative expression for the Brewster angle. It is important to note that Eq. (8-42) holds only when $\mathbf{H}_i$ is parallel to the boundary, and that there is a Brewster angle for any combination of ϵ_1 and ϵ_2.

The phenomenon described by the Brewster angle is used in some types of the gaseous laser, a device in which repeated transmissions through an active medium contained in a long tube are required. Reflection from the ends of the tube is minimized by using faces which are tilted at the Brewster angle.

8.6 Total Internal Reflection (the Critical Angle). Total reflection of the incident wave is experimentally known to occur under certain conditions for angles of incidence equal to, or greater than, a certain angle known as the *critical angle*. Under these conditions, the wave propagation (but not the electromagnetic field itself) remains entirely in the medium of incidence. For this reason, the total reflection is frequently called *total internal reflection*.

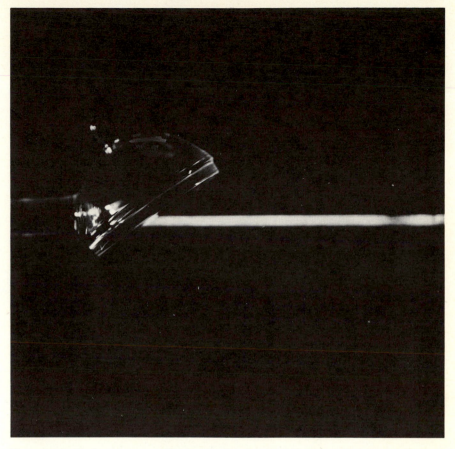

Photograph of a laser beam.   The beam exits through a Brewster window at the end of the laser tube.

To understand the phenomenon of total reflection, we must examine carefully the nature of the transmitted wave.   From Eqs. (8-28) and (8-33) we have

$$\mathbf{E}_t = \mathbf{E}_2 e^{j\beta_2(y\sin\theta_2 - z\cos\theta_2) + j\omega t} \qquad (8\text{-}43)$$

use this when
$\mathcal{E}_1 < \mathcal{E}_2$

where $\mathbf{E}_2$ is an arbitrary vector which lies in the plane $\mathbf{n}_2 \cdot \mathbf{r} = $ constant. Otherwise, the notation is that of Fig. 8-5.

If $\mu_1 = \mu_2 = \mu_0$, which for all practical purposes is true for most dielectric substances, Snell's law gives

$$\frac{\sin\theta_1}{\sin\theta_2} = \frac{\beta_2}{\beta_1} = \sqrt{\frac{\mu_2\epsilon_2}{\mu_1\epsilon_1}} = \sqrt{\frac{\epsilon_2}{\epsilon_1}} \qquad (8\text{-}44)$$

From this equation, we obtain

$$\cos \theta_2 = \sqrt{1 - \sin^2 \theta_2} = \sqrt{1 - \frac{\epsilon_1}{\epsilon_2} \sin^2 \theta_1} \qquad (8\text{-}45)$$

which allows us to express the space dependence in the exponent of Eq. (8-43) in the form

$$y \sin \theta_2 - z \cos \theta_2 = y \sqrt{\frac{\epsilon_1}{\epsilon_2}} \sin \theta_1 - z \sqrt{1 - \frac{\epsilon_1}{\epsilon_2} \sin^2 \theta_1} \qquad (8\text{-}46)$$

For $\epsilon_1 < \epsilon_2$, the quantity under the last radical on the right is always positive, and the radical itself is real. The transmitted wave, expressed by Eq. (8-43), in this case advances along the direction defined by the unit vector $\mathbf{n}_2$ at the phase velocity ω/β_2, and there is nothing new.

The phenomenon of total reflection occurs only if $\epsilon_1 > \epsilon_2$ and the angle of incidence θ_1 exceeds a certain value, which we shall now determine.

Once again, we examine the nature of the quantity under the last radical on the right of Eq. (8-46). We note that, even if $\epsilon_1 > \epsilon_2$, this quantity remains positive, provided the angle θ_1 is sufficiently small. However, as the angle of incidence is increased, the quantity under the radical becomes progressively smaller, decreasing to zero and ultimately becoming negative. That particular value of θ_1, call it θ_c, which makes

$$1 - \frac{\epsilon_1}{\epsilon_2} \sin^2 \theta_c = 0 \qquad (8\text{-}47)$$

is called the *critical angle*. It is explicitly given by the relation

$$\sin \theta_c = \sqrt{\frac{\epsilon_2}{\epsilon_1}} \qquad (8\text{-}48)$$

where $\epsilon_1 > \epsilon_2$.

It is obvious now that, when $\theta_1 = \theta_c$, Eq. (8-44) gives

$$\sin \theta_2 = \sqrt{\frac{\epsilon_1}{\epsilon_2}} \sin \theta_1 = \sqrt{\frac{\epsilon_1}{\epsilon_2}} \sin \theta_c = 1 \qquad (8\text{-}49)$$

which states that the angle of refraction is 90°, and thus implies that the transmitted wave is propagating parallel to the interface. This conclusion is also confirmed by Eq. (8-43), which, for $\theta_1 = \theta_c$, becomes

$$\mathbf{E}_t = \mathbf{E}_2 e^{j\beta_2 y + j\omega t} \qquad (8\text{-}50)$$

and clearly states that the planes of constant phase are perpendicular to the y axis. The complex Poynting vector is then

$$S_c = -\frac{E_2{}^2}{2\eta_2} \mathbf{a}_y \qquad (8\text{-}51)$$

which shows that the time-average power flow in the positive z direction, namely, into medium 2, is zero.

Now, the question that still remains is, what happens when the angle of incidence exceeds the critical angle θ_c? Obviously, the quantity under the last radical in Eq. (8-46) is then negative, and the radical itself is purely imaginary. Thus, when $\theta_1 > \theta_c$ and $\epsilon_1 > \epsilon_2$,

$$\sqrt{1 - \frac{\epsilon_1}{\epsilon_2}\sin^2\theta_1} = -j\sqrt{\frac{\epsilon_1}{\epsilon_2}\sin^2\theta_1 - 1} \tag{8-52}$$

where, as we shall presently see, the choice of negative sign is dictated by the condition that the field shall remain finite as $z \to \infty$. The amended form of Eq. (8-46) is then

$$y\sin\theta_2 - z\cos\theta_2 = y\sqrt{\frac{\epsilon_1}{\epsilon_2}}\sin\theta_1 + jz\sqrt{\frac{\epsilon_1}{\epsilon_2}\sin^2\theta_1 - 1} \tag{8-53}$$

so that, with $\mu_1 = \mu_2$,

$$j\beta_2(y\sin\theta_2 - z\cos\theta_2) = \omega\sqrt{\mu_1\epsilon_1}\left(-z\sqrt{\sin^2\theta_1 - \frac{\epsilon_2}{\epsilon_1}} + jy\sin\theta_1\right) \tag{8-54}$$

If we now let

$$\alpha = \omega\sqrt{\mu_1\epsilon_1}\sqrt{\sin^2\theta_1 - \frac{\epsilon_2}{\epsilon_1}} \tag{8-55}$$

$$\beta = \omega\sqrt{\mu_1\epsilon_1}\sin\theta_1 \tag{8-56}$$

where both α and β are real positive numbers, then we can write Eq. (8-43) as

when $\theta_1 > \theta_c$
$\epsilon_1 > \epsilon_2$

$$\boxed{E_t = E_2 e^{-\alpha z}e^{j(\omega t + \beta y)}} \quad \} \ \mu_1 = \mu_2 \tag{8-57}$$

From this expression, it is immediately apparent that the transmitted wave is *attenuated* [and this justifies the choice of the negative root in Eq. (8-52)] at a rate determined not only by the electromagnetic parameters of the medium, but also by the frequency and by the angle of incidence. According to Sec. 7.14, the electromagnetic disturbance represented by Eq. (8-57) is a *nonuniform* plane wave propagating in the negative y direction with a phase velocity

$$v_p = \frac{\omega}{\beta} = \frac{1}{\sin\theta_1\sqrt{\mu_1\epsilon_1}} \tag{8-58}$$

explicitly dependent upon the angle of incidence. As shown in Fig. 8-8, the planes of constant phase ($y = $ constant) are perpendicular to the planes of constant amplitude ($z = $ constant). Thus the wave is *guided* along the reflecting surface, with no average transport of energy into medium 2. This

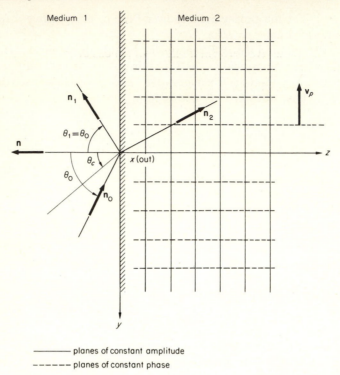

- planes of constant amplitude
- - - - - - planes of constant phase

FIGURE 8-8.   The case of total reflection ($\epsilon_1 > \epsilon_2$, $\theta_1 \geq \theta_c$).

also follows from a consideration of the expressions for Γ developed in Sec. 8.4.   Thus, from Eq. (8-31), we have

$$|\Gamma_\perp| = \left| \frac{\eta_2 \cos \theta_1 + j\eta_1 \sqrt{(\epsilon_1/\epsilon_2) \sin^2 \theta_1 - 1}}{\eta_2 \cos \theta_1 - j\eta_1 \sqrt{(\epsilon_1/\epsilon_2) \sin^2 \theta_1 - 1}} \right| = 1 \qquad (8\text{-}59)$$

when $\mathbf{n} \cdot \mathbf{E}_i = 0$, and similarly for Eq. (8-35) when $\mathbf{n} \cdot \mathbf{H}_i = 0$.   This means that *the intensity of the reflected wave is exactly equal to the intensity of the incident wave.*   Moreover, it means that the power flow into medium 2 is purely imaginary.   To prove this statement, we note that, since

$$\mathbf{S}_i = \tfrac{1}{2}\mathbf{E}_i \times \mathbf{H}_i^* = \frac{1}{2} \frac{E_i^2}{\eta_1} \mathbf{n_0}$$

$$\mathbf{S}_t = \tfrac{1}{2}\mathbf{E}_t \times \mathbf{H}_t^* = \frac{1}{2} \frac{E_t^2}{\eta_2} \mathbf{n_2}$$

$$(8\text{-}60)$$

the ratio

$$\frac{\mathbf{n} \cdot \mathbf{S}_t}{\mathbf{n} \cdot \mathbf{S}_i} = \frac{E_t^2/\eta_2}{E_i^2/\eta_1} \frac{\mathbf{n} \cdot \mathbf{n}_2}{\mathbf{n} \cdot \mathbf{n}_0} = \frac{E_t^2/\eta_2 \cos \theta_2}{E_i^2/\eta_1 \cos \theta_1} \tag{8-61}$$

is a purely imaginary quantity, by virtue of the imaginary character [Eqs. (8-45) and (8-52)] of $\cos \theta_2$; all other quantities on the right of Eq. (8-61) are obviously real.

Now let us look at the reflected wave. Examination of Eqs. (8-31) and (8-35) shows that the phase of the reflected wave is different for the two types of incidence. Thus, when $\mathbf{E}_i$ is parallel to the reflecting plane, the phase difference between incident and reflected waves is

$$\phi = 2 \tan^{-1} \frac{\sqrt{\sin^2 \theta_1 - (\epsilon_2/\epsilon_1)}}{\cos \theta_1} \tag{8-62}$$

and when $\mathbf{H}_i$ is parallel to the reflecting plane,

$$\phi = 2 \tan^{-1} \frac{\sqrt{\sin^2 \theta_1 - (\epsilon_2/\epsilon_1)}}{(\epsilon_2/\epsilon_1) \cos \theta_1} \tag{8-63}$$

In general, then, the reflected wave will be elliptically polarized.

To sum up, total reflection occurs when two lossless dielectrics meet at a plane interface, and the angle of incidence θ_0 exceeds the critical angle $\theta_c = \sin^{-1} \sqrt{\epsilon_2/\epsilon_1}$, where $\epsilon_1 > \epsilon_2$. In practice, this phenomenon is used in the design of optical instruments, such as the prism binocular.

8.7 Conducting Media; Reflection and Refraction. The results of Secs. 8.3 and 8.4, though written from the point of view of plane interfaces between two lossless dielectrics, are valid in the general case when one or both media are conducting.

Let us examine first the simplest case, namely, that of a wave incident normal to a plane boundary separating a dielectric and a perfect conductor. To fix ideas, we suppose that, in Fig. 8-3, medium 1 is a perfect dielectric and that $\sigma_2 \to \infty$. Then Eqs. (8-22) and (8-23) apply, but $\mathbf{E}_t = \mathbf{H}_t = 0$ in this case. Therefore, at the boundary ($z = 0$), we must have a zero tangential $\mathbf{E}$ field, from which it follows that $\mathbf{E}_r = -\mathbf{E}_i$. This says that $\Gamma = -1$, $T = 0$, and that total reflection occurs in the case of infinite conductivity.

The case of finite conductivity and arbitrary angles of incidence is more complicated, and is treated next.

As indicated in Fig. 8-9, we suppose that the wave is incident from medium 1, which is still a perfect dielectric, and that the refracting medium 2

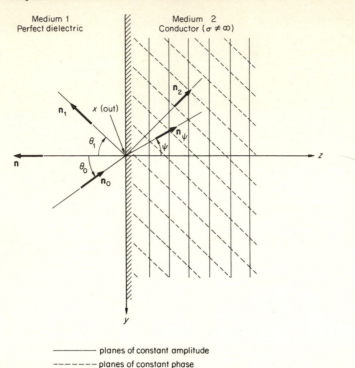

FIGURE 8-9.   Space relations applicable to the study of reflection
and refraction at a dielectric-conductor plane interface.

is now conducting, but σ is finite.   The propagation constants are defined by

$$\gamma_1 = j\beta_1 = j\omega\sqrt{\mu_1\epsilon_1} \tag{8-64}$$

$$\gamma_2 = \alpha_2 + j\beta_2 = (-\omega^2\mu_2\epsilon_2 + j\omega\mu_2\sigma_2)^{1/2} \tag{8-65}$$

In the last equation, α_2 and β_2 are expressed in terms of the parameters of the
medium by Eqs. (7-93) and (7-94), which we repeat here for ready reference:

$$\alpha_2 = \omega\sqrt{\mu_2\epsilon_2}\left\{\frac{1}{2}\left[\sqrt{1 + \left(\frac{\sigma_2}{\omega\epsilon_2}\right)^2} - 1\right]\right\}^{1/2} \tag{8-66}$$

$$\beta_2 = \omega\sqrt{\mu_2\epsilon_2}\left\{\frac{1}{2}\left[\sqrt{1 + \left(\frac{\sigma_2}{\omega\epsilon_2}\right)^2} + 1\right]\right\}^{1/2} \tag{8-67}$$

Within the conducting medium the transmitted wave is represented by an
expression analogous to Eq. (8-43).

$$\mathbf{E}_t = \mathbf{E}_2 e^{-\gamma_2(\mathbf{n}_2 \cdot \mathbf{r})} = \mathbf{E}_2 e^{\gamma_2(y\sin\theta_2 - z\cos\theta_2)} \tag{8-68}$$

This expression must now be expanded to take into account the complex nature of γ_2.

By Snell's laws, Eqs. (8-18) and (8-19),

$$\sin \theta_2 = \frac{\gamma_1}{\gamma_2} \sin \theta_0 = \frac{j\beta_1}{\alpha_2 + j\beta_2} \sin \theta_1 \tag{8-69}$$

so that

$$\cos \theta_2 = \sqrt{1 - \sin^2 \theta_2} = \sqrt{1 - \left(\frac{j\beta_1}{\alpha_2 + j\beta_2}\right)^2 \sin^2 \theta_1} \tag{8-70}$$

It is convenient to set

$$\cos \theta_2 = re^{j\delta} \tag{8-71}$$

in terms of which the space exponent in Eq. (8-68) becomes

$$\gamma_2(y \sin \theta_2 - z \cos \theta_2) = (\alpha_2 + j\beta_2)\left(y \frac{j\beta_1}{\alpha_2 + j\beta_2} \sin \theta_1 - zre^{j\delta}\right)$$

$$= j(\beta_1 \sin \theta_1)y - r(\alpha_2 \cos \delta - \beta_2 \sin \delta)z$$

$$- jr(\alpha_2 \sin \delta + \beta_2 \cos \delta)z$$

$$= -pz + j[(\beta_1 \sin \theta_1)y - qz] \tag{8-72}$$

where

$$p = r(\alpha_2 \cos \delta - \beta_2 \sin \delta)$$
$$q = r(\alpha_2 \sin \delta + \beta_2 \cos \delta) \tag{8-73}$$

are real quantities, dependent exclusively on the parameters of the medium and θ_1. Substituting Eq. (8-72) into Eq. (8-68), we find

$$\mathbf{E}_t = \mathbf{E}_2 e^{-pz + j[(\beta_1 \sin \theta_1)y - qz + \omega t]} \tag{8-74}$$

As in the case of total reflection, the refracted wave is nonuniform. The planes of constant amplitude ($z =$ constant) are parallel to the boundary plane, while the planes of constant phase $[(\beta_1 \sin \theta_1)y - qz =$ constant] are generally inclined relative to its normal, at an angle which no longer is specified by θ_2. In fact, the true angle of refraction ψ is the one defined by the imaginary part of the exponent in Eq. (8-74):

$$(\beta_1 \sin \theta_1)y - qz + \omega t = -\sqrt{(\beta_1 \sin \theta_1)^2 + q^2}$$

$$\times \left[\frac{\beta_1 \sin \theta_1}{\sqrt{(\beta_1 \sin \theta_1)^2 + q^2}}(-y) + \frac{qz}{\sqrt{(\beta_1 \sin \theta_1)^2 + q^2}}\right] + \omega t \tag{8-75}$$

If in this expression we let

$$\sin \psi = \frac{\beta_1 \sin \theta_1}{\sqrt{(\beta_1 \sin \theta_1)^2 + q^2}} \qquad \cos \psi = \frac{q}{\sqrt{(\beta_1 \sin \theta_1)^2 + q^2}} \tag{8-76}$$

we can then write

$$-\sqrt{(\beta_1 \sin \theta_1)^2 + q^2}(-y \sin \psi + z \cos \psi) + \omega t$$
$$= -\sqrt{(\beta_1 \sin \theta_1)^2 + q^2}\, \mathbf{n}_\psi \cdot \mathbf{r} + \omega t \qquad (8\text{-}77)$$

where $\mathbf{n}_\psi = -\sin \psi\, \mathbf{a}_y + \cos \psi\, \mathbf{a}_z$

At the same time we find that the true angle of refraction is

$$\psi = \tan^{-1} \frac{\beta_1 \sin \theta_1}{q} \qquad (8\text{-}78)$$

To gain further insight, let us consider the special, but important, practical case of a good conductor.

By definition, $\sigma_2/\omega\epsilon_2 \gg 1$, and therefore Eqs. (8-66) and (8-67) give

$$\alpha_2 \approx \beta_2 \approx \sqrt{\frac{\omega\mu_2\sigma_2}{2}} \qquad (8\text{-}79)$$

Under this condition (σ_2 large), and with $\mu_1 \approx \mu_2$, which is the practical case, Eq. (8-69) gives

$$\frac{\sin \theta_2}{\sin \theta_1} = \frac{j\omega\sqrt{\mu_1\epsilon_1}}{\sqrt{\omega\mu_2\sigma_2/2(1+j1)}} = \sqrt{\frac{\omega\epsilon_1}{\sigma_2}}\, e^{j\pi/4} \to 0 \qquad (8\text{-}80)$$

so that $\theta_2 = 0$, and in Eq. (8-71) $r = 1$ and $\delta = 0$. In view of this, we find

$$p \approx \alpha_2 \approx \sqrt{\frac{\omega\mu_2\sigma_2}{2}} \qquad q \approx \beta_2 \approx \sqrt{\frac{\omega\mu_2\sigma_2}{2}} \qquad (8\text{-}81)$$

and hence $$\psi = \tan^{-1} \frac{\omega\sqrt{\mu_1\epsilon_1} \sin \theta_1}{\sqrt{\omega\mu_2\sigma_2/2}} \to 0 \qquad (8\text{-}82)$$

Therefore, as the conductivity increases, the true angle of refraction tends to zero, and the planes of constant phase tend to orient themselves parallel to the reflecting plane and to the planes of constant amplitude. Typically, in the case of copper ($\sigma = 5.8 \times 10^7$ ℧/m) and at a frequency as high as 100 GHz, which is well above the range of practical microwave frequencies,

$$\psi = \tan^{-1} \frac{\omega\sqrt{\mu_1\epsilon_1} \sin \theta_1}{\sqrt{\omega\mu_2\sigma_2/2}} \leq \tan^{-1} \sqrt{\frac{\omega\epsilon_1}{\sigma_2/2}} \approx \tan^{-1} 10^{-3} \approx 10^{-3}\ \text{rad}$$

which fixes $\mathbf{n}_\psi$ virtually perpendicular to the boundary plane.

To the same approximation, Eq. (8-74) becomes

$$\mathbf{E}_t = \mathbf{E}_2 e^{-\alpha_2 z + j[(\beta_1 \sin \theta_1)y - \beta_2 z + \omega t]} \qquad (8\text{-}83)$$

and makes it evident that, at a distance

$$\delta = \frac{1}{\alpha_2} = \sqrt{\frac{2}{\omega\mu_2\sigma_2}} \qquad (8\text{-}84)$$

from the surface of the conductor, the amplitude of the electric vector is equal to 0.368 of its value at the surface. The factor δ is the skin depth, or the depth of penetration, encountered previously, in Eq. (7-96). Note that the inverse dependence of δ upon σ_2 dictates the use of very low frequencies when communication with a station submerged under water is considered. Since δ is very small, large amounts of power are usually required for this purpose.

Continuing our remarks, we should point out that, according to Eq. (8-77), the phase velocity in the conducting medium, measured along the $\mathbf{n}_\psi$ axis, is

$$v_p = \mathbf{n}_\psi \cdot \frac{d\mathbf{r}}{dt} = \frac{\omega}{\sqrt{(\beta_1 \sin \theta_1)^2 + q^2}} \approx \frac{\omega}{\sqrt{(\beta_1 \sin \theta_1)^2 + \beta_2^2}} \tag{8-85}$$

and exhibits a clear dependence on θ_1 and on the parameters of *both* media.

Finally, turning our attention to the reflected wave in medium 1, we note from Eqs. (8-31) and (8-35) that the reflection coefficient Γ is complex and in general dependent on the polarization of the incident wave. Therefore the reflected wave will be out of phase with respect to the incident wave and, more specifically, will be elliptically polarized if the incident wave is linearly polarized and if neither $\mathbf{E}_i$ nor $\mathbf{H}_i$ is entirely parallel to the reflecting plane.

For sufficiently large conductivities, $\theta_2 \approx 0$, as noted previously, and therefore

$$\Gamma \approx \begin{cases} \dfrac{\cos \theta_1 - (\eta_1/\eta_2)}{\cos \theta_1 + (\eta_1/\eta_2)} & \mathbf{E}_i \parallel S \\[3mm] \dfrac{(\eta_1/\eta_2) \cos \theta_1 - 1}{(\eta_1/\eta_2) \cos \theta_1 + 1} & \mathbf{H}_i \parallel S \end{cases} \tag{8-86}$$

On the other hand, for good conductors,

$$\eta = \sqrt{\frac{j\omega\mu}{\sigma + j\omega\epsilon}} \approx \sqrt{\frac{j\omega\mu}{\sigma}} \tag{8-87}$$

so that

$$\frac{\eta_1}{\eta_2} \approx \frac{\sqrt{\mu_1/\epsilon_1}}{\sqrt{j\omega\mu_2/\sigma_2}} \approx \sqrt{\frac{\sigma_2}{j\omega\epsilon_1}} \gg 1 \geq \cos \theta_1 \tag{8-88}$$

This means that *the reflection coefficient is approximately equal to 1 for all good conductors regardless of angle of incidence.*

We close our discussion by mentioning a few facts about the surface of the earth.

First, below about 1 MHz, sea water ($\sigma = 3$ ℧/m) behaves like a fairly good conductor. Its reflecting properties are nevertheless very much dependent upon the angle of incidence and the type of polarization, especially when $\mathbf{H}_i$ is parallel to the sea surface.   It turns out that, in this case, when the angle of incidence is near grazing, only a small portion of the incident wave is reflected, and the rest is absorbed.   For frequencies above 1 MHz, however, sea water, like freshwater and dry earth, acts like a lossy dielectric insofar as the reflection and refraction phenomena are concerned.

The reflection properties of sea water, freshwater, and dry earth can be used effectively in the design of Polaroid glasses to cut down glare.   This cause of discomfort to the eyes is often remedied by use of colored glasses; however, these are not altogether satisfactory, since they suppress reflections from objects indiscriminately, including those which are dimly lit.   The design of Polaroid glasses is based on the fact that sea water and freshwater and dry earth discriminate against the component of $\mathbf{H}_i$ which is parallel to the surface of the earth or the body of water.   Thus, after reflection, an incident wave is polarized predominantly with the electric vector parallel to the interface.   By rejecting principally this (horizontally polarized) component of the reflected light, it is possible to cut down glare effectively.

Summarizing, we have examined the reflection and refraction at a conductor-dielectric interface, with major emphasis on the case of good conductors.   We saw that good conductors represent an extreme case of the general behavior, and that their behavior is different from that of good dielectrics.

8.8   Multiple Interfaces.   Consider the situation shown in Fig. 8-10, where two plane interfaces separate three regions of space.   Two cases are of practical interest: one where all three regions are lossless, and one where a

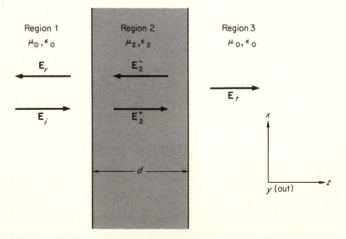

FIGURE 8-10.   A dielectric slab in free space.   (The y axis is positive out of the plane of the paper.)

lossy region separates two lossless dielectrics.† For simplicity, we shall consider only normal incidence and linearly polarized plane waves.

Consider the case of an infinite plane slab of dielectric material of thickness d, separating two semi-infinite regions of free space, as shown in Fig. 8-10. The incident plane wave $\mathbf{E}_i$ is partially reflected and partially transmitted, as shown. For concreteness, assume that the waves are x-polarized. Then we can write for the fields in the three regions

$$\mathbf{E}_1 = \mathbf{E}_i + \mathbf{E}_r = (E_i e^{-j\beta_0 z} + E_r e^{j\beta_0 z})\mathbf{a}_x$$

$$\mathbf{H}_1 = \mathbf{H}_i + \mathbf{H}_r = \left(\frac{E_i}{\eta_0} e^{-j\beta_0 z} - \frac{E_r}{\eta_0} e^{j\beta_0 z}\right)\mathbf{a}_y$$

$$\mathbf{E}_2 = \mathbf{E}_2^+ + \mathbf{E}_2^- = (E_2^+ e^{-j\beta_2 z} + E_2^- e^{j\beta_2 z})\mathbf{a}_x$$

$$\mathbf{H}_2 = \mathbf{H}_2^+ + \mathbf{H}_2^- = \left(\frac{E_2^+}{\eta_2} e^{-j\beta_2 z} - \frac{E_2^-}{\eta_2} e^{j\beta_2 z}\right)\mathbf{a}_y \qquad (8\text{-}89)$$

$$\mathbf{E}_3 = \mathbf{E}_t = E_t e^{-j\beta_0 z}\mathbf{a}_x$$

$$\mathbf{H}_3 = \mathbf{H}_t = \frac{E_t}{\eta_0} e^{-j\beta_0 z}\mathbf{a}_y$$

Recall that each of the amplitudes in these equations, in general, is complex. That is, the amplitudes can contain a relative phase factor. This allows use of a certain amount of leeway in choosing the z-coordinate origin. In particular, we can use the left interface as $z = 0$ for region 1, the right interface as $z = 0$ for region 3, and either (but not both) interface as $z = 0$ for region 2. Having made a coordinate-origin choice, we require the continuity of E_{tan} and H_{tan} at the interfaces, and obtain from Eqs. (8-89) four simultaneous equations in the four unknowns (E_1 is presumed known). From these we can obtain E_r and E_t, and hence have the solution to the problem.

If we use the right-hand interface for $z = 0$ in region 2 (Fig. 8-10) and the origins for the other two regions at their respective interfaces, as described above, the equations are

$$E_i + E_r = E_2^+ e^{j\beta_2 d} + E_2^- e^{-j\beta_2 d}$$

$$\frac{E_i}{\eta_0} - \frac{E_r}{\eta_0} = \frac{E_2^+}{\eta_2} e^{j\beta_2 d} - \frac{E_2^-}{\eta_2} e^{-j\beta_2 d} \qquad (8\text{-}90)$$

$$E_2^+ + E_2^- = E_t$$

$$\frac{E_2^+}{\eta_2} - \frac{E_2^-}{\eta_2} = \frac{E_t}{\eta_0}$$

† A practical example is a *radome*. This is an enclosure designed to protect an antenna from the undesirable effects of its physical environment and to leave its electrical performance essentially unaffected (Prob. 8-17).

Rather than solving these equations by routine methods, it is advantageous to recast them into different form, defining a few new concepts in the process.

First write the ratio of E_{tan} to H_{tan} at the two interfaces. This is just the ratio of the first pair of Eqs. (8-90) and the ratio of the second pair of Eqs. (8-90). We obtain for the first pair

$$\eta_0 \frac{E_i + E_r}{E_i - E_r} = \eta_2 \frac{E_2^+ e^{j\beta_2 d} + E_2^- e^{-j\beta_2 d}}{E_2^+ e^{j\beta_2 d} - E_2^- e^{-j\beta_2 d}} \tag{8-91}$$

or by factoring out E_i from numerator and denominator on the left and $E_2^+ e^{j\beta_2 d}$ from the right and defining,

$$\frac{E_r}{E_i} = \Gamma_1 \qquad \frac{E_2^-}{E_2^+} = \Gamma_2 \tag{8-92}$$

we obtain

$$\eta_0 \frac{1 + \Gamma_1}{1 - \Gamma_1} = \eta_2 \frac{1 + \Gamma_2 e^{-j2\beta_2 d}}{1 - \Gamma_2 e^{-j2\beta_2 d}} \tag{8-93}$$

Similarly, from the ratio of the second pair of Eqs. (8-90), we obtain

$$\eta_2 \frac{1 + \Gamma_2}{1 - \Gamma_2} = \eta_0 \tag{8-94}$$

From the definitions we see that the Γ's are the ratios of the complex magnitudes of the negative traveling wave to the positive traveling wave at the chosen origin in the particular region. Obviously, knowing the Γ's solves the original problem.

Example 8-3 A Dielectric Slab in Free Space. As an example, let us solve the problem of transmission through a slab which has a dielectric constant of 4 and a thickness d.

For $\epsilon_2 = 4\epsilon_0$, we have

$$\eta_2 = \sqrt{\frac{\mu_0}{4\epsilon_0}} = \frac{\eta_0}{2}$$

and from Eq. (8-94), $\Gamma_2 = \frac{1}{3}$. Inserting this value and $\eta_2 = \eta_0/2$ into Eq. (8-93), we find

$$\frac{1 + \Gamma_1}{1 - \Gamma_1} = \frac{1}{2} \frac{1 + \frac{1}{3} e^{-j2\beta_2 d}}{1 - \frac{1}{3} e^{-j2\beta_2 d}}$$

from which it is obvious that Γ_1 is complex unless $e^{-j2\beta_2 d}$ is real. Since solution for the complex case is really complex, we shall examine the real case first.

Note that $e^{-j2\beta_2 d}$ is real when $2\beta_2 d = n\pi$, and it becomes $+1$ when $2\beta_2 d = n\pi$, with n even, and -1 when $2\beta_2 d = n\pi$, with n odd. For n even, we have

$$\frac{1 + \Gamma_1}{1 - \Gamma_1} = \frac{1}{2} \frac{1 + \frac{1}{3}}{1 - \frac{1}{3}} = 1$$

and $\Gamma_1 = 0$. Thus $E_r/E_i = \Gamma_1 = 0$, and hence $E_r = 0$; there is no reflection when n is even. Since $\beta_2 = 2\pi/\lambda_2$, we see that $2\beta_2 d = n\pi$ gives

$$d = n\frac{\lambda_2}{4} \qquad n = 0, 2, 4, \ldots$$

Thus there is 100 percent transmission if the slab is an integral number of half wavelengths thick.

Similarly, for n odd, we obtain

$$\frac{1 + \Gamma_1}{1 - \Gamma_1} = \frac{1}{2}\frac{1 - \frac{1}{3}}{1 + \frac{1}{3}} = \frac{1}{4}$$

and $\Gamma_1 = -\frac{3}{5}$.

From Poynting's theorem, the ratio of the reflected to incident power is

$$\frac{E_r^2}{E_i^2} = \left(\frac{3}{5}\right)^2 = \frac{9}{25} \qquad \text{or} \qquad 36 \text{ percent}$$

Hence we have 36 percent of the power reflected and 64 percent of it transmitted for n odd, which means for $d = n(\lambda_2/4)$, with n odd.

A careful examination of the equation for thickness d other than an integral number of quarter wavelengths reveals that the power transmitted varies smoothly from 64 percent to 100 percent as the thickness varies. Additionally, the complex nature of Γ_1 means that the reflected wave is not in time phase with the incident wave at the interface.

The most general case of reflection and transmission through a slab with infinite transverse dimensions can be treated in much the same way as the simplified case we have considered thus far. Analysis shows that, for (generally) lossy substances and normal incidence,

$$\frac{E_r}{E_i} = \frac{(1 - Z_{12})(1 + Z_{23}) + (1 + Z_{12})(1 - Z_{23})e^{-2\gamma_2 d}}{(1 + Z_{12})(1 + Z_{23}) + (1 - Z_{12})(1 - Z_{23})e^{-2\gamma_2 d}} \qquad (8\text{-}95)$$

$$\frac{E_t}{E_i} = \frac{4}{(1 - Z_{12})(1 - Z_{23})e^{-\gamma_2 d} + (1 + Z_{12})(1 + Z_{23})e^{\gamma_2 d}} \qquad (8\text{-}96)$$

where

$$Z_{jk} = \frac{\mu_j \gamma_k}{\mu_k \gamma_j} \qquad j, k = 1, 2, 3, \ldots \qquad (8\text{-}97)$$

and the γ's are the familiar complex propagation constants. These formulas apply to both lossless and lossy substances.

Example 8-4 Radiation Pressure. It was shown in Sec. 6.8 that the general expression for force is

$$\mathbf{F} = \oint_\Sigma [\epsilon \mathbf{E}(\mathbf{E} \cdot \mathbf{n}) - \frac{\epsilon}{2} E^2 \mathbf{n}]\, da + \oint_\Sigma [\mu \mathbf{H}(\mathbf{H} \cdot \mathbf{n}) - \frac{\mu}{2} H^2 \mathbf{n}]\, da - \epsilon \int_V \frac{\partial}{\partial t}(\mathbf{E} \times \mathbf{B})\, dv$$

Let us use this formula to calculate the pressure exerted by a wave impinging upon a slab in free space.

Anechoic chamber.   (*Courtesy of Emerson and Cuming, Inc.*)

To this end, we assume that the slab is lossless and that $\mu_2 = \mu_0$ (Fig. 8-10). For simplicity, we let $d = \lambda_2/4$, so that, according to the results of the preceding example $\Gamma_1 = -\frac{3}{5}$. The first pair of Eqs. (8-89) becomes

$$\mathbf{E}_1 = E_i(e^{-j\beta_0 z} - \tfrac{3}{5}e^{j\beta_0 z})\mathbf{a}_x$$

$$\mathbf{H}_1 = \frac{E_i}{\eta_0}(e^{-j\beta_0 z} + \tfrac{3}{5}e^{j\beta_0 z})\mathbf{a}_y$$

If we now consider that the surface Σ in the force expression is a plane surface parallel to and just barely to the left of the left boundary, the volume V will consist of the slab itself and the free space on the right.   If, furthermore, we let the z axis originate at the boundary on the left, then at every point on Σ,

$$\mathbf{E}_1 = E_i(1 - \tfrac{3}{5})\mathbf{a}_x = \tfrac{2}{5}E_i\mathbf{a}_x$$

$$\mathbf{H}_1 = \frac{E_i}{\eta_0}(1 + \tfrac{3}{5})\mathbf{a}_y = \frac{8}{5}\frac{E_i}{\eta_0}\mathbf{a}_x$$

and $\mathbf{n} = -\mathbf{a}_z$.   This means that $\mathbf{E} \cdot \mathbf{n} = 0$, $\mathbf{H} \cdot \mathbf{n} = 0$, and therefore the component of the pressure (force per unit area) expressed by the surface integrals alone is

$$\tfrac{1}{2}\epsilon_0 E^2 + \tfrac{1}{2}\mu_0 H^2 = \tfrac{1}{2}\epsilon_0(E^2 + \eta_0^2 H^2) = \tfrac{1}{2}\epsilon_0 E_i^2[(\tfrac{2}{5})^2 + (\tfrac{8}{5})^2] = \tfrac{34}{25}\epsilon_0 E_i^2$$

which is quite small and negligible from a mechanical point of view.   Actually, this *is* the total pressure, since the volume integral averages out to zero over a complete cycle of the wave.   The same conclusion is drawn from a consideration of the fact that, for sinusoidal time variations, the quantity

$$\frac{\partial}{\partial t}(\mathbf{E} \times \mathbf{B}) = j2\omega \mathbf{E} \times \mathbf{B}$$

is a purely imaginary vector for uniform plane waves.

The theory developed in this section belongs to a much broader theory, where the problem is that of extremizing the reflection properties of a body. It is easy to see intuitively that if the thickness and the electromagnetic properties of a slab of material are properly chosen, the electromagnetic energy reflected by the slab will be minimized, and the body will be camouflaged (in the radar sense) at least over a narrow band of frequencies.   With this property in mind, entire rooms, known as *microwave dark rooms*, or *anechoic chambers*, are constructed with walls which are completely covered with special absorbing materials (some in corrugated form) designed to minimize reflections and thus to make the rooms suitable for antenna testing and for other similar activities.

8.9 Scattering.   The preceding section and, in particular, Example 8-3 are extreme simplifications of the general theory of scattering.   The effect of a material body, of any description, upon an impinging wave is to produce a *secondary* field which combines with the *primary* field, or the incident wave, in such a way that all boundary conditions are met at the surface of the body.

The secondary field, which arises from a system of secondary sources—charges and currents—induced on the surface (and in the interior) of the body is called the *scattered field*, and the phenomenon itself is called *scattering*. The incident and scattered fields combine to form the *diffracted field*, which is the total field in the presence of the object:

$$\text{Diffracted field} = \text{scattered field} + \begin{array}{l} \text{field that would exist} \\ \text{if object were} \\ \text{absent, but with} \\ \text{sources unchanged} \end{array} \qquad (8\text{-}98)$$

Clearly, then, the terms scattering and diffraction are associated with the same physical phenomenon.†

A prime example of scattering is the case of a *receiving antenna*. Whatever its size, weight, and configuration, and whatever the frequency of operation, an antenna receives by the same physical process; namely, the incident wave causes a constrained motion of charges in the antenna. The resulting distribution of charge and current gives rise to a secondary, or scattered, field, which, together with the impinging field, forms the diffracted field. Since the induced sources vary in synchronism with the incident field, the desired information can be extracted from the incident wave through proper coupling of the antenna to a suitable receiving apparatus.

Scattering is an everyday experience. We see things as a result of the scattering of light (either sunlight or artificial light) from objects around us. We control flying aircraft, even under cloudy conditions, by receiving backscattered energy from the aircraft, initially transmitted from a ground-based search radar. Inversely, the altitude of an aircraft relative to the terrain below can be measured by a self-contained radar altimeter, which translates into distance (distance = velocity of propagation × time) the time it takes for a wave to travel from the aircraft to ground and, after reflection (scattering), back to the airplane.

Yet, as simple as it is conceptually, and as often as it occurs, scattering remains, in practical terms, a formidable problem, beyond the scope of existing analytical tools. Only in a negligible number of extremely simple cases, where some sort of geometrical symmetry prevails, is it feasible to calculate the scattered field exactly in a closed mathematical form. Although computers may be used to obtain numerical results, the vast majority of cases, and those with the greatest practical interest, still remain unsolved. For this reason, radar targets are usually characterized by an experimentally observable quantity, known as the *radar cross section σ*. The magnitude of this single parameter depends not only on the electromagnetic properties of

---

† It should be emphasized that Eq. (8-98) represents an arbitrary way of decomposing the total field.

the object, but also on the target geometry, on the properties of the surrounding medium, on the frequency and the state of polarization, on the aspect angle (unsymmetrical objects look different from different angles), and even on whether or not the transmitter and the receiver are located at the same site (*monostatic* versus *bistatic* radars). Hence σ does not provide all the information required to characterize an object electromagnetically.

Quantitatively, the radar cross section of an object has the dimensions of an area (usually meters squared); *at a distance R, it is equal to $4\pi R^2$ times the ratio of the power density (power per unit area), $S_r = \frac{1}{2}|E_r \times H_r^*|$, of the scattered wave at the receiver site, to the power density, $S_i = \frac{1}{2}|E_i \times H_i^*|$, of the incident wave at the scatterer.*

$$\sigma = 4\pi R^2 \frac{S_r}{S_i} \qquad (8\text{-}99)$$

Generally, both S_r and S_i are complex. From Eq. (8-99) it is clear that σ is the area intercepting that amount of power which, when scattered isotropically (uniformly in all directions), sends a signal back to the receiver exactly equal in intensity to that observed in the case of the actual object. If we put Eq. (8-99) in the form

$$S_r = \frac{\sigma S_i}{4\pi R^2} \qquad (8\text{-}100)$$

we see immediately that the numerator is the total power intercepted by the scatterer, and that the denominator is the area of a spherical surface with radius R, centered about the scattering object; hence the notion of isotropic backscattering.

Example 8-5 Scattering by a Sphere. The simplest scattering object is a sphere. Returns from raindrops, almost spherical particles, can be used effectively to detect the arrival of storms either by means of ground-based or airborne *weather radars*. Given the advance warning, an aircraft can then maneuver around a dangerous area.

Consider a linearly polarized plane wave incident upon a perfectly conducting sphere, of radius a. The principles required for an exact solution to this problem are well known. Find a solution to Maxwell's equations which, far from the sphere, represents the incident plane wave, and which on the surface of the sphere satisfies the boundary conditions $E_{tan} = 0$, $H_{normal} = 0$. However, since an exact solution is quite lengthy, though it can be calculated with complete rigor, and since it has the form of an infinite series, thus providing little insight into the physical aspects of the problem, we shall concern ourselves only with a graphical display of the results.

Figure 8-11 shows the normalized radar cross section of a perfectly conducting sphere of radius a, expressed as a function of its radius measured in wavelengths. It is seen that, for sufficiently small a/λ, the normalized cross section increases with increasing a/λ. Analysis shows that in this region, called the *Rayleigh region*, the cross section varies as λ^{-4}; the behavior in this region is described by the approximate formula

$$\frac{\sigma}{\pi a^2} = 9\left(2\pi \frac{a}{\lambda}\right)^4 \qquad (8\text{-}101)$$

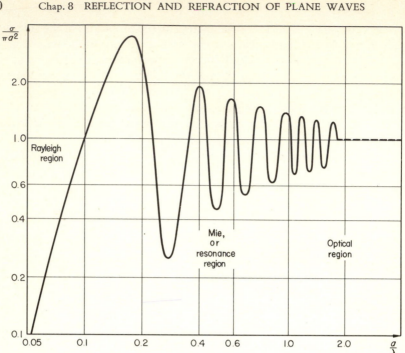

FIGURE 8-11.   The backscattering cross section of a perfectly conducting sphere of radius a, normalized with respect to its geometrical cross section (πa^2).

As the frequency increases in relation to the radius, the cross section begins to oscillate about the value $\sigma/\pi a^2 = 1$. This region is known as the *Mie*, or *resonance, region*; it precedes the so-called *optical region*, where $a \gg \lambda$ and the radar cross section approaches the optical cross section πa^2.

Example 8-6   Scattering by a Cylinder. In this example we illustrate a method by which a scattering problem may be approached analytically, and we point out the complexity of the resulting expressions. We purposely consider a configuration which, though idealized, provides answers in a closed analytical form.

We suppose that a z-polarized uniform plane wave is incident upon a perfectly conducting wire of radius a, which stretches in free space along the z axis (Fig. 8-12) and extends to infinity in both the positive and negative z directions.

Our plan of attack is first to express the incident wave in terms of functions peculiar to the coordinates used. It was pointed out in Sec. 3.12 that the Bessel functions play this role in the case of cylindrical coordinates. By virtue of their orthogonality and completeness, these functions can be used to represent an arbitrary function of the coordinates. Thus it can be shown that

$$e^{-j\beta x} = \sum_{n=-\infty}^{\infty} j^{-n} J_n(\beta r) e^{jn\varphi} \tag{8-102}$$

where J_n is the Bessel function of the first kind, and r and φ are the usual cylindrical coordinates. In view of this expansion, the incident field can be written

$$E_z{}^i = E_0 \sum_{n=-\infty}^{\infty} j^{-n} J_n(\beta r) e^{jn\varphi} \qquad (8\text{-}103)$$

Next, we represent the scattered field as an infinite series of *Hankel functions* of the second kind, defined by

$$H_n^{(2)}(r) = J_n(r) - jY_n(r)$$

We then express the scattered field as

$$E_z{}^s = E_0 \sum_{n=-\infty}^{\infty} A_n j^{-n} H_n^{(2)}(\beta r) e^{jn\varphi} \qquad (8\text{-}104)$$

where the amplitude constants A_n are yet to be determined. The total field is the sum of the incident and the reflected fields.

$$E_z = E_z{}^i + E_z{}^s = E_0 \sum_{n=-\infty}^{\infty} (-j)^n [J_n(\beta r) + A_n H_n^{(2)}(\beta r)] e^{jn\varphi} \qquad (8\text{-}105)$$

To determine the unknown constants A_n, we now require that E_{tan} be zero at $r = a$. This condition is automatically satisfied when the bracketed expression on the right is set equal to zero at $r = a$.

$$J_n(\beta a) + A_n H_n^{(2)}(\beta a) = 0$$

Then

$$A_n = - \frac{J_n(\beta a)}{H_n^{(2)}(\beta a)}$$

and

$$E_z = E_0 \sum_{n=-\infty}^{\infty} j^{-n} [J_n(\beta r) - \frac{J_n(\beta a)}{H_n^{(2)}(\beta a)} H_n^{(2)}(\beta r)] e^{jn\varphi} \qquad (8\text{-}106)$$

This is the required result. The complexity of what initially appeared to be a simple problem is undoubtedly overwhelming, since the formula for the diffracted field is far

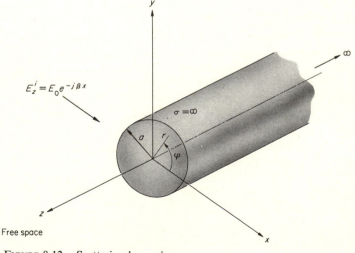

FIGURE 8-12. Scattering by a wire.

too involved to be understood and, what is even more important, to be used in practice unless the first few terms in the series suffice to give a reasonably accurate representation of the actual field.

Fortunately, we can gain some insight into the diffracted field from an examination of the scattered field at remote points of observation. If $\beta r \gg 1$, then

$$H_n^{(2)} \approx \sqrt{\frac{2}{\pi \beta r}} \exp \left[-j \left(\beta r - \frac{2n+1}{4} \pi \right) \right]$$

and the expression for the scattered field becomes

$$E_z^s = E_0 \sqrt{\frac{2}{\pi \beta r}} \left\{ \sum_{n=-\infty}^{\infty} A_n j^{-n} \exp \left[j \left(n\varphi + \frac{2n+1}{4} \pi \right) \right] \right\} e^{j(\omega t - \beta r)} \quad (8\text{-}107)$$

Clearly, this represents a z-polarized nonuniform cylindrical wave propagating in the positive r direction (outward), with a constant phase velocity $v_p = \omega/\beta$.

It should be intuitively evident that, because of symmetry, the cross section of a sphere will not be aspect-sensitive. However, the cross section of other objects, such as the infinite wire, will depend on the aspect angle and upon the polarization of the incident wave. Analysis shows that five independent radar measurements are required in order to describe the scattering properties of an object for a given aspect angle (Prob. 8-13).

To summarize, a wave incident upon a material body gives rise to the scattering phenomenon by causing a constrained movement of free and bound charges in synchronism with the applied field. The scattered field is a function of the properties and geometric configuration of the object; of the frequency, the polarization, and the aspect angle of the incident wave; and of the properties of the surrounding medium. The determination of the scattered field from Maxwell's equations can be accomplished only for the simplest of shapes. For this reason, an experimentally observable quantity, the radar cross section, is normally used in practice.

8.10 Inhomogeneous Media and Geometrical Optics.

The earth's atmosphere is a mixture of nitrogen, oxygen, a few other gases, and water vapor. The density of this gaseous layer is found to decrease, though not always monotonically, with increasing height above the surface of the earth. As a result, the index of refraction changes from about 1.0003 at zero altitude to about 1.00004 at 50,000 ft. In a very real sense, then, the atmosphere is a horizontally stratified inhomogeneous medium with continuously variable properties, and therefore refractive. To study the phenomena of propagation through such a medium it is convenient now to develop some of the basic aspects of the theory of geometrical optics.

We consider a linear and isotropic medium of infinite extent with a (possibly complex) dielectric constant which is a function of position. If the medium is also charge-free, Maxwell's equations in the sinusoidal steady state are

$$\mathbf{\nabla} \times \mathbf{E} = -j\omega\mu\mathbf{H}$$
$$\mathbf{\nabla} \times \mathbf{H} = j\omega\epsilon\mathbf{E}$$
$$\mathbf{\nabla} \cdot \mathbf{H} = 0 \tag{8-108}$$
$$\mathbf{\nabla} \cdot \epsilon\mathbf{E} = 0$$

If we take the curl of the first equation and substitute in the resulting expression the second Maxwell equation, we find

$$\mathbf{\nabla} \times \mathbf{\nabla} \times \mathbf{E} = \omega^2\mu\epsilon\mathbf{E} \tag{8-109}$$

By identity (1-84), the left side of Eq. (8-109) can be expanded to give

$$\mathbf{\nabla} \times \mathbf{\nabla} \times \mathbf{E} = \mathbf{\nabla}(\mathbf{\nabla} \cdot \mathbf{E}) - \mathbf{\nabla}^2\mathbf{E}$$

while by identity (1-79), the last Maxwell equation becomes

$$\mathbf{\nabla} \cdot \epsilon\mathbf{E} = \mathbf{E} \cdot \mathbf{\nabla}\epsilon + \epsilon\mathbf{\nabla} \cdot \mathbf{E}$$

Combining the last pair of relations gives

$$\mathbf{\nabla} \times \mathbf{\nabla} \times \mathbf{E} = -\mathbf{\nabla}\left(\mathbf{E} \cdot \frac{\mathbf{\nabla}\epsilon}{\epsilon}\right) - \mathbf{\nabla}^2\mathbf{E}$$

and transforms Eq. (8-109) to

$$\mathbf{\nabla}^2\mathbf{E} + \beta^2\mathbf{E} = -\mathbf{\nabla}\left(\mathbf{E} \cdot \frac{\mathbf{\nabla}\epsilon}{\epsilon}\right) \tag{8-110}$$

where $\beta^2 = \omega^2\mu\epsilon$.

Suppose now that $|\mathbf{\nabla}\epsilon|/\epsilon\beta^2$ is so small that the right side of Eq. (8-110) can be neglected by comparison with the second term on the left. This amounts to saying that *the spatial change of the refractive index per unit wavelength is small*, or in mathematical terms,

$$\frac{|\mathbf{\nabla}\epsilon|}{\epsilon\beta^2} \leq \frac{|\mathbf{\nabla}\epsilon|}{\epsilon_0\beta^2} = \frac{|\mathbf{\nabla}\epsilon_r|}{\beta^2} = \frac{\lambda^2|\mathbf{\nabla}n^2|}{(2\pi)^2} \ll 1 \tag{8-111}$$

which is true for air, where, on the average, the refractive index changes only 1 part in 10^8 in the course of one wavelength at 1 GHz. Under these conditions

$$\mathbf{\nabla}^2\mathbf{E} + \beta^2\mathbf{E} \approx 0 \tag{8-112}$$

and if ψ is any rectangular component of $\mathbf{E}$,

$$\mathbf{\nabla}^2\psi + n^2\beta_0^2\psi = 0 \tag{8-113}$$

where $\beta_0 = \omega\sqrt{\mu_0\epsilon_0}$, with $\mu \approx \mu_0$ for air; also, $n^2 = \epsilon_r$.

Let us seek solutions of Eqs. (8-113) in the form

$$\psi = \psi_0 e^{-j\beta_0 L} e^{j\omega t} \tag{8-114}$$

where ψ_0 and L are real functions of position. Substituting Eq. (8-114) into Eq. (8-113) and dividing out the common time factor, we obtain

$$\nabla^2(\psi_0 e^{-j\beta_0 L}) + n^2 \beta_0^2 \psi_0 e^{-j\beta_0 L} = 0 \tag{8-115}$$

Now

$$\nabla^2(\psi_0 e^{-j\beta_0 L}) = \psi_0 \, \nabla^2 e^{-j\beta_0 L} + e^{-j\beta_0 L} \, \nabla^2 \psi_0 + 2 \nabla \psi_0 \cdot \nabla e^{-j\beta_0 L}$$

But

$$\nabla^2 e^{-j\beta_0 L} = \nabla \cdot \nabla e^{-j\beta_0 L} = \nabla \cdot [-j\beta_0 (\nabla L) e^{-j\beta_0 L}]$$
$$= [-\beta_0^2 (\nabla L)^2 - j\beta_0 \, \nabla^2 L] e^{-j\beta_0 L}$$

where

$$|\nabla L|^2 = \nabla L \cdot \nabla L = \left(\frac{\partial L}{\partial x}\right)^2 + \left(\frac{\partial L}{\partial y}\right)^2 + \left(\frac{\partial L}{\partial z}\right)^2 \tag{8-116}$$

Hence

$$\nabla^2(\psi_0 e^{-j\beta_0 L}) = \psi_0(-\beta_0^2 \, |\nabla L|^2 - j\beta_0 \, \nabla^2 L) e^{-j\beta_0 L} + e^{-j\beta_0 L} \, \nabla^2 \psi_0$$
$$- j2\beta_0 e^{-j\beta_0 L} \, \nabla L \cdot \nabla \psi_0 \tag{8-117}$$

Substituting Eq. (8-117) into Eq. (8-115) and dividing out the common exponential, we find

$$\psi_0(-\beta_0^2 \, |\nabla L|^2 - j\beta_0 \, \nabla^2 L) + \nabla^2 \psi_0 - j2\beta_0 \, \nabla L \cdot \nabla \psi_0 + n^2 \beta_0^2 \psi_0 = 0$$

Finally, equating the real and imaginary parts of this equation, we obtain

$$-\psi_0 \beta_0^2 \, |\nabla L|^2 + \nabla^2 \psi_0 + n^2 \beta_0^2 \psi_0 = 0$$
$$\psi_0 \, \nabla^2 L + 2 \nabla L \cdot \nabla \psi_0 = 0$$

or rearranging the first and noting in the second that $\nabla \psi_0 / \psi_0 = \nabla(\ln \psi_0)$,

$$|\nabla L|^2 - \frac{\nabla^2 \psi_0}{\beta_0^2 \psi_0} = n^2 \tag{8-118}$$

$$\nabla^2 L + 2 \nabla L \cdot \nabla(\ln \psi_0) = 0 \tag{8-119}$$

Equations (8-118) and (8-119) represent an exact solution of Eq. (8-113).

To proceed further now it is necessary to introduce one additional approximation. Let us suppose that the frequency is sufficiently high, and β_0 correspondingly large, so that $\nabla^2 \psi_0 / \beta_0^2 \psi_0$ may be neglected. The phase function L is then determined approximately by

$$|\nabla L|^2 = n^2 \tag{8-120}$$

which is known as the equation of the *eikonal*. Letting **n** represent a vector in the direction of the wave normals ∇L (or *rays*) with a magnitude equal to $n = \sqrt{\epsilon_r}$, we obtain

$$\nabla L = \mathbf{n} \tag{8-121}$$

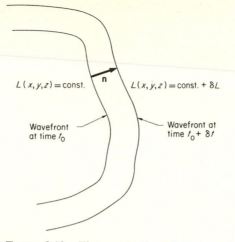

$L(x,y,z) = $ const.

$L(x,y,z) = $ const. $+ \delta L$

Wavefront at time t_0

Wavefront at time $t_0 + \delta t$

FIGURE 8-13. The propagation of a wavefront in geometrical optics.

This equation describes propagation along rays normal to families ($L =$ constant) of equiphase wavefronts (Fig. 8-13) and forms the theoretical basis of *geometrical optics*.

Example 8-7 Effective (⁴⁄₃) Earth Radius. Let us assume that the earth's atmosphere consists of a large number of thin but discrete layers, each layer having a constant dielectric constant that differs from that of the adjacent layers by a discernible amount. Generally speaking, the dielectric constant decreases with increasing altitude. Therefore a wave launched from any point in the atmosphere at an angle exceeding grazing will follow a curved path which tends to bend back toward the surface of the earth (Fig. 8-14).

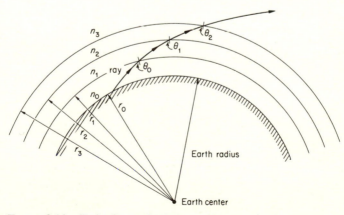

FIGURE 8-14. Path of a ray in the atmosphere.

Although the radius of curvature of the curved path generally varies with time and geographical location, for a *standard atmosphere* which fits *average* conditions, it is about four times the radius of the earth. The net effect is an apparent increase in the radius of the earth to an *effective* value of $\frac{4}{3}$ times the actual radius. When this radius is used, the atmosphere is considered to be homogeneous, and propagation is considered to occur along *straight* lines.

The systematic bending of the waves in the atmosphere thus accounts for the phenomenon of communication beyond the horizon. Taking the $\frac{4}{3}$ earth radius as $R = 27.9 \times 10^6$ ft, the apparent radio horizon from a point h ft above the surface of the earth is approximately equal to $\sqrt{2h}$ statute miles. This follows from a consideration of the simple geometrical relation

$$\sqrt{(R + h)^2 - R^2} = \sqrt{2hR + h^2} \approx \sqrt{2hR}$$

With both h and R expressed in feet, this answer comes out in feet. However, since $R \approx (5.28 \times 10^3)^2$, it is evident that the factor R may be dropped under the radical, and the answer would then be in miles, with the height expressed in feet.

From Eq. (8-121) it is apparent that if $d\mathbf{r}$ is a vector element of length in the direction of a ray, then

$$\phi = L(x,y,z) - L(x_0,y_0,z_0) = \int_{(x_0,y_0,z_0)}^{(x,y,z)} \nabla L \cdot d\mathbf{r} = \int_{(x_0,y_0,z_0)}^{(x,y,z)} \mathbf{n} \cdot d\mathbf{r} \quad (8\text{-}122)$$

is the phase difference at two points (x,y,z) and (x_0,y_0,z_0) along the ray. Since the phase velocity in the medium is

$$v = \frac{c}{n} = \frac{dr}{dt} \tag{8-123}$$

it is obvious that in Eq. (8-122) the phase ϕ is directly proportional to the total time elapsed during a displacement of the phase front. Since this displacement occurs along the direction of maximum phase change per unit distance, the time of the arrival at (x,y,z) of a disturbance originating at (x_0,y_0,z_0) is a minimum. This is *Fermat's principle*. In a homogeneous medium, this principle implies rays which are straight lines; in an inhomogeneous medium, they are curved (Prob. 8-15).

In summary, the theory of geometrical optics is the theory of ray tracings. Rays are normals to equiphase fronts. When the change in ϵ_r per wavelength is small and the dimensions of neighboring bodies are large compared with a wavelength, this theory predicts the propagation of waves through an inhomogeneous medium, according to simple geometrical principles (Snell's laws of reflection and refraction). When either condition is violated, the solution is obtained from the complete wave equation, in accordance with the principles of *physical optics*. In contrast to geometrical optics, physical optics predicts (see Huygens' principle in Chap. 10) the penetration of waves into shadow regions consistent with physical reality.

Read sometime for own use

8.11 Reflection and Refraction of Waves by an Ionized Gas.

Let us consider the behavior of an electromagnetic wave incident upon an ionized gas, such as the ionosphere. This is a region, extending, roughly, from 40 to 250 miles above the earth, in which the constituent gases are ionized as a result of ultraviolet radiation from the sun. At these heights the air pressure is so low that free electrons and ions can exist for a long time without recombining into neutral atoms. However, their density is not changing uniformly with height. Instead, four layers, called D, E, F_1, and F_2, quite thick vertically, occur in that order, at rather well defined heights. In each of these layers the ionization is nonuniform and varies with the hour of the day, the season, the geographical region, and the 11-year sunspot cycle. Typical free-electron density profiles range from 10^{10} to 10^{12} per cubic meter.

On the whole, the ionosphere tends to bend back waves impinging from the earth. Some of the gross features of this behavior are depicted in Fig. 8-15, where a strong dependence on the angle of incidence is seen.

To examine these and other related features of ionospheric reflection and refraction, let us consider the idealized problem of an electromagnetic wave, obliquely incident upon an ionized region of semi-infinite extent. We assume, as in Example 7-5, that the wave is incident from free space, which meets with the ionized region at a plane interface (Fig. 8-16). We also assume a distribution of charged particles, with mass m, charge e, and density N, throughout the half space $z > 0$. Finally, we assume that collisions between particles and the effect of the earth's magnetic field are to be neglected. All assumptions—clear-cut boundary, semi-infinite region, zero magnetic field, no collisions—are extreme simplifications, to be sure, but the errors introduced are unimportant here, since we are interested only in gross behavior.

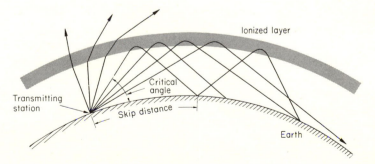

FIGURE 8-15. Ionospheric reflection and refraction.

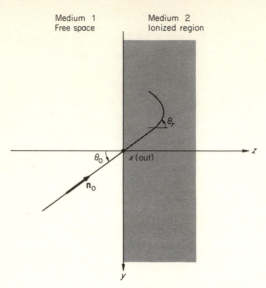

FIGURE 8-16. A wave incident on an ionized region.

Setting $B_0 = 0$ makes $\omega_g = 0$, and Eqs. (7-155) and (7-157) reduce to

$$v = \frac{1/\sqrt{\mu_0 \epsilon_0}}{\sqrt{1 - (\omega_0/\omega)^2}} \tag{8-124}$$

where $\omega_0{}^2 = Ne^2/\epsilon_0 m$ is the plasma frequency. In effect, then, the presence of the charged particles can be accounted for by an equivalent permittivity

$$\epsilon = \epsilon_0\left(1 - \frac{Ne^2}{\epsilon_0 m\omega^2}\right) = \epsilon_0\left(1 - \frac{81N}{f^2}\right) \tag{8-125}$$

insofar as propagation of TEM waves through the ionized medium is concerned.

Now, as formal solutions to Maxwell's equations, Snell's laws can be applied even in this case, if the ionized region is considered to act as a dispersive dielectric. This is particularly true if the frequency is sufficiently low so that, within a distance of one wavelength, the change in ionization is great enough to create effectively an abrupt discontinuity in the properties of the medium. By treating the ionized gas as a perfect dielectric with a dielectric constant (and hence index of refraction) of *less than unity*, the formulas developed in Secs. 8.2 to 8.6 can be applied in this case to study the reflection and refraction properties of the gas.

The situation is not as simple, however, at high frequencies. When the wavelength is sufficiently small so that only a slight change in ionization density occurs in the course of a wavelength, the ionosphere has to be treated by different methods, such as the method of ray tracing developed in the preceding section. The ionized medium is then considered as a dielectric with a continuously variable dielectric constant, or index of refraction, which causes the wave to follow a curved path, as indicated in Fig. 8-16, away from the region of higher electron density N. This curved path is such that, at every point along the path, the angle of refraction θ_r is related to the angle of incidence θ_0 by

$$n \sin \theta_r = \sin \theta_0 \qquad (8\text{-}126)$$

where
$$n = \sqrt{\epsilon_r} = \sqrt{1 - \frac{81N}{f^2}} \qquad (8\text{-}127)$$

is the local value of the refractive index. As N increases with distance into the ionized medium, the refractive index decreases, and θ_r increases in such a way that the product $n \sin \theta_r$ remains fixed. If at some point along the path $n = \sin \theta_0$, then, obviously, $\sin \theta_r = 1$, $\theta_r = 90°$, and the wave is reflected. The ionization density at this point is then obtained from the relation

$$\sqrt{1 - \frac{81N}{f^2}} = \sin \theta_0 \qquad (8\text{-}128)$$

Thus, if

$$N \geq \frac{f^2 \cos^2 \theta_0}{81} \qquad (8\text{-}129)$$

at any point in the ionized layer, the wave will be reflected; otherwise it will penetrate deeper into the ionized medium, possibly going completely through a layer, as suggested by the rays on the left in Fig. 8-15.

Inspection of Eq. (8-129) shows that the ionization density required for complete reflection increases with increasing frequency f and with decreasing angle of incidence θ_0. At a fixed frequency, N is maximum for $\theta_0 = 0°$. Inversely, for a given state of ionization, waves tend to be bent back toward the earth when they impinge at low angles of incidence relative to the horizon. The highest angle at which the ray just manages to be reflected back at a given frequency is the *critical angle* for the layer (Fig. 8-15).

From another point of view, reflection, or complete transmission, through a layer at a given angle of incidence is frequency-dependent. From Eq. (8-129) it is apparent that, as the angle of incidence increases above the critical value (θ_0 decreases), the frequency which will be reflected back tends to decrease, reaching a minimum value

$$f_c = \sqrt{81N} \qquad (8\text{-}130)$$

when the angle of incidence θ_0 is zero. Thus the *critical frequency* f_c is the highest frequency which will be reflected back for normal incidence.

A concept closely allied to the critical angle is the *skip distance*. As shown in Fig. 8-15, this is the distance from the transmitter to a point on the ground where the wave reflected from the ionosphere is first received. The wave then makes one *hop*. Both single-hop and multihop communication can be achieved by *bouncing* a wave off the ionosphere.

Of greater practical interest than the critical frequency (which by definition implies normal incidence) is the *maximum usable frequency* (abbreviated muf), which is a function of the locations of the transmitting and receiving points, the season of the year, and the time of day at which communication is desired. From Eqs. (8-129) and (8-130) it is apparent that

$$\text{muf} = \frac{f_c}{\cos \theta_0} \qquad (8\text{-}131)$$

which shows an intimate relation between the critical frequency and the muf. In fact, when $\theta_0 = 0$, muf $= f_c$. It is evident, then, that transmission must be virtually at a grazing angle if it is to occur at the highest possible frequency (which results in the lowest absorption by the ionized gas).

Absorption of radio waves in the ionosphere is a severe problem at the standard broadcast frequencies. During the daylight hours, all useful propagation is in the *surface-wave mode*, because the *sky wave* is attenuated so highly as to be negligible after reflection. At night, however, ionospheric absorption decreases radically, and it is not uncommon to hear, clearly, a broadcast station many hundreds of miles away.

Summarizing, waves normally incident on an ionized layer will be reflected only if the frequency of the wave is below the critical frequency of that ionized layer. This critical frequency is primarily dependent upon the degree of ionization present. Since the electron density in the ionospheric layers is a function of a large number of parameters such as sunspot cycle, time of day, season, etc., the critical frequencies of the various layers of the ionosphere are only coarsely predictable.

8.12   Summary.   In this chapter we have discussed the reflection and refraction of plane waves incident on the interface between two different media. The laws of reflection and refraction, expressed by Snell's law and the Fresnel equations, were derived from the requirement of continuity of the tangential components of **E** and **H** across the interface. Snell's law states that the product of the (complex) propagation constant times the sine of the angle of the direction of the wave relative to the normal to the interface is a constant.

At the Brewster angle there is no reflected wave when the incident wave is polarized with the incident $\mathbf{H}_i$ vector parallel to the boundary between two dielectrics. In contrast, total reflection occurs when the angle of incidence is equal to, or greater than, the critical angle defined by $\sin \theta_c = \sqrt{\epsilon_2/\epsilon_1}$. In this case, as in the case of incidence upon a conducting medium, the transmitted wave has the properties of a nonuniform wave, with the planes of constant amplitude generally not coincident with the planes of constant phase. There is, however, this difference. When total reflection occurs for two dielectrics, the time-average power transmitted into the refractive medium is equal to zero. In contrast, the amount of real power transmitted into the conducting medium is always greater than zero, unless the conductivity is infinitely large.

The skin depth for a conducting medium, or the depth of penetration, is the distance from the boundary of a point in the conductor at which the amplitude of the wave has decreased to 36.8 percent of its value at the surface. From the formula $\delta = \sqrt{2/\omega\mu\sigma}$, it is seen that the skin depth varies inversely with frequency and conductivity, becoming zero when $\sigma = \infty$.

Scattering occurs when a wave, incident upon a material body, causes a synchronous movement of free and bound charges. The scattered field is set up as a result of this movement and, in combination with the incident field, defines the total, or diffracted, field.

The theory of uniform plane waves can be extended to linear, isotropic, and nonconducting media, with a dielectric constant which is a function of position. When the wavelength is sufficiently small relative to all linear dimensions of nearby objects, and when ϵ_r changes by a very small amount in the course of a wavelength, the theory of geometrical optics, which predicts propagation along rays (normals to equiphase fronts), provides a very convenient tool. A study of the reflection and refraction phenomena in the earth's atmosphere and ionosphere furnishes excellent examples for the use of this theory. The theory of physical optics is based on the complete wave equations.

Problems

8-1 Reflection Coefficients for Oblique Incidence. For the problem of Example 8-2, obtain the expressions for the reflection coefficients in terms of the angle of incidence θ_1.

8-2 Conducting Media. Suppose that, in Fig. 8-3, both media 1 and 2 are conducting and that $\sigma_1/\omega\epsilon_1 = \sqrt{3}$, with $\omega = 3 \times 10^8$ rad/s. Also suppose that $\epsilon_1 = \epsilon_0$, $\mu_1 = \mu_0$, $\epsilon_2 = 2\epsilon_0$, $\mu_2 = \mu_0$, and $\sigma_2 = 2\sigma_1$. Then consider a plane wave propagating along the z axis, and determine the transmitted wave at $z = 1$ in both amplitude and phase, if at $z = -1$ the incident wave is $100e^{j0}$ μV/m.

8-3 Reflection from a Lossy Slab. In Fig. 8-10, let $\mu_2 = \mu_0$, $\sigma \neq 0$, and $d = 0.1\lambda_0$, where λ_0 is the wavelength of the incident wave measured in free space. Then obtain (most readily with a computer) a family of curves showing the variation of the magnitude and phase angle of the reflection coefficient in medium 1, Γ_1, as a function of ϵ_r in the range from $\epsilon_r = 2.0$ to $\epsilon_r = 5.0$ for fixed $\tan \delta = \sigma/\omega\epsilon$ values of 0, 10^{-2}, 5×10^{-2}, 10^{-1}, 2×10^{-1}.

8-4   Measurement of Dielectric Constants.   One of several methods of measuring the properties of dielectrics consists in measuring the transmission (power) loss and phase of a wave incident normally on, and transmitted through, a large sheet of the material to be tested.   The sheet is so located as to interfere with propagation from a transmitting to a receiving antenna, and the phase shift and attenuation in the sample are measured by comparison with a controlled phase shift and attenuation.

Derive expressions for the dielectric constant and loss tangent $(\sigma/\omega\epsilon)$ of the sample if the thickness is d and measurements are conducted in free space.

8-5   Total Reflection.   Derive Eqs. (8-62) and (8-63).

8-6   Multiple Dielectric Slabs.   A plane wave is incident normally on a pair of dielectric slabs joined together as shown in the figure.   Determine a condition between ϵ_1, ϵ_2, d_1, and d_2 for which all the incident power will be transmitted.

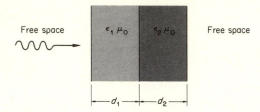

PROBLEM 8-6

8-7   Snell's Laws.   A plane wave is incident at an angle θ_0 upon an infinite slab of dielectric in free space.   Show that, in this case, Snell's laws are

$$\sin \theta_1 = \sin \theta_0 \qquad \gamma_2 \sin \theta_2 = \gamma_1 \sin \theta_0 \qquad \sin \theta_3 = \sin \theta_0$$

A complete mathematical development is required.

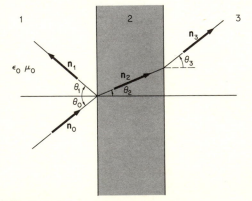

PROBLEM 8-7

8-8   Snell's Law for Spherical Boundaries.   With reference to Fig. 8-14, show that Snell's law for a spherically stratified medium gives

$$n_0 r_0 \sin \theta_0 = n_1 r_1 \sin \theta_1 = n_2 r_2 \sin \theta_2 = \cdots$$

8-9   Radiation Pressure. Show that

$$P = \frac{2\,|\tfrac{1}{2}\mathbf{E} \times \mathbf{H^*}|}{v} = 2\sqrt{\mu\epsilon}\,|\mathbf{S}_c|$$

is an expression for radiation pressure of a uniform plane wave traveling along an arbitrary axis in a medium of unlimited extent.

To get a feeling for the orders of magnitude, calculate the pressure associated with sunlight, for which $|\mathbf{S}_c| = 1.4\ kW/m^2$ near the earth's surface.

8-10   Geometrical Optics. Show that, if β_0 is large enough so that $|\nabla L|^2 = n^2$, then the amplitude function ψ_0 satisfies Laplace's equation.

8-11   The Doppler Effect. When a periodic disturbance is emitted by a source which is moving relative to an observer, the disturbance detected by the observer generally is different in frequency. This phenomenon, of frequency shift as a result of relative motion between source and observer, is called the *Doppler effect*, and the change in frequency is called the *Doppler frequency shift*, or simply, the *Doppler frequency*.

The most familiar manifestation of the Doppler effect occurs when one observer standing beside a railroad track detects a change in frequency, or pitch, of a train speeding by. The frequency of the sound waves emitted by the train decreases with the distance of the approaching train, and becomes still lower as it recedes away from the stationary observer.

A practical application of this phenomenon is in the measurement of the ground speed of airborne vehicles by means of *Doppler radars*. The most elementary aspect of this measurement can be understood with the aid of the accompanying figure. An electromagnetic wave transmitted from a vehicle at T, moving with a velocity $\mathbf{v}$ relative to earth, is backscattered toward the transmitter/receiver system with an upward-shifted frequency. Prove that this Doppler shift is given by

$$\Delta f = \frac{2v \cos \theta}{\lambda}$$

where λ represents the wavelength of the transmitted radar signal.

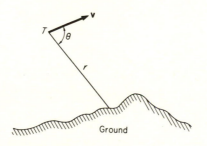

PROBLEM 8-11

8-12   Measurement of Dielectric Constants. A method for measuring the properties of low-loss dielectrics is founded on the reflection properties of relatively large samples at incidence near the Brewster angle. This has the desirable effect of minimizing interactions between sample and antenna. As in the case of normal incidence, described in Prob. 8-4, the dielectric constant and loss tangent are determined from phase and power-loss measurements.

Derive expressions for ϵ_r and $\tan \delta = \sigma/\omega\epsilon$ in terms of the experimentally observed quantities.

8-13 Scattering. Prove that the scattering properties of an object at a given aspect angle can be described by a single *scattering matrix* of the form

$$\begin{bmatrix} A & Be^{j\beta} \\ Be^{j\beta} & De^{j\delta} \end{bmatrix}$$

which includes five independent quantities—three amplitudes and two phase angles.

8-14 Scattering. Should we be able to see a collimated beam of light in vacuum if we were standing away from its direct path?

8-15 Geometrical Optics. Prove that in an inhomogeneous medium the curvature of the rays is given by

$$\frac{1}{\rho} = \mathbf{N} \cdot \nabla (\ln n)$$

where $\mathbf{N}$ is a unit vector in the direction of the radius of curvature, ρ, and n is the index of refraction. Then show that the rays bend toward the region where the index of refraction is higher.

8-16 Reflection from a Perfect Conductor. A uniform plane wave is incident upon the plane boundary of a semi-infinite, perfectly conducting slab. The angle of incidence θ_0 is arbitrary.

Prove that the reflected field is related to the incident field by

$$\mathbf{E}_r = (\mathbf{n} \cdot \mathbf{E}_i)\mathbf{n} - (\mathbf{n} \times \mathbf{E}_i) \times \mathbf{n}$$

$$\mathbf{n} \times \mathbf{H}_r = \mathbf{n} \times \mathbf{H}_i$$

and that, consequently, the surface current density is $\mathbf{K} = 2(\mathbf{n} \times \mathbf{H}_i)$. (*Note:* These formulas are sometimes used in practice, to analyze scattering problems.)

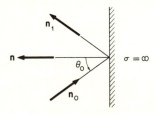

PROBLEM 8-16

8-17 Radome Analysis. A transverse electromagnetic wave is incident at some arbitrary angle θ_0 upon a dielectric sandwich of N layers. For each layer the thickness d_i, the permittivity ϵ_i, and the loss tangent $\tan \delta_i$ are arbitrary. The permeability is that of free space. Consider two cases:

Case I.   The incident electric field intensity $\mathbf{E}_0$ perpendicular to the plane of incidence
Case II.   The incident electric field intensity $\mathbf{E}_0$ parallel to the plane of incidence

In either case, define reflection coefficient R as the ratio of reflected (E_r) to incident (E_0)

field intensities in the free space to the left of the first plane boundary; also define transmission coefficient as the ratio of the transmitted (E_t) field intensity in the free space to the right of the last plane boundary to the incident field intensity (E_0). Thus

$$R_\perp = \left(\frac{E_r}{E_0}\right)_\perp \qquad T_\perp = \left(\frac{E_t}{E_0}\right)_\perp \qquad \text{for case I}$$

$$R_\parallel = \left(\frac{E_r}{E_0}\right)_\parallel \qquad T_\parallel = \left(\frac{E_t}{E_0}\right)_\parallel \qquad \text{for case II}$$

Then show that

$$R_\perp = \frac{B_0}{A_0} \qquad T_\perp = \frac{1}{A_0}$$

$$R_\parallel = \frac{G_0}{F_0} \qquad T_\parallel = \frac{1}{F_0}$$

where A_0, B_0, G_0, H_0 are determined from the recursion formulas

$$A_i = \frac{e^{k_i}}{2}\,[A_{i+1}(1 + Y_{i+1}) + B_{i+1}(1 - Y_{i+1})]$$

$$B_i = \frac{e^{-k_i}}{2}\,[A_{i+1}(1 - Y_{i+1}) + B_{i+1}(1 + Y_{i+1})]$$

$$F_i = \frac{e^{k_i}}{2}\,[F_{i+1}(1 + W_{i+1}) + G_{i+1}(1 - W_{i+1})]$$

$$G_i = \frac{e^{-k_i}}{2}\,[F_{i+1}(1 - W_{i+1}) + G_{i+1}(1 + W_{i+1})]$$

and $A_{N+1} = 1$, $B_{N+1} = 0$, $F_{N+1} = 1$, $B_{N+1} = 0$. Here

$$W_{i+1} = \frac{\cos\theta_{i+1}}{\cos\theta_i}\sqrt{\frac{\epsilon_i(1 - j\tan\delta_i)}{\epsilon_{i+1}(1 - j\tan\delta_{i+1})}}$$

$$Y_{i+1} = \frac{\cos\theta_{i+1}}{\cos\theta_i}\sqrt{\frac{\epsilon_{i+1}(1 - j\tan\delta_{i+1})}{\epsilon_i(1 - j\tan\delta_i)}}$$

$$k_i = d_i\gamma_i\cos\theta_i$$

$$\gamma_i{}^2 = -\omega^2\mu_0\epsilon_i + j\omega\mu_0\sigma_i$$

and θ_i is the complex angle of refraction in ith layer, and the factor $e^{j\omega t}$ is understood. [*Hint:* With reference to Fig. 8-5 show that, for $\mathbf{E}_i$ perpendicular to the plane of incidence,

$$\mathbf{E}_i + \mathbf{E}_r = \mathbf{E}_t$$

$$\cos\theta_1\,\mathbf{E}_i - \cos\theta_1\,\mathbf{E}_r = \frac{\eta_1}{\eta_2}\cos\theta_2\mathbf{E}_t$$

and that, for $\mathbf{H}_i$ perpendicular to the plane of incidence,

$$\mathbf{H}_i + \mathbf{H}_r = \mathbf{H}_t$$

$$\cos\theta_1\mathbf{H}_i - \cos\theta_1\mathbf{H}_r = \frac{\eta_2}{\eta_1}\cos\theta_2\mathbf{H}_t$$

Then apply a properly modified version of these formulas to the interface between the ith and $(i + 1)$st layer.]

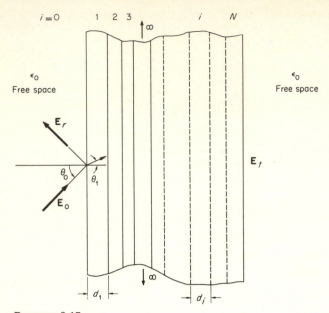

PROBLEM 8-17

8-18 Numerical Analysis of a Radome Wall. Implement a computer program to calculate $R_\perp^2$, $T_\perp^2$, $R_\parallel^2$, $T_\parallel^2$ of Prob. 8-17 for a five-layer radome wall having the following parameters:

Layer number	Dielectric constant	Loss tangent	d_i/λ_0
1	4.30	0.015	0.0406
2	1.15	0.005	0.1355
3	4.30	0.015	0.0406
4	3.15	0.035	0.0180
5	7.30	0.270	0.0027

Allow the angle of incidence to vary from 0 to 80° in 10° increments.   Include $\theta_0 = 88°$. *{Hint:* If $z = x + jy$, then

$$\sin^{-1} z = k\pi + (-1)^k \sin^{-1} \beta + (-1)^k j \ln [\alpha + (\alpha^2 - 1)^{\frac{1}{2}}]$$

where k is any positive integer and

$$\alpha = \tfrac{1}{2}[(x+1)^2 + y^2]^{\frac{1}{2}} + \tfrac{1}{2}[(x-1)^2 + y^2]^{\frac{1}{2}}$$
$$\beta = \tfrac{1}{2}[(x+1)^2 + y^2]^{\frac{1}{2}} - \tfrac{1}{2}[(x-1)^2 + y^2]^{\frac{1}{2}}$$

In this problem, set $k = 0$.}

8-19 Atmospheric Refraction of Radio Waves. Refraction of radio waves is caused by variations of the atmospheric composition which determines the radio refractive index.

Although variations of the refractive index are small, they do cause the rays to bend. Therefore their effect must be taken into account in precision (electronic) distance measurements.

For purposes of this exercise, let the index of refraction of the atmosphere be represented by the approximate formula

$$n(h) = 1 + 0.0003e^{-h/7}$$

where h is the height above the ground, in kilometers.

(a) Consider a ground-based radar tracking an airplane 100 miles away from the radar site and at an altitude of 5 miles. What error would be made in the distance measurement if the variations of the refractive index were ignored?

(b) Give an estimate of the lowest frequency at which the theory of geometrical optics is valid in this medium.

CHAPTER 9

TRANSMISSION LINES AND WAVEGUIDES

9.1 Introduction. At low frequencies we consider that lumped-circuit elements are connected together with conductors which have no electrical or circuit properties other than good conductivity. At higher frequencies the length of these conductors, or more specifically, the dimensions of the system, become nonnegligible fractions of a wavelength, and the simple lumped-circuit idealization leads to serious errors. The lumped-circuit model can be "patched up" in some cases by assigning "lead inductance" and "distributed capacitance" to the interconnections and to various parts of the system. These concepts have some utility, but they must be used with care, since, in reality, the problem has become an electromagnetic field problem. Even so, it is usually possible to preserve some of the simplicity of lumped-circuit analysis by use of *lumped impedances* connected to *uniform transmission lines*.

A *uniform transmission line* is a system of two or more conductors which has identical cross-sectional configurations for all positions along its length. Practical examples of uniform transmission lines are parallel wires, coaxial lines, and strip lines. It is our purpose in this chapter to present the basic features of transmission line analysis.

9.2 Existence and Definition of the Transmission Line Mode. The electromagnetic field configuration for a transmission line is in principle given by the solution of Maxwell's equations subject to the boundary conditions on the transmission line conductors. We do not present the solution of this problem at this time, but rather state as facts a few pertinent features of the solution.

Lossless transmission lines *Lossless* means that the conductors have infinite conductivity and that the dielectric surrounding them is perfect. The following features of the general lossless solution are pertinent.

1. A denumerable infinity of field configurations is possible. Each possible field configuration is called a *mode*.

2. If we focus our attention on the sinusoidal steady state, we find that every mode has a z dependence (z is the axial coordinate; that is, z measures distance along the line) of the form $e^{-\gamma z}$ and $e^{+\gamma z}$. Thus the modes are in the general form of waves traveling in the $\pm z$ direction.

3. The propagation constant γ, in general, is a function of the geometry of the transmission line and of the sinusoidal steady-state frequency. This function is different for each mode.

4. For any particular transmission line and mode, there is a frequency (usually called the *cutoff frequency*) at which $\gamma = 0$. Below this frequency, γ is positive real. Above this frequency, γ is positive imaginary. Thus true wave propagation is possible only above the cutoff frequency. Modes which have a cutoff frequency greater than the operating frequency can be ignored in elementary transmission line analysis.

5. For every lossless transmission line there is one special mode for which the **E** and **H** fields are transverse to the transmission line axis. This mode is called the TEM, or *transmission line mode*. It has a zero cutoff frequency.

Lossy transmission lines   A lossy transmission line is one where the conductors have finite conductivity or where the dielectric is not perfect. In the usual practical case, both conditions prevail.

The exact solution of the boundary-value problem for lossy transmission lines is very involved, and in most cases cannot be obtained in closed analytical form. The usual method of attack on the problem of lossy transmission lines is to use what amounts to a perturbation approach. The technique used is to solve for the fields as though the line were lossless and then to calculate the losses which would occur if the lossless fields were the exact fields.

9.3   Characteristics of the Transmission Line Mode.   The transmission line mode is a transverse electromagnetic (TEM) mode. This means that both the **E** and the **H** fields are transverse to the axial direction of the transmission line. If this axial direction is the z direction, $H_z = E_z = 0$.

We begin by writing Maxwell's equations for a nonconducting source-free region. Although other coordinate systems may be convenient later, there is no loss in generality at this point in using rectangular coordinates. In component form Maxwell's curl equations for $E_z = H_z = 0$ are

$$\frac{\partial E_y}{\partial z} = \mu \frac{\partial H_x}{\partial t} \qquad -\frac{\partial H_y}{\partial z} = \epsilon \frac{\partial E_x}{\partial t}$$

$$\frac{\partial E_x}{\partial z} = -\mu \frac{\partial H_y}{\partial t} \qquad \frac{\partial H_x}{\partial z} = \epsilon \frac{\partial E_y}{\partial t} \qquad (9\text{-}1)$$

$$\frac{\partial E_y}{\partial x} - \frac{\partial E_x}{\partial y} = 0 \qquad \frac{\partial H_y}{\partial x} - \frac{\partial H_x}{\partial y} = 0$$

These equations may be combined in various ways to give

$$\frac{\partial^2 E_x}{\partial z^2} - \mu\epsilon \frac{\partial^2 E_x}{\partial t^2} = 0$$

$$\frac{\partial^2 E_y}{\partial z^2} - \mu\epsilon \frac{\partial^2 E_y}{\partial t^2} = 0$$

$$\frac{\partial^2 H_x}{\partial z^2} - \mu\epsilon \frac{\partial^2 H_x}{\partial t^2} = 0$$

$$\frac{\partial^2 H_y}{\partial z^2} - \mu\epsilon \frac{\partial^2 H_y}{\partial t^2} = 0$$

(9-2)

As we have already seen in Chap. 7, these equations are wave equations whose solutions are each of the form

$$f(z,t) = f^+(z - vt) + f^-(z + vt) \tag{9-3}$$

where $v = 1/\sqrt{\mu\epsilon}$, and is to be interpreted as the z-directed velocity of propagation of the waves represented by f^+ and f^-.

Our first characteristic of the TEM mode is that it propagates along the transmission line with a velocity $v = 1/\sqrt{\mu\epsilon}$, which is independent of the geometry and frequency. It has a zero cutoff frequency.

Let us next look at the vector wave equations for source-free regions. From Chap. 7, they are

$$\nabla^2 \mathbf{E} = \mu\epsilon \frac{\partial^2 \mathbf{E}}{\partial t^2}$$

$$\nabla^2 \mathbf{H} = \mu\epsilon \frac{\partial^2 \mathbf{H}}{\partial t^2}$$

(9-4)

In the special case we are considering, where $E_z = H_z = 0$, these become

$$\nabla^2 E_x = \mu\epsilon \frac{\partial^2 E_x}{\partial t^2}$$

$$\nabla^2 E_y = \mu\epsilon \frac{\partial^2 E_y}{\partial t^2}$$

$$\nabla^2 H_x = \mu\epsilon \frac{\partial^2 H_x}{\partial t^2}$$

$$\nabla^2 H_y = \mu\epsilon \frac{\partial^2 H_y}{\partial t^2}$$

(9-5)

and, by substituting Eqs. (9-2) into these equations, we obtain

$$\frac{\partial^2 E_x}{\partial x^2} + \frac{\partial^2 E_x}{\partial y^2} = 0$$

$$\frac{\partial^2 E_y}{\partial x^2} + \frac{\partial^2 E_y}{\partial y^2} = 0$$

$$\frac{\partial^2 H_x}{\partial x^2} + \frac{\partial^2 H_x}{\partial y^2} = 0 \tag{9-6}$$

$$\frac{\partial^2 H_y}{\partial x^2} + \frac{\partial^2 H_y}{\partial y^2} = 0$$

Thus each of the field components obeys the two-dimensional Laplace's equation. Consequently, by Property 4 (p. 162), each of the field components may be written as the gradient of a scalar function. More usefully, however, for our present discussion, this means that the total **E** field and the total **H** field may each be written as the gradient of scalar functions. This allows us to define a potential difference

$$V(z,t) = \int_C \mathbf{E} \cdot d\mathbf{l} = \int_C (-\boldsymbol{\nabla}\phi) \cdot d\mathbf{l} = \phi_1 - \phi_2 \tag{9-7}$$

where the path of integration is any path in the transverse plane from conductor 1 to conductor 2 in Fig. 9-1. Although V depends upon the position z along the line and upon the time t at which the integration is performed, at a specific pair z and t its value is unique. We shall call $V(z,t)$ the *potential difference* between the conductors.

Similarly, we can define a unique current

$$I(z,t) = \oint_{C'} \mathbf{H} \cdot d\mathbf{l} \tag{9-8}$$

where C' is any contour lying in a transverse plane and encircling conductor 1. (For conductor 2 we obtain the negative of the defined current.) Notice that the displacement current is zero since $D_z \equiv 0$.

At this point we have shown only that it is possible to define a unique $V(z,t)$ and a unique $I(z,t)$. That they have the dimensions of voltage and current is rather obvious. However, their significance remains to be displayed.

As a preliminary step, we note that Eqs. (9-1) are, formally, those obeyed by the uniform plane waves of Chap. 7, with the additional condition imposed by the last two equations in the set (9-1), namely,

$$\frac{\partial E_x}{\partial y} = \frac{\partial E_y}{\partial x} \quad \text{and} \quad \frac{\partial H_x}{\partial y} = \frac{\partial H_y}{\partial x} \tag{9-9}$$

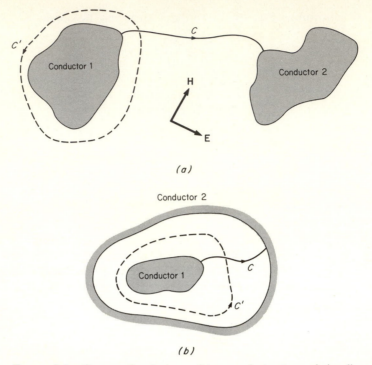

(a)

(b)

FIGURE 9-1.   Cross-sectional views of two-conductor transmission lines.   (a) Parallel conductors; (b) coaxial conductors.

which allows the fields to be nonuniform over transverse planes.   This additional degree of freedom is necessary so that the field can satisfy the boundary conditions without altering the implications of the first two pairs of Eqs. (9-1), which relate only the position and time behavior of the fields. Thus, using Eqs. (9-1), rewritten for the sinusoidal steady state, we can obtain an interpretation for our current problem similar to that obtained in Sec. 7-6.   This interpretation is

$$\mathbf{E}^+ = \mathbf{E}_T^+ e^{-j\beta z}$$

$$\mathbf{H}^+ = \frac{1}{\eta} (\mathbf{a}_z \times \mathbf{E}_T^+) e^{-j\beta z}$$

$$\mathbf{E}^- = \mathbf{E}_T^- e^{j\beta z} \qquad\qquad (9\text{-}10)$$

$$\mathbf{H}^- = -\frac{1}{\eta} (\mathbf{a}_z \times \mathbf{E}_T^-) e^{j\beta z}$$

where the factor $e^{j\omega t}$ is understood, and the subscript T is used to emphasize that the field is transverse.   In other words, these equations say that $\mathbf{E}^+$

and $\mathbf{H}^+$ are in time phase and orthogonal in space with a relative magnitude

$$\frac{|\mathbf{E}^+|}{|\mathbf{H}^+|} = \eta = \sqrt{\frac{\mu}{\epsilon}} \tag{9-11}$$

and similarly for the negatively traveling transverse fields.

Since the vectors $\mathbf{E}_T^+$ and $\mathbf{E}_T^-$ are solutions of the same boundary-value problem, the vectors $\mathbf{E}^+$ and $\mathbf{E}^-$ are collinear. This makes $\mathbf{H}^+$ and $\mathbf{H}^-$ also collinear. The actual directions and spatial dependence of fields are determined by the transmission line geometry. (An example is given below, in Sec. 9.4.)

We now wish to write the field vectors in a form which displays more clearly the transverse spatial dependence. To do this, we write

$$\mathbf{E}^+ = |\mathbf{E}^+| \, e^{j\phi^+} e^{-j\beta z} \mathbf{g}(x,y) = E_{T0}^+(z)\mathbf{g}(x,y) \tag{9-12}$$

where the scalar function $E_{T0}^+(z)$ contains all the axial position, phase, and magnitude information, and $\mathbf{g}(x,y)$ is a suitably normalized dimensionless vector function of the transverse coordinates (x,y) such that it contains the direction and relative-magnitude information. Similarly, we can write

$$\mathbf{E}^- = E_{T0}^- \mathbf{g}(x,y) \tag{9-13}$$

$$\mathbf{H}^+ = \frac{1}{\eta} E_{T0}^+ \mathbf{a}_z \times \mathbf{g}(x,y) \tag{9-14}$$

$$\mathbf{H}^- = -\frac{1}{\eta} E_{T0}^- \mathbf{a}_z \times \mathbf{g}(x,y) \tag{9-15}$$

Now we look at the previously defined $V(z,t)$ and $I(z,t)$. From Eq. (9-7) we have

$$V(z) = \int_C (E_{T0}^+ + E_{T0}^-)\mathbf{g}(x,y) \cdot d\mathbf{l} = E_T \int_C \mathbf{g}(x,y) \cdot d\mathbf{l} = E_T F_1 \tag{9-16}$$

where
$$E_T = E_{T0}^+ + E_{T0}^- \tag{9-17}$$

$$F_1 = \int_C \mathbf{g}(x,y) \cdot d\mathbf{l} \qquad \text{m} \tag{9-18}$$

Similarly, from Eq. (9-8), we have

$$I(z) = \oint_{C'} H_T[\mathbf{a}_z \times \mathbf{g}(x,y)] \cdot d\mathbf{l} = H_T F_2 \tag{9-19}$$

where
$$H_T = \frac{1}{\eta}(E_{T0}^+ - E_{T0}^-) \tag{9-20}$$

$$F_2 = \oint_{C'} [\mathbf{a}_z \times \mathbf{g}(x,y)] \cdot d\mathbf{l} \qquad \text{m} \tag{9-21}$$

Now, F_1 and F_2 are seen to depend only on the transmission line geometry. If we rewrite our results as

$$V(z) = V^+(z) + V^-(z) = [E_T^+(z) + E_T^-(z)]F_1$$

$$I(z) = I^+(z) + I^-(z) = \frac{1}{\eta} [E_T^+(z) - E_T^-(z)]F_2$$

these equations become

$$V(z) = V^+(z) + V^-(z)$$

$$I(z) = \frac{F_2}{\eta F_1} [V^+(z) - V^-(z)]$$

from which we see that

$$I^+(z) = \frac{F_2}{\eta F_1} V^+(z) = \frac{1}{Z_0} V^+(z) \tag{9-22}$$

$$I^-(z) = -\frac{F_2}{\eta F_1} V^-(z) = -\frac{1}{Z_0} V^-(z) \tag{9-23}$$

where $Z_0 = (F_1/F_2)\eta$ has the dimensions of an impedance, whose value is determined by the transmission line geometry and the dielectric medium. This impedance Z_0 is called the *characteristic impedance* of the transmission line. Its value for a coaxial transmission line is calculated in the next section.

9.4 The Coaxial Transmission Line. As a specific example of a transmission line, let us examine the TEM mode for the lossless coaxial structure shown in Fig. 9-2.

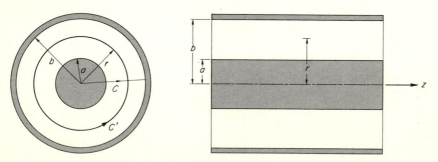

FIGURE 9-2. A cross-sectional and an axial view of a coaxial transmission line.

In cylindrical coordinates, Eqs. (9-1) are

$$\frac{\partial E_\varphi}{\partial z} = \mu \frac{\partial H_r}{\partial t} \qquad -\frac{\partial H_\varphi}{\partial z} = \epsilon \frac{\partial E_r}{\partial t}$$

$$\frac{\partial E_r}{\partial z} = -\mu \frac{\partial H_\varphi}{\partial t} \qquad \frac{\partial H_r}{\partial z} = \epsilon \frac{\partial E_\varphi}{\partial t} \qquad (9\text{-}24)$$

$$\frac{\partial}{\partial r}(rE_\varphi) - \frac{\partial E_r}{\partial \varphi} = 0 \qquad \frac{\partial}{\partial r}(rH_\varphi) - \frac{\partial H_r}{\partial \varphi} = 0$$

These equations can be combined in the same manner as was done for Eqs. (9-1) to give

$$\frac{\partial^2 E_r}{\partial z^2} - \mu\epsilon \frac{\partial^2 E_r}{\partial t^2} = 0$$

$$\frac{\partial^2 E_\varphi}{\partial z^2} - \mu\epsilon \frac{\partial^2 E_\varphi}{\partial t^2} = 0$$

$$\frac{\partial^2 H_r}{\partial z^2} - \mu\epsilon \frac{\partial^2 H_r}{\partial t^2} = 0 \qquad (9\text{-}25)$$

$$\frac{\partial^2 H_\varphi}{\partial z^2} - \mu\epsilon \frac{\partial^2 H_\varphi}{\partial t^2} = 0$$

showing that each of the field components individually obeys the wave equation, and hence each field component is of the form

$$f(z,t) = f^+(z - vt) + f^-(z + vt)$$

as was the case when expressed in rectangular coordinates.

Furthermore, our previous results show that the $\mathbf{E}^+$ and $\mathbf{H}^+$ fields are in time phase and orthogonal in space so that

$$\frac{E_T^+}{H_T^+} = \eta \qquad \frac{E_T^-}{H_T^-} = -\eta$$

Hence the only new part of the problem is to determine the transverse spatial behavior of the fields and the characteristic impedance of the line.

The spatial behavior can be determined by a direct solution of the set (9-24), which will show that, for fields with no φ dependence, the only allowable solution is of the form k/r, where k is a constant. This states that the fields are

$$E_r^+ = \frac{k_1}{r} \qquad H_r^+ = \frac{k_3}{r}$$

$$E_\varphi^+ = \frac{k_2}{r} \qquad H_\varphi^+ = \frac{k_4}{r} \qquad (9\text{-}26)$$

with similar expressions for the negative traveling fields. The boundary conditions $E_{tan} = 0$, $H_{normal} = 0$ demand that $k_2 = k_3 = 0$, resulting in the fields

$$E_r^+ = \frac{k^+}{r} \qquad H_\varphi^+ = \frac{1}{\eta} \frac{k^+}{r} \tag{9-27}$$

Now

$$V^+(z) = \int_C \frac{k^+}{r} \, dr = k^+ \ln \frac{b}{a} = E_T^+(z)F_1 \tag{9-28}$$

and

$$I^+(z) = \frac{1}{\eta} \oint_{C'} (\mathbf{a}_z \times \mathbf{E}_T) \cdot d\mathbf{l} = \frac{1}{\eta} \int_0^{2\pi} \frac{k^+}{r} r \, d\varphi = \frac{2\pi k^+}{\eta} = \frac{2\pi}{\eta} \frac{V^+(z)}{\ln (b/a)} \tag{9-29}$$

From this expression and Eq. (9-22) we obtain

$$\frac{F_2}{F_1} = \frac{2\pi}{\ln (b/a)} \tag{9-30}$$

and

$$Z_0 = \frac{\eta \ln (b/a)}{2\pi} \tag{9-31}$$

as the characteristic impedance of a lossless coaxial transmission line.

A circuit interpretation of Z_0 is sometimes desirable. If we recall that the capacitance per unit length of a coaxial structure is

$$C = \frac{2\pi\epsilon}{\ln (b/a)} \tag{9-32}$$

and the inductance per unit length is

$$L = \frac{\mu \ln (b/a)}{2\pi} \tag{9-33}$$

we see that

$$\sqrt{\frac{L}{C}} = \sqrt{\frac{[\mu \ln (b/a)]/2\pi}{2\pi\epsilon/[\ln (b/a)]}} = \sqrt{\frac{\mu}{\epsilon} \left[\frac{\ln (b/a)}{2\pi}\right]^2} = \sqrt{\frac{\mu}{\epsilon}} \frac{\ln (b/a)}{2\pi} = Z_0 \tag{9-34}$$

which interprets Z_0 as the ratio $\sqrt{L/C}$. In fact, if we had started our analysis by assuming that the line had a capacitance C and an inductance L per unit length and that a voltage and current V and I existed, and treated the problem as a distributed-circuits (Fig. 9-3) problem, we should have obtained the equations

$$\frac{\partial V}{\partial z} = -L \frac{\partial I}{\partial t}$$

$$\frac{\partial I}{\partial z} = -C \frac{\partial V}{\partial t} \tag{9-35}$$

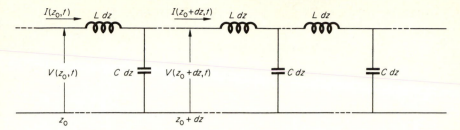

FIGURE 9-3. Distributed-circuit representation of a lossless transmission line.

These, in turn, can be combined to give

$$\frac{\partial^2 V}{\partial z^2} = LC\,\frac{\partial^2 V}{\partial t^2}$$

$$\frac{\partial^2 I}{\partial z^2} = LC\,\frac{\partial^2 I}{\partial t^2}$$

(9-36)

which are wave equations in V and I whose solutions are

$$V(z,t) = V^+(z - vt) + V^-(z + vt) \tag{9-37}$$

$$I(z,t) = I^+(z - vt) + I^-(z + vt) \tag{9-38}$$

where

$$v = \frac{1}{\sqrt{LC}} = \frac{1}{\sqrt{\dfrac{\mu \ln (b/a)}{2\pi}\,\dfrac{2\pi\epsilon}{\ln (b/a)}}} = \frac{1}{\sqrt{\mu\epsilon}}$$

as before. Direct substitution of these results into the original (9-35) yields

$$\frac{V^+(z,t)}{I^+(z,t)} = \sqrt{\frac{L}{C}} = Z_0$$

$$\frac{V^-(z,t)}{I^-(z,t)} = -Z_0$$

Note how minus sign appears in -z going wave.

as expected. The ease with which these results are obtained by the distributed-circuits method makes this approach advantageous for routine calculations. However, for actual lines, we need the insight provided by the fields approach to understand the frequency limitations imposed on the line by losses.

9.5 Transmission Line Analysis. Conventional transmission line analysis considers the behavior of the line in the sinusoidal steady state.

For time variations of the form $e^{j\omega t}$, Eqs. (9-36) become

$$\frac{d^2V}{dz^2} + \omega^2 LCV = 0$$

$$\frac{d^2I}{dz^2} + \omega^2 LCI = 0$$

(9-39)

and lead to the solutions

$$V(z) = V^+ e^{-j\beta z} + V^- e^{j\beta z}$$

(9-40)

$$I(z) = \frac{1}{Z_0}(V^+ e^{-j\beta z} - V^- e^{j\beta z})$$

(9-41)

↑ note minus

where the factor $e^{j\omega t}$ is understood and

$$\beta = \frac{2\pi}{\lambda} = \frac{\omega}{v} = \omega\sqrt{LC} = \omega\sqrt{\mu\epsilon}$$

(9-42)

The impedance at any point along the line is obtained by taking the ratio of Eqs. (9-40) and (9-41).

$$Z(z) = \frac{V(z)}{I(z)} = Z_0 \frac{V^+ e^{-j\beta z} + V^- e^{j\beta z}}{V^+ e^{-j\beta z} - V^- e^{j\beta z}}$$

(9-43)

It is convenient to change the form of this equation by dividing the numerator and the denominator by $V^+ e^{-j\beta z}$. It then becomes

$$Z(z) = Z_0 \frac{1 + (V^-/V^+)e^{j2\beta z}}{1 - (V^-/V^+)e^{j2\beta z}} = Z_0 \frac{1 + \Gamma_0 e^{j2\beta z}}{1 - \Gamma_0 e^{j2\beta z}}$$

(9-44)

where

$$\Gamma_0 = \frac{V^-}{V^+}$$

(9-45)

This definition implies that Γ_0 is the ratio of the reflected and incident voltage waves at $z = 0$, and is thus a characteristic of the load connected to the line. Since these waves are not necessarily in phase at $z = 0$, Γ_0 in general is complex and is of the form

$$\Gamma_0 = |\Gamma_0| e^{j\phi_0}$$

(9-46)

where ϕ_0 is the relative phase of V^- and V^+ at $z = 0$.

Transmission lines usually connect a generator to a load, as shown in Fig. 9-4. In these cases, it is almost always desirable to use the load position as the origin of the z-coordinate system, with the result that z is negative for all positions along the line. This causes no basic problem. However, the major portion of transmission line analysis is written in terms of the variable $s = -z$ so that we deal with $s = $ *positive distance from the load*. In terms of

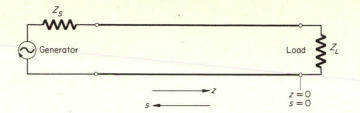

FIGURE 9-4. A transmission line connecting a generator to a load.

s, Eq. (9-44) becomes

$$Z(s) = Z_0 \frac{1 + \Gamma_0 e^{-j2\beta s}}{1 - \Gamma_0 e^{-j2\beta s}} \tag{9-47}$$

Manipulation and interpretation of Eq. (9-47) is the basic problem of transmission line analysis.   Since Z_0 is a constant for any particular line, it is convenient to define a normalized impedance $z_n(s)$ as

compare →

$$\frac{Z(s)}{Z_0} = z_n(s) = \frac{1 + \Gamma_0 e^{-j2\beta s}}{1 - \Gamma_0 e^{-j2\beta s}} \tag{9-48}$$

The advantage of using this normalized impedance is that it reduces all problems to a single class of problems.

First notice that $z_n(s)$ is complex, and can be written

$$z_n(s) = r(s) + jx(s) \tag{9-49}$$

Examination of Eq. (9-48) will show that, if all values of $|\Gamma_0|$ from zero to infinity are allowed, both $r(s)$ and $x(s)$ range from $-\infty$ to $+\infty$.   However, if we restrict our analysis to the case of a generator delivering power to a passive load, the net time-average power P_{av} delivered will be,† from Eqs. (9-40) and (9-41),

$$P_{av} = \text{Re} (\tfrac{1}{2}VI^*) = \frac{|V^+|^2}{2Z_0} - \frac{|V^-|^2}{2Z_0} \geq 0 \tag{9-50}$$

and therefore we shall have

$$|\Gamma_0| = \left| \frac{V^-}{V^+} \right| \leq 1$$

† The superposition of powers expressed by Eq. (9-50) is a significant phenomenon.   Not all systems, even linear ones, behave in this manner.   For example, the power delivered to a linear resistor by two voltage sources connected in series is not equal to the sum of the powers delivered by each source acting alone.

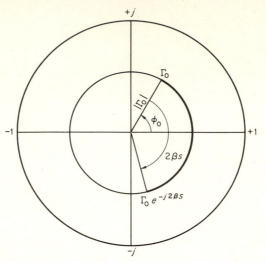

FIGURE 9-5. Γ in the complex plane ($|\Gamma_0| < 1$).

Under these conditions it is obvious that the so-called *generalized voltage reflection coefficient*

$$\Gamma = \Gamma_0 e^{-j2\beta s} = |\Gamma_0|\, e^{j\phi_0} e^{-j2\beta s} \tag{9-51}$$

is a quantity whose locus in the complex plane is a circle of radius $|\Gamma_0| \leq 1$. Figure 9-5 shows a plot of Γ.   Smith[†] showed that Eq. (9-48) permits the construction of a chart, known as the *Smith chart*, from which Eq. (9-48) can be solved graphically.

Consider Eq. (9-51), and let

$$\psi = \phi_0 - 2\beta s$$

Also, let p and q be the components of Γ.   Then

$$\boxed{\Gamma = |\Gamma_0|\, e^{j\psi} = p + jq} \tag{9-52}$$

and with the aid of (9-49), Eq. (9-48) becomes

$$r + jx = \frac{1 + \Gamma}{1 - \Gamma} = \frac{1 + (p + jq)}{1 - (p + jq)}$$

Separating the real and imaginary parts of this equation and performing some lengthy, but simple, algebraic manipulations (the details of which

† P. H. Smith, Transmission Line Calculator, *Electronics*, vol. 12, pp. 29–31, January, 1939.

are left as an exercise for the student), we obtain

$$\left(p - \frac{r}{r+1}\right)^2 + q^2 = \frac{1}{(r+1)^2}$$

$$\left(p - 1\right)^2 + \left(q - \frac{1}{x}\right)^2 = \frac{1}{x^2}$$

$(9\text{-}53)$

The Smith chart plots the loci of Γ for $r = $ constant and for $x = $ constant. It can be readily seen from Eqs. (9-53) that $r = $ constant results in loci of Γ which are circles of radius $1/(r+1)$ and whose centers are on the real axis at $r/(r+1)$. It can also be seen that $x = $ constant results in loci of Γ which are arcs of circles whose centers are located at $p = 1$, $q = 1/x$ and which have radii of $1/x$ such that the circles are tangent to the real axis at the point $1 + j0$.

These $r = $ constant and $x = $ constant lines are the lines plotted in the Smith chart, and are known as *Smith chart coordinates*. Figure 9-6 shows the Smith chart coordinates. Notice particularly that Γ is not plotted, but that it can be obtained graphically for any $r + jx$. On the actual Smith chart, the angle ψ is indicated around the unit circle in degrees and in wavelengths

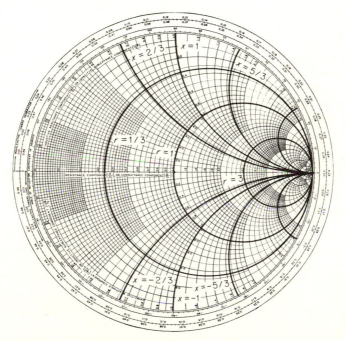

FIGURE 9-6. Smith chart coordinates.

$[2\beta s = (4\pi/\lambda)s]$ "toward the generator" and "toward the load." This facilitates certain calculations. A few examples will show how the Smith chart can be used. The student will need a Smith chart to follow these examples.

Example 9-1   Smith Chart Use.   For the equation

$$z_n = r + jx = \frac{1 + |\Gamma_0|\,e^{j\psi}}{1 - |\Gamma_0|\,e^{j\psi}}$$

(*a*) Given that $r = 2$, $x = 3$.   Find $|\Gamma_0|$ and ψ.

As shown in Fig. 9-7, locate the point $z_n = r + jx = 2 + j3$.   Read $\psi = 26°$, and by scaling find $|\Gamma_0| = 0.745$.

(*b*) Given $|\Gamma_0| = \frac{1}{3}$, $\psi = 90°$.   Find r and x.   Draw the circle $|\Gamma_0| = \frac{1}{3}$ and the $\psi = 90°$ line, as shown in Fig. 9-7.   At their intersection read $r = 0.8$, $x = 0.6$.

(*c*) For part (*b*) find the maximum and the minimum values of $|z_n|$ if all values of ψ are allowed.

From the equation, note that $|z_n|_{\max}$ occurs for $\psi = 0$ and $|z_n|_{\min}$ occurs for $\psi = \pi$.   Thus, analytically,

$$|z_n|_{\max} = \frac{1 + |\Gamma_0|}{1 - |\Gamma_0|} = \frac{1 + \frac{1}{3}}{1 - \frac{1}{3}} = 2$$

and

$$|z_n|_{\min} = \frac{1 - |\Gamma_0|}{1 + |\Gamma_0|} = \frac{1 - \frac{1}{3}}{1 + \frac{1}{3}} = \frac{1}{2}$$

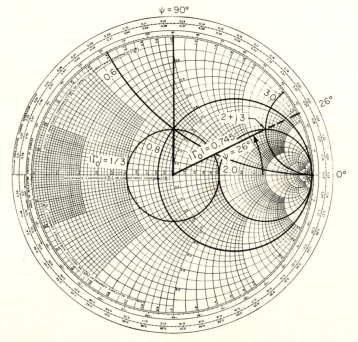

FIGURE 9-7.   Smith chart calculation of Example 9-1.

or graphically, complete the locus $\Gamma = \tfrac{1}{3}e^{j\psi}$ and read at $\psi = 0$, $z = 2 + j0 = |z_n|_{\max}$, and read at $\psi = \pi$, $z = \tfrac{1}{2} + j0 = |z_n|_{\min}$.

Note that the maximum and minimum values of the normalized impedance are both real.

Example 9-2 Smith Chart Use. Given a generator connected to a passive load by means of a transmission line whose characteristic impedance is 50 Ω. At a reference position $s = 0$ it is found that the line impedance, $Z(0) = V(0)/I(0)$, is given by

$$Z(0) = 100 + j50$$

What will the line impedance be 0.2λ toward the generator from the reference position?
First express the problem in Smith chart coordinates by writing

$$z_n(0) = \frac{Z(0)}{Z_0} = \frac{100 + j50}{50} = 2 + j1$$

Locate the Smith chart point $2 + j1$ as shown in Fig. 9-8. Note that the reference point is at 0.213λ "toward the generator," or $\psi_0 = 26.2°$, and that the origin for the two scales is not the same. The value 0.213λ has no significance in this problem, except that 0.2λ toward the generator from this reference point is at $0.213\lambda + 0.2\lambda = 0.413\lambda$ "toward the generator." ψ at this point is $-117.8°$.

Next use a compass to draw the (circular) locus of $\Gamma_0 e^{-j2\beta s}$; read the Smith chart coordinates at the point on this circle corresponding to $s = 0.2\lambda$, that is, at 0.413λ

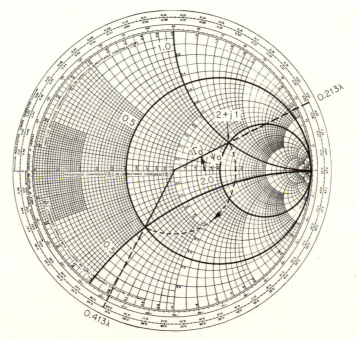

FIGURE 9-8. The solution of Example 9-2.

"toward the generator."   Obtain

$$z_n(s = 0.2\lambda) = 0.5 - j0.5$$

Hence $$Z(0.2\lambda) = (0.5 - j0.5)50 = 25 - j25 \ \Omega$$

is the line impedance 0.2λ toward the generator from the point where the line impedance is $100 + j50$.

9.6   Voltage Standing Wave Ratio (VSWR), Reflection Coefficient, and Matched Lines.

In terms of s, Eq. (9-43) is

$$Z(s) = \frac{V(s)}{I(s)} = Z_0 \frac{V^+e^{j\beta s} + V^-e^{-j\beta s}}{V^+e^{j\beta s} - V^-e^{-j\beta s}} \qquad (9\text{-}54)$$

Now consider a transmission line connecting a generator to an impedance, Z_L, as shown in Fig. 9-4.   At $s = 0$, Eq. (9-54) reduces to

$$Z_L = Z(0) = \frac{V(0)}{I(0)} = Z_0 \frac{V^+ + V^-}{V^+ - V^-}$$

Now for $Z(0) = Z_0$, we have

$$Z_0 = Z_0 \frac{V^+ + V^-}{V^+ - V^-} \qquad (9\text{-}55)$$

which has as its solution $V^- = 0$, giving $Z(s) = Z_0$ for all s.   Under these conditions, the line is said to be *matched*.   For all other load impedances, $V^- \neq 0$, and the line is said to be *mismatched*.

Since V^- is the complex value of a negative traveling wave whose origin is at a mismatch, it is desirable to view it as voltage reflected from the mismatch.   Furthermore, since the ratio $\Gamma_0 = V^-/V^+$ is determined by Z_L/Z_0, that is, by the mismatch, it is desirable to call Γ_0 the *voltage reflection coefficient* and to note that, given Γ_0, we can calculate Z_L/Z_0, or, given Z_L/Z_0, we can calculate Γ_0.

As a practical matter, it is rather difficult to measure the line impedance $Z(s)$ directly.   In some cases it is also difficult to measure the load impedance Z_L directly, although its location is usually known.   On the other hand, it is comparatively simple from a measurements point of view to obtain the (relative) value of $|V(s)|$ by using a *slotted line*.   This means that we need to be able to obtain Z_L and Γ_0 from measured values of $|V(s)|$.   The following discussion shows how this is accomplished.

First, with $s = -z$, Eq. (9-40) may be written

$$V(s) = V^+e^{j\beta s}(1 + \Gamma_0 e^{-j2\beta s})$$

and hence $$|V(s)| = |V^+e^{j\beta s}(1 + \Gamma_0 e^{-j2\beta s})|$$

which has a maximum value

$$|V(s)|_{\max} = |V^+| \, (1 + |\Gamma_0|)$$

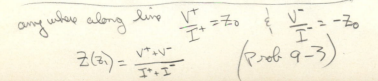

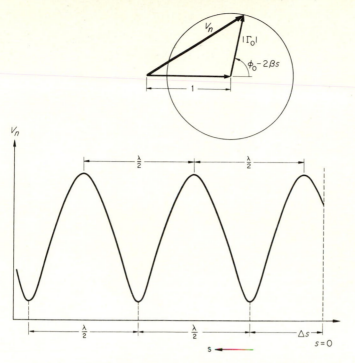

FIGURE 9-9.   A normalized voltage standing wave on a transmission line as a function of position.   The circle diagram shows the relationship of normalized voltage V_n to $\Gamma_0 = |\Gamma_0| e^{j\phi_0}$ and position along the line.

and a minimum value

$$|V(s)|_{min} = |V^+|\,(1 - |\Gamma_0|)$$

The ratio

$$S = \frac{|V(s)|_{max}}{|V(s)|_{min}} = \frac{1 + |\Gamma_0|}{1 - |\Gamma_0|} \tag{9-56}$$

is called the *voltage standing wave ratio*, usually abbreviated VSWR.   We can solve this equation for $|\Gamma_0|$ and obtain

$$|\Gamma_0| = \frac{S - 1}{S + 1} \tag{9-57}$$

A typical plot of normalized voltage $V_n = 1 + \Gamma_0 e^{-j2\beta s}$ is shown in Fig. 9-9. Notice that successive maxima (and minima) are one-half wavelength apart.

Now, having measured S and having calculated $|\Gamma_0|$, we can draw the circle on the Smith chart which is the locus of $\Gamma_0 e^{-j2\beta s}$.   However, it is expedient to note that, in terms of Smith chart coordinates, the normalized impedance

$$z_n = r + jx = \frac{1 + \Gamma_0 e^{-j2\beta s}}{1 - \Gamma_0 e^{-j2\beta s}} \tag{9-58}$$

has a maximum given by

$$|z_n|_{\max} = r_{\max} + j0 = \frac{1 + |\Gamma_0|}{1 - |\Gamma_0|} = S$$

Thus *the voltage standing wave ratio is numerically equal to the value of z_n at the point where the locus of $\Gamma_0 e^{-j2\beta s}$ crosses the positive real axis.* This point also corresponds to the position of $|V(s)|_{\max}$.

The minimum value of the normalized impedance is

$$|z_n|_{\min} = r_{\min} + j0 = \frac{1 - |\Gamma_0|}{1 + |\Gamma_0|} = \frac{1}{S}$$

and occurs at the point where the locus of $\Gamma_0 e^{-j2\beta s}$ crosses the negative real axis. This point also corresponds to the position of $|V(s)|_{\min}$.

The foregoing discussion shows that, if we measure the voltage standing wave ratio S, the half wavelength $\lambda/2$, and the distance Δs from the first minimum of $|V(s)|$ to the position $s = 0$, we can readily find the Smith chart

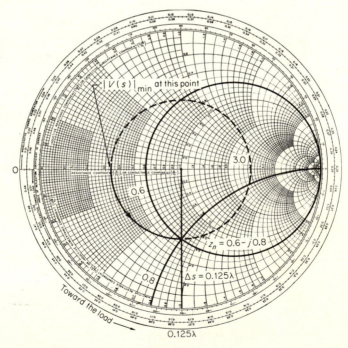

FIGURE 9-10.   The calculation of load impedance from VSWR and voltage minimum location.

coordinates for $s = 0$, and hence we can obtain the impedance $Z(0)$ at $s = 0$. If the load is at $s = 0$, then $Z(0) = Z_L$. An example will clarify the procedure.

Example 9-3 Use of VSWR to Calculate Load Impedance. Given that the VSWR is $S = 3$, that the first minimum of $|V(s)|$ is 5 cm from the load, and that the distance between successive minima is 20 cm, while the characteristic impedance of the line is 50 Ω. Find the load impedance.

The wavelength is 40 cm. The load is $\frac{5}{40} = 0.125\lambda$ "toward the load" from the first voltage minimum. With reference to Fig. 9-10, locate the point $r_{min} = 1/S = \frac{1}{3}$ and use a compass to draw the circle $\Gamma_0 e^{-j2\beta s}$. Draw the line $\Delta s = 0.125\lambda$ "toward the load." At the intersection of the circle and the line read the Smith chart coordinates

$$z_n(0) = r + jx = 0.6 - j0.8$$

Multiply by Z_0 and obtain

$$Z(0) = Z_L = Z_0 z_n(0) = 50(0.6 - j0.8) = 30 - j40 \qquad \Omega$$

The inverse problem of finding the VSWR for a given load impedance should be obvious.

We close this section with a summary of the most important formulas used in lossless transmission line analysis.

$$V = V^+ e^{-j\beta z} + V^- e^{j\beta z} = V^+ e^{j\beta s} + V^- e^{-j\beta s}$$

$$= V^+ e^{j\beta s}(1 + \Gamma_0 e^{-j2\beta s}) = V^+ e^{j\beta s}(1 + \Gamma)$$

$$I = \frac{1}{Z_0}(V^+ e^{-j\beta z} - V^- e^{j\beta z}) = \frac{1}{Z_0}(V^+ e^{j\beta s} - V^- e^{-j\beta s})$$

$$= \frac{V^+ e^{j\beta s}}{Z_0}(1 - \Gamma_0 e^{-j2\beta s}) = \frac{V^+ e^{j\beta s}}{Z_0}(1 - \Gamma)$$

$$\Gamma = \Gamma_0 e^{-j2\beta s} = \frac{Z - Z_0}{Z + Z_0}$$

See P 414

$$Z = \frac{V}{I} = Z_0 \frac{1 + \Gamma_0 e^{-j2\beta s}}{1 - \Gamma_0 e^{-j2\beta s}} = Z_0 \frac{1 + \Gamma}{1 - \Gamma}$$

At $z = 0$: $Z = Z_L$ $Z_L = Z_0 \frac{1 + \Gamma_0}{1 - \Gamma_0}$ $\Gamma_0 = \frac{Z_L - Z_0}{Z_L + Z_0}$

$$\text{VSWR} = \frac{1 + |\Gamma_0|}{1 - |\Gamma_0|}$$

This is magnitude! Remember!

$$|\Gamma_0| = \frac{\text{VSWR} - 1}{\text{VSWR} + 1}$$

An additional useful formula is given in Prob. 9-22.   It is

$$Z_L = Z_0 \frac{1 - j(\text{VSWR}) \tan \beta d_{\min}}{\text{VSWR} - j \tan \beta d_{\min}}$$

where $d_{\min}$ is the (positive) distance to the first voltage minimum from the load at $s = 0$.

9.7   Transmission Line Matching.   The time-average power delivered by a transmission line to a load is given by

$$P_{\text{av}} = \text{Re}\left(\frac{VI^*}{2}\right) = \text{Re}\left\{\frac{1}{2}\left[V^+e^{j\beta s}(1 + \Gamma_0 e^{-j2\beta s})\frac{V^{+*}e^{-j\beta s}}{Z_0}(1 - \Gamma_0^* e^{j2\beta s})\right]\right\}$$

$$= \frac{V^+V^{+*}}{2Z_0}(1 - \Gamma_0\Gamma_0^*) = \frac{|V^+|^2}{2Z_0}(1 - |\Gamma_0|^2) \qquad (9\text{-}59)$$

From this equation it is apparent that the condition $|\Gamma_0| \neq 0$ results in a reduced power to the load, assuming the voltage $|V^+|$ is held fixed.   This is nearly true for many practical situations.   Therefore, neglecting line losses, the power delivered to the load is a maximum, provided $|\Gamma_0| = 0$, that is, provided the load is *matched* to the line.

This section deals with the methods used to match a given load impedance to a transmission line.   This is accomplished by connecting additional lossless impedances to the line in such a manner that total line impedance makes $|\Gamma_0| = 0$.

From a practical point of view the only simple connections we can make to the line are parallel connections.   Parallel connections are best dealt with by using admittance concepts rather than impedance concepts.   The line admittance is given by

$$Y(s) = \frac{1}{Z(s)} = \frac{1}{Z_0}\left(\frac{1 - \Gamma_0 e^{-j2\beta s}}{1 + \Gamma_0 e^{-j2\beta s}}\right) = Y_0(g + jb) \qquad (9\text{-}60)$$

where Y_0 is the reciprocal of Z_0, and g and b are the normalized conductance and susceptance of the line.   Equation (9-60) shows that, by making the proper interpretation of the results, we can solve admittance problems using the same procedure on the Smith chart that we used for impedance problems. Specifically, we can readily convert from Smith chart impedance coordinates, $r + jx$, to Smith chart admittance coordinates, $g + jb$, and from line impedance to line admittance, by a simple angular shift of 180° around the $\Gamma_0 e^{-j2\beta s}$ circle, as the following example will show.   This is so because the normalized admittance, which is just the fraction within parentheses in Eq. (9-60), differs from the normalized impedance, Eq. (9-58), only in the interchange of signs, which can readily be taken into account by adding $j\pi$ to the exponentials.

Example 9-4 Transformation from Impedance to Admittance Using the Smith Chart.   Given that, at a specified point on a transmission line, the line impedance is $50 + j50$ Ω.   The characteristic impedance of the line is 50 Ω.   What is the line admittance at this point?

By direct calculation

$$Y = \frac{1}{Z} = \frac{1}{50 + j50} = (10 - j10) \times 10^{-3} \quad \mho$$

By use of the Smith chart, Fig. 9-11,

$$Z = Z_0(r + jx) = 50(1 + j1)$$

Locate the Smith chart impedance coordinate point $1 + j1$.   Draw the circle $\Gamma_0 e^{-j\beta s}$ through $1 + j1$.   At 180° along this circle from $1 + j1$ read the Smith chart admittance coordinates $0.5 - j0.5$.   Now

$$Y = Y_0(g + jb) = \tfrac{1}{50}(0.5 - j0.5) = (10 - j10) \times 10^{-3} \quad \mho$$

The steps are illustrated in Fig. 9-11.

Although the Smith chart procedure seems more involved than direct calculation, it actually facilitates most calculations of this type.

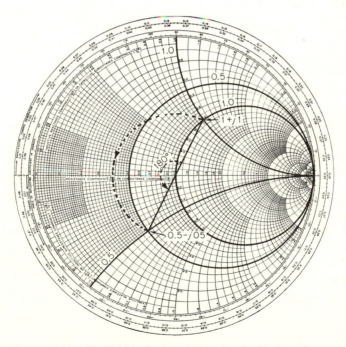

FIGURE 9-11.   The Smith chart operations involved in impedance-to-admittance transformations.

Appropriate lossless matching admittances are usually short-circuited or open-circuited sections of the same transmission line which is to be matched.

For a short-circuited stub, $\Gamma_0 = -1$, as may be seen by setting $V(0) = 0$ in Eq. (9-40); hence the locus of $\Gamma_0 e^{-j2\beta s}$ is a circle of radius 1, which is the outer circle of the Smith chart. Additionally, the expression for the stub input admittances, from Eq. (9-60), becomes

$$Y(s) = Y_0 \frac{1 + e^{-j2\beta s}}{1 - e^{-j2\beta s}} = Y_0(0 - j\cot\beta s) \qquad (9\text{-}61)$$

from which it is clear that

$$Y(0) = Y_0(0 - j\infty)$$

To find the Smith chart admittance coordinates of any length s of a short-circuited line, we enter the Smith chart at $-j\infty$ (which is the extreme point on the right in the chart), and we go around the outer circle "toward the generator" a distance s, measured in wavelengths; we then read the Smith chart coordinate $0 + jb$.

For open-circuited stubs the same procedure is followed, except that, for an open circuit, $\Gamma_0 = 1$, and from Eq. (9-60),

$$Y(s) = Y_0 \frac{1 - e^{-j2\beta s}}{1 + e^{-j2\beta s}} = Y_0(0 + j\tan\beta s) \qquad (9\text{-}62)$$

In this case the distance s is measured "toward the generator" from the point $0 + j0$ (which is the extreme point on the left in the Smith chart).

We are now in a position to perform matching calculations.

Single-stub matching   A load impedance of any arbitrary finite value can be matched to a transmission line by use of a single (open- or short-circuited) matching stub of appropriate length in parallel with the line at the proper distance from the load. Figure 9-12 shows the line positions and the Smith chart operations involved in single-stub matching.

For a given load impedance Z_L, we use the Smith chart to obtain Y_L and then construct the admittance circle, from which we can obtain

$$Y(s) = Y_0[g(s) + jb(s)] \qquad (9\text{-}63)$$

for any desired distance s from the load. We are already familiar with this operation. Now, if at s we parallel the line with a stub whose admittance is given by

$$Y_s = Y_0(0 + jb)$$

the parallel combination at s will have an admittance given by

$$Y_{\parallel} = Y(s) + Y_s = Y_0[g(s) + jb(s) + jb] \qquad (9\text{-}64)$$

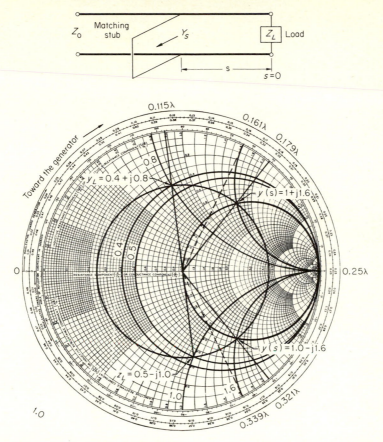

FIGURE 9-12. Single-stub matching.

from which it will be seen that the line will be matched if we choose s such that $g(s) = 1$, and also choose a stub length such that $b = -b(s)$.

It is clear that, for any finite nonzero load impedance, there are two values of s, every half wavelength, which make $g(s) = 1$. We usually choose to match at the position closest to the load, or at the position which requires the shorter stub. Usually, but not always, we avoid open-circuited stubs.

An example of single-stub matching follows.

Example 9-5   Single-stub Matching.   A 50-Ω transmission line is connected to a load impedance $Z_L = 25 - j50$. Find the position and length of the short-circuited stub required to match the line.

Since $Z_0 = 50\ \Omega$, the normalized load impedance is

$$z_L = 0.5 - j1.0$$

From the Smith chart (Fig. 9-12) find

$$y_L = 0.4 + j0.8$$

at 0.115λ "toward the generator" (or $\phi_0 = 97°$), and

$$y(s) = 1.0 + j1.6$$

at 0.179λ "toward the generator," giving $s = (0.179 - 0.115)λ$, or 0.064λ, as the matching point nearest the load.

Notice that there is a second matching point, where

$$y(s) = 1.0 - j1.6$$

at 0.321λ "toward the generator." This point is $(0.321 - 0.115)λ$, or 0.206λ from the load.

The first matching point requires a stub whose normalized susceptance b is -1.6. Reading around the outer circle from $-j\infty$, we find that the length of the short-circuited stub with this value of b is $(0.339 - 0.250)λ$, or 0.089λ long.

The second matching point would require a stub whose normalized susceptance b is $+1.6$. This stub would have a length of 0.161λ if it were open-circuited or a length of $(0.161 + 0.250)λ = 0.411λ$ if it were short-circuited.

Double-stub matching The necessity of providing an adjustable location for the matching stub of single-stub matching causes some difficulty, particularly in permanent installations. In such cases two (or more) matching stubs may be built into the line at fixed positions. Three fixed stubs, properly located, can in principle match any arbitrary load. However, in practice, the stub adjustments required to produce a match is a trial-and-error procedure, and three stubs are exasperatingly difficult to adjust. For this reason, most fixed-location stub arrangements are double-stub. This results in some loss of versatility in that there are some impedance values which cannot be matched with two stubs at fixed locations. With some a priori knowledge of the load impedance, these unmatchable impedances can be avoided by using a fixed line length between the load and the position of the first matching stub.

To be specific, let us examine the case of a load impedance Z_L in parallel with a stub, and with a second stub located a distance d toward the generator from the load. The configuration is shown in Fig. 9-13. The basic problem is the same as in the case of single-stub matching, except that we now can adjust the imaginary part of the effective load admittance $(Y_L + Y_1)$ but we cannot adjust the location of the matching stub.

In terms of Smith chart coordinates, the line will be matched if we make

$$y(d) + y_2 = 1 + j0 \tag{9-65}$$

Since $y_2 = 0 + jb_2$, we must make $y(d) = 1 + jb$ and $b_2 = -b$. For a graphical analysis of the problem, we ask this question: If the normalized admittance at $s = d$ is to be

$$y(d) = 1 + jb = \frac{1 + \Gamma_0 e^{-j2\beta d}}{1 - \Gamma_0 e^{-j2\beta d}} \tag{9-66}$$

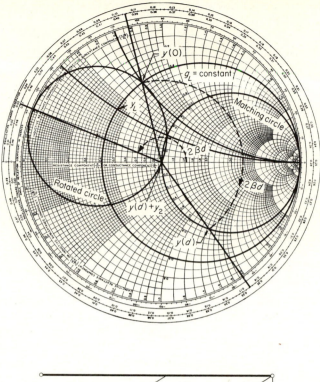

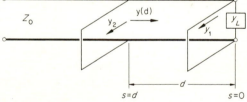

FIGURE 9-13. Double-stub matching.

while at $s = 0$

$$y(0) = g_L + jb_L + jb_1 = \frac{1 + \Gamma_0}{1 - \Gamma_0} \tag{9-67}$$

how is the locus of $y(0)$ related graphically to the locus of $y(d)$? First, notice that the locus $1 + jb$ is the unit, or *matching*, circle centered at the point $0.5 + j0$ in the complex plane. (This is the admittance-coordinate circle $g = 1$ on the Smith chart.) Also, notice that, point by point, the locus (9-67) is obtained from the locus (9-66) by moving along the complex plane circle $|\Gamma_0| = $ constant by an angle $2\beta d$ *toward the load*. Thus the locus

(9-67) is still a circle, of radius $\frac{1}{2}$, whose center is at an angle $2\beta d$ from the positive real axis.

Now we are in a position to outline the step-by-step procedure for double-stub matching. Refer to Fig. 9-13.

1. Draw the rotated circle on the Smith chart. $2\beta d°$ from $0°$
2. Locate $y_L = g_L + jb_L$ on the Smith chart.
3. Follow the $g_L = $ constant Smith chart coordinate to the point where it intersects† the rotated circle, and read $y(0) = g_L + jb_c$, where the normalized susceptance $b_c = b_L + b_1$. This is the required effective load admittance coordinate.
4. Since b_c and b_L are known, the required stub admittance coordinate is $b_1 = b_c - b_L$.
5. From $y(0)$ follow the $|\Gamma_0| = $ constant circle a distance $2\beta d$. It will intersect the $1 + jb$, or matching, circle at this point. This is $y(d)$.
6. Read the value of b.
7. The required susceptance of the matching stub at $s = d$ is $b_2 = -b$.

Example 9-6 Double-stub Matching. Given that

$$Y_L = (8 + j8) \times 10^{-3} \quad \mho$$
$$Y_0 = 20 \times 10^{-3} \quad \mho$$

Use one short-circuited stub in parallel with the load and one short-circuited stub in parallel with the line at $d = 0.22\lambda$; match the line and find the length of the required stubs.

With reference to Fig. 9-13:

1. Draw the rotated circle.
2. Locate $y_L = 0.4 + j0.4$.
3. Follow the $g_L = 0.4$ curve to the point $y(0) = 0.4 + j0.7$, where it intersects the rotated circle.
4. Calculate $b_1 = 0.7 - 0.4 = 0.3$.
5. From $y(0) = 0.4 + j0.7$, follow the $|\Gamma_0| = $ constant circle a distance $2\beta d = 0.22\lambda$ "toward the generator" to the point $y(d)$ on the matching circle.
6. Read $y(d) = 1 - j1.45$.
7. The required normalized susceptance of the matching stub at $s = d$ is 1.45.

The corresponding short-circuited stub lengths are

For $b_1 = 0.3$: $\qquad\qquad\qquad l_1 = 0.297\lambda$

For $b_2 = 1.45$: $\qquad\qquad\qquad l_2 = 0.404\lambda$

As an exercise, the student should choose the other point of intersection on the rotated circle and find the new matching-stub lengths.

---

† There are two points of intersection. The entire procedure is valid for either point of intersection, but it generally leads to different matching-stub lengths.

9.8 Lossy Transmission Lines. The distributed-circuit approach to lossy transmission lines is fairly straightforward. The fields approach provides more insight and understanding of subtleties and limitations of lossy transmission line analysis. However, it is rather involved.† In this section we present only the distributed-circuit approach to the problem.

An actual lossy transmission line can be adequately represented *at a single frequency* by the distributed circuit shown in Fig. 9-14, where R represents the series ohmic resistance of the conductors per unit length, L represents the series inductance of the line per unit length, G represents the shunt conductance per unit length resulting from the conductivity and dielectric losses of the medium, and C represents the shunt capacitance of the line per unit length. Unfortunately, these parameters are frequency-dependent. However, if we consider them to be constant, we can write the circuit equations

$$\frac{\partial V}{\partial z} = -RI - L\frac{\partial I}{\partial t}$$

$$\frac{\partial I}{\partial z} = -GV - C\frac{\partial V}{\partial t} \tag{9-68}$$

or, in the sinusoidal steady state,

$$\frac{dV}{dz} = -(R + j\omega L)I$$

$$\frac{dI}{dz} = -(G + j\omega C)V \tag{9-69}$$

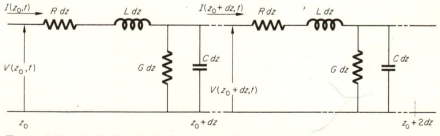

FIGURE 9-14. A distributed-circuit representation of a lossy transmission line.

† R. B. Adler, L. J. Chu, and R. M. Fano: "Electromagnetic Energy Transmission and Radiation," chap. 9, John Wiley & Sons, Inc., New York, 1960.

which can be differentiated and combined to give

$$\frac{d^2V}{dz^2} = (R + j\omega L)(G + j\omega C)V$$

$$\frac{d^2I}{dz^2} = (R + j\omega L)(G + j\omega C)I$$

(9-70)

If we define

$$\gamma = [(R + j\omega L)(G + j\omega C)]^{\frac{1}{2}}$$

(9-71)

the solutions to Eqs. (9-70) are

$$V(z) = V^+e^{-\gamma z} + V^-e^{+\gamma z}$$

$$I(z) = I^+e^{-\gamma z} + I^-e^{+\gamma z}$$

(9-72)

where V^+, V^-, I^+, I^- are the complex amplitudes of positive and negative traveling voltages and currents.

If Eqs. (9-72) are substituted into Eqs. (9-69), we find that

$$\frac{V(z)}{I(z)} = Z_0 \frac{V^+e^{-\gamma z} + V^-e^{+\gamma z}}{V^+e^{-\gamma z} - V^-e^{+\gamma z}}$$

(9-73)

where

$$Z_0 = \left(\frac{R + j\omega L}{G + j\omega C}\right)^{\frac{1}{2}}$$

(9-74)

which shows that, although it is still possible to define a characteristic impedance for the line, it is now a complex function of frequency. Furthermore, as shown by Eq. (9-71), the propagation constant γ is also complex. If we separate γ into its real and imaginary parts by writing

$$\gamma = \alpha + j\beta$$

(9-75)

where α and β are defined by

$$\alpha = \text{Re } [(R + j\omega L)(G + j\omega C)]^{\frac{1}{2}}$$

$$\beta = \text{Im } [(R + j\omega L)(G + j\omega C)]^{\frac{1}{2}}$$

(9-76)

we should find that V^+ (and I^+) is attenuated by a factor $e^{-\alpha z}$ while V^- (and I^-) is attenuated by a factor $e^{-\alpha s} = e^{+\alpha z}$ as they propagate with a phase velocity

$$v_p = \frac{\omega}{\beta}$$

(9-77)

Note that both α and β are functions of frequency, in addition to depending on R, L, G, and C, which are also functions of frequency. This emphasizes that any lossy transmission line analysis is valid only at a single frequency. Fortunately, in many problems of interest, the range of frequencies involved is small enough so that α, β, and Z_0 may be considered constant (measured, of course, at the frequency of interest).

A particularly interesting special case of a lossy line is the case where the line parameters R, L, G, and C have been adjusted, at least as nearly as possible, so that

$$\frac{L}{R} = \frac{C}{G} \qquad (9\text{-}78)$$

for all frequencies. Such a line is called *distortionless*, for reasons which will become apparent. For this situation, we have

$$Z_0 = \left(\frac{R + j\omega L}{G + j\omega C}\right)^{\frac{1}{2}} = \sqrt{\frac{L}{C}}\left(\frac{R/L + j\omega}{G/C + j\omega}\right)^{\frac{1}{2}} = \sqrt{\frac{L}{C}} \qquad (9\text{-}79)$$

which is independent of frequency and, in fact, is just the value which it would have if the line were lossless. Also, for this distortionless line, we have

$$\gamma = [(R + j\omega L)(G + j\omega C)]^{\frac{1}{2}} = [RG + j\omega RC + j\omega LG + (j\omega)^2 LC]^{\frac{1}{2}}$$

But since $RC = LG$, this can be written

$$\gamma = [RG + j2\omega\sqrt{RG}\sqrt{LC} + (j\omega)^2 LC]^{\frac{1}{2}} = \sqrt{RG} + j\omega\sqrt{LC}$$

which says that

$$\alpha = \sqrt{RG} \qquad \beta = \omega\sqrt{LC} \qquad (9\text{-}80)$$

Recalling that the phase velocity is given by

$$v_p = \frac{\omega}{\beta} = \frac{1}{\sqrt{LC}}$$

we see that the phase velocity is independent of frequency and has the same value as that of a lossless line. This means that pulses which, as implied by Fourier analysis, contain a wide range of frequencies will be propagated without shape distortion. This is the basis for the term *distortionless line*.

A second interesting and useful special case is the *low-loss line*. A precise definition of low loss is of course arbitrary. If we use as a criterion

$$R \ll \omega L \qquad G \ll \omega C \qquad (9\text{-}81)$$

We shall see that the line is also distortionless over the range of frequencies for which our low-loss criteria are satisfied. In this case we have, to a zero-order approximation,

$$Z_0 = \left(\frac{R + j\omega L}{G + j\omega C}\right)^{\frac{1}{2}} \approx \sqrt{\frac{L}{C}} \qquad (9\text{-}82)$$

and to a first-order approximation,

$$\gamma = [(R + j\omega L)(G + j\omega C)]^{\frac{1}{2}}$$

$$= j\omega\sqrt{LC}\left[\left(1 + \frac{R}{j\omega L}\right)\left(1 + \frac{G}{j\omega C}\right)\right]^{\frac{1}{2}}$$

$$= j\omega\sqrt{LC}\left[1 + \frac{R}{j\omega L} + \frac{G}{j\omega C} + \frac{RG}{(j\omega)^2 LC}\right]^{\frac{1}{2}}$$

$$\approx j\omega\sqrt{LC}\left(1 + \frac{R}{j\omega L} + \frac{G}{j\omega C}\right)^{\frac{1}{2}}$$

$$\approx j\omega\sqrt{LC}\left(1 + \frac{R}{2j\omega L} + \frac{G}{2j\omega C}\right)$$

$$= j\omega\sqrt{LC} + \frac{R}{2\sqrt{L/C}} + \frac{G}{2\sqrt{C/L}} \qquad (9\text{-}83)$$

or

$$\gamma \approx \frac{R}{2Z_0} + \frac{G}{2Y_0} + j\omega\sqrt{LC} \qquad (9\text{-}84)$$

which is similar to the truly distortionless case, and in fact reduces to it if $R/L = G/C$.

The foregoing remarks show that, if a line is made as nearly distortionless as possible and at the same time as low-loss as possible, an analysis based upon the distortionless-line assumption will *probably* be satisfactory. One must, of course, be wary in borderline cases.

Lossy-transmission-line analysis can be accomplished by use of the Smith chart at a single frequency regardless of whether it is low loss distortionless or not. The basic mathematical problem which must be solved is

$$Z(s) = Z_0 \frac{1 + \Gamma_0 e^{-2\alpha s} e^{-j2\beta s}}{1 - \Gamma_0 e^{-2\alpha s} e^{-j2\beta s}} \qquad (9\text{-}85)$$

or

$$\frac{Z(s)}{Z_0} = \frac{1 + \Gamma_0 e^{-2\alpha s} e^{-j2\beta s}}{1 - \Gamma_0 e^{-2\alpha s} e^{-j2\beta s}} \qquad (9\text{-}86)$$

from which it is seen that the locus of

$$\Gamma(s) = \Gamma_0 e^{-2\alpha s} e^{-j2\beta s} \qquad (9\text{-}87)$$

which was a circle for a lossless line, is now a logarithmic spiral. The Smith chart still gives the real and imaginary parts of $Z(s)/Z_0$ for $|\Gamma(s)| \leq 1$. The extension $|\Gamma(s)| > 1$ is treated in the literature.†

Some examples of Smith chart calculations for lossy lines will help clarify the procedures.

† *Ibid.*, chap. 5.

Example 9-7 Lossy Line. (*a*) Suppose that a 58-cm length of lossy line which is known to be less than $\lambda/2$ long is open-circuited at one end and that the input impedance $Z(s)$ at the other end is measured to be

$$\frac{Z(s)}{Z_0} = 0.20 + j0.25$$

Find α and β.

SOLUTION

Let d be the length of the line. Then we have

$$\frac{Z(0)}{Z_0} \to \infty \qquad \text{at open end}$$

$$\frac{Z(s)}{Z_0} = 0.20 + j0.25 \qquad \text{at input}$$

$$\frac{1 + 1e^{-2\alpha d}e^{-j2\beta d}}{1 - 1e^{-2\alpha d}e^{-j2\beta d}} = 0.20 + j0.25$$

As shown in Fig. 9-15, we locate the point $0.2 + j0.25$ and draw the line from the center to this point. By measurement, we find that its length is 0.685 of the chart's outer radius, which means that $|\Gamma(d)| = 0.685$. Extending this line to the scale around the circumference, we see that $d = 0.04\lambda + 0.25\lambda = 0.29\lambda$.

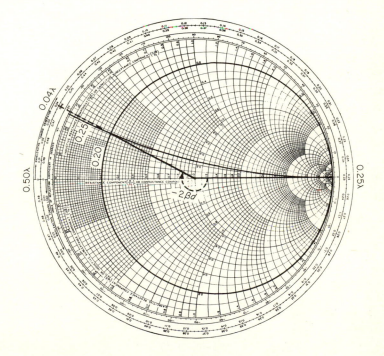

FIGURE 9-15. Smith chart calculations for lossy lines.

From these measurements and the length $d = 0.58$ m, we obtain $e^{-2\alpha d} = 0.685$, or $2\alpha d = 0.375$, and

$$\alpha = \frac{0.375}{1.16} = 0.323 \text{ Np/m}$$

Actually, since it is the factor $2\alpha d$ which appears in the exponential, and since d is frequently specified in wavelengths, it is convenient at times to calculate

$$2\alpha = \frac{0.375}{0.29} = 1.295 \text{ Np/wavelength}$$

(*b*) Assume that this same transmission line is terminated with a load impedance such that

$$\frac{Z(0)}{Z_0} = 1.0 + j2.0$$

and that $Z_0 = 50 + j5 \ \Omega$.   Find the input impedance.

SOLUTION

Refer to Fig. 9-16.   Locate the point $1.0 + j2.0$ as shown.   Rotate 0.29λ "toward the generator" and draw the radial line shown.   Then $Z(0.29\lambda)/Z_0$ lies on this line.   Using a compass and a linear scale, find $|\Gamma| = 0.7$.   Using $e^{-2\alpha d} = 0.685$, obtain $|\Gamma| e^{-2\alpha d} = 0.48$.   Again, using a compass and the linear scale, locate the point

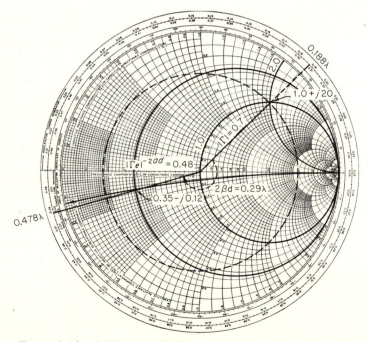

FIGURE 9-16.   Smith chart calculations for a lossy transmission line.

$|\Gamma|e^{-2\alpha d}e^{-j2\beta d}$ along the previously located radial line, as shown in Fig. 9-16. Read

$$\frac{Z(0.29\lambda)}{Z_0} = 0.35 - j0.12$$

and since $Z_0 = 50 + j5$, calculate

$$Z(0.29\lambda) = (50 + j5)(0.35 - j0.12) = 18.1 - j4.25 \quad \Omega$$

9.9 Transients on Lossless Transmission Lines.

In this section we shall deal with wave motion in the transient state along a lossless trans-mission line, terminated† at $z = l$ in a linear passive load (Fig. 9-17a). The input to the line is at $z = 0$.

The partial differential equations of a lossless line are given by Eqs. (9-35) and (9-36), and lead to general solutions of the form given in the following equations,

$$V(z,t) = f_+\left(t - \frac{z}{v}\right) + f_-\left(t + \frac{z}{v}\right)$$

$$I(z,t) = \frac{1}{Z_0}\left[f_+\left(t - \frac{z}{v}\right) - f_-\left(t + \frac{z}{v}\right)\right] \tag{9-88}$$

where $v = 1/\sqrt{LC}$ is the velocity at which waves travel.   At $z = 0$, this pair

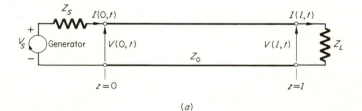

(a)

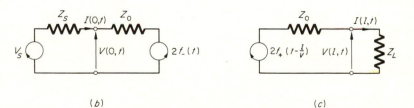

(b)   (c)

FIGURE 9-17.   Circuit characterization of a lossless transmission line. (a) A finite line; (b) sending-end equivalent circuit; (c) receiving-end equivalent circuit.

† The change of coordinates in this section is deliberate, and is made to accommodate our conventional ideas about positively traveling waves, left to right in the direction of increasing coordinate.

becomes

$$V(0,t) = f_+(t) + f_-(t)$$

$$I(0,t) = \frac{1}{Z_0}[f_+(t) - f_-(t)] \tag{9-89}$$

and provides the basis for deriving an equivalent circuit at the transmitting end of the line. Thus, by eliminating the function $f_+(t)$, we find

$$V(0,t) - Z_0 I(0,t) = 2f_-(t)$$

and deduce the equivalent circuit of Fig. 9-17b. From this circuit the sending-end current $I(0,t)$ can be found, once the function $f_-(t)$ is given; then the function $f_+(t)$ can be determined easily, since from Eqs. (9-89) it is clear that

$$f_+(t) = Z_0 I(0,t) + f_-(t) \tag{9-90}$$

At the receiving end ($z = l$) the development follows a similar course. Thus

$$V(l,t) = f_+\left(t - \frac{l}{v}\right) + f_-\left(t + \frac{l}{v}\right)$$

$$I(l,t) = \frac{1}{Z_0}\left[f_+\left(t - \frac{l}{v}\right) - f_-\left(t + \frac{l}{v}\right)\right] \tag{9-91}$$

and by eliminating the function $f_-(t + l/v)$,

$$V(l,t) + Z_0 I(l,t) = 2f_+\left(t - \frac{l}{v}\right)$$

which leads to the equivalent circuit of Fig. 9-17c. Thus, if we know f_+, we can solve for $I(l,t)$ and, ultimately, for f_-.

Let us examine the transient behavior of a lossless line by considering some specific examples.

Example 9-8 Pulses on Transmission Lines. Let both Z_S and Z_L in Fig. 9-17a be pure resistances. In particular, let $Z_S = 150\ \Omega$, $Z_L = 50\ \Omega$. Also, let $Z_0 = 50\ \Omega$. Assume that the line length is 800 m, and that the velocity of propagation on the line is 200 m/μs. At $t = 0$, a single pulse of 16 V and 1-μs duration is transmitted by the generator. Discuss the pulse behavior.

At the initial instant, $t = 0$, the line is completely at rest, and both the voltage and the current are zero for $0 < z \le l$. More specifically, $f_-(0) = 0$, so that, in Fig. 9-17b, the generator feeds a series combination of two resistances, Z_S and Z_0. Hence, by simple voltage division, we find that the pulse amplitude at the input terminals to the line is

$$V(0,t) = \frac{16 \times 50}{150 + 50} = 4\ \text{V} \qquad 0 < t < 1\ \mu s$$

This 4-V pulse arrives at the load $800/200 = 4\ \mu s$ later.

From Fig. 9-17c, the load voltage $V(l,t)$ is seen to be equal to $f_+(t - l/v)$ when $Z_L = Z_0$. Additionally, the voltage equation in Eqs. (9-91) predicts $f_-(t + l/v) \equiv 0$ when $V(l,t) = f_+(t - l/v)$. Therefore, in this case, there is no reflection, and all the pulse power is adsorbed in the load. [Notice that the reflection coefficient at the load is $\Gamma_L = (50 - 50)/(50 + 50) = 0$, which also means that there is no reflection. However, we cannot use this approach when reactive elements are included in the load.]

We see, then, that at the load, we have a pulse of 4 V magnitude, which begins at $t = 4$ μs and lasts for 1 μs.

Example 9-9 Pulses on Transmission Lines. Let the conditions and line length be the same as in Example 9-8, except that

$$Z_S = 150\ \Omega \qquad Z_0 = 50\ \Omega \qquad Z_L = 30\ \Omega$$

with $V_s = 16$ V as before.

Since there are no reactive elements, either at the generator or at the load, we can make effective use of the reflection coefficients

$$\Gamma_s = \frac{150 - 50}{150 + 50} = \frac{1}{2}$$

$$\Gamma_L = \frac{30 - 50}{30 + 50} = -\frac{1}{4}$$

We have already seen that this gives an initial f_+ of 4 V, which arrives at the load end at $t = 4$ μs, where it produces a reflected pulse of $(-\frac{1}{4})(4) = -1$ V. At $t = 8$ μs, this pulse arrives at the generator and produces a reflected pulse of $(\frac{1}{2})(-1) = -\frac{1}{2}$ V. At $t = 12$ μs, this pulse arrives at the load end and produces a reflected pulse of $(-\frac{1}{2})(-\frac{1}{4}) = \frac{1}{8}$ V. This goes on forever. After a few round trips, however, the pulse is of negligible amplitude.

Example 9-10. Reactive Termination As a last example, let us consider a lossless transmission line terminated in a capacitor. Initially both the line and the load capacitor are uncharged. As shown in Fig. 9-18a, the source voltage is a unit step and the internal impedance of the generator is equal to the characteristic (resistive) impedance of the line.

Let $T = l/v$ be the time required for a wave to travel the length of the line. In the time interval $0 \leq t < T$, there is no wave traveling in the negative z direction; hence $f_-(t + z/v) \equiv 0$, and the sending-end equivalent circuit of Fig. 9-17b reduces to the simpler circuit shown in Fig. 9-18b. From this circuit, we find $V(0,t) = \frac{1}{2}u_{-1}(t)$, so that a rectangular voltage wave, described by

$$f_+\left(t - \frac{z}{v}\right) = \frac{1}{2}u_{-1}\left(t - \frac{z}{v}\right)$$

will travel down the line accompanied by a current step of magnitude $1/2Z_0$. At $t = T$ the waves reach the load where the conditions are described by the equivalent circuit of Fig. 9-17c; here the equivalent generator voltage is

$$2f_+\left(t - \frac{l}{v}\right) = 2\left[\frac{1}{2}u_{-1}\left(t - \frac{l}{v}\right)\right] = u_{-1}(t - T)$$

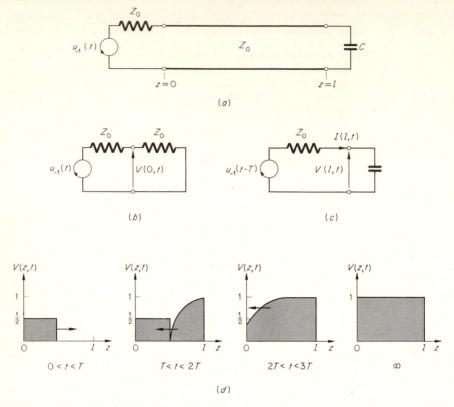

FIGURE 9-18. The details of Example 9-10. (a) A finite line terminated in an initially uncharged capacitor; (b) sending-end equivalent circuit; (c) receiving-end equivalent circuit; (d) line voltage at various instants of time.

The specific circuit applying to this example is shown in Fig. 9-18c, and leads to the solution

$$V(l,t) = u_{-1}(t - T)(1 - e^{-(t-T)/Z_0 C})$$

The reflected wave is determined from Eqs. (9-91).   Thus

$$f_-\left(t + \frac{l}{v}\right) = V(l,t) - f_+\left(t - \frac{l}{v}\right) = u_{-1}(t - T)(\tfrac{1}{2} - e^{-(t-T)/Z_0 C})$$

At any point along the line, this wave is delayed by an amount $(l - z)/v$.   Therefore, for $T < t < 2T$, there will be a reflected voltage wave

$$f_-\left(t + \frac{z}{v}\right) = u_{-1}\left(t - 2T + \frac{z}{v}\right)(\tfrac{1}{2} - e^{-(t-2T+z/v)/Z_0 C})$$

which will reach the sending end at $t = 2T$.   The terminating impedance at this end is equal to the characteristic impedance of the line.   Therefore there will be no reflection at this end.   At each moment, the voltage at any point along the line will be given by the sum of the incident and reflected waves. Thus

$$V(z,t) = \tfrac{1}{2}u_{-1}\left(t - \frac{z}{v}\right) + u_{-1}\left(t - 2T + \frac{z}{v}\right)(\tfrac{1}{2} - e^{-(t-2T+z/v)/Z_0C})$$

which shows that, as time goes on, the entire line will be charged up to 1 V, as expected. A few pictorial stages along the way are given in Fig. 9-18d.

9.10 Waveguides. As the frequency increases, the performance of ordinary transmission lines becomes progressively less desirable. In particular, the losses become prohibitively large. Fortunately, it is possible to transmit electromagnetic energy by using structures, known as *waveguides*, whose performance becomes optimum at about the same frequency for which ordinary transmission line performance becomes seriously degraded. This section discusses the elementary aspects of waveguide theory.

For our purposes *a waveguide is a hollow conducting tube of any arbitrary uniform cross section.* As a practical matter, a rectangular cross section with a width about twice the height is the most common waveguide. To show that this waveguide will propagate electromagnetic waves, we solve Maxwell's equations subject to the boundary conditions and show that a nontrivial solution in fact exists.

Consider the waveguide shown in Fig. 9-19. The interior of the guide is presumed to be a linear, homogeneous, and isotropic dielectric, in which case Maxwell's equations, as shown in Sec. 7.5, may be combined to yield the vector Helmholtz equations in **E** and **H**. Thus, for a lossless medium, Eqs. (7-38) give

$$\nabla^2\mathbf{E} + \omega^2\mu\epsilon\mathbf{E} = 0 \qquad\qquad (9\text{-}92)$$

$$\nabla^2\mathbf{H} + \omega^2\mu\epsilon\mathbf{H} = 0 \qquad\qquad (9\text{-}93)$$

In rectangular coordinates, these equations are actually a set of six identical

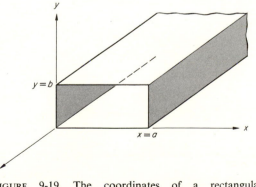

FIGURE 9-19. The coordinates of a rectangular waveguide.

scalar Helmholtz equations in the **E**- and **H**-field components. Let us write out explicitly the equation for H_z:

$$\frac{\partial^2 H_z}{\partial x^2} + \frac{\partial^2 H_z}{\partial y^2} + \frac{\partial^2 H_z}{\partial z^2} + \omega^2 \mu \epsilon H_z = 0 \qquad (9\text{-}94)$$

It is well known that a solution to this equation may be obtained by assuming a product function of the form

$$H_z(x,y,z) = X(x)\,Y(y)Z(z) \qquad (9\text{-}95)$$

where $\qquad\qquad\quad X(x) = $ a function of x only

$$Y(y) = \text{a function of } y \text{ only}$$

$$Z(z) = \text{a function of } z \text{ only}$$

If we use this information, we see that Eq. (9-94) can be written

$$\frac{d^2 X}{dx^2}\,(YZ) + \frac{dY^2}{dy^2}\,(XZ) + \frac{d^2 Z}{dz^2}\,(XY) + \omega^2 \mu \epsilon XYZ = 0$$

or upon dividing by XYZ,

$$\frac{1}{X}\frac{d^2 X}{dx^2} + \frac{1}{Y}\frac{d^2 Y}{dy^2} + \frac{1}{Z}\frac{d^2 Z}{dz^2} = -\omega^2 \mu \epsilon \qquad (9\text{-}96)$$

In this equation, the variables are separated. Hence each term on the left must be equal to a constant. Accordingly, Eq. (9-96) may be written

$$-A^2 - B^2 + \gamma^2 = -\omega^2 \mu \epsilon \qquad (9\text{-}97)$$

where the constants† A^2, B^2, γ^2 separate the original partial differential equation into three total differential equations, of the form

$$\frac{d^2 X}{dx^2} + A^2 X = 0$$

$$\frac{d^2 Y}{dy^2} + B^2 Y = 0 \qquad (9\text{-}98)$$

$$\frac{d^2 Z}{dz^2} - \gamma^2 Z = 0$$

The solutions of these equations should be well known to the student. They

† We anticipated the final result when we chose to write the constants as $-A^2$, $-B^2$, and γ^2. It will be seen that this combination of algebraic signs best fits the rest of the problem.

are

$$X = C_1 \sin Ax + C_2 \cos Ax$$
$$Y = C_3 \sin By + C_4 \cos By \qquad (9\text{-}99)$$
$$Z = C_5 e^{\gamma z} + C_6 e^{-\gamma z}$$

which means that

$$H_z = XYZ = (C_1 \sin Ax + C_2 \cos Ax)$$
$$\times (C_3 \sin By + C_4 \cos By)(C_5 e^{\gamma z} + C_6 e^{-\gamma z}) \quad (9\text{-}100)$$

Examination of the original vector Helmholtz equations in **E** and **H** [Eqs. (9-92) and (9-93)] shows that E_x, E_y, E_z, H_x, and H_y, all yield solutions identical in form with Eq. (9-100), and hence the solutions can differ only in the—as yet undetermined—multiplicative constants C_1, C_2, C_3, C_4, C_5, C_6. To find the relationship between the field components, we need to return to Maxwell's equations

$$\mathbf{\nabla} \times \mathbf{H} = j\omega\epsilon\mathbf{E} \qquad \mathbf{\nabla} \times \mathbf{E} = -j\omega\mu\mathbf{H}$$

which, when written in rectangular components, yield

$$\frac{\partial H_z}{\partial y} - \frac{\partial H_y}{\partial z} = j\omega\epsilon E_x \qquad \frac{\partial E_z}{\partial y} - \frac{\partial E_y}{\partial z} = -j\omega\mu H_x$$

$$\frac{\partial H_x}{\partial z} - \frac{\partial H_z}{\partial x} = j\omega\epsilon E_y \qquad \frac{\partial E_x}{\partial z} - \frac{\partial E_z}{\partial x} = -j\omega\mu H_y \qquad (9\text{-}101)$$

$$\frac{\partial H_y}{\partial x} - \frac{\partial H_x}{\partial y} = j\omega\epsilon E_z \qquad \frac{\partial E_y}{\partial x} - \frac{\partial E_x}{\partial y} = -j\omega\mu H_z$$

At this point we return to Eqs. (9-99) and note that the function Z is the sum of the two traveling waves, the first of which is traveling in the $-z$ direction and the second of which is traveling in the $+z$ direction. Because of linearity, it is legitimate to consider the two directions of travel separately. Accordingly, we shall deal with the $+z$ traveling wave, and write the z dependence simply as $e^{-\gamma z}$ for all the field components. Having done this, $\partial/\partial z$ becomes $-\gamma$, and the set (9-101) transforms to

$$\frac{\partial H_z}{\partial y} + \gamma H_y = j\omega\epsilon E_x \qquad \frac{\partial E_z}{\partial y} + \gamma E_y = -j\omega\mu H_x$$

$$-\gamma H_x - \frac{\partial H_z}{\partial x} = j\omega\epsilon E_y \qquad -\gamma E_x - \frac{\partial E_z}{\partial x} = -j\omega\mu H_y$$

$$\frac{\partial H_y}{\partial x} - \frac{\partial H_x}{\partial y} = j\omega\epsilon E_z \qquad \frac{\partial E_y}{\partial x} - \frac{\partial E_x}{\partial y} = -j\omega\mu H_z$$

By properly combining these equations we obtain

$$E_x = -\frac{\gamma}{h^2}\frac{\partial E_z}{\partial x} - \frac{j\omega\mu}{h^2}\frac{\partial H_z}{\partial y}$$

$$E_y = -\frac{\gamma}{h^2}\frac{\partial E_z}{\partial y} + \frac{j\omega\mu}{h^2}\frac{\partial H_z}{\partial x}$$

$$H_x = -\frac{\gamma}{h^2}\frac{\partial H_z}{\partial x} + \frac{j\omega\epsilon}{h^2}\frac{\partial E_z}{\partial y}$$

$$H_y = -\frac{\gamma}{h^2}\frac{\partial H_z}{\partial y} - \frac{j\omega\epsilon}{h^2}\frac{\partial E_z}{\partial x}$$

(9-102)

where
$$h^2 = \gamma^2 + \omega^2\mu\epsilon = A^2 + B^2 \qquad (9\text{-}103)$$

From these equations it can be seen that two special classes of solutions are possible:

1. Those solutions for which $E_z = 0$, $H_z \neq 0$. Since the remaining **E**-field components are transverse to z, these solutions are called *transverse electric modes*, or TE *modes*.
2. Those solutions for which $H_z = 0$, $E_z \neq 0$. For reasons parallel to those stated above, these solutions are called *transverse magnetic modes*, or TM modes.

Obviously, the most general solution† is the simultaneous existence of TE and TM modes. However, it is advantageous to examine each case separately.

9.11 TM Modes in Rectangular Waveguides. In this case $H_z = 0$. The boundary conditions are that $E_{\text{tan}} = 0$ at $x = 0$, $y = 0$, $x = a$, and $y = b$. Moreover E_z is of the form given by Eq. (9-100), specialized for a positive traveling wave, that is,

$$E_z = (C_1 \sin Ax + C_2 \cos Ax)(C_3 \sin By + C_4 \cos By)e^{-\gamma z} \quad (9\text{-}104)$$

Using the boundary condition $E_z = 0$ at $x = 0$ and also at $y = 0$, we see immediately that the cosine terms must be zero, and that E_z must be given by

$$E_z = C \sin Ax \sin By\, e^{-\gamma z} \qquad (9\text{-}105)$$

where $C = C_1 C_3$. Now, using the boundary condition $E_z = 0$ at $x = a$ and also at $y = b$, we see that we must have

$$\sin Aa = 0$$
$$\sin Bb = 0$$

---

† Note that a TEM wave cannot exist inside a waveguide, because setting $E_z = H_z = 0$ in Eqs. (9-102) would make all the other components of the field equal to zero.

These two equations require that

$$Aa = m\pi \qquad m = 1, 2, 3, \ldots$$
$$Bb = n\pi \qquad n = 1, 2, 3, \ldots \tag{9-106}$$

Neither m nor n can be zero since zero for either one would make E_z identically zero, and there would be no field.

Having completely determined the x and y dependence of E_z, we could insert E_z into Eqs. (9-102) to obtain the other field components. However, since γ appears explicitly in Eqs. (9-102), it is expedient to examine the nature of γ before proceeding.

In terms of the explicitly determined values of A and B, Eq. (9-103) gives

$$\gamma = \sqrt{\left(\frac{m\pi}{a}\right)^2 + \left(\frac{n\pi}{b}\right)^2 - \omega^2\mu\epsilon} \tag{9-107}$$

Thus we see that γ is a real number for

$$\omega^2\mu\epsilon < \left(\frac{m\pi}{a}\right)^2 + \left(\frac{n\pi}{b}\right)^2$$

and that γ is a pure imaginary number for

$$\omega^2\mu\epsilon > \left(\frac{m\pi}{a}\right)^2 + \left(\frac{n\pi}{b}\right)^2$$

This means that true wave propagation occurs only for ω greater than a *cutoff (angular) frequency* ω_c, defined by

$$\omega_c{}^2\mu\epsilon = \left(\frac{m\pi}{a}\right)^2 + \left(\frac{n\pi}{b}\right)^2 \tag{9-108}$$

Since we are primarily concerned with the waveguide as a wave-propagating structure, we note explicitly the expected pure imaginary form of γ by writing $\gamma = j\beta$, where

$$\beta = \sqrt{\omega^2\mu\epsilon - \left(\frac{m\pi}{a}\right)^2 - \left(\frac{n\pi}{b}\right)^2} \tag{9-109}$$

is the phase-shift constant for the particular field configuration implied by the integers m and n.

If we now insert all our results into Eqs. (9-102), we shall have the final complete set of equations which describe the field vector for the TM modes.

The resulting equations are

$$E_z = C \sin Ax \sin By \, e^{-j\beta z}$$

$$E_x = \frac{-j\beta A}{h^2} C \cos Ax \sin By \, e^{-j\beta z}$$

$$E_y = \frac{-j\beta B}{h^2} C \sin Ax \cos By \, e^{-j\beta z}$$

$$H_x = \frac{j\omega\epsilon B}{h^2} C \sin Ax \cos By \, e^{-j\beta z} \qquad (9\text{-}110)$$

$$H_y = -\frac{j\omega\epsilon A}{h^2} C \cos Ax \sin By \, e^{-j\beta z}$$

$$H_z = 0$$

where
$$A = \frac{m\pi}{a} \qquad m = 1, 2, 3, \ldots$$
$$\qquad (9\text{-}111)$$
$$B = \frac{n\pi}{b} \qquad n = 1, 2, 3, \ldots$$

The wave associated with the integers m and n is designated as the TM_{mn} mode.

9.12 TE Modes in Rectangular Waveguides. We derive the field equations for the TE modes ($E_z = 0$) in a similar manner, except that, since $E_z = 0$, we must look to E_x and E_y to satisfy the boundary condition $E_{\tan} = 0$. We leave it as an exercise for the student to go through the detailed steps in the derivation, and assert that the complete set of equations for the field vectors which define the TE_{mn} modes are

$$H_z = C \cos Ax \cos By \, e^{-j\beta z}$$

$$H_x = \frac{j\beta A}{h^2} C \sin Ax \cos By \, e^{-j\beta z}$$

$$H_y = \frac{j\beta B}{h^2} C \cos Ax \sin By \, e^{-j\beta z} \qquad (9\text{-}112)$$

$$E_x = \frac{j\omega\mu B}{h^2} C \cos Ax \sin By \, e^{-j\beta z}$$

$$E_y = -\frac{j\omega\mu A}{h^2} C \sin Ax \cos By \, e^{-j\beta z}$$

$$E_z = 0$$

where β, A, B, and h^2 are the same wave parameters as defined for TM modes.

Another group of wave parameters which apply to both types of modes are:

1. The *cutoff wavelength*

$$\lambda_c = \frac{c}{f_c} = \frac{1}{\sqrt{\mu\epsilon}\, f_c} = \frac{2}{[(m/a)^2 + (n/b)^2]^{1/2}} \qquad (9\text{-}113)$$

where the last equality follows from Eq. (9-108). This is the longest wavelength (measured in an unbounded medium with the same μ and ϵ) at which true wave propagation occurs. For longer wavelengths the modes become *evanescent*.

In terms of f_c,

$$\beta = \omega\sqrt{\mu\epsilon}\sqrt{1 - \left(\frac{f_c}{f}\right)^2} = \frac{2\pi}{\lambda}\sqrt{1 - \left(\frac{f_c}{f}\right)^2} \qquad (9\text{-}114)$$

where $\lambda = c/f$ is the wavelength measured in the same unbounded medium.

2. The *phase velocity*

$$v_p = \frac{\omega}{\beta} = \frac{\omega}{\sqrt{\omega^2\mu\epsilon - (m\pi/a)^2 - (n\pi/b)^2}} = \frac{c}{\sqrt{1 - (f_c/f)^2}} \qquad (9\text{-}115)$$

This is the velocity of propagation of a constant phase front. Notice that since the phase velocity is not linearly dependent upon frequency, a broad-band signal will be dispersed as it travels down a waveguide. For narrow-band signals the notion of *group velocity* was found to be meaningful. From Eq. (7-127) we obtain

$$v_g = \frac{1}{d\beta/d\omega} = \frac{\sqrt{\omega^2\mu\epsilon - (m\pi/a)^2 + (n\pi/b)^2}}{\omega\mu} = \frac{c^2}{v_p} \qquad (9\text{-}116)$$

Notice that $v_p \geq c$ for all $f \geq f_c$.

3. The *guide wavelength*

$$\lambda_g = \frac{v_p}{f} = \frac{2\pi}{\beta} = \frac{2\pi}{\sqrt{\omega^2\mu\epsilon - (m\pi/a)^2 - (n\pi/b)^2}} = \frac{\lambda}{\sqrt{1 - (f_c/f)^2}} \qquad (9\text{-}117)$$

Notice that $\lambda_g \geq \lambda$ for all $f \geq f_c$. Also, notice that

$$\frac{1}{\lambda^2} = \frac{1}{\lambda_g{}^2} + \frac{1}{\lambda_c{}^2} \qquad (9\text{-}118)$$

Now, there is one significant difference between TE and TM modes. Recall that, for TM modes, either condition $m = 0$ or $n = 0$ resulted in a zero field. However, inspection of Eqs. (9-112) shows that both

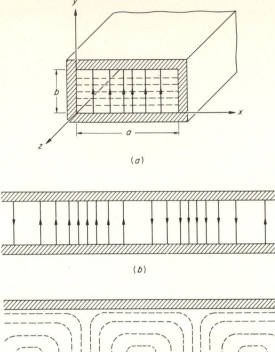

(a)

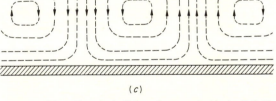

(b)

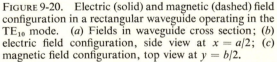

(c)

FIGURE 9-20.   Electric (solid) and magnetic (dashed) field configuration in a rectangular waveguide operating in the TE$_{10}$ mode.   (a) Fields in waveguide cross section; (b) electric field configuration, side view at $x = a/2$; (c) magnetic field configuration, top view at $y = b/2$.

$m = 0$, $n \neq 0$ and $m \neq 0$, $n = 0$ yield nonvanishing TE-mode field configurations which are considerably simpler than the general TE$_{mn}$ configurations.   For example, $m = 1$, $n = 0$ yields

$$H_z = C \cos \frac{\pi x}{a} e^{-j\beta z}$$

$$H_x = \frac{j\beta a}{\pi} C \sin \frac{\pi x}{a} e^{-j\beta z}$$

$$E_y = \frac{-j\omega\mu a}{\pi} C \sin \frac{\pi x}{a} e^{-j\beta z}$$

$$H_y = E_x = E_z = 0$$

(9-119)

This special case is of particular practical importance because it is possible to design a waveguide such that all the wave propagation is in this mode. This mode is called the *dominant mode*, designated as the TE_{10} mode. The corresponding field configuration is shown in Fig. 9-20.

As can be verified from the appropriate equations, the wave parameters of the dominant TE_{10} mode are

$$\beta = \sqrt{\omega^2\mu\epsilon - \left(\frac{\pi}{a}\right)^2} = \frac{2\pi}{\lambda}\sqrt{1 - \left(\frac{f_c}{f}\right)^2}$$

$$f_c = \frac{c}{2a}$$

$$\lambda_c = 2a \tag{9-120}$$

$$v_p = \frac{\omega}{\sqrt{\omega^2\mu\epsilon - (\pi/a)^2}} = \frac{c}{\sqrt{1 - (f_c/f)^2}}$$

$$\lambda_g = \frac{2\pi}{\sqrt{\omega^2\mu\epsilon - (\pi/a)^2}} = \frac{\lambda}{\sqrt{1 - (f_c/f)^2}}$$

Wave propagation is restricted to the TE_{10} mode by designing waveguides with width-to-height ratios a/b of about 2 and using an operating frequency which is above the TE_{10} cutoff frequency but below the first higher cutoff frequency.

Example 9-11 Design of a Rectangular Waveguide. Design a rectangular waveguide which, at 10 GHz, will operate in the TE_{10} mode with 25 percent safety factor ($f \geq 1.25 f_c$) when the interior of the guide is filled with air. It is required that the mode with the next higher cutoff will operate at 25 percent below its cutoff frequency.
For the TE_{10} mode, $f_c = c/2a$. Therefore the lower bound for the dimension a is

$$\frac{3 \times 10^{10}}{2 \times 10^{10}} = 1.5 \text{ cm}$$

The upper bound is found from the condition $f \geq 1.25 f_c$. Thus

$$a \leq \frac{1.25c}{2f} = \frac{1.25 \times 3 \times 10^{10}}{2 \times 10^{10}} = 1.875 \text{ cm}$$

so that a must be chosen such that $1.5 \leq a \leq 1.875$.
The wave with the next higher cutoff frequency is the TE_{01} mode. By analogy, $f_c' = c/2b$, and

$$f \leq 0.75 f_c' = 0.75 \frac{c}{2b}$$

from which we find that the choice of b must be such that

$$b \leq \frac{0.75c}{2f} = \frac{0.75 \times 3 \times 10^{10}}{2 \times 10^{10}} = 1.125 \text{ cm}$$

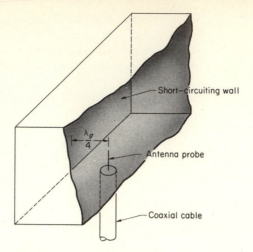

Short-circuiting wall

$\frac{\lambda_g}{4}$

Antenna probe

Coaxial cable

FIGURE 9-21. Excitation method for TE_{10} mode.

If we choose the upper bounds $a = 1.875$ cm, $b = 1.125$ cm, the cutoff frequency of the next higher mode, the TE_{11} mode, will be 15.52 GHz, well beyond the operating frequency.   Therefore wave propagation will be confined to the TE_{10} mode.

One method of exciting the TE_{10} mode is shown in Fig. 9-21.

9.13 Power Calculations in Waveguides. The time-average power transmitted by a waveguide can be expressed in terms of the complex Poynting vector.   Thus the time-average power per unit area transmitted in the positive z direction is given by

$$P_z = \tfrac{1}{2}\, \text{Re}\, (E_x H_y^* - E_y H_x^*) \tag{9-121}$$

It is a relatively easy task to show that this expression reduces to

$$P_z = \tfrac{1}{2} Z_0 (|H_x|^2 + |H_y|^2) \tag{9-122}$$

where the vertical bars indicate amplitudes of complex quantities, and Z_0 represents a wave impedance which is defined thus:

$$Z_{0_{TE}} = \frac{\omega \mu}{\beta} \quad \text{for TE modes}$$

$$\tag{9-123}$$

$$Z_{0_{TM}} = \frac{\beta}{\omega \epsilon} \quad \text{for TM modes}$$

Notice that

$$Z_{0_{TE}} Z_{0_{TM}} = \frac{\mu}{\epsilon} = (\text{characteristic impedance of medium})^2 \tag{9-124}$$

Now, the total time-average power propagated in the z direction is given by

$$P_T = \int_0^b \int_0^a \tfrac{1}{2} Z_0(|H_x|^2 + |H_y|^2)\, dx\, dy$$

The details of the integrations are lengthy, but the results are simple. We have, for TM modes,

$$P_T = \frac{ab}{8} \sqrt{\frac{\epsilon}{\mu}} \left(\frac{\lambda_c}{\lambda}\right)^2 C^2 \sqrt{1 - \left(\frac{\lambda}{\lambda_c}\right)^2} \tag{9-125}$$

and for TE modes,

$$P_T = \frac{ab}{8} \sqrt{\frac{\mu}{\epsilon}} \left(\frac{\lambda_c}{\lambda}\right)^2 C^2 \sqrt{1 - \left(\frac{\lambda}{\lambda_c}\right)^2} \qquad \text{except } \text{TE}_{10} \text{ and } \text{TE}_{01} \text{ modes} \tag{9-126}$$

$$P_T = \frac{ab}{4} \sqrt{\frac{\mu}{\epsilon}} \left(\frac{\lambda_c}{\lambda}\right)^2 C^2 \sqrt{1 - \left(\frac{\lambda}{\lambda_c}\right)^2} \qquad \text{TE}_{10} \text{ and } \text{TE}_{01} \text{ modes} \tag{9-127}$$

These equations show that P_T is independent of the z coordinate, as expected, since the waveguide was assumed to be lossless. However, even if the guide is completely evacuated, the finite, though very high, conductivity of the walls will have some effect on the power transmitted, causing it to decrease with increasing distance from the source. The reduction in the total power transmitted is due to the heat generated by currents flowing in the waveguide walls. As shown in Fig. 9-22, the wall current flow is nonuniform, but the configuration repeats itself every full wavelength. The actual power loss per unit area in the walls can be computed by multiplying the current density squared by the *surface resistance*† R_s of the walls. The power lost per unit length of guide, P_L, can then be obtained by integrating

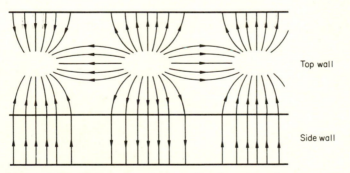

FIGURE 9-22.   Current flow in the walls of a rectangular waveguide operating in the TE_{10} mode.

† This is a quantity defined as $R_s = 1/\sigma\delta$, where δ is the depth of penetration, discussed in Sec. 8.7. Surface resistance is measured in ohms per square.

this power-loss density over the wall surface of a unit length of guide. For a guide operating in the TE_{10} mode, this gives

$$P_L = R_s C^2 \left(b + \frac{a}{2} + \frac{a}{2} \frac{\beta^2 a^2}{\pi^2} \right) \tag{9-128}$$

A convenient way of accounting for wall losses is to attach the exponential factor $e^{-\alpha z}$ to all field expressions. The so-called *attenuation constant* α can then be determined by noting, first, that the z variation of power transmitted will be of the form $e^{-2\alpha z}$. Since the power loss per unit length of guide is equal to the space rate of decrease of power flow, the relation of P_L to P_T is, simply,

$$P_L = -\frac{\partial P_T}{\partial z} = -(-2\alpha)P_T = 2\alpha P_T \tag{9-129}$$

from which it is clear that

$$\alpha = \frac{P_L}{2P_T} \tag{9-130}$$

For the TE_{10} mode, this definition gives

$$\alpha = \frac{R_s(1 + 2bf_c^2/af^2)}{b\sqrt{\mu/\epsilon}\,\sqrt{1 - (f_c/f)^2}} \tag{9-131}$$

and is seen to be a frequency-dependent quantity.

Example 9-12   Power Transmission. At 9 GHz the inside dimensions of the standard waveguide used in practice are 0.9 by 0.4 in. Find the maximum power that can be transmitted in the TE_{10} mode, assuming that the air dielectric will break down when the electric field intensity exceeds 3×10^6 V/m.

From Eq. (9-119), the maximum value of the transverse electric field intensity occurs at $x = a/2$ and is equal to

$$\frac{\omega\mu aC}{\pi} = \frac{2\pi \times 9 \times 10^9 \times 4\pi \times 10^{-7} \times 0.9 \times 0.0254 \times C}{\pi} \quad \text{V/m}$$

This quantity is not to exceed 3×10^6 V/m. Therefore the maximum value of the constant C is 5.79×10^3 A/m.

On the other hand,

$$\lambda_c = 2a = 2 \times 0.9 \times 0.0254 = 0.0457 \quad \text{m}$$

$$\lambda = \frac{c}{f} = \frac{3 \times 10^8}{9 \times 10^9} = 0.0333 \quad \text{m}$$

$$\sqrt{\frac{\mu}{\epsilon}} = 377 \quad \Omega$$

Substituting these quantities in Eq. (9-127) gives

$$P_T = \frac{ab}{4}\sqrt{\frac{\mu}{\epsilon}}\left(\frac{\lambda_c}{\lambda}\right)^2 C^2 \sqrt{1 - \left(\frac{\lambda}{\lambda_c}\right)^2}$$

$$= \frac{0.9 \times 0.4 \times (0.0254)^2}{4} \; (377) \left(\frac{0.0457}{0.0333}\right)^2 (5.79 \times 10^3)^2 \sqrt{1 - \left(\frac{0.0333}{0.0457}\right)^2}$$

$$= 0.944 \times 10^6 \quad \text{W}$$

as the maximum amount of power that can be transmitted without causing breakdown of the air dielectric.

9.14 Cylindrical Waveguides. The analysis of cylindrical waveguides follows exactly the same route. The results are summarized below.

TM modes In the usual cylindrical coordinate system notation, the expressions for the field components in this mode are

$$E_r = \frac{\beta}{\omega\epsilon} H_\varphi$$

$$E_\varphi = -\frac{\beta}{\omega\epsilon} H_r$$

$$E_z = CJ_n(rh) \cos n\varphi \, e^{-j\beta z} \qquad\qquad (9\text{-}132)$$

$$H_r = -\frac{j\omega\epsilon n}{h^2 r} CJ_n(rh) \sin n\varphi \, e^{-j\beta z}$$

$$H_\varphi = -\frac{j\omega\epsilon}{h} CJ_n'(rh) \cos n\varphi \, e^{-j\beta z}$$

$$H_z = 0$$

where the factor $e^{j\omega t}$ is understood and

where C = a constant proportional to the strength of the field

J_n = Bessel function of first kind and order n

J_n' = derivative of J_n with respect to its argument (rh)

$h^2 = \omega^2\mu\epsilon - \beta^2$, and necessarily less than $\omega^2\mu\epsilon$ for transmission to occur

The magnitude of h is determined from the boundary condition

$$J_n(ha) = 0 \qquad\qquad (9\text{-}133)$$

which makes E_z identically zero at $r = a$. To avoid the use of relatively high frequencies, h must be small. Therefore only the first few roots of

Eq. (9-133) are of practical interest.   They are

$n = 0$	$n = 1$	$n = 2$
$(ha)_{0,1} = 2.405$	$(ha)_{1,1} = 3.832$	$(ha)_{2,1} = 5.136$
$(ha)_{0,2} = 5.520$	$(ha)_{1,2} = 7.016$	$(ha)_{2,2} = 8.417$

The first subscript refers to the value of n, and with reference to Eqs. (9-132), it specifies the number of full-cycle variations of the field components in the peripheral direction.   The second subscript refers to the roots in ascending order of magnitude, and specifies the number of half-cycle variations (Fig. 3-13) of the field components in the radial direction.

Corresponding to the ordered values of h, that is, $h_{nm} = (ha)_{n,m}/a$, the various TM waves are referred to as TM_{01}, TM_{12}, and in general TM_{nm}.   It will be noted that

$$f_c = \frac{h_{nm}}{2\pi\sqrt{\mu\epsilon}}$$

$$\beta = \sqrt{\omega^2\mu\epsilon - h_{nm}{}^2} = \omega\sqrt{\mu\epsilon}\sqrt{1 - \left(\frac{f_c}{f}\right)^2}$$

$$v_p = \frac{\omega}{\beta} = \frac{\omega}{\sqrt{\omega^2\mu\epsilon - h_{nm}{}^2}} = \frac{c}{\sqrt{1 - (f_c/f)^2}} \qquad (9\text{-}134)$$

$$\lambda_g = \frac{v_p}{f} = \frac{2\pi}{\sqrt{\omega^2\mu\epsilon - h_{nm}{}^2}} = \frac{\lambda}{\sqrt{1 - (f_c/f)^2}}$$

TE modes   Expressions for the field components are

$$E_r = \frac{\omega\mu}{\beta} H_\varphi$$

$$E_\varphi = -\frac{\omega\mu}{\beta} H_r$$

$$E_z = 0$$

$$H_r = -\frac{j\beta}{h} CJ'_n(hr) \cos n\varphi \, e^{-j\beta z} \qquad (9\text{-}135)$$

$$H_\varphi = \frac{jn\beta}{h^2 r} CJ_n(hr) \sin n\varphi \, e^{-j\beta z}$$

$$H_z = CJ_n(hr) \cos n\varphi \, e^{-j\beta z}$$

The requirement at the boundary $r = 1$ is given by

$$J'_n(ha) = 0 \qquad (9\text{-}136)$$

the first few roots of which are

$n = 0$	$n = 1$	$n = 2$
$(ha)'_{0,1} = 3.832$	$(ha)'_{1,1} = 1.841$	$(ha)'_{2,1} = 3.054$
$(ha)'_{0,2} = 7.016$	$(ha)'_{1,2} = 5.331$	$(ha)'_{2,2} = 6.706$

The corresponding TE waves are referred to as TE_{01}, TE_{11}, and in general TE_{nm}. The equations for the wave parameters f_c, β, v_p, and λ_g are the same as for TM modes, and are expressed by Eqs. (9-134).

Figure 9-23 shows a field plot for the TE_{11} mode, in which the cutoff frequency

$$f_c = \frac{c}{2\pi} \frac{(ha)'_{1,1}}{a} = \frac{0.293}{a\sqrt{\mu\epsilon}} \tag{9-137}$$

is the lowest. Thus the TE_{11} mode is the dominant mode in cylindrical waveguides.

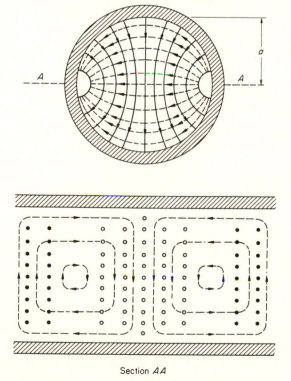

Section AA

FIGURE 9-23. Electric (solid) and magnetic (dashed) field configuration in a cylindrical waveguide operating in the dominant TE_{11} mode.

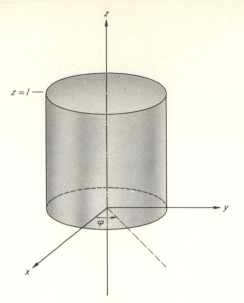

FIGURE 9-24.   A cylindrical cavity resonator.

A practical application of cylindrical waveguides is in the construction of *resonant cavities*, suitable for use, for example, in direct-reading frequency meters.   One of the simplest forms of a resonant cavity is a finite section of cylindrical waveguide terminated at both ends by flat conducting plates (Fig. 9-24).   Waves traveling in both the positive z direction and negative z direction can exist simultaneously in such a structure, thus giving rise to a standing wave.   The requirement that $H_z = 0$ at $z = 0$ leads to

$$H_z = -j2CJ_n(hr) \cos n\varphi \sin \beta z \tag{9-138}$$

for all TE$_{nm}$ modes.   The requirement $H_z = 0$ at $z = l$ leads to

$$\sin \beta l = 0$$

from which

$$\beta = \frac{k\pi}{l} \qquad k = 1, 2, 3, \ldots \tag{9-139}$$

The symbol k stands for the number of half-cycle variations in the z direction. Since $\beta^2 = \omega^2 \mu \epsilon - h_{nm}{}^2$, it follows that the frequency at which a cavity operating in the TE$_{nmk}$ mode will resonate is

$$f_r = \frac{1}{2\pi\sqrt{\mu\epsilon}} \sqrt{\left(\frac{k\pi}{l}\right)^2 + h_{nm}{}^2} \tag{9-140}$$

where
$$h_{nm} = \frac{(ha)'_{n,m}}{a}$$

Practical frequency meters operate in the TE_{111} mode.

We could obtain many more interesting and practical deductions and extensions of our introduction to waveguides. However, a more complete discussion would be too lengthy to include in this text. Accordingly, we conclude the discussion with a suggestion that the student consult the literature.†

9.15 Summary.   This chapter began with a treatment of the TEM mode on a general transmission line from an electromagnetic field point of view. By proper manipulation and interpretation of the general results, it was shown that they could be cast in the more conventional form, which expresses transmission line behavior in terms of a line voltage, a line current, and a line characteristic impedance, where line voltage and line current were seen to be field integrals, and the characteristic impedance was seen to express the line geometry. The results were then applied to the particular case of a coaxial transmission line.

Following this development, basic classical transmission line analysis using Smith chart calculations was presented for both lossless and lossy lines, and then a brief presentation of transients on lossless lines was given.

The chapter concluded with an introduction to waveguide theory, with considerable emphasis placed on the dominant modes.

Problems

9-1 **Characteristic Impedance.** Calculate the characteristic impedance of a lossless transmission line, which, when terminated in a resistive load of 100 Ω, presents an impedance of $40 - j30\ \Omega$ at a point $\lambda/8$ away from the load.

9-2 **Characteristic Quantities of the Standing Wave Distribution.** On a certain lossless transmission line with a characteristic impedance of 50 Ω, the measured VSWR is 3. The first voltage minimum occurs at $\lambda/8$ away from the load. Find the terminating impedance.

9-3 **Characteristic Quantities of the Standing Wave Distribution.** Given that

$$V^+(z_1) = 30\underline{/30°}\quad \text{V}$$

$$I^-(z_1) = -1\underline{/-30°}\quad \text{A}$$

Determine the impedance and the instantaneous current at the input terminals.

† R. E. Collin, "Foundations for Microwave Engineering," McGraw-Hill Book Company, New York, 1966.

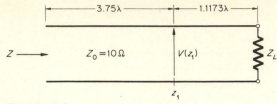

PROBLEM 9-3

9-4 **Standing Wave Ratio.** A lossless transmission line having $Z_0 = 100\ \Omega$ is terminated with an impedance of $Z_L = 100 + j50\ \Omega$. What will the voltage standing wave ratio be?

9-5 **Smith Chart Calculations.** The VSWR on a lossless line is 5. At a certain point on the line, within $\lambda/4$ m from the load, the impedance has an angle $45°$ and has a normalized value greater than 1. The load has a normalized magnitude of 1. How far is the point from the load, and what is the actual value of the load, if $Z_0 = 100\ \Omega$?

9-6 **Impedance.** Starting with Eqs. (9-40) and (9-41), prove that the input impedance of an open-circuited, lossless transmission line of length L is

$$Z_{oc} = -jZ_0 \cot \beta L$$

Also prove that the impedance of a short-circuited lossless transmission line of length L is

$$Z_{sc} = jZ_0 \tan \beta L$$

9-7 **Single-stub Matching.** A line with a normalized load admittance of $0.55 + j0.27$ is to be matched, using a single short-circuited stub. What is the minimum distance of the stub from the load and the required stub length?

9-8 **Impedance Determination and Matching.** A lossless transmission line of $Z_0 = 50\ \Omega$ is terminated in an unknown impedance Z_L. The standing wave ratio on the line is 3. Successive voltage minima are 20 cm apart, and the first minimum is 5 cm from the load.

(a) What is the terminating impedance?

(b) Find the location and length of the short-circuited stub required to match the line.

9-9 **Double-stub Matching.** Before the matching section is inserted, the VSWR is 4. Choose the shortest distance d such that there is only one solution for l_1 and l_2 (*not two*).

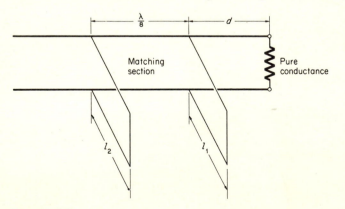

PROBLEM 9-9

9-10 Double-stub Tuning. A lossless transmission line of $Z_0 = 200$ Ω has an unknown load impedance Z_L and a standing wave ratio of 5. The first voltage minimum is 4 cm from the load, and the minima are 20 cm apart. We wish to match the line by placing a short-circuited stub in parallel with the load and a second stub 10 cm from the load.

Find the required stub lengths.

9-11 Series-stub Matching. A short-circuited series stub is to be used for matching. What value of d will yield the smallest length l?

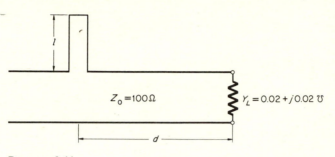

PROBLEM 9-11

9-12 Matching. Determine the capacitive susceptance of C needed to effect a match.

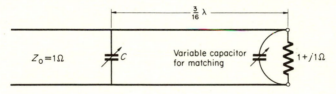

PROBLEM 9-12

9-13 Matching. What values of inductive and capacitive reactance (in ohms) must be used to obtain a match?

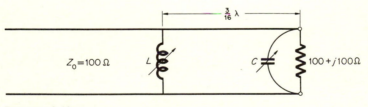

PROBLEM 9-13

9-14 Lossy Line. On a transmission line operating at $\omega = 1.5 \times 10^7$ rad/s, the following constants are observed:

$$\alpha = 0.866 \times 10^{-1} \text{ Np/m}$$
$$\beta = 0.5 \times 10^{-1} \text{ rad/m}$$
$$Z_0 = 100 \ \Omega$$

What are R, G, L, C for the line?

9-15 Lossy Line. A lossy transmission line has constants per meter of $R = 10^{-2}\ \Omega$, $G = 10^{-6}\ \mho$, $L = 10^{-6}\ H$, $C = 10^{-9}\ F$. At a frequency of 1590 Hz, find

    (a) The characteristic impedance of the line

    (b) The velocity of wave propagation

    (c) The percentage to which the absolute magnitude of the voltage of a traveling wave decreases in 1 km

9-16 Lossy Line. At point A the instantaneous voltage is

$$V_A = 100 \sin\left(\omega t + \frac{\pi}{6}\right) + 5 \sin\left(\omega t - \frac{\pi}{6}\right)$$

At point B the instantaneous voltage is

$$V_B = 100 e^{-1} \sin\left(\omega t + \frac{\pi}{6} - \frac{1}{\sqrt{3}}\right) + \cdots \sin(\cdots)$$

    (a) Calculate the values of α and β.

    (b) Fill in the $\cdots$ blanks.

    (c) What is the voltage reflection coefficient Γ_A?

    (d) What is the instantaneous current at A?

    (e) What is $R + jX$ and $G + jB$ for the line?

    (f) If $\Gamma_0 = 0.5\underline{/30°}$, what is Z_L?

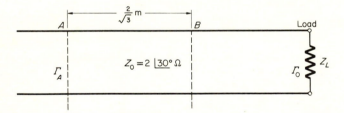

PROBLEM 9-16

9-17 Transients. The open-circuit voltage at the generator terminals is

$$V_g = 2u_{-1}(t) + 2u_{-1}(t - 10^{-4}) - 4u_{-1}(t - 2 \times 10^{-4})$$

Sketch and label, where appropriate, the voltage on the line at $t = 1.25T$, where $T = 10^{-4}\ s$ is the time for signal to travel the length of the line. The line is l m long.

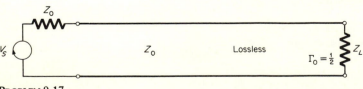

PROBLEM 9-17

9-18 General. A load on a transmission line, with a characteristic impedance of 1 Ω, consists of a variable L and R series combination such that $|Z_L| = 5$.

(a) As one proceeds toward the generator, which does he encounter first, a voltage maximum or a voltage minimum? Show why.

(b) For all possible variations of the load, what are the least and greatest distances to the first voltage minimum?

(c) A single short-circuited stub is to be used for matching. What value of load will require the shortest distance to the stub? What will be the length of the stub?

9-19 Transmission Line Theory and the Laplace Transform. An initially quiescent transmission line d m long is energized by an arbitrary voltage $f(t)$. Expressed in the complex frequency variable s, the arbitrary load, source, and characteristic impedances are $Z_L(s)$, $Z_S(s)$, and $Z_0(s)$, respectively.

Develop a general expression for $V(z,s)$, that is, $\mathscr{L}V(z,t)$, which applies for any source voltage. [*Hint:* Develop the first three terms of $V(z,s)$ by the following procedure. Assume that the line is lossless, that it is terminated in pure resistances at both input and output, and that the source voltage is a unit step, that is, $f(t) = u_{-1}(t)$. Obtain, in sequence, $V(z,t)$ and $V(z,s)$ for this case. Then "upgrade" $V(z,s)$ to get its form for the general case.]

9-20 Standing Wave. A lossless transmission line whose characteristic impedance is Z_0 Ω is terminated in a mismatched load. When the load is replaced by a short circuit, it is found that no shift occurs in the position of a certain minimum in the standing wave pattern. Show that:

(a) The load reflection coefficient is a purely real number.

(b) The load is a resistance of Z_0/VSWR Ω.

9-21 Power. Prove that the instantaneous power transmitted by a lossless line to a purely resistive load is given by

$$P(z,t) = \frac{1}{2}\frac{V^{+2}}{Z_0}\left[(1 - \Gamma_0{}^2) + \cos 2(\omega t - \beta z) - \Gamma_0{}^2 \cos 2(\omega t + \beta z)\right]$$

Note that the time-average power is equal to $\frac{1}{2}(V^{+2}/Z_0)(1 - \Gamma_0{}^2)$, as expected.

9-22 Impedance. Let $d_{\min}$ be the distance from the load at $s = 0$ to the first voltage minimum. (Note that $d_{\min}$ is a positive quantity.) If S is the VSWR, prove that

$$Z_L = Z_0 \frac{1 - jS \tan \beta d_{\min}}{S - j \tan \beta d_{\min}}$$

9-23 Transients. For the open-circuited line shown, plot $V(0,t)$ and $I(0,t)$ and include their asymptotic values.

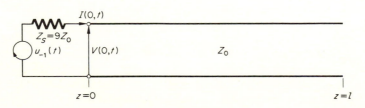

PROBLEM 9-23

9-24 Transients. Solve the problem of Example 9-10 when the line is terminated in an inductance L.

9-25 Rectangular Waveguide Modes. The inside dimensions of a waveguide, widely used at frequencies around 10 GHz (X-band), are 0.4 by 0.9 in. Find the cutoff frequency of the dominant mode and the maximum frequency that can be used to the exclusion of all other modes.

9-26 Field Plots. For the TM_{11} mode in a rectangular guide draw sketches of (a) E_x versus x, (b) E_y versus z, (c) E_z versus y, for the same instant of time.

9-27 Rectangular Waveguide Modes. Is it possible for the waveguide structure (0.4 by 0.9 in.) shown to allow only dominant-mode propagation? If so, over what frequency range?

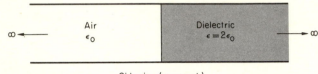

Air
ϵ_0

Dielectric
$\epsilon = 2\epsilon_0$

Side view ($x =$ const.)

PROBLEM 9-27

9-28 Waveguide Impedance. A section of 0.4- by 0.9-in. rectangular waveguide, 0.2 m long, is operated at 10 GHz. The VSWR is 6, and the distance to the first minimum is $0.10\lambda_g$. What will be the normalized input impedance of this waveguide?

9-29 Delay Line. What length of 0.4- by 0.9-in. waveguide is required to introduce a signal delay of 2 μs at 10 GHz?

9-30 Power in Waveguides. A rectangular waveguide is designed to have a ratio $\lambda/\lambda_c = 0.8$ at a frequency of 5000 MHz in the TE_{10} mode. The guide has a height-to-width ratio of 0.5. The time-average power flow is 1 kW. Compute the maximum values of electric and magnetic intensities in the guide, and indicate where these occur.

9-31 Standing Waves in Waveguides. One-half of a very long section of X-band waveguide (0.4 by 0.9 in.) is filled with a dielectric whose relative permittivity is 1.5. The guide is designed to operate in the dominant mode at a frequency of 9350 MHz. Draw a sketch showing the variation of the E field with z in the plane $x = a/2$. Label your sketch with all pertinent information. Assume that the wave is launched in the dielectric and that the load is perfectly matched to the air-filled guide.

9-32 Attenuation. The power level at a certain point in a rectangular waveguide operating in the TE_{10} mode is P_0. What is the power level at a second point whose distance is 1.5 m from the first toward the load? The attenuation constant α is 0.0328 dB/m. [*Hint:* To convert decibels (dB) to nepers multiply by 0.1151.]

9-33 Cutoff Frequencies in Cylindrical Waveguides. An air-filled cylindrical waveguide has a dominant-mode cutoff frequency of 9 GHz. What is its inside diameter? Determine the cutoff frequencies for the next three lowest-order modes.

9-34 Power Loss in Cylindrical Waveguides. Derive an expression for the instantaneous current density in the walls of an infinitely long circular waveguide excited in the TM_{01} mode. Next, evaluate the power loss due to the finite conductivity of the walls per unit guide length in terms of the surface resistance R_s.

9-35 Rectangular Resonator. Show that the resonant frequency of a rectangular resonator is given by

$$f_r = \frac{1}{2\sqrt{\mu\epsilon}} \sqrt{\left(\frac{m}{a}\right)^2 + \left(\frac{n}{b}\right)^2 + \left(\frac{k}{l}\right)^2}$$

where l is the length of the resonator, and $k = 1, 2, 3, \ldots$.

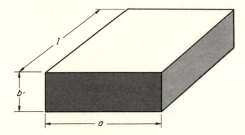

PROBLEM 9-35

9-36 Cylindrical Resonator Design. It is desired to build a cylindrical cavity resonator at 9000 MHz to operate in the TE_{011} mode. If the frequency of operation is to be 10 percent above cutoff, determine the cavity dimensions.

9-37 Frequency Calibration of a Cavity Resonator. The diameter of a cylindrical resonator is 4 cm. Draw a calibration curve showing the variation of resonant frequency with cavity length in the TE_{111} mode. The dielectric is air.

$$N = \frac{m}{3}$$

CHAPTER 10

RADIATION AND ANTENNAS

10.1 Introduction. Any system of conductors and material media which is connected to a power source so as to produce a time-varying electromagnetic field in an external region will radiate energy. When the system is arranged so as to optimize or accentuate the radiation of energy from some portion of the system while at the same time minimizing or suppressing radiation from the rest of the system, that portion of the system which radiates energy is called an *antenna*.

Thus antenna theory tacitly assumes that the antenna is connected to a nonradiating power source by means of a nonradiating transmission line. This idealization can usually be achieved in practice, and although, in some practical antenna problems, achieving this idealization may be the most difficult part of the problem, in this chapter we presume that it has been solved, and we concern ourselves only with the antenna.

10.2 The Radiation Problem. In Chap. 2 we showed that we could cast Maxwell's equations into a form involving a scalar wave equation and a vector wave equation plus some subsidiary equations. In particular, we were able to show that the set of equations

$$\nabla^2 \phi - \mu\epsilon \frac{\partial^2 \phi}{\partial t^2} = -\frac{\rho}{\epsilon} \tag{10-1}$$

$$\nabla^2 \mathbf{A} - \mu\epsilon \frac{\partial^2 \mathbf{A}}{\partial t^2} = -\mu\mathbf{J} \tag{10-2}$$

where
$$\mathbf{E} = -\nabla\phi - \frac{\partial \mathbf{A}}{\partial t} \qquad \mathbf{B} = \nabla \times \mathbf{A} \tag{10-3}$$

with $\mathbf{A}$ and ϕ connected by the Lorentz condition

$$\nabla \cdot \mathbf{A} = -\mu\epsilon \frac{\partial \phi}{\partial t} \tag{10-4}$$

was an alternative statement of Maxwell's equations. This formulation is particularly useful for radiation problems in that it directly relates the scalar

458

and vector potentials to the sources of the fields. The scalar potential ϕ is not really necessary in antenna problems since $\mathbf{B}$ can be obtained from $\mathbf{A}$, and then Maxwell's equation

$$\nabla \times \mathbf{H} = \epsilon \frac{\partial \mathbf{E}}{\partial t} \tag{10-5}$$

can be integrated with respect to time to give

$$\mathbf{E} = \frac{1}{\epsilon} \int \nabla \times \mathbf{H} \, dt \tag{10-6}$$

Thus it is evident that what we need is the solution to Eq. (10-2).

Although a rigorous solution† of this equation is possible, the details are involved, and hence we present only arguments which make the result seem logical. With this in mind, we note that, in rectangular coordinates, Eq. (10-2) can be expressed as three scalar equations in the three components of $\mathbf{A}$, and that each of these scalar equations is of the same mathematical form as Eq. (10-1). In source-free regions, $\rho = 0$, and Eq. (10-1) is the scalar wave equation in ϕ whose general solution is a completely arbitrary, analytic function of the arguments $t - r/v$ and $t + r/v$, where r denotes distance measured along the direction of propagation and $v = 1/\sqrt{\mu\epsilon}$. Also, for time-independent source distributions, Eq. (10-1) is just Poisson's equation, whose solution is

$$\phi = \frac{1}{4\pi\epsilon} \int_V \frac{\rho}{r} \, dv \tag{10-7}$$

We should expect the nonhomogeneous time-dependent case to incorporate the features of both types of solutions, since these are just special cases of the general solution. More precisely, it seems intuitively reasonable that the correct solution would be obtained by simply substituting $t - r/v$ for t in the integral of Eq. (10-7).

$$\phi = \frac{1}{4\pi\epsilon} \int_V \frac{\rho(t - r/v)}{r} \, dv \tag{10-8}$$

That this expression is indeed a valid solution of the time-dependent non-homogeneous wave equation can be shown by direct substitution. However, the computations are rather involved, and are not presented at this time.

Notice that the time t in Eq. (10-8) is the time at the point of observation. On the other hand, $t' = t - r/v$ is the time at the source point. Thus the equation says that sources which had the configuration ρ at $t' = t - r/v$

† J. A. Stratton, "Electromagnetic Theory," chap. 8, McGraw-Hill Book Company, New York, 1941.

produce a potential ϕ at a time t which is later† than the time t' by an amount that takes into account the finite velocity of propagation of waves in the medium. Because of this time-delay aspect of the solution, the potential ϕ is known as the *retarded potential*, and the phenomenon itself, *retardation*.

In antenna problems it is convenient to eliminate the scalar potential ϕ and to cast the entire problem in terms of the vector potential $\mathbf{A}$. In rectangular coordinates the time-dependent nonhomogeneous vector wave equation in $\mathbf{A}$ can be written as three simultaneous scalar wave equations, namely,

$$\nabla^2 A_x - \mu\epsilon \frac{\partial^2 A_x}{\partial t^2} = -\mu J_x \tag{10-9}$$

$$\nabla^2 A_y - \mu\epsilon \frac{\partial^2 A_y}{\partial t^2} = -\mu J_y \tag{10-10}$$

$$\nabla^2 A_z - \mu\epsilon \frac{\partial^2 A_z}{\partial t^2} = -\mu J_z \tag{10-11}$$

Since these are mathematically the same equations as the equation in ϕ, we can write down their solutions, by inspection, as

$$A_x = \frac{\mu}{4\pi} \int_V \frac{J_x(t - r/v)}{r} \, dv \tag{10-12}$$

$$A_y = \frac{\mu}{4\pi} \int_V \frac{J_y(t - r/v)}{r} \, dv \tag{10-13}$$

$$A_z = \frac{\mu}{4\pi} \int_V \frac{J_z(t - r/v)}{r} \, dv \tag{10-14}$$

or more compactly, in vector notation,

$$\mathbf{A} = \frac{\mu}{4\pi} \int_V \frac{\mathbf{J}(t - r/v)}{r} \, dv \tag{10-15}$$

The problem of calculation of the field of an antenna of known current distribution thus reduces essentially to the evaluation of Eq. (10-15).

10.3   The Field of a Current Element (Hertzian Dipole).

A large class of antennas consists of conducting wires arranged so as to produce desired radiation properties. In most cases the cross-sectional size of the wires can be neglected, and the wires can be treated as perfectly conducting filamentary conductors. With this idealization, Eq. (10-15) can be written

$$\mathbf{A} = \frac{\mu}{4\pi} \int_C \frac{I(t - r/v)}{r} \, d\mathbf{l} \tag{10-16}$$

---

† The alternative argument, $t + r/v$, represents advanced time, implying that the phenomenon represented by the quantity ϕ can be observed before it has been generated by the sources. This is physically inconceivable, and that part of the solution which depends on $t + r/v$ is henceforth discarded.

where $I(t - r/v)$ is the current carried by the wire along the contour C, and $d\mathbf{l}$ is a vector element of length in the direction of the wire.   An *isolated* infinitesimal section of the wire is known as a *current element*, or *Hertzian dipole*.   Although, obviously, a current element cannot be isolated from the rest of the antenna, it is still very useful to calculate the fields which an isolated current element would produce.   The fields of an actual antenna can be calculated from the fields of a current element by integration.

In this section we propose to calculate the field of the current element $I\,d\mathbf{l}$.

From Eq. (10-16) we have that the vector potential $\mathbf{A}$ is

$$\mathbf{A} = \frac{\mu}{4\pi r} I\left(t - \frac{r}{v}\right) d\mathbf{l} \qquad (10\text{-}17)$$

or if I is a sinusoidal current,

$$\mathbf{A} = \frac{\mu}{4\pi r} I \cos \omega\left(t - \frac{r}{v}\right) d\mathbf{l} \qquad (10\text{-}18)$$

From this expression it is apparent that the *phase delay*, corresponding to a time delay of r/v, is $\omega r/v$ rad.   It will be convenient to use the spherical geometry of Fig. 10-1, where $d\mathbf{l}$ is in the z direction, and for notational

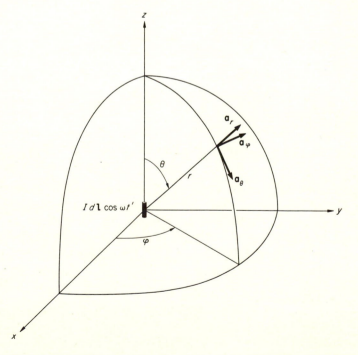

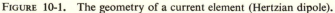

FIGURE 10-1.   The geometry of a current element (Hertzian dipole).

convenience we shall write $t - r/v$ as t' in the final results.   In this notation, **A** has only a z component, which is given by

$$A_z = \frac{\mu I \, dl \cos \omega t'}{4\pi r} \tag{10-19}$$

However, the **E** and **H** fields are more useful when expressed in spherical coordinates.   Using Eqs. (1-13), we can write

$$A_r = A_z \cos \theta \qquad A_\theta = -A_z \sin \theta \qquad A_\varphi = 0 \tag{10-20}$$

and using $\mathbf{B} = \mathbf{\nabla} \times \mathbf{A}$, we find

$$B_r = (\mathbf{\nabla} \times \mathbf{A})_r = 0 \tag{10-21}$$

$$B_\theta = (\mathbf{\nabla} \times \mathbf{A})_\theta = 0 \tag{10-22}$$

$$
\begin{aligned}
B_\varphi &= \frac{1}{r} \left[\frac{\partial}{\partial r}(r A_\theta) - \frac{\partial A_r}{\partial \theta} \right] \\
&= \frac{\mu I \, dl}{4\pi r} \left\{ \frac{\partial}{\partial r}\left[-\sin \theta \cos \omega\left(t - \frac{r}{v} \right) \right] - \frac{\partial}{\partial \theta}\left[\frac{\cos \theta}{r} \cos \omega\left(t - \frac{r}{v} \right) \right] \right\} \\
&= \frac{\mu I \, dl \sin \theta}{4\pi} \left[\frac{-\omega \sin \omega(t - r/v)}{rv} + \frac{\cos \omega(t - r/v)}{r^2} \right]
\end{aligned}
\tag{10-23}
$$

To find the **E** field, we use Eq. (10-6), and obtain

$$E_\theta = \frac{I \, dl \sin \theta}{4\pi \epsilon} \left(\frac{-\omega \sin \omega t'}{rv^2} + \frac{\cos \omega t'}{r^2 v} + \frac{\sin \omega t'}{\omega r^3} \right) \tag{10-24}$$

$$E_r = \frac{2I \, dl \cos \theta}{4\pi \epsilon} \left(\frac{\cos \omega t'}{r^2 v} + \frac{\sin \omega t'}{\omega r^3} \right) \tag{10-25}$$

Lastly, dividing Eq. (10-23) by μ, we have

$$H_\varphi = \frac{I \, dl \sin \theta}{4\pi} \left(\frac{-\omega \sin \omega t'}{rv} + \frac{\cos \omega t'}{r^2} \right) \tag{10-26}$$

We see from these equations† that, even for a simple current element, the exact total field is complicated.   Fortunately, we seldom need to consider the exact total field.   There are two reasons for this.   First, we notice that the terms involve inverse r, r^2, and r^3 terms.   For large distances from the current element we can neglect the higher-order terms.   For instance, in E_θ and H_φ the inverse r and inverse r^2 terms are equal in magnitude when

$$\frac{\omega}{v} = \frac{1}{r}$$

† The corresponding forms in the frequency domain are given in Prob. 10-3.

that is, for
$$r = \frac{v}{\omega} = \frac{\lambda}{2\pi} \approx \frac{\lambda}{6} \qquad (10\text{-}27)$$

The second reason is more fundamental. Since the antenna's primary function is to radiate energy, it will be possible most of the time to ignore those terms which do not contribute to energy radiation. The next section will show that only the inverse r terms contribute to the time-average radiated power.

Accordingly, it is customary in practice to call the field represented by the $1/r$ terms the *radiation field* and, in so doing, to distinguish it from the *induction field*, which is represented by the $1/r^2$ terms and which predominates at small distances r. Note that, aside from a time dependence, the induction field in Eq. (10-26) is predictable from the Biot-Savart law. Note also that the $1/r^3$ term in Eq. (10-24) is just the electric field intensity of an electric dipole if the time dependence were to be suppressed. Accordingly, the $1/r^3$ term is sometimes called the *electrostatic field* term.

10.4 Power Radiated by a Current Element.

In order to calculate the power radiated by a current element, we need to calculate Poynting's vector. The instantaneous Poynting's vector is given by $\mathscr{P} = \mathscr{E} \times \mathscr{H}$. For the fields given by Eqs. (10-24) to (10-26), $\mathscr{P}$ has a θ component and an r component, namely,

$$P_\theta = -E_r H_\varphi \qquad P_r = E_\theta H_\varphi \qquad (10\text{-}28)$$

It is obvious that the radial component is the only component which contributes to the net outward power flow. Thus

$$P_r = \frac{I^2 \, dl^2 \sin^2 \theta}{16\pi^2 \epsilon} \left(\frac{\omega^2 \sin^2 \omega t'}{r^2 v^3} - \frac{\omega \sin \omega t' \cos \omega t'}{r^3 v^2} \right.$$
$$\left. - \frac{\sin^2 \omega t'}{r^4 v} - \frac{\omega \sin \omega t' \cos \omega t'}{r^3 v^2} + \frac{\cos^2 \omega t'}{r^4 v} + \frac{\sin \omega t' \cos \omega t'}{\omega r^5} \right) \qquad (10\text{-}29)$$

or after application of some trigonometric identities,

$$P_r = \frac{I^2 \, dl^2 \sin^2 \theta}{16\pi^2 \epsilon} \left[\frac{\sin 2\omega t'}{2\omega r^5} + \frac{\cos 2\omega t'}{r^2 v^2} \right.$$
$$\left. - \frac{\omega \sin 2\omega t'}{r^3 v^2} + \frac{\omega^2 (1 - \cos 2\omega t')}{2r^2 v^3} \right] \qquad (10\text{-}30)$$

Noting that the time average of both $\sin 2\omega t'$ and $\cos 2\omega t'$ is zero, we can write the time average of P_r as

$$P_{r(\text{av})} = \frac{\omega^2 I^2 \, dl^2 \sin^2 \theta}{32\pi^2 \epsilon r^2 v^3} \qquad (10\text{-}31)$$

The important feature of this result, for our present use, is that the time-average value of the radial component of Poynting's vector is one-half times the product of the inverse r terms in $\mathbf{E}$ and $\mathbf{H}$. Hence the *far field* of our isolated current element, which is specified by

$$\mathbf{E} = -\frac{\omega I\, dl \sin\theta}{4\pi\epsilon rv^2} \sin\omega t'\, \mathbf{a}_\theta$$

$$\mathbf{H} = -\frac{\omega I\, dl \sin\theta}{4\pi rv} \sin\omega t'\, \mathbf{a}_\varphi$$

(10-32)

is all that is needed for calculation of radiated power and is also a valid approximation to the total field for large distances. This is true for the far field of any antenna. For this reason the far field is frequently called the *radiation field*.

Before we calculate the total radiated power from our current element, let us examine the radiation field further. First, we note that the $\mathbf{E}$ and the $\mathbf{H}$ fields are in time phase and normal to each other. Second, we note that

$$|\mathbf{E}| = \eta\, |\mathbf{H}| \tag{10-33}$$

where $\eta = \sqrt{\mu/\epsilon}$. Thus, except for a $(\sin\theta)/r$ term in both $\mathbf{E}$ and in $\mathbf{H}$, the radiation field has the properties of a uniform plane wave. For spherical surfaces of large r and for regions on the surface which are small enough so that $\sin\theta$ can be considered constant, the far field appears to be a uniform plane wave.

Returning to the problem of calculation of the total radiated power, we see that

$$\text{Power radiated} = \oint_\Sigma P_{r(\text{av})}\, da = \oint_\Sigma \frac{\eta\, |H_\varphi|^2}{2}\, da$$

$$= \frac{\eta\omega^2 I^2\, dl^2}{32\pi^2 v^2} \int_0^{2\pi} \int_0^\pi \frac{\sin^2\theta}{r^2}\, r^2 \sin\theta\, d\theta\, d\varphi$$

$$= \frac{\eta\omega^2 I^2\, dl^2}{12\pi v^2} \tag{10-34}$$

Generally, it is useful to assume that the antenna is in free space, for which η has the value of 120π, and to note that $\omega/v = \beta = 2\pi/\lambda$ and that $I^2/2 = I_{\text{rms}}^2$. With these substitutions we obtain

$$\text{Power radiated} = 80\pi^2\left(\frac{dl}{\lambda}\right)^2 I_{\text{rms}}^2 \tag{10-35}$$

By analogy with circuit theory we like to write power $= I_{\text{rms}}^2\, R$, and we define

$$R_{\text{rad}} = 80\pi^2\left(\frac{dl}{\lambda}\right)^2 \tag{10-36}$$

as the *radiation resistance* of a current element.

10.5 The General Nature of the Far Field of an Antenna. In general terms, we can write for the field of a current element

$$E_\theta = \frac{E_0}{r} \sin \omega t'$$

$$H_\varphi = \frac{E_0}{r\eta} \sin \omega t'$$

(10-37)

where E_0 contains all the amplitude factors in Eqs. (10-32), or if we absorb a 90° phase factor into E_0, we can write

$$E_\theta = \text{Re}\left(\frac{E_0 e^{j(\omega t - \beta r)}}{r}\right)$$

$$H_\varphi = \text{Re}\left(\frac{E_0 e^{j(\omega t - \beta r)}}{\eta r}\right)$$

(10-38)

which suggests that we can make use of the simplification of manipulation afforded by working with the complex fields†

$$E_\theta = \frac{E_0 e^{-j\beta r}}{r}$$

$$H_\varphi = \frac{E_0 e^{-j\beta r}}{\eta r}$$

(10-39)

For most antennas, the far field is of a similar form, such that we can, in general, write the far field

$$\mathbf{E} = \mathbf{E}_0 \frac{e^{-j\beta r}}{r}$$

(10-40)

$$\mathbf{H} = \frac{1}{\eta} \mathbf{a}_r \times \mathbf{E}$$

(10-41)

with $\mathbf{E}$ perpendicular to $\mathbf{H}$, and both $\mathbf{E}$ and $\mathbf{H}$ perpendicular to $\mathbf{a}_r$, where $\mathbf{a}_r$ is the radius vector from the *phase center* (at a given observation point this is defined‡ as the center of that sphere on which the plane of the field vectors exhibits the least local variation) of the antenna, which usually coincides with its physical center. We should note that some complicated antennas do not have a true phase center, and this simplification is not valid.

In simple cases, the field is linearly polarized. Since the general case of elliptical polarization can usually be treated as superposition of two linear polarizations, in the rest of this chapter we assume linear polarization.

† See Prob. 10-3, previously cited.

‡ For standard definitions see Test Procedures for Antennas, *IEEE Trans. on Antennas and Propagation*, vol. AP-13, pp. 464–466, May, 1965.

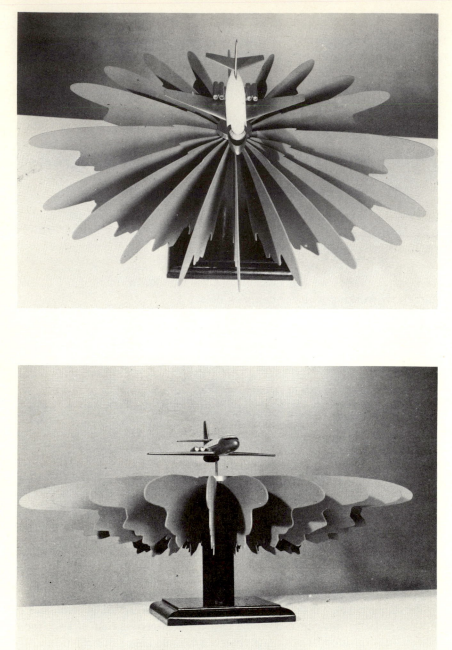

Three-dimensional displays of an aircraft antenna pattern.
(*Courtesy of Lockheed-Georgia Company.*)

Usually, of course, E_0 will be a function of angular position, just as it was for a current element. Additionally, the general nature of the far field of an antenna is such that at any point in space it behaves locally as a uniform plane wave.

10.6 Antenna Patterns.

An antenna pattern is a three-dimensional plot which shows the antenna's characteristics as a radiator of energy. Three types of antenna patterns are in general use which show the relative angular distribution of (1) field intensity, (2) power density, or (3) radiation intensity.

The E-field pattern A plot of $|E|$ as a function of θ and φ is called the *E-field pattern* (three-dimensional). As a practical matter, it is of course impossible to present a complete three-dimensional plot. In most cases, a plot of $|E|$ as a function of θ for some particular value of φ plus a plot of $|E|$ as a function of φ for some particular value of θ give most of the useful information.

Example 10-1 Field Patterns. Given that the E field has only a θ component

$$E_\theta = \frac{E_0 \sin \theta}{r} e^{-j\beta r} \tag{10-42}$$

Plot the E-field pattern for $\theta = $ constant and for $\varphi = $ constant. These plots are shown in Fig. 10-2a and b.

Usually, the patterns are normalized so that the maximum magnitude is 1, and are then called *normalized E-field patterns*.

The power pattern A plot of the time-average Poynting's vector is called the *power pattern*. The power pattern may be thought of as a plot of

$$\text{Re}\,(S_c) = \tfrac{1}{2}\,\text{Re}\,(\mathbf{E} \times \mathbf{H^*})$$

Since the complex Poynting's vector for the far field is real, this gives for our example

$$\text{Re}\,(S_c) = S_c = \frac{1}{2}\frac{E_0{}^2 \sin^2 \theta}{\eta r^2} \qquad \text{W/m}^2$$

This pattern is shown plotted in Fig. 10-2c. The half-power points in this figure specify the *beamwidth*. This is the angular distance (90° in Fig. 10-2c) between the directions at which the power is one-half the maximum power.

The radiation intensity pattern If we multiply $\text{Re}\,(S_c)$ by r^2, we obtain the *radiation intensity U*. Thus

$$U = r^2 \,\text{Re}\,(S_c) \qquad \text{W/unit solid angle}$$

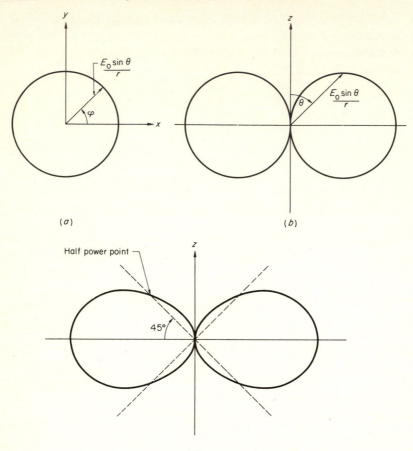

FIGURE 10-2.   The field patterns of $E_\theta = (1/r)E_0 \sin \theta \, e^{-j\beta r}$.   (a) E_θ for $\theta = $ constant; (b) E_θ for $\varphi = $ constant; (c) power pattern for $\varphi = $ constant.

A plot of the radiation intensity is called the *radiation intensity pattern*.   In our example

$$U = \frac{1}{2} \frac{E_0{}^2 \sin^2 \theta}{\eta} \qquad \text{W/unit solid angle}$$

Both the power patterns and the radiation intensity patterns are usually normalized to unity by dividing by the maximum value at a particular r. Thus

$$\text{Re} \, (S_{c_n}) = \sin^2 \theta \qquad \qquad (10\text{-}43)$$

$$U_n = \sin^2 \theta \qquad \qquad (10\text{-}44)$$

The normalized patterns are obviously identical, and it is customary in practice to refer to either of them as *the* pattern.   Use of this imprecise terminology is also extended by the methods generally employed to measure patterns experimentally. Most recording techniques actually present the logarithm of the power per unit area relative to an arbitrary reference, and the usual reference is the maximum power per unit area. The quantity

$$10 \log \frac{\text{Re } (S_c)}{\text{Re } (S_{c_{max}})} \tag{10-45}$$

expresses the result in decibels (dB).

10.7 Directivity and Gain.

10.7 Directivity and Gain. Before proceeding further with our discussion, we need precise definitions of two terms which, because of their similarity, are frequently misused.

The *directivity D* of an antenna is defined as the ratio of the maximum radiation intensity to the average radiation intensity. Put mathematically,

$$D = \frac{U_{max}}{U_{av}} \tag{10-46}$$

where U_{av} is the average radiation intensity. Notice that, by use of the general method of obtaining an average, we have

$$U_{av} = \frac{1}{4\pi} \oint_\Sigma U(\theta,\varphi) \, d\Omega \tag{10-47}$$

or

$$4\pi U_{av} = \oint_\Sigma U(\theta,\varphi) \, d\Omega = P_{rad} \tag{10-48}$$

Equation (10-48) states that 4π times the average radiation intensity equals the total power radiated by the antenna. Hence we can write Eq. (10-46) as

$$D = \frac{4\pi U_{max}}{4\pi U_{av}} = \frac{4\pi U_{max}}{P_{rad}} \tag{10-49}$$

which says that we can calculate the directivity by taking the ratio of 4π times the maximum radiation intensity to the total power radiated.

Equation (10-49) is the IEEE standard definition of directivity, and it is frequently the more convenient equation to use for calculation.

A secondary concept which we need is that of a *lossless isotropic radiator* (antenna). For a *lossless antenna*, the power input equals the power radiated. For an *isotropic antenna*, the radiation intensity is the same in all directions, and hence $U(\theta,\varphi) = \text{constant} = U_{av}$. Thus, for a lossless isotropic antenna, we have

$$P_{input} = P_{radiated} = 4\pi U_{av} \tag{10-50}$$

Such an antenna, although conceptually very useful, cannot be achieved in practice.†

Now, the *gain G* of an antenna is defined as the ratio of the maximum radiation intensity from the antenna, U_{max}, to the maximum radiation intensity from a reference antenna, $(U_{\text{max}})_r$, with the same power *input*. Thus

$$G = \frac{U_{\text{max}}}{(U_{\text{max}})_r} \qquad (10\text{-}51)$$

The gain of an antenna is a relative quantity. It involves the use of a reference antenna and takes into consideration the *efficiency* of both antennas. Efficiency is defined as the ratio of total power radiated by an antenna to the net power accepted by the antenna.

To standardize the gain specification, two reference antennas are commonly used. They are the lossless half-wave dipole (see next section) and the lossless isotropic antenna. For these two cases, we use the terms *gain over a dipole* and *gain over isotropic*. Gain over isotropic is used often enough so that a special symbol G_0 is defined for that case.

$$G_0 = \frac{U_{\text{max}}}{U_0} \qquad (10\text{-}52)$$

where U_0 is the (constant) radiation intensity of a lossless isotropic radiator with the same power input. From these definitions, we have $U_{\text{max}} = kU'_{\text{max}}$, where k denotes the efficiency of the antenna, and U'_{max} is the maximum value which the radiation intensity would have if the antenna were lossless. Notice that, for a lossless antenna, $k = 1$ and $G_0 = D$; that is, for a lossless antenna, the gain over isotropic is exactly equal to the directivity. Since many antennas have relatively small losses, there is a tendency to be rather lax in distinguishing between gain and directivity.

Frequently, it is convenient to specify gain and directivity in decibels by giving 10 times the logarithm to the base 10 of the actual ratio.

Gain in decibels = $10 \log G_0$

For example, a gain $G_0 = 20$ would be given as 13 dB.

In concluding our present set of definitions, we should point out that the concepts of gain and directivity are frequently generalized to include the concept of gain and directivity as a function of direction. Specifically,

$$G_0(\theta, \varphi) = \frac{U(\theta, \varphi)}{U_0} \qquad (10\text{-}53)$$

---

† It has been shown, however, that one can approach arbitrarily close to this idealization. See W. K. Saunders, On the Unity Gain Antenna, "Symposium on Electromagnetic Theory and Antennas," Copenhagen, June 25–30, 1962 (Pergamon Press).

is the gain over isotropic as a function of direction (θ, φ), and

$$D(\theta, \varphi) = \frac{U(\theta, \varphi)}{U_{av}} \qquad (10\text{-}54)$$

is the directivity as a function of direction (θ, φ). These expressions are really just normalized radiation intensity patterns. The gain $G_0(\theta, \varphi)$ has been normalized by dividing the radiation intensity $U(\theta, \varphi)$ by the radiation intensity U_0 of a lossless isotropic antenna of the same input power, and $D(\theta, \varphi)$ has been normalized by dividing by the average radiation intensity U_{av}. Note that, for a lossless antenna, $U_{av} = U_0$ and $D(\theta, \varphi) = G_0(\theta, \varphi)$, so that for low-loss antennas the difference between gain and directivity is small, and one tends to be lax about making a distinction.

Example 10-2   Calculation of Directivity.   Suppose an antenna has a power input of 40π W and an efficiency of 98 percent.   Also, suppose that the radiation intensity has been found to have a maximum value of 200 W/unit solid angle.   Find the directivity and gain of the antenna.
   We have

$$U_{av} = \frac{P_{rad}}{4\pi} = \frac{(0.98)(40\pi)}{4\pi} = 9.8 \text{ W/sr}$$

Hence

$$D = \frac{200}{9.8} = 20.4, \text{ or } 13.1 \text{ dB}$$

Also,

$$U_0 = \frac{P_{in}}{4\pi} = \frac{40\pi}{4\pi} = 10 \text{ W/sr}$$

and

$$G_0 = 200/10 = 20, \text{ or } 13.0 \text{ dB}$$

10.8   Linear Dipole Antennas.   A *linear dipole antenna* is a straight-wire antenna, usually *center-fed* (Fig. 10-3a). A *linear monopole antenna* is a straight-wire antenna *fed against* a ground plane (Fig. 10-3b).   It is fairly obvious that a monopole antenna differs structurally from a dipole antenna. However, the electrical problem of a monopole antenna is basically the same as that of a dipole antenna, and is best handled by the method of images. This method implies that the vector potential and the field intensity of a monopole in the region above the ground plane are exactly the same as those of a center-fed dipole with the same current and an *overall* length which is twice the monopole length.   It is clear, then, that analysis of the center-fed dipole includes analysis of the monopole.

   Let us fix our attention on the *short dipole*, defined as a center-fed antenna having a length that is very short compared with a wavelength.   If the overall length L of a center-fed dipole is short enough so that the contributions to the far field from each infinitesimal element of its length are in time phase with each other, the total fields can be calculated by simple scalar addition of the

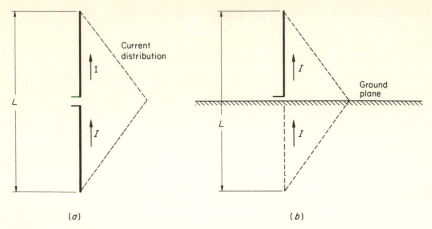

FIGURE 10-3.   The current distribution on linear antennas.   (a) Short dipole; (b) short monopole.

infinitesimal fields produced by a chain of Hertzian dipoles.   This approximation is valid for $L \leq \lambda/10$, roughly speaking.   For such an antenna we have

$$E_\theta = \int_{-L/2}^{L/2} \frac{j\omega I \sin \theta}{4\pi\epsilon v^2 r} e^{-j\beta r} \, dl$$

Our approximations here mean that we can consider the distance r from the point of observation to the source point (position along the antenna) to be constant, allowing us to write

$$E_\theta = \frac{j\omega \sin \theta e^{-j\beta r}}{4\pi\epsilon v^2 r} \int_{-L/2}^{L/2} I \, dl = \frac{j\omega \sin \theta e^{-j\beta r}}{4\pi\epsilon v^2 r} I_{av} L \qquad (10\text{-}55)$$

$$H_\varphi = \frac{E_\theta}{\eta} \qquad (10\text{-}56)$$

This result simply means that the fields of a short dipole are obtained from the expressions for an infinitesimal dipole by simple substitution of L for dl and I_{av} for I, where

$$I_{av} = \frac{1}{L} \int_{-L/2}^{L/2} I \, dz \qquad (10\text{-}57)$$

The current along a short dipole and along a short monopole varies nearly linearly (Sec. 10.9) from I_0 at the center to zero at the end, as shown in Fig. 10-3.   From this it is obvious that

$$I_{av} = \frac{I_0}{2} \qquad (10\text{-}58)$$

We can readily calculate the total radiated power from a short antenna by substituting $I_{av} = I_0/2$ and $L = dl$ into Eq. (10-35) and noting that, for a monopole, the radiated power is just $\frac{1}{2}$ that of a dipole (it radiates only into the upper half space). We obtain, for the dipole,

$$P_{rad} = 20\pi^2 \left(\frac{L}{\lambda}\right)^2 I_{rms}^2 \qquad (10\text{-}59)$$

and for the monopole,

$$P_{rad} = 10\pi^2 \left(\frac{L}{\lambda}\right)^2 I_{rms}^2 \qquad (10\text{-}60)$$

where, as in Eq. (10-35), I_{rms}^2 denotes the root-mean-square value of I.

We can also define a radiation resistance for each case by

$$R_{rad} \text{ (dipole)} = 20\pi^2 \left(\frac{L}{\lambda}\right)^2 \quad \Omega \qquad (10\text{-}61)$$

$$R_{rad} \text{ (monopole)} = 10\pi^2 \left(\frac{L}{\lambda}\right)^2 \quad \Omega \qquad (10\text{-}62)$$

Some numerical values are of interest. For $L = \lambda/10$ the two resistances are approximately 2 and 1 Ω, respectively. These values are very small for transmission line loads and cause rather severe problems of transmission line matching. The matching networks required frequently have large losses, with the result that the overall system efficiency is small.

A quantity which is often specified for short dipoles is the *effective length*, defined by the relation

$$L_{eff} = \frac{1}{I_0} \int_{-L/2}^{L/2} I(z) \, dz \qquad (10\text{-}63)$$

where I_0 is the current fed to the antenna which extends from $z = -L/2$ to $z = L/2$. From this definition it is apparent that the effective length of an antenna is that length which, by supporting the feed current I_0 of the actual antenna throughout the entire length, has the same overall effectiveness as the original antenna.

10.9 Current Distribution on a Linear Antenna.

We saw that for short antennas we had to know the current distribution on the antenna before we could finish the problem. For longer antennas it is necessary to know it before we start. In principle, we can find the current distribution by solving Maxwell's equations subject to the boundary conditions along the antenna. In practice, it turns out that this is such a formidable problem that it has been

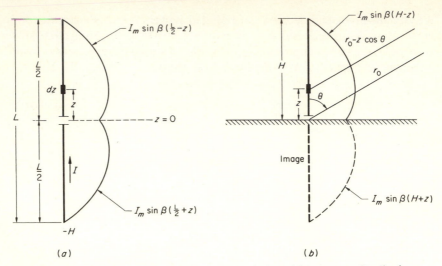

FIGURE 10-4. The geometry and the assumed sinusoidal current distribution for a center-fed dipole and a monopole. (*a*) Dipole; (*b*) monopole ($H = L/2$).

solved rigorously for only one case.† We are therefore faced with the necessity of assuming a distribution, hoping it is correct. Intuitively, we might expect the current to have the standing wave distribution characteristic of an open-circuited transmission line (see previous chapter.) This assumption proves to be correct, at least as a valid engineering approximation, in that antenna calculations based upon this assumption yield quite accurate results.

The standing wave current distribution in the z direction has a magnitude

$$
I = \begin{cases} I_m \sin \beta\left(\dfrac{L}{2} - z\right) & z > 0 \\[3mm] I_m \sin \beta\left(\dfrac{L}{2} + z\right) & z < 0 \end{cases} \tag{10-64}
$$

and hence is called a *sinusoidal current distribution* (Fig. 10-4).

10.10 The Longer Linear Dipole and Monopole. When a center-fed dipole or a monopole, base-fed against a ground plane, has a half length which exceeds approximately $\lambda/10$, the simple analysis of Sec. 10.8 for short linear antennas is no longer a valid approximation, and we must use a more

† R. W. P. King, The Linear Antenna: Eighty Years of Progress, *Proc. IEEE*, vol. 55, pp. 2–16, January, 1967.

exact analysis.   The usual procedure is to calculate the vector potential, and the far-field electric and magnetic fields from the vector potential.

If we use the geometry and assume a sinusoidal current distribution, as shown in Fig. 10-4, we shall find that the vector potential has only a z component, given in complex notation by

$$A_z = \int_{-H}^{0} \frac{\mu I_m \sin \beta(H+z) e^{-j\beta r}}{4\pi r} \, dz + \int_{0}^{H} \frac{\mu I_m \sin \beta(H-z) e^{-j\beta r}}{4\pi r} \, dz$$

where $H = L/2$.   In these integrands we set $r = r_0 - z \cos \theta$ in the exponential factors and $r = r_0$ in the denominator; we then factor the constant terms, and obtain

$$A_z = \frac{\mu I_m e^{-j\beta r_0}}{4\pi r_0} \int_{-H}^{0} \sin \beta(H+z) e^{j\beta z \cos \theta} \, dz$$

$$+ \frac{\mu I_m e^{-j\beta r_0}}{4\pi r_0} \int_{0}^{H} \sin \beta(H-z) e^{j\beta z \cos \theta} \, dz \quad (10\text{-}65)$$

Noting that in the first integrand z is negative allows us to change signs on z and change limits so as to obtain

$$A_z = \frac{\mu I_m e^{-j\beta r_0}}{4\pi r_0} \int_{0}^{H} \sin \beta(H-z)(e^{-j\beta z \cos \theta} + e^{j\beta z \cos \theta}) \, dz$$

$$= \frac{\mu I_m e^{-j\beta r_0}}{2\pi r_0} \int_{0}^{H} \sin \beta(H-z) \cos (\beta z \cos \theta) \, dz \quad (10\text{-}66)$$

and after integration,

$$A_z = \frac{\mu I_m e^{-j\beta r_0}}{2\pi \beta r_0} \frac{\cos \beta(H \cos \theta) - \cos \beta H}{\sin^2 \theta} \quad (10\text{-}67)$$

From A_z we obtain the far field in a manner similar to that employed for the infinitesimal dipole.   The details are not displayed here.   The results are

$$H_\varphi = \frac{j I_m e^{-j\beta r_0}}{2\pi r_0} \frac{\cos (\beta H \cos \theta) - \cos \beta H}{\sin \theta} \quad (10\text{-}68)$$

$$E_\theta = \eta H_\varphi \quad (10\text{-}69)$$

When the half length H is equal to a quarter wavelength, the antenna is known as a *half-wave dipole* (or a *quarter-wave monopole*, if fed against a ground plane).   These antennas are of particular practical importance because they have desirable input characteristics and also have desirable

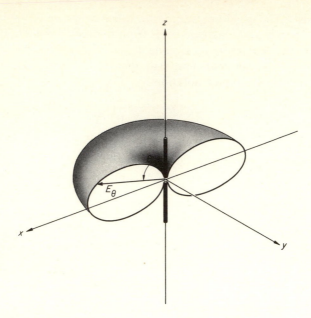

FIGURE 10-5. The normalized E-field pattern for a half-wave dipole.   (A quarter-wave monopole has the same pattern for $\theta \leq \pi/2$, but the fields are zero for $\theta > \pi/2$.)

radiation patterns.   Substituting $H = \lambda/4$ into the general expressions, Eqs. (10-68) and (10-69), gives the results

$$E_\theta = \frac{j60I_m e^{-j\beta r}}{r} \frac{\cos\left(\dfrac{\pi}{2}\cos\theta\right)}{\sin\theta} \tag{10-70}$$

$$H_\varphi = \frac{jI_m e^{-j\beta r}}{2\pi r} \frac{\cos\left(\dfrac{\pi}{2}\cos\theta\right)}{\sin\theta} \tag{10-71}$$

where the subscript on r has been dropped in accordance with usual notation.

The E-field patterns and the power, or radiation intensity, patterns for this special case are relatively easy to plot.   A three-dimensional cutaway view of the normalized E-field pattern is shown in Fig. 10-5 for the $\lambda/2$ dipole.   The $\lambda/4$ monopole pattern is the same as the $\lambda/2$ dipole in the upper half space, and is zero in the lower half space ($\theta > \pi/2$).

The total time-average power radiated by a half-wave dipole can be calculated by integrating the complex Poynting's vector over a sphere of radius r. We have

$$P_{\text{rad}} = \oint_{\Sigma} \tfrac{1}{2} E_{\theta} H_{\varphi}^{*} \, da$$

$$= \frac{30}{2\pi} I_m^2 \int_0^{2\pi} d\varphi \int_0^{\pi} \frac{\cos^2 \left(\dfrac{\pi}{2} \cos \theta \right)}{\sin \theta} \, d\theta$$

$$= 30 \, I_m^2 \int_0^{\pi} \frac{\cos^2 \left(\dfrac{\pi}{2} \cos \theta \right)}{\sin \theta} \, d\theta \qquad (10\text{-}72)$$

The remaining integral in Eq. (10-72) is not easy to evaluate. It may be attacked by a change-of-variable technique and eventually cast into a slowly convergent infinite series, or it may be programmed on a digital computer. The result, in any event, is that its value to four significant figures is 1.2186, and we obtain the numerical result that the radiated power is given by

$$P_{\text{rad}} = 73 \frac{I_m^2}{2} \qquad \text{half-wave dipole}$$

Using our previous definition of radiation resistance, we see that

$$R_{\text{rad}} = 73 \, \Omega \qquad \text{half-wave dipole} \qquad (10\text{-}73)$$

Since the fields of a quarter-wave monopole for $\theta < \pi/2$ are exactly the same as those of a half-wave dipole and zero for $\theta > \pi/2$, and since the half-wave dipole fields are symmetrical about $\theta = \pi/2$, we should have just one-half the radiated power of a dipole for the monopole case. That is,

$$P_{\text{rad}} = 36.5 \frac{I_m^2}{2} \qquad \text{quarter-wave monopole}$$

and the radiation resistance would be

$$R_{\text{rad}} = 36.5 \, \Omega \qquad \text{quarter-wave monopole} \qquad (10\text{-}74)$$

A look at Fig. 10-4, specialized to $H = \lambda/4$, shows that the driving-point current is given by

$$I = I_m \sin \beta (H - 0) e^{j\omega t} = I_m e^{j\omega t}$$

In this special case, the resistive part of the driving-point impedance of the antenna is just the radiation resistance. It is beyond the scope of our presentation to derive the driving-point reactance, but it turns out to be approximately zero. It is exactly zero for H slightly less than $\lambda/4$. Most $\lambda/2$ dipoles and $\lambda/4$ monopoles are adjusted to make the driving-point reactance zero.   When the length is adjusted, the driving-point resistance is slightly less than 73, or 36.5 Ω.   The resulting driving-point impedances of

$$Z_{in} \approx \begin{cases} 73 + j0 & \text{half-wave dipole} \hspace{3em} \text{(10-75)} \\ 36.5 + j0 & \text{quarter-wave monopole} \hspace{2em} \text{(10-76)} \end{cases}$$

are comparatively easy to match to transmission lines, and in a large measure account for the popularity of these antennas.

We complete our discussion of the two special cases by calculating their directivity.   From the definition of directivity we obtain, for a $\lambda/2$ dipole,

$$D = \frac{4\pi U_{max}}{P_{rad}} = \frac{4\pi(\frac{1}{2}E_\theta H_\varphi^* r^2)_{max}}{R_{rad}(I_m^2/2)} = \frac{2\pi(60 I_m^2/2\pi)}{73(I_m^2/2)} = 1.64, \text{ or } 2.15 \text{ dB}$$

$$\text{(10-77)}$$

and similarly, for a $\lambda/4$ monopole,

$$D = 3.28, \text{ or } 5.15 \text{ dB} \hspace{4em} \text{(10-78)}$$

10.11 Antenna Arrays.   When two or more antennas are located in a common region of space and driven either directly or indirectly from a common generator, we have an *antenna array*.   In principle, the general antenna-array problem is handled by superposition.   That is, the resulting **E** and **H** fields can, in principle at least, be found by writing the vector-phasor sum of the fields produced by the individual antennas.

To obtain specific results, one must consider specific arrays.   The simplest array consists of two identical antennas and, more specifically, of two identical short dipoles oriented in space, as shown in Fig. 10-6, and driven with in-phase currents of equal magnitude.

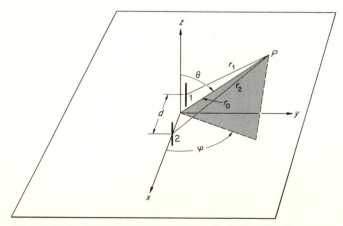

FIGURE 10-6.   The geometry of two identical short dipoles.

The far fields of the individual antennas at an ordinary far-field point will be in the θ direction, and will be given by the sum of

$$E_1 = E_m \sin \theta \frac{e^{-j\beta r_1}}{r_1} \tag{10-79}$$

and

$$E_2 = E_m \sin \theta \frac{e^{j\beta r_2}}{r_2} \tag{10-80}$$

Thus

$$E = E_1 + E_2 = E_m \sin \theta \left(\frac{e^{-j\beta r_1}}{r_1} + \frac{e^{-j\beta r_2}}{r_2} \right) \tag{10-81}$$

In the far-field region, the lines r_1, r_0, r_2 are essentially parallel, and we have

$$r_1 \approx r_0 + \frac{d}{2} \cos \varphi$$

$$\tag{10-82}$$

$$r_2 \approx r_0 - \frac{d}{2} \cos \varphi$$

In the far field $r_0 \gg d/2$, and we can use $r_1 \approx r_2 \approx r_0$ in the denominator. However, because of the periodic nature of the exponential, r_1 and r_2 must be expressed as in Eqs. (10-82). This states that the variation of r affects the phase in the integrand but has little effect on the magnitude. Accordingly,

$$E = \frac{E_m \sin \theta}{r_0} (e^{-j\beta r_0 - j\beta(d/2)\cos\varphi} + e^{-j\beta r_0 + j\beta(d/2)\cos\varphi})$$

$$= \frac{E_m \sin \theta e^{-j\beta r_0}}{r_0} (e^{-j\psi/2} + e^{j\psi/2}) \tag{10-83}$$

where $\psi = \beta d \cos \varphi$, and finally,

$$E = E_0 \left(2 \cos \frac{\psi}{2} \right) \tag{10-84}$$

where

$$E_0 = \frac{E_m \sin \theta e^{-j\beta r_0}}{r_0} \tag{10-85}$$

is the field of the individual short dipole.

To interpret the $2 \cos (\psi/2)$ factor, we note that the pattern of two in-phase isotropic radiators of unity magnitude separated by a distance d would be

$$E_i = \left(\frac{e^{-j\beta r_1}}{r_1} + \frac{e^{+j\beta r_2}}{r_2} \right) = \frac{e^{-j\beta r_0}}{r_0} \left(2 \cos \frac{\psi}{2} \right)$$

In view of this expression, we see that the pattern of the two identical in-phase dipole antennas can be expressed in normalized form as

$$E_n = (\sin \theta) \left(\cos \frac{\psi}{2} \right)$$

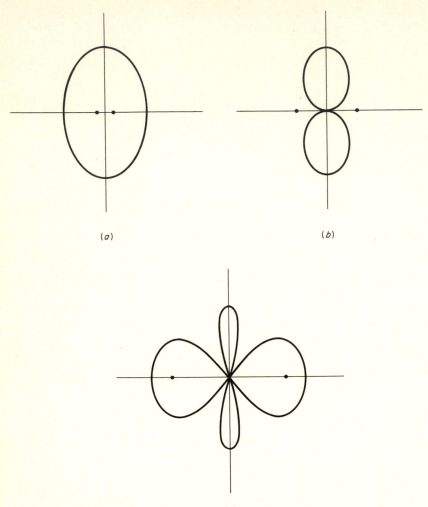

FIGURE 10-7. The array factor for two isotropic in-phase point sources for three different spacings.   (a) $d = \lambda/4$; (b) $d = \lambda/2$; (c) $d = \lambda$.

where the first term is the normalized pattern of the individual dipole, and the second term is the normalized pattern of the array of isotropic point sources.   This result is general for arrays of identical antennas.   It is common practice to analyze arrays of identical antennas by first finding the *array pattern* (frequently called the *array factor*) and multiplying it by the pattern of the individual antenna.

Analysis of arrays of nonidentical antennas is usually a formidable problem, which we do not present in this brief treatment of antennas.

Example 10-3 Arrays of Isotropic In-phase Point Sources. In this example we give the array factor (that is, the normalized E-field pattern) of two isotropic in-phase point sources.

As derived in this section, the array factor is $AF = \cos{(\psi/2)}$, where $\psi = \beta d \cos{\varphi}$.

CASE I. $d = \lambda/4$

$$\psi = \beta d \cos{\varphi} = \frac{2\pi}{\lambda}\frac{\lambda}{4}\cos{\varphi} = \frac{\pi}{2}\cos{\varphi}$$

and

$$AF = \cos\left(\frac{\pi}{4}\cos{\varphi}\right)$$

CASE II. $d = \lambda/2$

$$AF = \cos\left(\frac{\pi}{2}\cos{\varphi}\right)$$

CASE III. $d = \lambda$

$$AF = \cos{(\pi \cos{\varphi})}$$

These array factors are plotted in Fig. 10-7.

If the two individual antennas of an array are not in phase, the previous results will be modified to include their relative phase. In particular, if we let α be the relative phase by which antenna 2 leads antenna 1, we can let the phase reference of the array be the centerpoint of the array, and let antenna 1 lag this point by $\alpha/2$, and antenna 2 lead this point by $\alpha/2$. Then the result will be

$$E = E_0(e^{-j\beta(d/2)\cos{\varphi}-j\alpha/2} + e^{j\beta(d/2)\cos{\varphi}+j\alpha/2}) = E_0\left(2\cos\frac{\psi}{2}\right) \quad (10\text{-}86)$$

where

$$\psi = \beta d \cos{\varphi} + \alpha \quad (10\text{-}87)$$

which shows that the array factor is still $\cos{(\psi/2)}$. But in this case ψ includes the relative phase α.

We now present three examples to show typical results.†

Example 10-4 Specific Array Factors. We consider three cases.

CASE I. $d = \lambda/4, \alpha = \pi/2$

$$\frac{\psi}{2} = \frac{\beta d}{2}\cos{\varphi} + \frac{\alpha}{2} = \frac{2\pi}{2\lambda}\frac{\lambda}{4}\cos{\varphi} + \frac{\pi}{4} = \frac{\pi}{4}(1 + \cos{\varphi})$$

$$AF = \cos\left[\frac{\pi}{4}(1 + \cos{\varphi})\right]$$

† A rather extensive set of array factors for two isotropic point sources is given in J. D. Kraus, "Antennas," chap. 11, McGraw-Hill Book Company, New York, 1950.

CASE II. $d = \lambda/4, \alpha = \pi$

$$\frac{\psi}{2} = \frac{\pi}{4} \cos \varphi + \frac{\pi}{2} = \frac{\pi}{4} (2 + \cos \varphi)$$

$$AF = \cos \left[\frac{\pi}{4} (2 + \cos \varphi) \right]$$

CASE III. $d = \lambda, \alpha = \pi/2$

$$\frac{\psi}{2} = \pi \cos \varphi + \frac{\pi}{4} = \frac{\pi}{4} (1 + 4 \cos \varphi)$$

$$AF = \cos \left[\frac{\pi}{4} (1 + 4 \cos \varphi) \right]$$

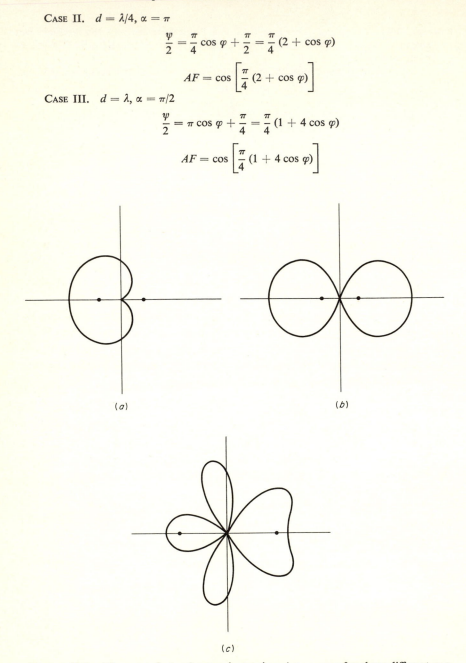

(a) (b)

(c)

FIGURE 10-8.   The array factor for two isotropic point sources for three different conditions of spacing and phase. (a) $d = \lambda/4$, $\alpha = \pi/2$; (b) $d = \lambda/4$, $\alpha = \pi$; (c) $d = \lambda$, $\alpha = \pi/2$.

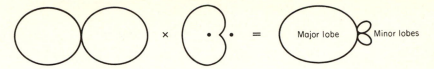

FIGURE 10-9.   The individual antenna pattern, the array factor, and the resultant E-field array pattern in the plane of the array for two identical half-wave dipoles spaced $\lambda/4$ apart and fed in phase quadrature.   Individual dipole $\times$ array factor = resultant pattern.

These array factors are shown plotted in Fig. 10-8.   Examination of this figure and Fig. 10-7, which is for $\alpha = 0$, shows that a wide range of array factors can be obtained by adjusting the spacing and phasing of the array.   Arrays are commonly used to produce some desired modification of the pattern of the individual antenna. Figure 10-9 shows an example of this for two half-wave dipoles spaced $\lambda/4$ apart and fed with the right-hand antenna leading the left-hand antenna by 90°.   Notice that a fairly accurate sketch of the resultant pattern can be made by geometrically multiplying the individual pattern and the array factor.

The example given in Fig. 10-9 shows only the pattern in the plane of the dipoles.   The total pattern of the array of course is three-dimensional. A fairly accurate visualization of its three-dimensional behavior can be obtained by applying the same technique in the plane perpendicular to the plane of the dipoles ($\theta = 90°$ plane), and then in the plane perpendicular to the line joining the two dipoles.   This is left as an exercise for the student.

10.12   Uniform Linear Arrays.   A *uniform linear array* is an array of identical antennas uniformly spaced along a straight line, fed with currents of equal magnitude and having a uniform progressive phase shift.   Figure 10-10 shows an n-element uniform linear array.   Using the methods already developed, we can write the total radiation **E** field as

$$E = E_0(1 + e^{j\psi} + e^{j2\psi} + e^{j3\psi} + \cdots + e^{j(n-1)\psi}) \qquad (10\text{-}88)$$

where
$$\psi = \beta d \cos \varphi + \alpha \qquad (10\text{-}89)$$

and E_0 is the individual antenna's E-field pattern.   Since we are primarily interested in the array factor, we suppress E_0 and, after an algebraic manipulation, write

$$|E| = \left| \frac{1 - e^{jn\psi}}{1 - e^{j\psi}} \right| = \left| \frac{\sin (n\psi/2)}{\sin (\psi/2)} \right| \qquad (10\text{-}90)$$

This result has several interesting properties, which we now examine.

1.   The angle $\psi/2$ has a maximum value of $(\beta d + \alpha)/2$ at $\varphi = 0$ and a minimum value of $(-\beta d + \alpha)/2$ at $\varphi = \pi$.   At $\varphi = 2\pi$, $\psi/2$ returns to

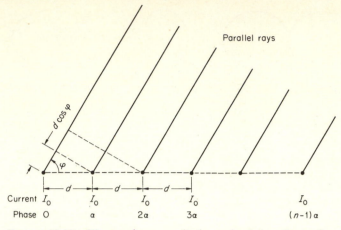

FIGURE 10-10.   The spacing, magnitude, and relative phase of an *n*-element uniform linear array.

its maximum value.   Examination of this behavior and a little thought show that the array factor, Eq. (10-90), is symmetrical about the line of the array ($\varphi = 0$, $\varphi = \pi$ line).

2. By differentiation and inspection we find that the *principal maximum* of the array factor occurs at $\psi = 0$ and that the magnitude of this principal maximum is equal to n.

3. The secondary maxima occur at, approximately,

$$\sin \frac{n\psi}{2} = 1$$

which means $$\frac{n\psi}{2} = \pm(2k + 1)\frac{\pi}{2} \qquad k = 1, 2, 3, \ldots$$

In particular, the *first secondary maximum* is at

$$\frac{\psi}{2} = \frac{3\pi}{2n}$$

and has a magnitude of $|\sin (3\pi/2n)|^{-1}$.   Notice that this has a limiting value, for n large,

$$\lim_{n \to \infty} \frac{1}{|\sin (3\pi/2n)|} = \frac{2n}{3\pi} = 0.212n$$

Since the magnitude of the principal maximum is n, we have that the ratio of the first secondary maximum to the principal maximum is 0.212, or 13.5 dB.

4. The array factor has zero nulls when the numerator is zero; that is, when

$$\sin \frac{n\psi}{2} = 0 \qquad \text{except } \psi = 0$$

which means $\qquad \dfrac{n\psi}{2} = \pm k\pi \qquad k = 1, 2, 3, \ldots$

Two special cases of uniform linear arrays are of particular interest and practicality.

Case I. Broadside array If all the elements of the array are in phase ($\alpha = 0$), the array is called a *broadside array*, for reasons which are obvious from an examination of the location of the principal maxima. These maxima are at $\psi/2 = 0$ because, for $\alpha = 0$,

$$\psi = \beta d \cos \varphi = 0 \qquad \text{at } \varphi = \pm \frac{\pi}{2}$$

Thus we have the principal maxima perpendicular to the line of the array (or broadside). The angle φ, in Fig. 10-10, is actually an angle of revolution with only positive values from 0 to π. In many practical antenna situations it is convenient to view φ as a planar angle whose range is from 0 to 2π.

Case II. End-fire array If the progressive phase shift α is related to the spacing by $\alpha = -\beta d$, the principal maximum is at $\varphi = 0$, and the array is called an *end-fire array*.

There are several other classes of wire antennas and arrays of linear elements.† However, at frequencies approaching, roughly, 1 GHz, the wavelength is only a fraction of 1 m, and the size and power-handling capability of wire antennas are correspondingly small. For applications in which microwave frequencies are employed, it is necessary to use reflector-type antennas, like those shown in Fig. 10-11, with large physical dimensions. The next section analyzes the radiation from such antennas.

10.13 Huygens' Principle and Aperture Antennas. In our discussion of linear dipole antennas we assumed a well-defined, known current distribution, and we were able to calculate the far field of the antenna by rather straightforward methods. In the case of microwave antennas we do not have a well-defined, known current distribution. However, in most cases of interest, we are able to determine the E- and H-field distributions over a

† H. Jasik (ed.), "Antenna Engineering Handbook," McGraw-Hill Book Company, New York, 1961.

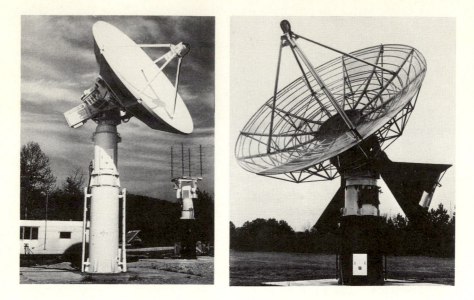

FIGURE 10-11.   Reflector-type antennas.   (*Courtesy of Scientific-Atlanta, Inc.*)

finite open surface located in front of the antenna.   In such cases we call this surface the *aperture*, we call the antenna an *aperture antenna*, and we call the method of calculation of the far-field pattern of the antenna the *aperture field method*.

A basic postulate in the analysis of aperture antennas is the validity of *Huygens' principle*.

Let Σ be a closed surface consisting of a perfect screen S_2, on which the tangential component of the electric field intensity and the normal component of the magnetic field intensity are both zero, and an aperture S_1 which is bounded by a closed contour C (Fig. 10-12).   For harmonic ($e^{j\omega t}$) fields, Maxwell's equations predict† that the field at every interior point (x,y,z) of a linear, homogeneous, isotropic, and source-free medium bounded by Σ is given in terms of the *aperture field* $\mathbf{E}_1$, $\mathbf{H}_1$ by the relations

$$\mathbf{E}(x,y,z) = \frac{j}{4\pi\omega\epsilon} \int_{S_1} \{[(\mathbf{n} \times \mathbf{H}_1) \cdot \nabla']\nabla'\phi + \beta^2\phi(\mathbf{n} \times \mathbf{H}_1)$$
$$+ j\omega\epsilon(\mathbf{n} \times \mathbf{E}_1) \times \nabla'\phi\} \, da \quad (10\text{-}91)$$

$$\mathbf{H}(x,y,z) = \frac{-j}{4\pi\omega\mu} \int_{S_1} \{[(\mathbf{n} \times \mathbf{E}_1) \cdot \nabla']\nabla'\phi + \beta^2\phi(\mathbf{n} \times \mathbf{E}_1)$$
$$- j\omega\mu(\mathbf{n} \times \mathbf{H}_1) \times \nabla'\phi\} \, da \quad (10\text{-}92)$$

† S. Silver (ed.), "Microwave Antenna Theory and Design," chap. 5, McGraw-Hill Book Company, New York, 1949.   (Note opposite direction of the vector **n**.)

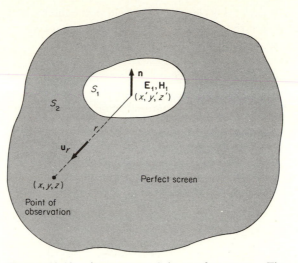

FIGURE 10-12.   An aperature S_1 in a perfect screen.   The
primary sources are outside the closed surface defined by S_1
and S_2.

where

$$[(\mathbf{n} \times \mathbf{H}_1) \cdot \nabla']\nabla'\phi \triangleq \mathbf{a}_x\left[(\mathbf{n} \times \mathbf{H}_1) \cdot \frac{\partial}{\partial x'} (\nabla'\phi)\right] + \mathbf{a}_y\left[(\mathbf{n} \times \mathbf{H}_1) \cdot \frac{\partial}{\partial y'} (\nabla'\phi)\right]$$

$$+ \mathbf{a}_z\left[(\mathbf{n} \times \mathbf{H}_1) \cdot \frac{\partial}{\partial z'} (\nabla'\phi)\right] \quad (10\text{-}93)$$

and where $\phi = e^{-j\beta r}/r$, $\beta = \omega\sqrt{\mu\epsilon}$, and r is the distance from the fixed point
of observation (x,y,z) to the variable source point (x',y',z') on the aperture
S_1.   Furthermore, ∇' denotes differentiation with respect to the primed
variables, and $\mathbf{n}$ is the unit outward vector normal to the surface Σ.

Equations (10-91) and (10-92) represent a general mathematical statement of Huygens' principle for harmonic fields and state that, if the field can
be described on the boundary, it can be found at any point inside.   By means
of lengthy calculations requiring the introduction of several simplifying
assumptions, this pair of expressions can be reduced to the forms most
suitable for analysis and physical interpretation.

Thus, for $r \gg \lambda/2\pi$, the transformed expressions for the field intensities
are

$$\mathbf{E}(x,y,z) = \frac{j\beta}{4\pi} \int_{S_1} \mathbf{u}_r \times \left[(\mathbf{n} \times \mathbf{E}_1) - \sqrt{\frac{\mu}{\epsilon}} \mathbf{u}_r \times (\mathbf{n} \times \mathbf{H}_1)\right]\frac{e^{-j\beta r}}{r} \, da \quad (10\text{-}94)$$

$$\mathbf{H}(x,y,z) = \frac{j\beta}{4\pi} \int_S \mathbf{u}_r \times \left[(\mathbf{n} \times \mathbf{H}_1) - \sqrt{\frac{\epsilon}{\mu}} \mathbf{u}_r \times (\mathbf{n} \times \mathbf{E}_1)\right]\frac{e^{-j\beta r}}{r} \, da \quad (10\text{-}95)$$

From these expressions it is clear that the field in the interior can be evaluated from a knowledge of the *tangential components of the field* on the aperture S_1. It is also clear that both integrands are transverse to $\mathbf{u}_r$. Therefore, for all points of observation such that $r \gg \lambda/2\pi$, *the contribution to the field from each infinitesimal Huygens' source on S_1 is perpendicular to $\mathbf{u}_r$, the direction of wave travel.*

An application of the results obtained so far is illustrated in the following example.

Example 10-5   The Pyramidal Horn.   Electromagnetic horns, in general, comprise an important class of aperture antennas.   Figure 10-13 shows a commonly used type of horn, the *pyramidal horn*, which is fed by a rectangular waveguide in the dominant mode.   It is known† that, for all practical purposes, the mouth of the horn is uniformly polarized in one direction (the y direction in this case) and that the *aperture illumination function* is given by

$$\mathbf{E}_1 = E_0 \cos \frac{\pi x}{a} \exp \left[-j\beta \left(\frac{x^2}{2l_H} + \frac{y^2}{2l_E} \right) \right] \mathbf{a}_y \tag{10-96}$$

where E_0 is a constant.   As indicated in Fig. 10-13, the distances l_H and l_E in Eq. (10-96) define the flare of the horn.   The aperture S_1 is the mouth of the horn, and the surface S_2 is the rest of the xy plane.   Also, $\mathbf{n} = -\mathbf{a}_z$, and

$$\mathbf{u}_r = u_x \mathbf{a}_x + u_y \mathbf{a}_y + u_z \mathbf{a}_z = \frac{x - x'}{r} \mathbf{a}_x + \frac{y - y'}{r} \mathbf{a}_y + \frac{z}{r} \mathbf{a}_z$$

where u_x, u_y, u_z are the direction cosines of the unit vector $\mathbf{u}_r$.

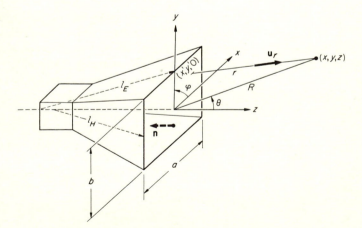

FIGURE 10-13.   A pyramidal horn.

† S. A. Schelkunoff and H. T. Friis, "Antennas," chap. 16, John Wiley & Sons, Inc., New York, 1952.

Since the aperture field is nearly a uniform plane wave, the magnetic field intensity is

$$\mathbf{H}_1 = -\frac{E_0}{120\pi} \cos \frac{\pi x}{a} \exp\left[-j\beta\left(\frac{x^2}{2l_H} + \frac{y^2}{2l_E}\right)\right]\mathbf{a}_x \qquad (10\text{-}97)$$

where 120π represents the intrinsic impedance $\sqrt{\mu_0/\epsilon_0}$.

Setting

$$\delta = \beta\left[r + \frac{1}{2}\left(\frac{x^2}{l_H} + \frac{y^2}{l_E}\right)\right]$$

then reduces Eqs. (10-94) and (10-95) to

$$E(x,y,z) = \frac{j}{2\lambda}\int_{-a/2}^{a/2}\int_{-b/2}^{b/2} E_0 \cos\frac{\pi x}{a}$$
$$\times\left[-u_x u_y \mathbf{a}_x + (u_z + u_x^2 + u_z^2)\mathbf{a}_y - u_y(1 + u_z)\mathbf{a}_z\right]\frac{e^{-j\delta}}{r}\,dy\,dx \quad (10\text{-}98)$$

$$H(x,y,z) = \frac{j}{240\pi\lambda}\int_{-a/2}^{a/2}\int_{-b/2}^{b/2} E_0 \cos\frac{\pi x}{a}$$
$$\times\left[-(u_z + u_y^2 + u_z^2)\mathbf{a}_x + u_x u_y \mathbf{a}_y + u_x(1 + u_z)\mathbf{a}_z\right]\frac{e^{-j\delta}}{r}\,dy\,dx \quad (10\text{-}99)$$

These integrals are best evaluated on the digital computer. For numerical results, we consider the case of a standard horn, for which $a = 5.984$ in., $b = 4.908$ in., $l_H = 14.333$ in., $l_E = 13.633$ in. For this horn, the power patterns calculated at 16 GHz are shown plotted in Fig. 10-14, and are seen to agree well with the measured patterns throughout the dynamic range of the recording apparatus.

The complexity of the use of the vector equations (10-94) and (10-95) was largely hidden in the statement of the preceding example that calculations are best made on the digital computer. The computational difficulties can be greatly reduced, and the working equations considerably simplified, by introducing several additional approximations. These approximations reduce the analysis of aperture antennas to a scalar problem. The final result is given by Eq. (10-113), below. The algebraic details follow.†

Let us consider a TEM wave impinging upon a *plane* aperture S_1, along a direction $\mathbf{p}$, as indicated in Fig. 10-15. The source field is zero everywhere except over S_1. Then

$$\mathbf{H}_1 = \sqrt{\frac{\epsilon}{\mu}}\,(\mathbf{p}\times\mathbf{E}_1)\qquad \mathbf{n}\times\mathbf{H}_1 = -\mathbf{a}_z\times\left[\sqrt{\frac{\epsilon}{\mu}}\,(\mathbf{p}\times\mathbf{E}_1)\right]\quad (10\text{-}100)$$

and with $\beta = 2\pi/\lambda$, Eq. (10-94) becomes

$$E(x,y,z) = \frac{j}{2\lambda}\int_{S_1}\mathbf{u}_r\times\{(-\mathbf{a}_z\times\mathbf{E}_1) + \mathbf{u}_r\times[\mathbf{a}_z\times(\mathbf{p}\times\mathbf{E}_1)]\}\frac{e^{-j\beta r}}{r}\,da$$
$$(10\text{-}101)$$

† Use of the final result, Eq. (10-113), does not require a full understanding of these algebraic details. They are essential only for a full realization of the implications of the approximations.

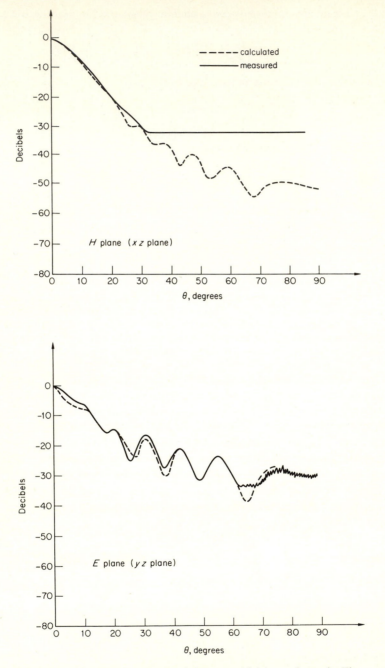

Figure 10-14. Power patterns of the pyramidal horn shown in Fig. 10-13.

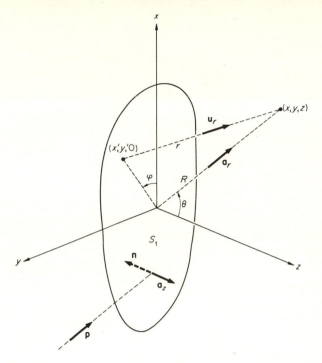

FIGURE 10-15. Geometry of a plane aperture. The wave is incident from $z < 0$, giving rise to a diffraction field in the region $z > 0$.

Over the plane aperture S_1 the direction of the source field vector $\mathbf{E}_1$ is arbitrary. However, no generality will be lost if, for the purposes of this analysis, we consider it to be uniformly polarized in the x direction, namely,

$$\mathbf{E}_1 = F(x',y',0)\mathbf{a}_x = A(x',y')e^{-j\beta L(x',y')}\mathbf{a}_x \qquad (10\text{-}102)$$

Here A and L denote amplitude and phase functions, respectively. Then

$$\mathbf{a}_z \times (\mathbf{p} \times \mathbf{E}_1) = \mathbf{a}_z \times (\mathbf{p} \times \mathbf{a}_x)F = [\mathbf{p}(\mathbf{a}_z \cdot \mathbf{a}_x) - \mathbf{a}_x(\mathbf{p} \cdot \mathbf{a}_z)]F$$

and since $\mathbf{a}_z \cdot \mathbf{a}_x \equiv 0$, the vector integrand $\mathbf{I}$ in Eq. (10-101) transforms to

$$\mathbf{I} = \mathbf{u}_r \times \{-(\mathbf{a}_z \times \mathbf{a}_x) - \mathbf{u}_r \times [(\mathbf{p} \cdot \mathbf{a}_z)\mathbf{a}_x]\}F = -\mathbf{u}_r \times \{[\mathbf{a}_z + (\mathbf{p} \cdot \mathbf{a}_z)\mathbf{u}_r] \times \mathbf{a}_x\}F$$

$$= -[\mathbf{a}_z + (\mathbf{p} \cdot \mathbf{a}_z)\mathbf{u}_r](\mathbf{u}_r \cdot \mathbf{a}_x)F + \mathbf{a}_x[(\mathbf{a}_z \cdot \mathbf{u}_r) + (\mathbf{p} \cdot \mathbf{a}_z)(\mathbf{u}_r \cdot \mathbf{u}_r)]F$$

For points near the z axis, that is, for θ small, $\mathbf{u}_r \cdot \mathbf{a}_x \approx 0$, and the first term on the right vanishes. Then

$$\mathbf{I} \approx \mathbf{a}_x(\mathbf{a}_z \cdot \mathbf{u}_r + \mathbf{a}_z \cdot \mathbf{p})F$$

The amended expression for the **E** field is

$$E(x,y,z) = a_x \frac{j}{2\lambda} \int_{S_1} F(a_z \cdot u_r + a_z \cdot p) \frac{e^{-j\beta r}}{r} \, da \qquad (10\text{-}103)$$

This equation expresses the so-called *diffraction field* (Sec. 8.9). It is customary in practice to divide this diffraction field into three general zones:

1. The near-field zone
2. The Fresnel zone
3. The Fraunhofer, or far-field, zone

The near-field zone is the immediate neighborhood of the aperture. No further simplifying approximation can be made, and Eq. (10-103) applies.

The Fresnel region is far enough from the aperture so that in Eq. (10-103) $a_z \cdot u_r \approx a_z \cdot a_r = \cos \theta$, and $r \approx R$ in every term except in the phase factor $e^{-j\beta r}$. With these approximations, Eq. (10-103) becomes

$$E_x = \frac{j}{2\lambda R} \int_{S_1} F(a_z \cdot p + \cos \theta) e^{-j\beta r} \, da \qquad (10\text{-}104)$$

The variation of the phase factor $e^{-j\beta r}$ can be determined from a consideration of the distance

$$r = [(x - x')^2 + (y - y')^2 + z^2]^{\frac{1}{2}}$$

in terms of spherical variables

$$x = R \sin \theta \cos \varphi$$
$$y = R \sin \theta \sin \varphi$$
$$z = R \cos \theta$$

We have

$$r = [R^2 - 2R(x' \sin \theta \cos \varphi + y' \sin \theta \sin \varphi) + x'^2 + y'^2]^{\frac{1}{2}} \quad (10\text{-}105)$$

Adding and subtracting the term

$$T^2 = (x' \sin \theta \cos \varphi + y' \sin \theta \sin \varphi)^2$$

under the radical and, subsequently, factoring out the completed square, we obtain

$$r = (R - T)\left[1 + \frac{x'^2 + y'^2 - T^2}{(R - T)^2}\right]^{\frac{1}{2}} \qquad (10\text{-}106)$$

Now the earlier hypothesis $r \approx R$ carries with it the tacit implication that $R^2 \gg T^2$. Therefore the radical can be approximated by the first two terms of a binomial expansion, giving

$$r \approx (R - T)\left[1 + \frac{1}{2}\frac{x'^2 + y'^2 - T^2}{(R - T)^2}\right]$$

$$= R - T + \frac{x'^2 + y'^2 - T^2}{2(R - T)}$$

$$\approx R - T + \frac{x'^2 + y'^2 - T^2}{2R}$$

$$= R + r_1 \qquad\qquad (10\text{-}107)$$

where

$$r_1 = -(x' \sin \theta \cos \varphi + y' \sin \theta \sin \varphi)$$
$$+ \frac{x'^2 + y'^2 - (x' \sin \theta \cos \varphi + y' \sin \theta \sin \varphi)^2}{2R} \qquad (10\text{-}108)$$

The diffraction field, Eq. (10-104), for the Fresnel region then becomes

$$E_x = \frac{je^{-j\beta R}}{2\lambda R}\int_{S_1} Fe^{-j\beta r_1}(\mathbf{a}_z \cdot \mathbf{p} + \cos \theta)\, da \qquad (10\text{-}109)$$

In the Fraunhofer region, the second term on the right of Eq. (10-108) is neglected, and

$$r_1 \approx -(x' \sin \theta \cos \varphi + y' \sin \theta \sin \varphi) \qquad (10\text{-}110)$$

For apertures perpendicular to the direction of propagation, $\mathbf{p} = \mathbf{a}_z$, and

$$E_x = \frac{je^{-j\beta R}}{2\lambda R}(1 + \cos \theta)\int_{S_1} Fe^{j\beta \sin \theta\,(x' \cos \varphi + y' \sin \varphi)}\, da \qquad (10\text{-}111)$$

This equation is valid for small θ only.   Therefore

$$1 + \cos \theta \approx 2 \qquad (10\text{-}112)$$

so that, finally,

$$E_x = \frac{je^{-j\beta R}}{\lambda R}\int_{S_1} Fe^{j\beta \sin \theta\,(x' \cos \varphi + y' \sin \varphi)}\, da \qquad (10\text{-}113)$$

When the illumination function F can be represented as a product of two functions,

$$F = Ae^{-j\beta L} = F_1(x')F_2(y') \qquad (10\text{-}114)$$

the scalar wave function E_x is the product of two Fourier integrals,

$$E_x = \frac{je^{-j\beta R}}{\lambda R}\int_{x'} F_1(x')e^{j\beta x' \sin \theta \cos \varphi}\, dx' \int_{y'} F_2(y')e^{j\beta y' \sin \theta \sin \varphi}\, dy' \quad (10\text{-}115)$$

which usually provide the point of departure in the analysis of aperture antennas.

From the preceding derivation it is clear that Eq. (10-115) is subject to several restrictions, namely:

1. Harmonic time variations
2. Linear, homogeneous, isotropic, and source-free medium
3. Zero tangential field intensities over the complement of S_1
4. $r \gg \lambda/2\pi$
5. Plane aperture S_1
6. $\mathbf{u}_r \cdot \mathbf{a}_x \approx 0$
7. $\mathbf{a}_z \cdot \mathbf{u}_r \approx \mathbf{a}_z \cdot \mathbf{a}_r = \cos\theta$
8. $\dfrac{e^{-j\beta r}}{r} \approx \dfrac{e^{-j\beta R}}{R}\, e^{j\beta \sin\theta(x'\cos\varphi + y'\sin\varphi)}$
9. Incidence along the normal to the aperture ($\mathbf{a}_z \cdot \mathbf{p} = 1$)
10. $1 + \cos\theta \approx 2$
11. Separable illumination function

In closing, it is important to remark that the radial distance which marks the boundary between the Fresnel and Fraunhofer regions is taken in practice to be $2D^2/\lambda$, where D is the maximum linear dimension of the aperture. In terms of Eq. (10-108), this corresponds to a maximum phase deviation of $\pi/8$ deg. To prove this statement, we note that the second term in the numerator of the fraction may be neglected in comparison with the first, so that

$$\frac{2\pi}{\lambda}\frac{x'^2 + y'^2}{2R} \leq \frac{2\pi}{\lambda}\frac{(D/2)^2}{2R} < \frac{\pi}{8}$$

or
$$\frac{2D^2}{\lambda} < R \tag{10-116}$$

Let us now apply Eq. (10-113) to a specific problem.

Example 10-6   Uniformly Illuminated Rectangular Aperture. Let a rectangular aperture in the xy plane (Fig. 10-16) be centered about the origin, and suppose that the E field is uniform and is polarized in the x direction; that is, in Eq. (10-113), let

$$F(x',y') = \begin{cases} E_0 & -\dfrac{a}{2} \leq x' \leq \dfrac{a}{2}, \quad -\dfrac{b}{2} \leq y' \leq \dfrac{b}{2} \\ 0 & \text{elsewhere} \end{cases} \tag{10-117}$$

We wish to derive expressions for the scalar far field and to determine the 3-dB, or half-power, beamwidths.

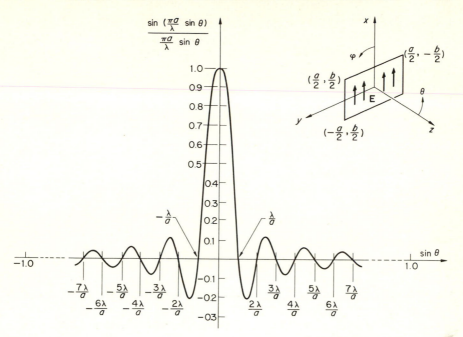

FIGURE 10-16.   Radiation from a uniformly illuminated rectangular aperture.

We have

$$
\begin{aligned}
E_x &= \frac{je^{-j\beta R}}{\lambda R}\int_{-b/2}^{b/2}\int_{-a/2}^{-a/2} E_0 e^{j\beta \sin\theta(x'\cos\varphi + y'\sin\varphi)}\,dx'\,dy' \\[2mm]
&= \frac{je^{-j\beta R}}{\lambda R}\int_{-a/2}^{a/2} E_0 e^{jx'\,\beta\sin\theta\cos\varphi}\,dx'\int_{-b/2}^{b/2} e^{jy'\,\beta\sin\theta\sin\varphi}\,dy' \\[2mm]
&= \frac{je^{-j\beta R}}{\lambda R}\left[aE_0\,\frac{\sin\left(\dfrac{a}{2}\beta\sin\theta\cos\varphi\right)}{\dfrac{a}{2}\beta\sin\theta\cos\varphi}\right]\left[b\,\frac{\sin\left(\dfrac{b}{2}\beta\sin\theta\sin\varphi\right)}{\dfrac{b}{2}\beta\sin\theta\sin\varphi}\right]
\end{aligned}
$$

$$(10\text{-}118)$$

In the $\varphi = 0°$ plane, which in this case is the E plane, this expression reduces to

$$
E_x(\varphi = 0°) = \frac{je^{-j\beta R}}{\lambda R}\,(ab)E_0\,\frac{\sin\left[(\pi a/\lambda)\sin\theta\right]}{(\pi a/\lambda)\sin\theta}
$$

while in the $\varphi = 90°$ plane, which in this case is the H plane, it becomes

$$
E_x(\varphi = 90°) = \frac{je^{-j\beta R}}{\lambda R}\,(ab)E_0\,\frac{\sin\left[(\pi b/\lambda)\sin\theta\right]}{(\pi b/\lambda)\sin\theta}
$$

In either plane, the radiation pattern exhibits a dependence of the type displayed in Fig. 10-16.

The 3-dB beamwidth in the $\varphi = 0°$ plane is obtained by setting

$$0.707 = \frac{\sin\left[(\pi a/\lambda)\sin\theta_0\right]}{(\pi a/\lambda)\sin\theta_0}$$

and solving for the angle θ_0. (For the $\varphi = 90°$ plane, a is simply replaced by b in this expression.)   For small values of the argument $x = (\pi a/\lambda)\sin\theta_0$,

$$\sin x \approx x - \frac{x^3}{6} + \frac{x^5}{120}$$

Therefore $$0.707 = \frac{x - x^3/6 + x^5/120}{x}$$

from which it follows that $x \approx 1.4$, or $\theta_0 \approx \sin^{-1}(1.4\,\lambda/\pi a)$.   Thus, in the $\varphi = 0°$ plane,

$$\text{3-dB beamwidth} = 2\sin^{-1}\left(0.445\frac{\lambda}{a}\right) \approx 51\frac{\lambda}{a} \quad \text{deg} \qquad (10\text{-}119)$$

while in the $\varphi = 90°$ plane,

$$\text{3-dB beamwidth} \approx 51\frac{\lambda}{b} \quad \text{deg} \qquad (10\text{-}120)$$

Clearly, the larger the aperture, the smaller the beamwidth.

The usual type of aperture-antenna problem consists in finding the aperture field which will optimize the far-field pattern to specific requirements. A few general characteristics are:

1. Symmetrical apertures with A and L [Eq. (10-102)] symmetrical about the center of the aperture produce symmetrical far-field patterns.
2. $A = $ constant, $L = 0$, and a symmetrical aperture produces zero nulls in the far-field pattern, and the main beam is the narrowest that can be obtained for the given aperture size.   However, the secondary maxima (*side lobes*) are high (see previous example).
3. Larger apertures give narrower beams.
4. $L = 0$ and A symmetrical but monotonically decreasing from the center of the aperture will, in general, decrease the secondary maxima.
5. In general, $L \neq 0$ will produce a moderate increase in main beamwidth compared with the uniform-phase case, and the pattern will have nonzero nulls.

Since narrow beamwidth and low side-lobe level are very common aperture-antenna requirements, a great deal of work has been done on defining the "best" amplitude distribution to optimize the conflicting requirements.

To sum up, this section has given a brief introduction to aperture antennas.   Such antennas may be analyzed with the aid of Eqs. (10-94) and (10-95) or, when more assumptions are allowed, with the aid of Eq. (10-113).

10.14   Aperture Synthesis.†   A field synthesis problem of great importance is that of designing radiating systems to produce desired radiation characteristics.   In most practical situations, the designer of a linear or planar array is required to solve the following problem.   Given the specified radiation pattern, how must the array be arranged, and how must the individual elements be excited in order to produce the best (in some prespecified sense) approximation to the given pattern?   One method of attacking this problem is known as the *Fourier synthesis method*, based on Schelkunoff's‡ early mathematical treatment of arrays of isotropic sources.

Recall that the array factor for uniform linear arrays is given by Eq. (10-88).   If the amplitude and phase progression are not constant, the corresponding expression for the radiation pattern factor will be

$$F(\psi) = B_0 + B_1 e^{j\psi} + B_2 e^{j2\psi} + \cdots + B_{n-1} e^{j(n-1)\psi} \qquad (10\text{-}121)$$

where $B_k = |B_k| \, e^{j\alpha k}$ is the complex amplitude and phase of the kth element, and

$$\psi = \beta d \cos \varphi \qquad (10\text{-}122)$$

as in Sec. 10.12.

In Eq. (10-121) the left-end element is the reference element.   If the center element is taken as the reference element, as in Fig. 10-17, the radiation pattern for an array with $2N + 1$ elements will be

$$F(\psi) = A_{-N} e^{-jN\psi} + \cdots + A_{-1} e^{-j\psi} + A_0 + A_1 e^{j\psi} + \cdots + A_N e^{jN\psi}$$

$$(10\text{-}123)$$

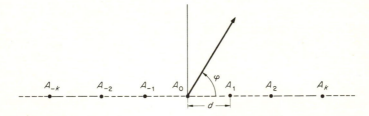

FIGURE 10-17.   Array of $2N + 1$ equally spaced isotropic point sources.

† This section was written in collaboration with G. W. Breland.   This section may be omitted with no loss in continuity.

‡ S. A. Schelkunoff, A Mathematical Theory of Linear Arrays, *Bell System Tech. J.*, vol. 22, pp. 80–107, 1943.

where $A_k = |A_k|\, e^{j\alpha k}$ is the complex amplitude of the kth element. In compact notation, this can be written

$$F(\psi) = A_0 + \sum_{k=1}^{N}(A_{-k}e^{-jk\psi} + A_k e^{jk\psi}) \qquad (10\text{-}124)$$

If, further, we demand that the excitations of elements equidistant from the center element be complex conjugate pairs, that is,

$$A_{-k} = A_k^* \qquad (10\text{-}125)$$

then the pattern factor of the array becomes purely real, since now

$$F(\psi) = a_0 + \sum_{k=1}^{N}(a_k + jb_k)(\cos k\psi - j\sin k\psi)$$

$$+ \sum_{k=1}^{N}(a_k - jb_k)(\cos k\psi + j\sin k\psi)$$

$$= a_0 + 2\sum_{k=1}^{N}(a_k \cos k\psi + b_k \sin k\psi) \quad (10\text{-}126)$$

where we have set

$$A_0 = a_0 + j0$$
$$\qquad\qquad\qquad\qquad (10\text{-}127)$$
$$A_{-k} = a_k + jb_k$$

We recognize Eq. (10-126) as having the appearance of a finite Fourier series for $F(\psi)$. It will actually be a Fourier series if $F(\psi)$ is a periodic function with a periodicity of 2π. From Eq. (10-122) it is apparent that

$$\psi_{\varphi=0} = \beta d$$
$$\qquad\qquad\qquad\qquad (10\text{-}128)$$
$$\psi_{\varphi=\pi} = -\beta d$$

We see that $F(\psi)$ satisfies the periodicity requirement only if $2\beta d = 2\pi$, or since $\beta = 2\pi/\lambda$, only if† $d = \lambda/2$. If $d = \lambda/2$, the excitation coefficients are

---

† The requirement that $d = \lambda/2$ can be changed to $d \leq \lambda/2$ by making $F(\psi)$ pseudo-periodic by use of a fill-in function. In this case the method leads to a nonunique solution in that each new choice of fill-in function gives a new solution.

calculated by the usual methods of Fourier analysis. They are thus given by

$$a_0 = \frac{1}{2\beta d} \int_{-\beta d}^{\beta d} F(\psi)\, d\psi \tag{10-129}$$

$$a_k = \frac{1}{2\beta d} \int_{-\beta d}^{\beta d} F(\psi) \cos k\psi\, d\psi \tag{10-130}$$

$$b_k = \frac{1}{2\beta d} \int_{-\beta d}^{\beta d} F(\psi) \sin k\psi\, d\psi \tag{10-131}$$

It is worth noting that, for a given number of equally spaced radiators, this finite Fourier series approximation for any given radiation pattern is optimum in the mean-square sense.

Example 10-7 Fourier Synthesis. As an example, let us approximate the idealized radiation pattern shown in Fig. 10-18a. It is clear that such a pattern is not exactly realizable with a finite number of radiators, but the mathematical simplicity of this type of pattern makes it attractive for illustrative purposes.

Noting that Fig. 10-18a specifies the pattern as a function of the angle φ, we apply the simple transformation $\psi = \beta d \cos \varphi$, in which case we obtain Fig. 10-18b, noting

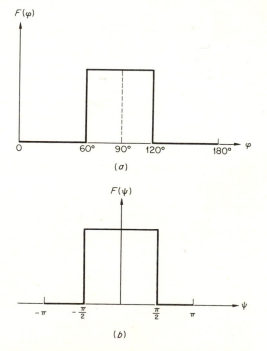

FIGURE 10-18. Example on Fourier synthesis. (a) Pattern to be approximated by Fourier synthesis method; (b) pattern after change of variable from $F(\varphi)$ to $F(\psi)$.

the change on the horizontal axis.   Applying the expressions for the Fourier coefficients, we obtain

$$a_0 = \frac{1}{2\pi} \int_{-\pi}^{\pi} F(\psi) \, d\psi$$

$$= \frac{1}{2\pi} \int_{-\pi/2}^{\pi/2} d\psi = 0.5$$

$$a_k = \frac{1}{2\pi} \int_{-\pi}^{\pi} F(\psi) \cos k\psi \, d\psi$$

(10-132)

$$= \frac{1}{2\pi} \int_{-\pi/2}^{\pi} \cos k\psi \, d\psi = \begin{cases} \dfrac{1}{k\pi} \sin \dfrac{k\pi}{2} & k \text{ odd} \\ 0 & k \text{ even} \end{cases}$$

$$b_k = \frac{1}{2\pi} \int_{-\pi}^{\pi} F(\psi) \sin k\psi \, d\psi = 0 \qquad \text{for all } k$$

Table 15-1 displays values of a_k and b_k for values of k to 15.   Figure 10-19 illustrates
the prescribed pattern and representative synthesized patterns for values of φ between
90° (broadside) and 180° (end-fire) for arrays having 3, 7, and 15 elements.   We note,
particularly, the results of increasingly larger numbers of elements in that the approximation to the desired pattern is improved.

Fourier synthesis has thus been demonstrated for aperture design where
discrete radiators are employed.   The task of presenting a complete survey

TABLE 15-1.   ELEMENT EXCITATIONS FOR FOURIER
SYNTHESIS EXAMPLE

Element	a (Real parts)	b (Imaginary parts)
0	0.5	0
1	0.3183	0
2	0	0
3	−0.1061	0
4	0	0
5	0.06332	0
6	0	0
7	−0.04547	0
8	0	0
9	0.03537	0
10	0	0
11	−0.02894	0
12	0	0
13	0.02448	0
14	0	0
15	−0.02122	0

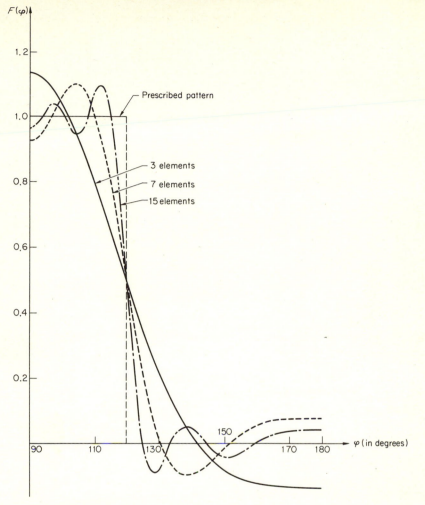

FIGURE 10-19. Example on Fourier synthesis.

of all the synthesis methods is prohibitive in length and completely out of place in such an introductory treatment as was intended here.   The interested reader is referred to the suggested references† for coverage of the numerous other techniques of antenna design.

† P. M. Woodward, A Method of Calculating the Field over a Plane Aperture Required to Produce a Given Polar Diagram, *J. IEE*, vol. 93, pt. III-A, pp. 1554–1558, 1947.

    C. L. Dolph, A Current Distribution Which Optimizes the Relationship between Beamwidth and Sidelobe Levels, *Proc. IRE*, vol. 34, pp. 335–348, June, 1946.

    T. T. Taylor, Design of Line Source Antennas for Narrow Beamwidths and Low Sidelobes, *IRE Trans. on Antennas and Propagation*, vol. AP-3, pp. 16–28, January, 1955.

10.15 Reciprocity. The central theme of this section is the reciprocity theorem, a theorem which follows from the symmetry of Maxwell's equations, providing a powerful tool for the solution of electromagnetic field problems.

Let a system of sources J_1, ρ_1 produce a field E_1, H_1, and let a second system of sources J_2, ρ_2, operating at the same frequency, produce a second field E_2, H_2. If at all points of interest the medium is linear and isotropic, but not necessarily homogeneous, then

$$D = \epsilon E \qquad B = \mu H$$

where ϵ and μ are scalars possibly dependent on position, and

$$\nabla \times E_1 = -j\omega\mu H_1$$
$$\nabla \times H_1 = J_1 + j\omega\epsilon E_1 \tag{10-133}$$

while, similarly,

$$\nabla \times E_2 = -j\omega\mu H_2$$
$$\nabla \times H_2 = J_2 + j\omega\epsilon E_2 \tag{10-134}$$

Dot-multiplying the second equation in Eqs. (10-133) by E_2 and the second equation in Eqs. (10-134) by E_1 and subtracting, we obtain

$$E_2 \cdot (\nabla \times H_1) - E_1 \cdot (\nabla \times H_2) = E_2 \cdot J_1 - E_1 \cdot J_2 \tag{10-135}$$

By dot-multiplying the first equations in Eqs. (10-133) and (10-134) by H_2 and H_1, respectively, and subtracting, as before, we obtain

$$H_2 \cdot (\nabla \times E_1) - H_1 \cdot (\nabla \times E_2) = 0 \tag{10-136}$$

Adding Eqs. (10-135) and (10-136) and applying the identity

$$\nabla \cdot (a \times b) = b \cdot (\nabla \times a) - a \cdot (\nabla \times b) \tag{10-137}$$

to the left side of the resulting expression, gives

$$\nabla \cdot (E_1 \times H_2 - E_2 \times H_1) = E_2 \cdot J_1 - E_1 \cdot J_2 \tag{10-138}$$

This is the differential form of the general reciprocity theorem.   In a current-free region, $J_1 = J_2 = 0$, and

$$\nabla \cdot (E_1 \times H_2 - E_2 \times H_1) = 0 \tag{10-139}$$

This is the earliest form of the reciprocity theorem as derived by Lorentz.

The integral form of the reciprocity theorem is obtained by integrating both sides of Eq. (10-138) over a volume V, enclosed by a surface Σ, and then applying the divergence theorem to the left side of the resulting expression. We find

$$\oint_{\Sigma} (\mathbf{E}_1 \times \mathbf{H}_2 - \mathbf{E}_2 \times \mathbf{H}_1) \cdot \mathbf{n} \, da = \int_{V} (\mathbf{E}_2 \cdot \mathbf{J}_1 - \mathbf{E}_1 \cdot \mathbf{J}_2) \, dv \quad (10\text{-}140)$$

where $\mathbf{n}$, as usual, is the unit vector drawn in the outward direction at every point on the surface Σ. Although quite general, this form of the reciprocity theorem is not the most practical.

Let us assume that both systems of sources are within a finite distance from some arbitrary origin. If Σ is a large spherical surface centered about that origin, the fields $\mathbf{E}_1$, $\mathbf{H}_1$ and $\mathbf{E}_2$, $\mathbf{H}_2$ are both locally plane at every point on the spherical surface, where, as a result,

$$\mathbf{H}_1 = \sqrt{\frac{\epsilon}{\mu}} \, (\mathbf{n} \times \mathbf{E}_1) \qquad \mathbf{H}_2 = \sqrt{\frac{\epsilon}{\mu}} \, (\mathbf{n} \times \mathbf{E}_2)$$

Accordingly, the integrand on the left of Eq. (10-140) can be written

$$\mathbf{E}_1 \times \mathbf{H}_2 - \mathbf{E}_2 \times \mathbf{H}_1 = \sqrt{\frac{\epsilon}{\mu}} \, [\mathbf{E}_1 \times (\mathbf{n} \times \mathbf{E}_2) - \mathbf{E}_2 \times (\mathbf{n} \times \mathbf{E}_1)]$$

$$= \sqrt{\frac{\epsilon}{\mu}} \, [(\mathbf{E}_1 \cdot \mathbf{E}_2)\mathbf{n} - (\mathbf{E}_1 \cdot \mathbf{n})\mathbf{E}_2 - (\mathbf{E}_1 \cdot \mathbf{E}_2)\mathbf{n} + (\mathbf{E}_2 \cdot \mathbf{n})\mathbf{E}_1] \quad (10\text{-}141)$$

where the last equality follows from the identity

$$\mathbf{a} \times (\mathbf{b} \times \mathbf{c}) = (\mathbf{a} \cdot \mathbf{c})\mathbf{b} - (\mathbf{a} \cdot \mathbf{b})\mathbf{c}$$

Since the waves are locally plane at every point on the large sphere, it follows that

$$\mathbf{E}_1 \cdot \mathbf{n} = \mathbf{E}_2 \cdot \mathbf{n} = 0$$

The right side of Eq. (10-141) therefore vanishes, and in this case the reciprocity theorem becomes

$$\oint_{\Sigma} (\mathbf{E}_1 \times \mathbf{H}_2 - \mathbf{E}_2 \times \mathbf{H}_1) \cdot \mathbf{n} \, da = \int_{V} (\mathbf{E}_2 \cdot \mathbf{J}_1 - \mathbf{E}_1 \cdot \mathbf{J}_2) \, dv = 0$$

The results of this analysis, namely,

$$\oint_{\Sigma} (\mathbf{E}_1 \times \mathbf{H}_2 - \mathbf{E}_2 \times \mathbf{H}_1) \cdot \mathbf{n} \, da = 0 \qquad (10\text{-}142)$$

$$\int_{V} (\mathbf{E}_2 \cdot \mathbf{J}_1 - \mathbf{E}_1 \cdot \mathbf{J}_2) \, dv = 0 \qquad (10\text{-}143)$$

can now be developed into a number of reciprocity theorems of fundamental importance in circuit and in field theory. For example, Eq. (10-143) can

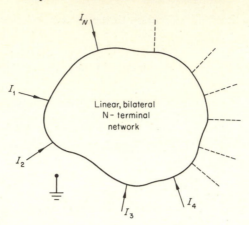

FIGURE 10-20.   Reciprocity in networks.

be used immediately to show that the impedance and admittance matrices of a linear bilateral network are both symmetric.

It will be demonstrated in Chap. 11 that a circuit is a configuration of electromagnetic structures which allow specialized fields to exist in confined regions of space, beyond which the field is everywhere essentially zero. Accordingly, the left-hand side of Eq. (10-140) is zero if Σ is a closed surface completely surrounding the entire circuit, and Eq. (10-143) holds identically. As in Sec. 6.7, each of the two terms in this integral can be transformed to give

$$\int_V (\mathbf{E}_2 \cdot \mathbf{J}_1 - \mathbf{E}_1 \cdot \mathbf{J}_2)\, dv = \sum_{k=1}^{N} (\phi_{2k} I_{1k} - \phi_{1k} I_{2k}) = 0 \qquad (10\text{-}144)$$

where the ϕ's and the I's denote the terminal variables of a network with N terminals (Fig. 10-20), and the subscripts 1 and 2 distinguish two solutions of the network problem. It is tacitly assumed that the potentials ϕ_{1k} and ϕ_{2k} are measured relative to some arbitrary ground.

Since Eq. (10-144) holds for any set of applied voltages at the terminals of the network, we can let

$$
\begin{aligned}
\phi_{1k} = 0 \quad k \neq i \qquad \phi_{1k} \neq 0 \quad k = i \\
\phi_{2k} = 0 \quad k \neq j \qquad \phi_{2k} \neq 0 \quad k = j
\end{aligned}
\qquad (10\text{-}145)
$$

where i and j are any integers between 1 and N. In this special case, the sum in Eq. (10-144) reduces to

$$\phi_{2j} I_{1j} - \phi_{1i} I_{2i} = 0 \qquad (10\text{-}146)$$

Since $\qquad\qquad\qquad I_{2i} = y_{ij}\phi_{2j} \qquad I_{1j} = y_{ji}\phi_{1i}$

where y_{ij} and y_{ji} denote transfer admittances, Eq. (10-146) simplifies to

$$y_{ij} = y_{ji} \qquad (10\text{-}147)$$

which satisfies the criterion of symmetry. In a similar way it can be shown that

$$z_{ij} = z_{ji} \qquad (10\text{-}148)$$

Thus *the admittance and impedance matrices of a linear bilateral network are always symmetrical.* Stated differently, the ratio of the voltage applied at one port of a network to the current flowing at a second, when all other terminals are open-circuited, remains unchanged if the voltage is applied to the second port, and the current is measured at the first.

It will be seen that this same notion can be extended to show that *the receiving antenna pattern is the same as the transmitting pattern.*

We recall that the pattern of a transmitting antenna is the aggregate of values of relative field (or power) intensity measured at all points on a large spherical surface centered about the phase center of the antenna. If we divide the spherical surface into a large number N of contiguous and infinitesimally small elements of surface, the pattern could be taken by moving a small exploring dipole from one small element of surface to the next, and measuring the current I flowing in the short dipole. The exploring dipole must be placed flat against each surface element and must be oriented at each stop along the way in both the θ and φ directions in order to take into account possible polarization effects. We can describe what is actually happening here by considering the transmitting antenna and the N positions on the surface as a single network with a total of $N + 1$ terminals, which, by our previous results, is therefore characterized by symmetrical impedance and admittance matrices. Consequently, for every location on the surface, the ratio of transmitted and received energy is the same regardless of whether the antenna is used as a transmitter and the dipole as a receiver, or vice versa. In practice, the antenna under test is usually rotated about two mutually orthogonal axes, while the source is transmitting a fixed amount of power from a distance normally equal to, or greater than, $2D^2/\lambda$.

It is worth noting in closing that the reciprocity theorem can also be used to prove that *the antenna impedance for transmitting is the same as the internal impedance of a generator representing the antenna for receiving.*

10.16 Summary. This chapter has presented an introduction to the important subject of radiation from antennas. The first topic discussed was the relation of the field to its sources. It was shown that considerable simplification of the problem results when only the far field is required. The concepts of antenna patterns, directivity, and gain were developed and defined.   As specific examples of antenna far-field calculations, the far fields of linear dipoles were obtained.   Antenna arrays were discussed and specific results obtained for uniform linear arrays.   Finally, an introduction to aperture antennas and the aperture field method for obtaining and synthesizing the far-field pattern of aperture antennas was presented.

Problems

10-1   Field of a Hertzian Dipole (Review).   For a current element, the θ component of the electric field intensity is given by

$$E_\theta = \frac{I\,dl\sin\theta}{4\pi\epsilon}\left(\frac{-\omega\sin\omega t'}{rv^2} + \frac{\cos\omega t'}{r^2 v} + \frac{\sin\omega t'}{\omega r^3}\right)$$

where $t' = t - r/v$.   Identify each term (in powers of $1/r$) in this expression by name, and single out the term which is most important in the study of antennas.

10-2   Charge Distribution.   In a quarter-wave monopole mounted over a perfect ground, the approximate current distribution is given by the expression $I_m \cos\beta z e^{j\omega t}$, where $z = 0$ coincides with the feed point of the antenna.   Determine the charge distribution in the antenna.   Is it in phase with the current distribution?

10-3   Complex Field of a Hertzian Dipole.   Prove that in complex notation the field of an elementary dipole is specified by

$$E_\theta = \frac{\eta I\,dl\sin\theta e^{-j\beta r}}{4\pi}\left(\frac{j\beta}{r} + \frac{1}{r^2} + \frac{1}{j\beta r^3}\right)$$

$$E_r = \frac{\eta I\,dl\cos\theta e^{-j\beta r}}{4\pi}\left(\frac{2}{r^2} + \frac{2}{j\beta r^3}\right)$$

$$H_\varphi = \frac{I\,dl\sin\theta e^{-j\beta r}}{4\pi r}\left(j\beta + \frac{1}{r}\right)$$

where the factor $e^{j\omega t}$ is understood.   Note that these are the frequency-domain representations of Eqs. (10-24) to (10-26).

10-4   Hertz Potential.   The radiation field is often derived from a single potential, known as the *Hertz potential*, defined by

$$\mathbf{A} = \mu\epsilon\frac{\partial\mathbf{\Pi}}{\partial t}\qquad \phi = -\nabla\cdot\mathbf{\Pi}$$

Prove that in a linear, homogeneous, isotropic, and source-free medium

$$\nabla^2\mathbf{\Pi} - \mu\epsilon\frac{\partial^2\mathbf{\Pi}}{\partial t^2} = 0$$

and that

$$\mathbf{E} = \nabla(\nabla\cdot\mathbf{\Pi}) - \mu\epsilon\frac{\partial^2\mathbf{\Pi}}{\partial t^2}\qquad \mathbf{B} = \mu\epsilon\nabla\times\frac{\partial\mathbf{\Pi}}{\partial t}$$

10-5   Polarization.   Two short-wire antennas are excited by currents of equal amplitude and phase.   A distant receiving antenna is located on the z axis.   If this is a short linear antenna, how should it be oriented to intercept maximum signal?

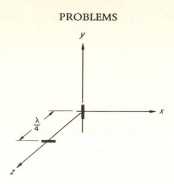

PROBLEM 10-5

10-6 Loop Antenna. A loop antenna consists of one or more complete turns of wire carrying, essentially, a uniform circulatory current and producing a radiation pattern approximating that of an elementary magnetic dipole. (An elementary magnetic dipole is a loop with a uniform in-phase current and linear dimensions which are very small compared with the wavelength.)

Show that the far field of a small loop antenna, the plane of which is coincident with the xy plane, is given by

$$E_\varphi = \frac{\beta^2}{4\pi} \sqrt{\frac{\mu}{\epsilon}} \frac{m \sin \theta \, e^{-j\beta r}}{r} \qquad H_\theta = -\frac{E_\varphi}{\eta}$$

where m is the strength of the dipole (current times area), and r, θ, and φ are the usual spherical coordinates. The time factor $e^{j\omega t}$ is understood.

10-7 Array Factor. A linear array is formed by spacing $\lambda/2$ apart three short dipoles. The currents fed into all three dipoles are equal in magnitude and in time phase. Derive an expression, and plot the array-factor pattern in a plane normal to the elements of the array.

10-8 Antenna Arrays. Two thin linear antennas are operated at the same frequency and are positioned a half-wavelength apart in the xy plane, as shown. The antennas are individually isotropic radiators in the xy plane. When the impressed terminal currents are $I_1 = I_2 = 1\underline{/0°}$ A, the field intensity at point P in the far field due to the array has a magnitude of 2 μV/m.

Determine the magnitude of the field intensity at the same point P when the impressed currents are $I_1 = e^{j\pi/2}$ A and $I_2 = 2e^{j0}$ A.

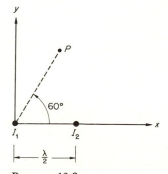

PROBLEM 10-8

10-9 Uniform Linear Arrays. Calculate the width between nulls for a six-element uniform linear array of spacing $d = \lambda/4$ (a) for a broadside array and (b) for an end-fire array.

10-10 Uniform Linear Arrays. Calculate the half-power (3-dB) beamwidth for the arrays of Prob. 10-9.

10-11 Aperture Antennas. A square aperture antenna of dimensions $a = b = 20\lambda$ is illuminated with an x-polarized uniform plane wave. Calculate the angular positions of the nulls of this pattern in the xz plane and in the yz plane.

10-12 Aperture Antennas. In Example 10-6, let

$$F(x',y') = \begin{cases} E_0 e^{j\beta x' \sin \theta_0} & -\dfrac{a}{2} \leq x' \leq \dfrac{a}{2}, \quad -\dfrac{b}{2} \leq y' \leq \dfrac{b}{2} \\ 0 & \text{elsewhere} \end{cases}$$

Derive an expression for and plot the scalar far field as a function of θ. The angle θ_0 is fixed, and $\beta = 2\pi/\lambda$.

10-13 Aperture Antennas. In Example 10-6, let

$$F(x',y') = \begin{cases} \dfrac{E_0}{2}\left(1 + \cos\dfrac{2\pi x'}{a}\right) & -\dfrac{a}{2} \leq x' \leq \dfrac{a}{2}, \quad -\dfrac{b}{2} \leq y' \leq \dfrac{b}{2} \\ 0 & \text{elsewhere} \end{cases}$$

Plot the θ variation of the far field in the xz plane, and determine the 3-dB beamwidth.

Note that the beamwidth of this pattern is larger than that of the uniformly illuminated aperture, but the side lobes are smaller in this case.

10-14 Electrodynamic Similitude. When antennas are located on large structures such as airplanes, ships, or satellites, it becomes unfeasible to test the antenna characteristics on the full-scale model. Instead, a scale model is normally constructed, and the tests are performed at a frequency different from the design frequency. The exact relationships of frequency to scaled dimensions can be determined directly from Maxwell's equations, and are expressible as a pair of equations

$$\mu\epsilon\left(\frac{l}{\tau}\right)^2 = \text{constant} \qquad \mu\sigma\frac{l^2}{\tau} = \text{constant}$$

where l and τ represent distance and time scale factors, and which together constitute a statement of the *principle of electrodynamic similitude*. In essence, this pair of equations requires that when the linear dimensions of the model are reduced to $1/n$ times those of the full-scale antenna, the test frequency and conductivity of the materials used in the model must be n times those of the full-scale antenna. (*Note:* Because of the high conductivity of the metals, the requirement for a change in conductivity is usually disregarded in practice.) At the same time, ϵ and μ of the materials used in the model must be the same as in the full-scale antenna.

Consider a 30-MHz antenna, the size of which is to be reduced by a factor of 4 for testing purposes.

(*a*) Determine the operating frequency.

(*b*) If the antenna includes a dielectric substance as a part of the radiating system, prove that the loss tangent ($\sigma/\omega\epsilon$) of the dielectric used in the test must be the same as in the full-scale antenna.

PROBLEM 10-14. Testing of scaled aircraft antennas. (*Courtesy of Lockheed-Georgia Company.*)

10-15 Aperture Synthesis. Show that the far field of a line source and the illumination of that line source form a Fourier transform pair.

10-16 Aperture Synthesis. Discuss in detail why Fourier synthesis is unique only for half-wavelength spacing of elements.

10-17 Aperture Synthesis. Approximate the following pattern with an array of seven elements, using Fourier synthesis.

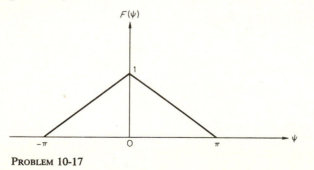

PROBLEM 10-17

10-18 Mature Approach to Practical Problems. The following case studies are intended to test your maturity in electromagnetics. Each case may well arise in actual engineering practice. This, however, does not imply that a unique solution exists for all. On the contrary, it is quite probable that many of the situations presented may not possess unique solutions, while others may have no apparent solution at all. Nevertheless, for purposes of this exercise, you are expected to put forth a reasonable effort in order to come up with a technically plausible solution.

(a) A vertically polarized uniform plane wave impinges upon a body whose physical dimensions and electromagnetic properties are entirely arbitrary. Is the reflected (backscattered) wave vertically polarized also? Explain, and if your answer happens to be negative, give a counterexample.

(b) The term *reflectivity* is occasionally used in practice to refer to such quantities as Γ, which was defined in Chap. 8. In some instances, however, when this term is applied, the physical medium under consideration is a slab which is finite in extent in all directions. The question, then, is whether a mere specification of a numerical value for Γ is sufficient to provide complete identification of the electromagnetic properties of the slab, assuming it to be linear, homogeneous, and isotropic.

(c) Consider a cloud of electrically short dipoles (short pieces of wire) randomly dispersed in free space. Is it sufficient to know only the backscattering properties of an isolated dipole as a function of its orientation in order to predict the overall behavior of the cloud, assuming that the relative orientation of the various dipoles in the cloud has in some way been established a priori?

(d) A uniform plane wave is incident normally upon a perfectly conducting screen extending to infinity in both (transverse) directions. A hole cut out in the screen with dimensions which are comparable with, but not very much larger than, the wavelength allows a diffracted field to exist in the half space on the opposite side of the screen. Does the relative field distribution across the aperture (and hence the relative diffracted field itself) remain unaltered if an object is placed in front of the aperture, but slightly away from the plane of the screen, so as to block partially the direct rays impinging upon the aperture?

(e) Is mechanical pressure exerted on a transmitting antenna while in operation? On a receiving antenna? Explain.

(f) In designing a 30- to 70-MHz antenna, a practicing antenna engineer decides, because of tight structural specifications, to use a single-turn half-loop antenna

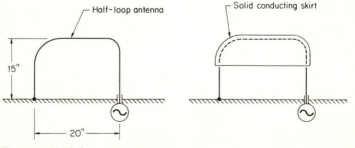

PROBLEM 10-18 f.

mounted over a ground plane, as indicated on the left of the figure.   Much to his disappointment, however, he discovers later that such an antenna has an extremely low radiation resistance, and on top of that, it presents rather severe matching limitations.   In order to circumvent the first problem, he decides to *top-load* the antenna by placing a conducting skirt over the loop, as shown on the right. Can you justify his action technically?   What would you have done in this case?

(g) Would you expect the jet exhaust from an aircraft engine to affect in any way the radiation characteristics of an antenna transmitting through, and in a direction normal to, the exhaust column?   Specifically, would you expect the antenna pattern to be altered in any way by the exhaust stream?

(h) Suppose you are flying above a layer of clouds and your problem is to detect a second object flying below the clouds but near enough to the earth's surface so that appreciable *clutter* background exists by virtue of the presence of the terrain below.   Give a preliminary design of a system to solve this problem, conceptually speaking.

(i) How would you communicate by wireless means in a thick forest if the distance separating you from a second party was less than, say, 10 miles?

CHAPTER 11

THE QUASISTATIC FIELD

11.1 Introduction. This chapter is devoted to a study of the field theory of lumped electric circuits and devices.

Early in our careers, most of us acquire the impression that an electric circuit is an interconnection of elements, identified through their terminal volt-ampere relationships: $V = IR$ for a resistor, $V = L\, dI/dt$ for an inductor, and $I = C\, dV/dt$ for a capacitor. Though this is all we need in an introductory course on electric circuits, the behavior of these elements at high frequencies is difficult to understand on the basis of classical circuit theory alone. It seems quite puzzling when later we discover that, for example, certain structures behave like inductors when their physical appearance is anything but that of an inductor. This is particularly true in the frequency range around 100 MHz and above.

In order to overcome such conceptual difficulties we need to realize that, with every type of circuit element, there is a *specific-type*, *frequency-dependent*, *localized electromagnetic field*.

A second point, often overlooked in basic presentations, is that the definition of *potential difference* is normally couched in terms of stationary electric field concepts. We saw in Sec. 3.3 that the value of the electrostatic potential at a point is unique and that potential difference between any two points is independent of connecting path. It is demonstrated in this chapter that under time-varying conditions the scalar potential cannot be defined uniquely as the line integral of electric field intensity. What, then, do we mean by potential difference in a circuit in which terminal variables change with time? We hope to give a satisfactory answer to this question, basing our arguments mainly on the fundamentals of field theory. Circuit theory alone cannot provide an answer.

Thus the major purpose of this chapter is to develop the ideas associated with electric circuits from the field theory point of view. A second, and equally important, purpose is to develop the mathematical formalism needed in the study of electromechanical devices and transformers. The common thread running through this chapter is the notion of *quasistatic* (slowly varying) fields, which we develop.

512

11.2 Fields in Lumped Circuit Elements. In previous chapters we have dealt essentially with two major topics, static fields and time-varying fields. Stated in terms of frequency, we have examined fields when the frequency was either zero or so high that the corresponding wavelength was small compared with linear dimensions of the structures considered. Thus we have left a gap in the frequency spectrum where wavelengths and linear dimensions are comparable with each other. What, then, are the field phenomena which take place at frequencies between these two extremes?

To begin answering this question, we consider the problem of a TEM wave guided by a pair of perfectly conducting parallel plates in a source-free nonconducting region characterized by the two scalar constants ϵ and μ. We suppose that the plates are large enough in the y direction (Fig. 11-1) so that, for all practical purposes, $\partial/\partial y = 0$. We also suppose that the field owes its existence to a set of harmonic sources connected to the plates at $z = -a$, and we require the solution to Maxwell's equations which satisfies the boundary condition $\mathbf{a}_z \times \mathbf{H} = 0$ at $z = 0$, and also the condition $\mathbf{a}_x \times \mathbf{E} = 0$ at $x = 0$ and at $x = d$.

A solution of Maxwell's equations which satisfies the second boundary condition is the familiar (from TEM waves) pair

$$E_x = E_+ e^{-j\beta z} + E_- e^{j\beta z} \tag{11-1}$$

$$H_y = \frac{1}{\eta} (E_+ e^{-j\beta z} - E_- e^{j\beta z}) \tag{11-2}$$

where $\beta = \omega\sqrt{\mu\epsilon}$, $\eta = \sqrt{\mu/\epsilon}$, and the factor $e^{j\omega t}$ is understood. Upon requiring that the tangential component of $\mathbf{H}$ vanish (open circuit) at $z = 0$,

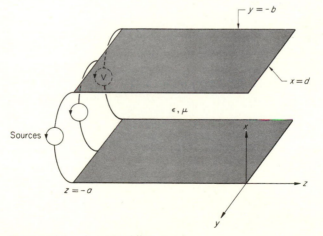

FIGURE 11-1. A waveguiding structure.

we obtain $E_+ = E_-$, and thus

$$E_x = E_+(e^{-j\beta z} + e^{j\beta z}) = 2E_+ \cos \beta z \tag{11-3}$$

$$H_y = \frac{E_+}{\eta}(e^{-j\beta z} - e^{j\beta z}) = -j2\frac{E_+}{\eta} \sin \beta z \tag{11-4}$$

We recognize Eqs. (11-3) and (11-4) as expressions for standing waves similar to those encountered in the study of transmission lines. Using the infinite series expansions

$$\cos x = 1 - \frac{x^2}{2!} + \frac{x^4}{4!} - \cdots + -\text{higher-order terms}$$

$$\sin x = x - \frac{x^3}{3!} + \frac{x^5}{5!} - \cdots + -\text{higher-order terms}$$

we find

$$E_x = 2E_+\left[1 - \frac{(\beta z)^2}{2!} + \frac{(\beta z)^4}{4!} - \cdots + -\text{higher-order terms}\right] \tag{11-5}$$

$$H_y = -j2\frac{E_+}{\eta}\left[\beta z - \frac{(\beta z)^3}{3!} + \frac{(\beta z)^5}{5!} - \cdots + -\text{higher-order terms}\right] \tag{11-6}$$

Since $\beta = \omega\sqrt{\mu\epsilon} = 2\pi/\lambda$, Eqs. (11-5) and (11-6) may be represented as power series expansions in ω.

$$E_x = 2E_+\left[1 - \frac{(z\sqrt{\mu\epsilon})^2}{2}\omega^2 + \frac{(z\sqrt{\mu\epsilon})^4}{4!}\omega^4 - \cdots + -\text{higher-order terms}\right] \tag{11-7}$$

$$H_y = -j2\frac{E_+}{\eta}\left[(z\sqrt{\mu\epsilon})\omega - \frac{(z\sqrt{\mu\epsilon})^3}{3!}\omega^3\right.$$
$$\left. + \frac{(z\sqrt{\mu\epsilon})^5}{5!}\omega^5 - \cdots + -\begin{array}{l}\text{higher-}\\\text{order terms}\end{array}\right] \tag{11-8}$$

Alternatively, using $\beta = 2\pi/\lambda$,

$$E_x = 2E_+\left[1 - \frac{(2\pi z/\lambda)^2}{2!} + \frac{(2\pi z/\lambda)^4}{4!} - \cdots + -\text{higher-order terms}\right] \tag{11-9}$$

$$H_y = -j2\frac{E_+}{\eta}\left[\frac{2\pi z}{\lambda} - \frac{(2\pi z/\lambda)^3}{3!} + \frac{(2\pi z/\lambda)^5}{5!} - \cdots + -\text{higher-order terms}\right] \tag{11-10}$$

Clearly, the argument $2\pi z/\lambda$ reaches a maximum (absolute) value at $z = -a$.

At low frequencies, a is small compared with wavelength, $2\pi a/\lambda \ll 1$, and in this case the field may be represented by the first few terms in the expansions.   Specifically, as the frequency approaches zero,

$$E_x = 2E_+ \qquad H_y = 0 \qquad\qquad (11\text{-}11)$$

where, it should be emphasized, the time dependence $e^{j\omega t}$ is understood. Since the magnetic field intensity is everywhere zero, no current will be drawn ($\mathbf{n} \times \mathbf{H} = \mathbf{K}$ at both plates) from the sources, and the field will be uniform. As a result, charge will appear on the surfaces of both plates with an absolute density $\rho_s = \epsilon E_x = 2\epsilon E_+$; the net charge on each plate will be

$$Q = \rho_s \times \text{area of plates} = \epsilon E_x(ab)$$

Since the field is uniform, the potential difference V between the plates will be unique, with $V = E_x d$.   Hence the capacitance seen by the sources will be

$$C = \frac{Q}{V} = \epsilon\,\frac{ab}{d} \qquad\qquad (11\text{-}12)$$

where ab is the area of the plates, d their separation, and ϵ the permittivity of the medium between the plates.

This is just the familiar expression for capacitance of a pair of parallel plates.   But inasmuch as the current drawn from the sources is zero, it is not at all clear how the capacitance C defined by Eq. (11-12) is in any way related to the capacitance C expressly defined by the volt-ampere relation $I = C\,dV/dt$.   Yet the equivalence of the two definitions is usually taken for granted.

To resolve this dilemma, we must take a closer look at Eqs. (11-7) and (11-8).   We assume that the frequency is higher, but is low enough so that, to a good approximation, the spatial dependence of the field is given by

$$E_x = 2E_+ \qquad H_y = -j2\,\frac{E_+}{\eta}\,(z\sqrt{\mu\epsilon})\omega \qquad\qquad (11\text{-}13)$$

In other words, for slow time variations the field is represented by terms up to and including the first power in ω in the two series expansions.   To the same approximation, potential difference is still defined by $V = 2E_+ d$. However, since $\mathbf{H}$ is no longer zero, current will be drawn from the sources, which, at $z = -a$, will be given by

$$I = \text{plate width} \times \text{surface current density} = b\left[j2\,\frac{E_+}{\eta}\,(a\sqrt{\mu\epsilon})\omega\right]$$

For harmonic time variations, the relation $I = C\,dV/dt$ reduces to $I = j\omega CV$. Hence the dynamic capacitance

$$C = \frac{I}{j\omega V} = \frac{ba\sqrt{\mu\epsilon}}{d\sqrt{\mu/\epsilon}} = \epsilon\,\frac{ab}{d} \qquad\qquad (11\text{-}14)$$

is obviously the same as the static capacitance.

Though in no way conclusive, the preceding development strongly suggests that under time-varying conditions the field in idealized lumped circuit elements is not purely periodic in space, like a wave, nor is it purely static. Indications are that it is nearly static, that is, quasistatic.

As the frequency is increased, more terms must be retained in the power series expansions. With an increasing number of terms, more questions arise as to their meaning from a circuits point of view. In due time, it will be seen that such terms can often be taken into account as equivalent inductances and capacitances interconnected to form circuits with equivalent terminal properties. Thus, as the frequency is allowed to increase, what appears at first to be a very simple structure eventually exhibits complex field and circuit properties.

11.3   An Iterative Solution for Time-varying Fields.†   To examine further the field basis of circuit theory, it is instructive at this point to introduce a series-expansion technique for field analysis. By this method, the solution to a dynamic-field problem is obtained as a power series in the frequency variable, each term of which is derived by iterating upon the preceding term of the series.

Let us assume that the field vectors contain the time only as a factor $e^{j\omega t}$, and write the field equations in the form

$$\nabla \times \mathbf{E} = -j\omega \mathbf{B} \tag{11-15}$$

$$\nabla \times \mathbf{H} = \mathbf{J} + j\omega \mathbf{D} \tag{11-16}$$

$$\nabla \cdot \mathbf{B} = 0 \tag{11-17}$$

$$\nabla \cdot \mathbf{D} = \rho \tag{11-18}$$

The medium is assumed to be linear, homogeneous, and isotropic. A natural consequence of the field equations is that the field vectors are generally dependent upon frequency, and also upon the space coordinates. This is quite obvious from inspection of Eqs. (11-15) and (11-16), where the angular frequency ω appears explicitly. It is reasonable, then, to expect that every field quantity can be expanded in a Taylor series about $\omega = 0$. Recalling the formula

$$f(x) = \sum_{n=0}^{\infty} \frac{(x - x_0)^n}{n!} \frac{d^n f(x_0)}{dx^n} \tag{11-19}$$

† R. M. Fano, L. J. Chu, and R. B. Adler: "Electromagnetic Fields, Energy and Forces," John Wiley & Sons, Inc., New York, 1960.

A. F. Stevenson, Solution of Electromagnetic Scattering Problems as Power Series in the Ratio (Dimensions of Scatterer)/Wavelength, *J. Appl. Phys.*, vol. 24, pp. 1134–1142, 1953.

we can write, in particular,

$$\mathbf{E}(x,y,z,\omega) = \sum_{n=0}^{\infty} \frac{\omega^n}{n!} \frac{\partial^n [\mathbf{E}(x,y,z,0)]}{\partial \omega^n} = \sum_{n=0}^{\infty} \omega^n \mathbf{E}_n \qquad (11\text{-}20)$$

where the coefficient of the nth power of ω, namely,

$$\mathbf{E}_n = \frac{1}{n!} \frac{\partial^n [\mathbf{E}(x,y,z,0)]}{\partial \omega^n} \qquad (11\text{-}21)$$

will be called the *nth-order* electric field intensity. It is recognized that all derivatives are to be evaluated at $\omega = 0$.   The development follows the same course for the other field quantities, with the corresponding results

$$\mathbf{B}(x,y,z,\omega) = \sum_{n=0}^{\infty} \omega^n \mathbf{B}_n \qquad \mathbf{B}_n = \frac{1}{n!} \frac{\partial^n [\mathbf{B}(x,y,z,0)]}{\partial \omega^n} \qquad (11\text{-}22)$$

$$\mathbf{D}(x,y,z,\omega) = \sum_{n=0}^{\infty} \omega^n \mathbf{D}_n \qquad \mathbf{D}_n = \frac{1}{n!} \frac{\partial^n [\mathbf{D}(x,y,z,0)]}{\partial \omega^n} \qquad (11\text{-}23)$$

$$\mathbf{H}(x,y,z,\omega) = \sum_{n=0}^{\infty} \omega^n \mathbf{H}_n \qquad \mathbf{H}_n = \frac{1}{n!} \frac{\partial^n [\mathbf{H}(x,y,z,0)]}{\partial \omega^n} \qquad (11\text{-}24)$$

$$\mathbf{J}(x,y,z,\omega) = \sum_{n=0}^{\infty} \omega^n \mathbf{J}_n \qquad \mathbf{J}_n = \frac{1}{n!} \frac{\partial^n [\mathbf{J}(x,y,z,0)]}{\partial \omega^n} \qquad (11\text{-}25)$$

$$\rho(x,y,z,\omega) = \sum_{n=0}^{\infty} \omega^n \rho_n \qquad \rho_n = \frac{1}{n!} \frac{\partial^n [\rho(x,y,z,0)]}{\partial \omega^n} \qquad (11\text{-}26)$$

Each of the six field quantities is thus expressed as a power series in ω.

Introducing Eqs. (11-21) and (11-22) into Eq. (11-15), we obtain

$$\nabla \times \left(\sum_{n=0}^{\infty} \omega^n \mathbf{E}_n \right) = -j\omega \sum_{n=0}^{\infty} \omega^n \mathbf{B}_n \qquad (11\text{-}27)$$

Interchanging the order of differentiation and summation and collecting terms, we find

$$\nabla \times \mathbf{E}_0 + \omega(\nabla \times \mathbf{E}_1 + j\mathbf{B}_0) + \omega^2(\nabla \times \mathbf{E}_2 + j\mathbf{B}_1) + \cdots$$
$$+ \text{ higher-order terms } = 0 \quad (11\text{-}28)$$

or in compact form,

$$\nabla \times \mathbf{E}_0 + \sum_{n=1}^{\infty} \omega^n (\nabla \times \mathbf{E}_n + j\mathbf{B}_{n-1}) = 0 \qquad (11\text{-}29)$$

If this equality is to hold for all ω, the coefficients of all the powers of ω must be equal to zero.   Thus

$$\nabla \times \mathbf{E}_0 = 0$$
$$\nabla \times \mathbf{E}_1 + j\mathbf{B}_0 = 0$$
$$\nabla \times \mathbf{E}_2 + j\mathbf{B}_1 = 0 \qquad (11\text{-}30)$$
$$\cdots \cdots \cdots \cdots \cdots \cdots$$
$$\nabla \times \mathbf{E}_n + j\mathbf{B}_{n-1} = 0$$

The other field equations can be treated in a similar manner.   We have

$$\nabla \times \mathbf{H}_0 = \mathbf{J}_0$$
$$\nabla \times \mathbf{H}_1 = \mathbf{J}_1 + j\mathbf{D}_0$$
$$\dots\dots\dots\dots\dots$$ \hfill (11-31)
$$\nabla \times \mathbf{H}_n = \mathbf{J}_n + j\mathbf{D}_{n-1}$$

Also

$$\nabla \cdot \mathbf{B}_n = 0 \qquad n = 0, 1, 2, 3, \dots \qquad (11\text{-}32)$$

$$\nabla \cdot \mathbf{D}_n = \rho_n \qquad n = 0, 1, 2, 3, \dots \qquad (11\text{-}33)$$

Finally, from the equation of continuity,

$$\nabla \cdot \mathbf{J} + j\omega\rho = 0 \qquad (11\text{-}34)$$

we obtain

$$\nabla \cdot \mathbf{J}_0 = 0$$
$$\nabla \cdot \mathbf{J}_1 = -j\rho_0$$
$$\dots\dots\dots\dots$$ \hfill (11-35)
$$\nabla \cdot \mathbf{J}_n = -j\rho_{n-1}$$

It is to be noted that Eqs. (11-32) and (11-33), unlike the rest, involve only field quantities of the *same order*.   This is because the angular frequency ω does not appear explicitly in Eqs. (11-17) and (11-18).   On the other hand, when ω does appear in an equation, as in Eq. (11-15), it brings about a coupling between field quantities of *sequential order*.   For example, the second equation in Eqs. (11-30) relates the *zero-order* **B** field to the *first-order* **E** field, while the third equation relates the *first-order* **B** field to the *second-order* **E** field.   This characteristic property allows us to calculate a field through a series of iterations.   Explicitly, we note that the *zero-order field* is a solution of

$$\nabla \times \mathbf{E}_0 = 0$$
$$\nabla \times \mathbf{H}_0 = \mathbf{J}_0$$
$$\nabla \cdot \mathbf{B}_0 = 0 \qquad (11\text{-}36)$$
$$\nabla \cdot \mathbf{D}_0 = \rho_0$$
$$\nabla \cdot \mathbf{J}_0 = 0$$

and that the *first-order field* is a solution of

$$\nabla \times \mathbf{E}_1 = -j\mathbf{B}_0$$
$$\nabla \times \mathbf{H}_1 = \mathbf{J}_1 + j\mathbf{D}_0$$
$$\nabla \cdot \mathbf{B}_1 = 0 \qquad (11\text{-}37)$$
$$\nabla \cdot \mathbf{D}_1 = \rho_1$$
$$\nabla \cdot \mathbf{J}_1 = -j\rho_0$$

while, in general, the *nth-order field* is a solution of

$$\nabla \times \mathbf{E}_n = -j\mathbf{B}_{n-1}$$

$$\nabla \times \mathbf{H}_n = \mathbf{J}_n + j\mathbf{D}_{n-1}$$

$$\nabla \cdot \mathbf{B}_n = 0 \qquad\qquad (11\text{-}38)$$

$$\nabla \cdot \mathbf{D}_n = \rho_n$$

$$\nabla \cdot \mathbf{J}_n = -j\rho_{n-1}$$

The method of successive iterations consists in solving, first, the set (11-36). This solution establishes the zero-order field. Next, the known zero-order field is introduced in the first-order equations (11-37), and this set is solved for the first-order field. The first-order field is then used to determine the second-order field; the latter is used to determine the third-order field; and in general the *n*th-order field is used to determine the $(n + 1)$st-order field. Thus all the terms in the series expansion for the dynamic field are determined in succession by solving a particular set of equations in which the known quantities are the previously derived results.

We hasten to add, however, that the power-series method of analysis, in general, does not enable us to obtain solutions to problems not amenable to solution by classical methods. It does, nevertheless, provide a convenient means of identifying the intrinsic nature of lumped circuit element fields and to predict their behavior with frequency variations. Generally speaking, the number of terms in each power series which will result in an approximate, yet sufficiently accurate, answer depends on the ratio of physical dimensions to wavelength.

Example 11-1 A Parallel-plate Waveguide. As an example, let us obtain the solution to the problem considered in Sec. 11.2 by the method of successive iterations. Specifically, let us seek a power-series solution to Maxwell's equations of the form

$$\mathbf{E} = E_x e^{j\omega t}\mathbf{a}_x \qquad \mathbf{H} = H_y e^{j\omega t}\mathbf{a}_y \qquad (11\text{-}39)$$

which is independent of the variable y and which, on the boundaries of the waveguide (Fig. 11-1), satisfies the conditions $\mathbf{a}_z \times \mathbf{E} = 0$ and $\mathbf{a}_z \cdot \mathbf{H} = 0$. It is furthermore required that, at $z = 0$, $H_y = 0$ (open circuit). Translated in terms of $\mathbf{E}$, the last requirement is $\nabla \times \mathbf{E} = 0$ at $z = 0$. Since, moreover, $\nabla \cdot \mathbf{E} = 0$ everywhere between the plates, it follows that, on the plane $z = 0$, the electric field intensity vector is

$$\mathbf{E}\big|_{z=0} = E_{0x} e^{j\omega t}\mathbf{a}_x \qquad (11\text{-}40)$$

In other words, when the E field is expanded in a power series of the form

$$E = E_0 + \omega E_1 + \omega^2 E_2 + \cdots + \text{higher-order terms}$$

the term $\mathbf{E}_0$ must be specified by Eq. (11-40), and all higher-order terms must be zero at $z = 0$. With this in mind, we can now proceed with the actual solution of the problem.

Zero-order Field

The zero-order electric field must be a solution of

$$\nabla \times \mathbf{E}_0 = 0 \qquad \nabla \cdot \mathbf{E}_0 = 0 \tag{11-41}$$

which is normal to the perfectly conducting plates and which, furthermore, assumes the value $E_{0x}\mathbf{a}_x$ for $z = 0$.   The solution is

$$\mathbf{E}_0 = E_{0x}\mathbf{a}_x \tag{11-42}$$

In this expression, E_{0x} is a scalar function completely independent of the space coordinates.   Its magnitude is determined by the sources at $z = -a$.

In order to find the zero-order magnetic field, we must first determine the zero-order current.   Since the plates are perfectly conducting, the current flow in the system is confined to the surface of the plates, and can therefore be represented by a surface current density $\mathbf{K}$.   According to Eqs. (11-36), the zero-order equation of continuity is simply $\nabla \cdot \mathbf{J}_0 = 0$.   Therefore the zero-order component of the vector $\mathbf{K}$ must be divergenceless:

$$\nabla \cdot \mathbf{K}_0 = 0 \tag{11-43}$$

or in expanded form,

$$\frac{\partial K_{0y}}{\partial y} + \frac{\partial K_{0z}}{\partial z} = 0 \tag{11-44}$$

Since by hypothesis $\partial/\partial y = 0$, it follows that $\mathbf{K}$ is independent of the coordinate z:

$$\frac{\partial K_{0z}}{\partial z} = 0 \tag{11-45}$$

Consequently,

$$K_{0z} = \text{constant} \qquad \text{(independent of } x, y, \text{ and } z)$$

Since the plates terminate at $z = 0$, no steady current can flow in the plates (direct currents cannot flow through an open circuit); as a result, $K_{0z} = 0$.   The y-directed zero-order component of $\mathbf{K}$ is also zero for a similar reason.

It follows now that the zero-order magnetic field must be a solution of

$$\nabla \times \mathbf{H}_0 = 0 \qquad \nabla \cdot \mathbf{H}_0 = 0 \tag{11-46}$$

such that $\mathbf{H}_0 \times \mathbf{a}_x = 0$ and $\mathbf{H}_0 \cdot \mathbf{a}_x = 0$ at $x = 0$ and at $x = d$.   In other words, both the tangential (no surface currents present) and the normal (perfect conductors) components of $\mathbf{H}_0$ must be zero at the plates.   Additionally, $\mathbf{H}_0 = 0$ at $z = 0$, by assumption.   We conclude that $\mathbf{H}_0 \equiv 0$ everywhere.   This is quite reasonable since there are no zero-order currents anywhere to support such a field.

Summarizing, the zero-order field is

$$\mathbf{E}_0 = E_{0x}\mathbf{a}_x \qquad \mathbf{H}_0 = 0 \tag{11-47}$$

where the factor $e^{j\omega t}$ is understood, and E_{0x} is a scalar quantity whose magnitude is a measure of the relative strength of the field.

First-order Field

In view of the fact that $\mathbf{H}_0 = 0$, the first-order electric field equations (11-37) reduce to the pair

$$\nabla \times \mathbf{E}_1 = 0 \qquad \nabla \cdot \mathbf{E}_1 = 0 \tag{11-48}$$

Expanded in rectangular coordinates, these become

$$\left(\frac{\partial E_{1z}}{\partial y} - \frac{\partial E_{1y}}{\partial z}\right)\mathbf{a}_x + \left(\frac{\partial E_{1x}}{\partial z} - \frac{\partial E_{1z}}{\partial x}\right)\mathbf{a}_y + \left(\frac{\partial E_{1y}}{\partial x} - \frac{\partial E_{1x}}{\partial y}\right)\mathbf{a}_z = 0$$

$$\frac{\partial E_{1x}}{\partial x} + \frac{\partial E_{1y}}{\partial y} + \frac{\partial E_{1z}}{\partial z} = 0$$

or since a solution of the form (11-39) is sought which has no y dependence,

$$\frac{\partial E_{1x}}{\partial z} = 0 \qquad \frac{\partial E_{1x}}{\partial x} = 0 \qquad (11\text{-}49)$$

From this it follows that

$$E_{1x} = \text{constant} \qquad \text{(independent of } x, y, \text{ and } z)$$

a constant which, by virtue of Eq. (11-40), must be equal to zero (no first-order or higher-order terms are allowed at $z = 0$). Hence the first-order electric field is zero throughout the region bounded by the plates.

As for the first-order magnetic field, we have from Eqs. (11-37)

$$\nabla \times \mathbf{H}_1 = j\mathbf{D}_0 = (j\epsilon E_{0x})\mathbf{a}_x \qquad \nabla \cdot \mathbf{H}_1 = 0 \qquad (11\text{-}50)$$

Since the required field is to obey Eqs. (11-39) and the condition $\partial/\partial y = 0$, the second equation immediately reduces to the identity $0 = 0$, while the first equation, after expansion, becomes

$$-\frac{\partial H_{1y}}{\partial z} = j\epsilon E_{0x} \qquad \frac{\partial H_{1y}}{\partial x} = 0$$

To satisfy both equations simultaneously, we must have

$$H_{1y} = -j\epsilon z E_{0x} + \text{constant} \qquad \text{(independent of } x \text{ and } z)$$

The requirement $H_y = 0$ at $z = 0$ can obviously be met only by setting the constant of of integration equal to zero.

In summary, then,

$$\mathbf{E}_1 = 0 \qquad \mathbf{H}_1 = (-j\epsilon z E_{0x})\mathbf{a}_x \qquad (11\text{-}51)$$

define the first-order electric and magnetic fields, with the factor $e^{j\omega t}$ again understood.

SECOND-ORDER FIELD

The second-order electric field obeys the relations

$$\nabla \times \mathbf{E}_2 = -j\mathbf{B}_1 \qquad \nabla \cdot \mathbf{D}_2 = 0 \qquad (11\text{-}52)$$

Upon substituting the expression for $\mathbf{H}_1$ from Eqs. (11-51), keeping in mind the conditions (11-39) and $\partial/\partial y = 0$, we find

$$\frac{\partial E_{2x}}{\partial z} = -\mu\epsilon z E_{0x} \qquad (11\text{-}53)$$

and after integration,

$$E_{2x} = -\tfrac{1}{2}\mu\epsilon z^2 E_{0x} + \text{constant} \qquad \text{(independent of } x, y, \text{ and } z)$$

The constant of integration must be zero, since at $z = 0$ the **E** field is to have a zero-order component only. Thus

$$E_{2x} = -\tfrac{1}{2}\mu\epsilon z^2 E_{0x} \qquad (11\text{-}54)$$

In exactly the same way, the second-order magnetic field is obtained from

$$\nabla \times \mathbf{H}_2 = 0 \qquad \nabla \cdot \mathbf{B}_2 = 0 \tag{11-55}$$

with the corresponding result, $\mathbf{H}_2 \equiv 0$.

Summarizing, the second-order field is

$$\mathbf{E}_2 = (-\tfrac{1}{2}\mu\epsilon z^2 E_{0x})\mathbf{a}_x \qquad \mathbf{H}_2 = 0 \tag{11-56}$$

HIGHER-ORDER FIELDS

Continuing, this process will show that the nth-order field is specified by

$$\mathbf{E}_n = \begin{cases} 0 & n \text{ odd} \\ \left[(-1)^{n/2}(\mu\epsilon)^{n/2}\dfrac{z^n}{n!}\,E_{0x} \right]\mathbf{a}_x & n \text{ even} \end{cases} \tag{11-57}$$

$$\mathbf{H}_n = \begin{cases} -\left[(-1)^{(n-1)/2}\,j\sqrt{\dfrac{\epsilon}{\mu}}\,(\mu\epsilon)^{n/2}\dfrac{z^n}{n!}\,E_{0x} \right]\mathbf{a}_y & n \text{ odd} \\ 0 & n \text{ even} \end{cases} \tag{11-58}$$

Hence the desired power series expansions for the electric and magnetic fields are

$$\mathbf{E} = \sum_{n=0}^{\infty} \omega^n E_n = \sum_{n \text{ even}} \left[(-1)^{n/2}(\omega\sqrt{\mu\epsilon})^n \dfrac{z^n}{n!}\,E_{0x} \right]\mathbf{a}_x \tag{11-59}$$

$$\mathbf{H} = \sum_{n=0}^{\infty} \omega^n H_n = \sum_{n \text{ odd}} \left[-j\sqrt{\dfrac{\epsilon}{\mu}}\,(-1)^{(n-1)/2}(\omega\sqrt{\mu\epsilon})^n \dfrac{z^n}{n!}\,E_{0x} \right]\mathbf{a}_y \tag{11-60}$$

or since $\beta = \omega\sqrt{\mu\epsilon}$,

$$E_x = E_{0x} \sum_{n \text{ even}} \left[(-1)^{n/2}\dfrac{(\beta z)^n}{n!} \right] = E_{0x}\cos\beta z \tag{11-61}$$

$$H_y = -j\sqrt{\dfrac{\epsilon}{\mu}}\,E_{0x} \sum_{n \text{ odd}} \left[(-1)^{(n-1)/2}\dfrac{(\beta z)^n}{n!} \right] = -j\dfrac{E_{0x}}{\eta}\sin\beta z \tag{11-62}$$

Except for a slight difference in multiplicative constants (E_{0x} in place of $2E_+$), this pair of expressions is clearly identical with the pair (11-3) and (11-4).

Although much longer and by far more tedious than the classical solution (Sec. 11.2), the power-series solution will enable us to point out the basic nature of fields found in lumped circuit elements and electric devices. In the following section, we shall see that the zero-order field and the first-order field give rise to the conventional idea of lumped capacitance. In so doing, we shall understand why at times we can solve (as is often done in actual practice) a static-field problem, and then tag on a time factor to the static solution in order to obtain a reasonable approximation to the actual field.

11.4 Variations of a Capacitor with Frequency. The power-series solution provides a ready means of analyzing the frequency dependence of a structure from a driving immittance (impedance or admittance) point of view. For this purpose, we can take one of two equivalent approaches: we can define impedance (admittance) as a ratio of voltage (current) to current (voltage) or we can define it as a ratio of complex power to current (voltage) squared. The latter approach is developed and exemplified in Sec. 11.5. The more familiar former approach can best be presented with the aid of an example.

Example 11-2 Frequency Behavior of a Capacitor. A very practical form of a capacitor is the *paper capacitor*. It can easily be constructed by sandwiching together a flexible dielectric film between a pair of aluminum foils. The resulting structure, rolled into a cylinder, can be connected to a circuit by means of a pair of leads permanently attached to the ends of the foils. The electromagnetic field in this structure is understandably similar in many respects to that of a simple parallel-plate capacitor of the type shown in Fig. 11-1.

It was previously shown that, neglecting fringing, the field in a capacitor, with plates of infinite extent, is defined by

$$E_x = E_{0x}\left[1 - \frac{(\beta z)^2}{2!} + \frac{(\beta z)^4}{4!} - + \cdots + \text{higher-order terms} \right] \tag{11-63}$$

$$H_y = -j\frac{E_{0x}}{\eta}\left[\beta z - \frac{(\beta z)^3}{3!} + \frac{(\beta z)^5}{5!} - + \cdots + \text{higher-order terms} \right] \tag{11-64}$$

At sufficiently low frequencies, all structural dimensions are small compared with wavelength, and in this case a reasonable approximation to the actual field is the zero-order field

$$\mathbf{E}^0 \approx \mathbf{E}_0 = E_{0x}\mathbf{a}_x \tag{11-65}$$

$$\mathbf{H}^0 \approx \mathbf{H}_0 = 0 \tag{11-66}$$

These are just the first terms in the expansions (11-63) and (11-64). The factor $e^{j\omega t}$ is understood. Clearly, the zero-order field has the same space variation as it would have for dc excitation, but it varies periodically in time. Since $\mathbf{K} = \mathbf{n} \times \mathbf{H}$ and $\mathbf{H} = 0$, no current will be drawn from the sources, and, to a zero-order approximation, the admittance seen by the sources at $z = -a$ is

$$Y^0 = \frac{I^0(-a)}{V^0(-a)} = 0 \tag{11-67}$$

as expected (capacitors are open circuits at zero frequency).

At slightly higher frequencies, the first-order field must be included.

$$\mathbf{E}^1 = \mathbf{E}_0 + \omega\mathbf{E}_1 = E_{0x}\mathbf{a}_x \tag{11-68}$$

$$\mathbf{H}^1 = \mathbf{H}_0 + \omega\mathbf{H}_1 = (-j\omega\epsilon z E_{0x})\mathbf{a}_y \tag{11-69}$$

The total current drawn from the sources at $z = -a$ may be obtained by first calculating the density of the current flowing on the surface of the lower plate.

$$\mathbf{K}^1 = \mathbf{a}_x \times \mathbf{H}^1 = (-j\omega\epsilon z E_{0x})\mathbf{a}_z \tag{11-70}$$

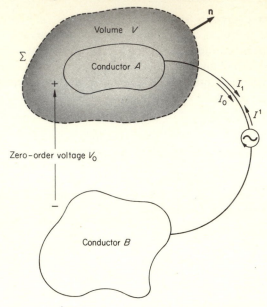

FIGURE 11-2.   The general definition of circuit capacitance.

Since $\mathbf{K}^1$ is independent of y, in full agreement with the initially imposed condition $\partial/\partial y = 0$, it follows that

$$I^1(-a) = \int_0^b K^1(-a)\,dy = bK^1(-a) = j\omega\epsilon abE_{0x} \qquad (11\text{-}71)$$

Accepting the definition of voltage,

$$V^1(-a) = \int_d^0 E^1(-a)\,dx = E^1(-a)\,d = E_{0x}d \qquad (11\text{-}72)$$

as an extension from the static case, we have for the first approximation to admittance

$$Y^1 = \frac{I^1(-a)}{V^1(-a)} = \frac{j\omega\epsilon abE_{0x}}{E_{0x}d} = j\omega\,\frac{\epsilon ab}{d} = j\omega C \qquad (11\text{-}73)$$

where C is just the electrostatic capacitance defined by Eq. (11-12), and $Y^1 = j\omega C$ is clearly the circuit definition of a capacitor for sinusoidal time variations.

It should be observed at this point that Eq. (11-73) was derived by retaining only the first two terms in the power-series solution for the field. The zero-order and first-order fields together define the so-called *quasistatic* field, and always lead to the accepted notion of lumped circuit elements. That this is true for a general capacitor is demonstrated as follows.

Figure 11-2 shows a capacitor driven by a sinusoidal source. Assuming that the dielectric separating the conductors is perfect ($\sigma = 0$), it is evident that $I_0 = 0$. Since the first-order equation of continuity is, simply,

$$\nabla \cdot \mathbf{J}_1 = -j\rho_0 \qquad (11\text{-}74)$$

the first-order component of the current is

$$I_1 = \oint_\Sigma \mathbf{J}_1 \cdot \mathbf{n} \, da = \int_V \mathbf{\nabla} \cdot \mathbf{J}_1 \, dv = -j \int_V \rho_0 \, dv = -jQ_0 \qquad (11\text{-}75)$$

where Q_0 is the total zero-order charge deposited on the surface of conductor 4. The relation of Q_0 to the zero-order component of voltage, V_0, is the definition $Q_0 = CV_0$ for electrostatic capacitance. Hence

$$I_1 = -jCV_0 \qquad (11\text{-}76)$$

and the total current drawn from the source, to a first approximation, is

$$I^1 = -(I_0 + \omega I_1) = j\omega C V_0 \qquad (11\text{-}77)$$

where the minus sign is the result of definitions of current directions in Fig. 11-2. To the same approximation, then,

$$\frac{I^1}{V_0} = j\omega C \qquad (11\text{-}78)$$

and this proves our assertion.

To continue, as the frequency becomes still higher and the structural dimensions become appreciable fractions of wavelength, the second-order field must be included. Thus

$$\mathbf{E}^2 = \mathbf{E}_0 + \omega\mathbf{E}_1 + \omega^2\mathbf{E}_2 = E_{0x}(1 - \tfrac{1}{2}\omega^2\mu\epsilon z^2)\mathbf{a}_x \qquad (11\text{-}79)$$

$$\mathbf{H}^2 = \mathbf{H}_0 + \omega\mathbf{H}_1 + \omega^2\mathbf{H}_2 = E_{0x}(-j\omega\epsilon z)\mathbf{a}_y \qquad (11\text{-}80)$$

and the second-order approximation to the admittance, seen at the generator terminals, is

$$Y^2 = \frac{I^2(-a)}{V^2(-a)} = \frac{bH^2(-a)}{dE^2(-a)} = \frac{b(j\omega\epsilon a)}{d(1 - \tfrac{1}{2}\omega^2\mu\epsilon a^2)} = \frac{j\omega(\epsilon ab/d)}{1 - \tfrac{1}{2}\omega^2\mu\epsilon a^2} \qquad (11\text{-}81)$$

or

$$Y^2 = \frac{j\omega C}{1 + (j\omega)^2[(a/v)^2/2]} \qquad (11\text{-}82)$$

Here $v = 1/\sqrt{\mu\epsilon}$ is the velocity of TEM wave propagation in the medium filling the space between the plates. Dividing numerator and denominator by the factor $j\omega C$ leads to

$$Y^2 = \frac{1}{1/j\omega C + j\omega[(a/v)^2/2C]} = \frac{1}{1/j\omega C + j\omega L} \qquad (11\text{-}83)$$

where we have set

$$L = \frac{(a/v)^2}{2C} = \mu\frac{ad}{2b} \qquad (11\text{-}84)$$

As shown in Fig. 11-3c, the admittance Y^2 can be synthesized readily by connecting a capacitor and an inductor in series.

At frequencies such that $\omega \ll \sqrt{2}\,v/a$, Eq. (11-81) reduces to $Y^2 = j\omega C$, and the admittance is just that of a capacitance C. From the equivalent relation

$$a \ll \frac{\sqrt{2}\,v}{\omega} = \frac{\sqrt{2}\,\lambda f}{2\pi f} = \frac{\lambda}{\sqrt{2}\,\pi} \approx \frac{\lambda}{4} \qquad (11\text{-}85)$$

it becomes immediately clear that the quasistatic approximation to a field is generally valid when the dimension a is much less than a quarter wavelength. This is a useful rule of thumb to remember.

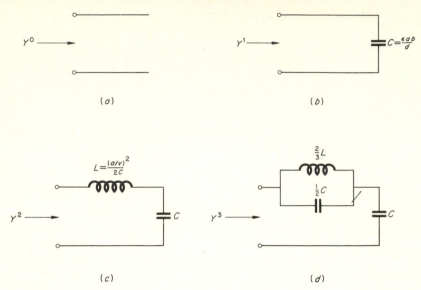

FIGURE 11-3. Equivalent-circuit representation of a capacitor as a function of frequency.

(a) $Y^0 = 0$

(b) $Y^1 = j\omega C$

(c) $Y^2 = \dfrac{j\omega C}{1 + (j\omega)^2 \dfrac{(a/v)^2}{2}}$

(d) $Y^3 = \dfrac{j\omega C\left[1 + (j\omega)^2 - \dfrac{(a/v)^2}{6}\right]}{1 + (j\omega)^2 \dfrac{(a/v)^2}{2}}$

By now we are probably startled, having discovered that a common, ordinary, parallel-plate capacitor is not just a capacitor, but in fact a structure capable of acting like a series combination of a capacitor and (of all things) an inductor! Strangely enough, this is not all. We shall discover, as we go on, that more and more elements will be needed to represent a single capacitor.

To a third-order approximation, the field within the capacitor plates is given by

$$\mathbf{E}^3 = \mathbf{E}_0 + \omega\mathbf{E}_1 + \omega^2\mathbf{E}_2 + \omega^3\mathbf{E}_3 = E_{0x}(1 - \tfrac{1}{2}\omega^2\mu\epsilon z^2)\mathbf{a}_x \tag{11-86}$$

$$\mathbf{H}^3 = \mathbf{H}_0 + \omega\mathbf{H}_1 + \omega^2\mathbf{H}_2 + \omega^3\mathbf{H}_3 = E_{0x}(-j\omega\epsilon z + j\tfrac{1}{6}\omega^3\mu\epsilon^2 z^3)\mathbf{a}_y \tag{11-87}$$

and the driving-point admittance by

$$Y^3 = \frac{bH^3(-a)}{dE^3(-a)} = \frac{j\omega C\left[1 + (j\omega)^2 \dfrac{(a/v)^2}{6}\right]}{1 + (j\omega)^2 \dfrac{(a/v)^2}{2}} \tag{11-88}$$

An equivalent representation for Y^3 is

$$Y^3 = \cfrac{1}{\cfrac{1}{j\omega C} + \cfrac{1}{j\omega \cfrac{C}{2} + \cfrac{1}{j\omega\,(2L/3)}}} \tag{11-89}$$

and in this form it is easily recognized as a driving-point admittance for the circuit shown in Fig. 11-3d.

It is easy to see that as the process of successive approximations is continued, more and more elements will be needed to synthesize an equivalent network. In the limit as the structural dimensions become very large compared with wavelength, the driving-point admittance will approach the ratio of E_x to H_y, as expressed by Eqs. (11-61) and (11-62). Thus

$$Y = \frac{bH(-a)}{dE(-a)} = j\frac{b}{\eta d}\tan\beta a \tag{11-90}$$

which is just the familiar expression for the admittance of an open-circuited transmission line. For βa very small, $\tan\beta a \approx \beta a$, and $Y = j\omega\epsilon ab/d = j\omega C$, as expected.

It is customary in practice to speak of *stray*, or *distributed*, effects when the behavior of a circuit or a device cannot be predicted on the basis of ordinary network theory. One distinct advantage of the field approach is that it provides the basis for a clearer understanding of these effects at the higher frequencies.

The term *distributed network* can now be appreciated fully in the light of the preceding development.

In summary, the power-series method of analysis provides the means for understanding the frequency dependence of lumped circuit elements. At sufficiently high frequencies a simple capacitor, for example, exhibits the properties of an equivalent circuit comprised of two or more ideal circuit elements.

11.5 Immittance from Power Considerations.

We saw in Chap. 7 that the complex Poynting's theorem is expressed by

$$-\oint_\Sigma \tfrac{1}{2}(\mathbf{E}\times\mathbf{H}^*)\cdot\mathbf{n}\,da = \int_V \tfrac{1}{2}\mathbf{E}\cdot\mathbf{J}^*\,dv + j\omega\int_V (\tfrac{1}{2}\mu H^2 - \tfrac{1}{2}\epsilon E^2)\,dv \tag{11-91}$$

The surface integral on the left represents the complex power flowing *into* the region bounded by the closed surface Σ. Let this power be denoted by the symbol P_c. If the development in Sec. 6.7 is repeated for sinusoidal time variations, it will be found that complex power in the circuit sense is given by the following equation:

$$P_c = \tfrac{1}{2}VI^* \tag{11-92}$$

Let the volume V enclosed by the surface Σ contain a linear passive two-terminal network. Then

$$\langle P_d\rangle = \int_V \tfrac{1}{2}\mathbf{E}\cdot\mathbf{J}^*\,dv \tag{11-93}$$

is the power dissipated in the volume V, averaged over a complete cycle, and

$$\langle W_m \rangle = \int_V \tfrac{1}{4}\mu H^2 \, dv \tag{11-94}$$

$$\langle W_e \rangle = \int_V \tfrac{1}{4}\epsilon E^2 \, dv \tag{11-95}$$

are the time-average magnetic and electric energies, respectively, stored in the field. For such a circuit, the equation

$$P_c = \tfrac{1}{2}VI^* = \langle P_d \rangle + j2\omega(\langle W_m \rangle - \langle W_e \rangle) \tag{11-96}$$

provides a convenient means for defining immittance in terms of power and energy, even when the network cannot be identified in terms of conventional resistors, inductors, and capacitors. For, by definition,

$$V = ZI = \frac{I}{Y} \tag{11-97}$$

and hence

$$P_c = \tfrac{1}{2}VI^* = \tfrac{1}{2}(ZI)I^* = \tfrac{1}{2}Z|I|^2 \tag{11-98}$$

where $|I|$ is the magnitude of the current flowing through the network terminals. Introducing Eq. (11-98) into (11-96) and dividing through by $|I|^2/2$, we obtain the desired result:

$$Z = \frac{P_c}{|I|^2/2} = \frac{1}{|I|^2/2}\,[\langle P_d \rangle + j2\omega(\langle W_m \rangle - \langle W_e \rangle)] \tag{11-99}$$

The corresponding expression for input admittance can be obtained in a similar manner. We have

$$P_c^* = \tfrac{1}{2}V^*I = \tfrac{1}{2}V^*(VY) = \tfrac{1}{2}|V|^2Y \tag{11-100}$$

and therefore

$$Y = \frac{P_c^*}{|V|^2/2} = \frac{1}{|V|^2/2}\,[\langle P_d \rangle - j2\omega(\langle W_m \rangle - \langle W_e \rangle)] \tag{11-101}$$

The validity of this result is contingent only upon our being able to define the voltage V at the terminals of the network.

Now let us apply these results to the power-series representation of a field. Let

$$S_c = \tfrac{1}{2}(\mathbf{E} \times \mathbf{H}^*) \qquad S_c^* = \tfrac{1}{2}(\mathbf{E}^* \times \mathbf{H}) \tag{11-102}$$

and

$$P_c = -\oint_\Sigma S_c \cdot \mathbf{n} \, da \qquad P_c^* = -\oint_\Sigma S_c^* \cdot \mathbf{n} \, da \tag{11-103}$$

By virtue of its direct dependence upon $\mathbf{E}$ and $\mathbf{H}$, the Poynting vector can be expressed as a power series in ω. Thus, by direct substitution of the series representations of $\mathbf{E}$ and $\mathbf{H}$ in the expression for S_c^*, we find

$$\mathbf{S}_c^* = \tfrac{1}{2}(\mathbf{E}_0^* + \omega\mathbf{E}_1^* + \omega^2\mathbf{E}_2^* + \cdots) \times (\mathbf{H}_0 + \omega\mathbf{H}_1 + \omega^2\mathbf{H}_2 + \cdots)$$

$$(11\text{-}104)$$

so that

$$Y = \frac{P_c^*}{|V|^2/2} = \frac{-\oint_\Sigma \mathbf{S}_c^* \cdot \mathbf{n}\, da}{|V|^2/2} = \frac{-\oint_\Sigma \left[\frac{1}{2}\left(\sum_{n=0}^N \omega\mathbf{E}_n^*\right) \times \left(\sum_{n=0}^\infty \omega\mathbf{H}_n\right)\right] \cdot \mathbf{n}\, da}{\frac{1}{2}\left(\sum_{n=0}^\infty \omega V_n\right)\left(\sum_{n=0}^\infty \omega V_n^*\right)}$$

$$(11\text{-}105)$$

Example 11-3   Frequency Behavior of a Capacitor.   Let us apply the foregoing results to the parallel-plate capacitor of Example 11-2.   Let Σ be the infinite plane $z = -a$. Since $\mathbf{n} = -\mathbf{a}_z$ and the direction of the Poynting vector is along the positive z axis, the numerator in Eq. (11-105) is, simply,

$$P_c^* = \oint_\Sigma [\tfrac{1}{2}E(-a)H(-a)]\, da = [\tfrac{1}{2}E(-a)H(-a)]\, bd$$

Taking into account the fact that $\mathbf{E}$ is real, we may write

$$|V|^2 = VV^* = VV = [E(-a)d][E(-a)d]$$

Hence the admittance expression

$$Y = \frac{P_c^*}{|V|^2/2} = \frac{[\tfrac{1}{2}E(-a)H(-a)]bd}{\tfrac{1}{2}[E(-a)d]^2} = \frac{bH(-a)}{dE(-a)} \qquad (11\text{-}106)$$

is the same as that found in Example 11-2, and the same answers apply.

11.6   Basic Concepts of the Calculus of Variations.

Preparatory to a full discussion of electric networks from the field-theory point of view, we now consider the simplest problem of the calculus of variations.   The preceding sections were primarily concerned with single lumped devices; the following three sections are devoted to a study of the field aspects of groups of such devices connected in networks.

In the basic calculus, the ordinary derivative is a measure of the local rate of change in functional value, which in turn fixes the sensitivity of the latter to small variations of the independent variable(s) about the point of observation.   This is illustrated in Fig. 11-4 for a function of one variable. The changes $\Delta f(x_3)$, $\Delta f(x_2)$, $\Delta f(x_1)$ in functional value at three points x_3, x_2, x_1 are seen to become correspondingly smaller, in keeping with the magnitude of the local derivative.   For a given change Δx, the corresponding change in the value of the function is $\Delta f(x)$.   It is evident that, for sufficiently small Δx, the change in the value of the function is least at points

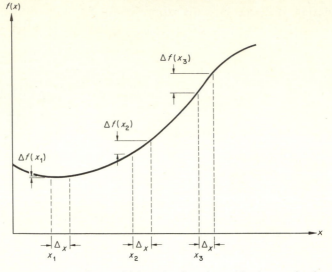

FIGURE 11-4.   The sensitivity of a function with changes in the absolute magnitude of its local derivative.

along the x axis where the derivative is zero.   This can also be seen from the Taylor series [Eq. (11-19)] expansion of the function f.   If x_0 denotes any point along the x axis, than at any other point

$$f(x) = f(x_0) + (x - x_0)\frac{df(x_0)}{dx}$$
$$+ \frac{(x - x_0)^2}{2!}\frac{d^2f(x_0)}{dx^2} + \frac{(x - x_0)^3}{3!}\frac{d^3f(x_0)}{dx^3} + \cdots$$

and therefore the functional change $\Delta f(x) = f(x) - f(x_0)$ due to a change $\Delta x = (x - x_0)$ is

$$\Delta f(x) = (x - x_0)\frac{df(x_0)}{dx} + \frac{(x - x_0)^2}{2!}\frac{d^2f(x_0)}{dx^2} + \frac{(x - x_0)^3}{3!}\frac{d^3f(x_0)}{dx^3} + \cdots$$
$$(11\text{-}107)$$

For Δx sufficiently small, the first term on the right is the dominant one. Therefore the absolute magnitude of $\Delta f(x)$ is primarily determined by the magnitude of the first derivative, and has its minimum when $df(x_0)/dx \to 0$.

The simplest problem of the calculus of variations is concerned with the properties of a function of a *function*, or a *functional*.   The independent variable, in other words, is a function $f(x)$, and not x itself.   The problem is then to determine *that function* which, among a class of *admissible comparison functions*, makes the value of the prescribed functional least sensitive to changes in the independent function.   We say that, for that function, the functional is *stationary in the variational sense*.

Consider the *basic integral*

$$I = \int_{x_1}^{x_2} f(x,y,y')\, dx \qquad (11\text{-}108)$$

where the integrand is a continuous function of its arguments, namely, the variable x, the function $y(x)$, and its derivative $y'(x)$, and has continuous second derivatives either mixed or unmixed, with respect to all its arguments, for all y' and for x and y in the region defined by

$$x_1 \leq x \leq x_2$$

$$y_1(x) < y(x) < y_2(x)$$

The problem is to determine $y(x)$ such that the integral (11-108) has a *relative extremum*, either maximum or minimum, in this region, subject to boundary conditions

$$y(x_1) = K_1 \qquad y(x_2) = K_2$$

where K_1, K_2 are given constants. For our purposes, let it be required that the basic integral I is to be minimized. Then the question is, among all the admissible comparison functions $y(x)$, is there a particular function $y_0(x)$ such that

$$I[y_0] \leq I[y] \qquad (11\text{-}109)$$

for all $y(x)$? The brackets in Eq. (11-109) signify that I is a functional of y.

Suppose $y_0(x)$ is, indeed, the minimizing function. Then

$$\int_{x_1}^{x_2} f(x,y_0,y_0')\, dx \leq \int_{x_1}^{x_2} f(x,y,y')\, dx \qquad (11\text{-}110)$$

for every admissible comparison function $y(x)$. Let $\eta(x)$ be a given function which has continuous second partial derivatives on $[x_1, x_2]$ and which vanishes at x_1 and x_2. Then, for a suitably chosen constant α, the sum $y_0(x) + \alpha\eta(x)$ is also an admissible comparison function because, first of all, it has continuous second partial derivatives, being the sum of two functions satisfying this condition; second, $y_0(x_1) + \alpha\eta(x_1) = K_1$ and $y_0(x_2) + \alpha\eta(x_2) = K_2$; and finally, $y_1(x) < y_0(x) + \alpha\eta(x) < y_2(x)$ when α is sufficiently small. As a result,

$$\int_{x_1}^{x_2} f(x,y_0,y_0')\, dx \leq \int_{x_1}^{x_2} f(x,\, y_0 + \alpha\eta,\, y_0' + \alpha\eta')\, dx \qquad (11\text{-}111)$$

The right-hand side of Eq. (11-111) is a function of α, once y_0 and η are assigned, and has a minimum when $\alpha = 0$. Thus, letting $I(0)$ and $I(\alpha)$ denote, respectively, the left and right members of Eq. (11-111), we obtain

$$I(0) \leq I(\alpha) \qquad (11\text{-}112)$$

and therefore

$$\frac{dI(\alpha)}{d\alpha} = 0 \qquad \text{when } \alpha = 0 \qquad (11\text{-}113)$$

That is,

$$\frac{dI(0)}{d\alpha} = \int_{x_1}^{x_2} \left[\frac{\partial f(x,y_0,y_0')}{\partial x} \frac{dx}{d\alpha} + \frac{\partial f(x,y_0,y_0')}{\partial y_0} \frac{d(y_0 + \alpha\eta)}{d\alpha} \right.$$

$$\left. + \frac{\partial f(x,y_0,y_0')}{\partial y_0'} \frac{d(y_0' + \alpha\eta')}{d\alpha} \right] dx$$

$$= \int_{x_1}^{x_2} \left[\eta \frac{\partial}{\partial y_0} f(x,y_0,y_0') + \eta' \frac{\partial}{\partial y_0'} f(x,y_0,y_0') \right] dx = 0 \quad (11\text{-}114)$$

Integrating by parts, we find

$$\int_{x_1}^{x_2} \eta' \frac{\partial}{\partial y_0'} f(x,y_0,y_0') \, dx = \left[\eta \frac{\partial}{\partial y_0'} f(x,y_0,y_0') \right]_{x_1}^{x_2} - \int_{x_1}^{x_2} \eta \frac{d}{dx} \left[\frac{\partial}{\partial y_0'} f(x,y_0,y_0') \right] dx$$

$$(11\text{-}115)$$

Since, by assumption, $\eta(x_1) = \eta(x_2) = 0$, the first term on the right vanishes. So Eq. (11-114) may now be written

$$\frac{dI(0)}{d\alpha} = \int_{x_1}^{x_2} \eta \left\{ \frac{\partial}{\partial y_0} f(x,y_0,y_0') - \frac{d}{dx} \left[\frac{\partial}{\partial y_0'} f(x,y_0,y_0') \right] \right\} dx = 0 \quad (11\text{-}116)$$

We now invoke the *fundamental lemma* of the calculus of variations. Let x_1 and x_2 be real constants, with $x_1 < x_2$. If $F(x)$ is continuous on $[x_1,x_2]$, and if $\int_{x_1}^{x_2} \eta(x)F(x) \, dx = 0$ for every function $\eta(x)$ which has continuous second partial derivatives on $[x_1,x_2]$ and for which $\eta(x_1) = \eta(x_2) = 0$, then the function $F(x)$ is identically zero on $[x_1,x_2]$. This lemma applied to Eq. (11-116) gives

$$\frac{\partial}{\partial y_0} f(x,y_0,y_0') - \frac{d}{dx} \left[\frac{\partial}{\partial y_0'} f(x,y_0,y_0') \right] = 0 \qquad (11\text{-}117)$$

This means that a function y_0 which satisfies Eq. (11-110) is a solution of the differential equation

$$\frac{\partial}{\partial y} f(x,y,y') - \frac{d}{dx} \left[\frac{\partial}{\partial y'} f(x,y,y') \right] = 0 \qquad (11\text{-}118)$$

In the calculus of variations, Eq. (11-118) is known as *Euler's equation*.

Example 11-4 Shortest Distance between Two Points. As an example, let us show that the shortest distance between any two points in a plane is a straight line (Fig. 11-5).

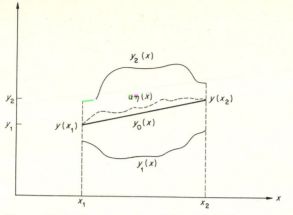

FIGURE 11-5.   The parameters of Example 11-4.

Let (x_1, y_1) and (x_2, y_2) be any two points in the xy plane, and let $y(x)$ represent the set of admissible comparison functions. The integral to be extremized is

$$I[y] = \int_{x_1}^{x_2} \sqrt{1 + y'^2}\, dx$$

and the corresponding Euler's equation is

$$\frac{d}{dx}\left(\frac{\partial}{\partial y'} \sqrt{1 + y'^2} \right) = 0$$

The solution of this differential equation is an elementary task. We obtain

$$y' = C_1$$

and therefore

$$y = C_1 x + C_2 \tag{11-119}$$

Clearly, this is the equation of a straight line. The constants of integration C_1, C_2 may be evaluated from the boundary conditions $y(x_1) = y_1$ and $y(x_2) = y_2$.

The term $\alpha\eta(x)$, which was introduced in the preceding development, changes a given function $y(x)$ into a new function $y(x) + \alpha\eta(x)$. By convention, the quantity

$$\delta y = \alpha\eta(x) \tag{11-120}$$

is called the *variation of* $y(x)$. Corresponding to this change in $y(x)$, a given function $f(x, y, y')$ varies by an amount which, neglecting higher-order terms, is defined to be the *variation* of f, and is accordingly expressed by

$$\delta f = \frac{\partial f}{\partial y} \alpha\eta + \frac{\partial f}{\partial y'} \alpha\eta' \tag{11-121}$$

Evidently, the integrand in Eq. (11-114) is equal to the variation of $f(x,y,y')$ to within the multiplicative constant α. In view of this, the preceding development of the simplest problem of the calculus of variations may be stated thus: A necessary condition that the integral (11-108) be *stationary* is that the *first variation* vanish:

$$\delta I = \int_{x_1}^{x_2} \delta f(x,y,y')\, dx = 0 \qquad (11\text{-}122)$$

If the basic integral involves m independent variables $x, y, z, \ldots$ and n dependent variables $y, u, v, \ldots$, a similar development leads to one Euler equation for each of the n dependent variables. The interested reader is referred to the literature.†

Summarizing, the simplest problem of the calculus of variations is to find a function which extremizes the value of a given functional.

11.7 A Variational Principle for Electromagnetic Fields.

It is often advantageous to examine field and circuit problems from an energy point of view, particularly when we seek information about their fundamental nature and a deeper understanding of their behavior. For example, through energy considerations we are able to define impedance for a given electromagnetic structure, even though a clear-cut identification of its component parts in terms of lumped circuit elements is impossible. Similarly, we often prefer to think of resonance as the phenomenon which occurs when a system stores, on a time-average basis, equal amounts of electric and magnetic energies. A variational principle,‡ discussed in this section, expresses from an energy point of view what is perhaps one of the least appreciated properties of fields and circuits. In essence, the principle states that *the electromagnetic state of a system at every instant of time is at the lowest possible level consistent with the constraints of the system.*

The principal objectives of this discussion are, first, to gain insight into the properties of electromagnetic fields through study of the variational principle, and second, to establish the conceptual framework for deriving Kirchhoff's voltage law, using this principle as a basis. In the process, some related topics will be singled out for special attention.

The variational principle is the statement of Hamilton's principle for the electromagnetic case. In mathematical form and for a linear, homogeneous, and isotropic medium, this principle is

$$\delta \int_{t_0}^{t} \int_{V} \left(\tfrac{1}{2}\epsilon E^2 - \tfrac{1}{2}\mu H^2 + \mathbf{J} \cdot \mathbf{A} - \phi\rho \right) dv\, dt = 0 \qquad (11\text{-}123)$$

† F. B. Hildebrand, "Methods of Applied Mathematics," Prentice-Hall, Inc., Englewood Cliffs, N.J., 1952.

‡ D. T. Paris and F. K. Hurd: Relaxation Properties of Fields and Circuits, *Proc. IEEE*, vol. 53, pp. 150–156, February, 1965.

where the integrations extend over arbitrary intervals in the three-dimensional space and time domains. The variations are with respect to three mutually orthogonal space variables, and time. Stated in words, Eq. (11-123) asserts that the behavior of electromagnetic fields, generated by currents and charges which are distributed in a prescribed system of conductors, is characterized by the fact that the time integral of the energy function

$$L = \int_V (\tfrac{1}{2}\epsilon E^2 - \tfrac{1}{2}\mu H^2 + \mathbf{J} \cdot \mathbf{A} - \phi\rho)\, dv \qquad (11\text{-}124)$$

has a stationary value when compared with all nearby varied fields which satisfy the requisite boundary conditions for the prescribed system of bodies.

For a proof of the validity of Eq. (11-123), the reader is referred to the original article cited.

The statement of Eq. (11-123) for fields produced by charges at rest or in uniform motion is

$$\delta \int_V (\tfrac{1}{2}\epsilon E^2 - \tfrac{1}{2}\mu H^2 + \mathbf{J} \cdot \mathbf{A} - \phi\rho)\, dv = 0 \qquad (11\text{-}125)$$

Specialized for the case of a stationary distribution of charge, this equation assumes the form

$$\delta \int_V (\tfrac{1}{2}\epsilon E^2 - \phi\rho)\, dv = 0 \qquad (11\text{-}126)$$

Let us consider Eq. (11-126) as it relates to a closed system. For such a system the integration extends over the entire domain of the field, including the regions of its sources. Then, as in Chap. 6,

$$\int_V \tfrac{1}{2}\epsilon E^2\, dv = \int_V \tfrac{1}{2}(-\nabla\phi) \cdot \mathbf{D}\, dv$$

$$= \int_V \tfrac{1}{2}(-\nabla \cdot \phi\mathbf{D} + \phi\nabla \cdot \mathbf{D})\, dv$$

$$= -\oint_\Sigma \tfrac{1}{2}\phi\mathbf{D} \cdot \mathbf{n}\, da + \int_V \tfrac{1}{2}\phi\rho\, dv \qquad (11\text{-}127)$$

where $\mathbf{n}$ is the unit normal vector drawn outward to the surface enclosing the domain of the field. The closed surface integral on the right-hand side of Eq. (11-127) vanishes identically as the boundary surface is allowed to recede to infinity; the second integrand on the right vanishes outside of the region of the sources. It follows that, for closed systems,

$$\int_V \tfrac{1}{2}\epsilon E^2\, dv = \int_V \tfrac{1}{2}\phi\rho\, dv \qquad (11\text{-}128)$$

and that, consequently, Eq. (11-126) becomes

$$\delta \int_V \tfrac{1}{2}\epsilon E^2\, dv = 0 \qquad (11\text{-}129)$$

We saw in Chap. 6 that the requirement imposed by Eq. (11-129) is usually known as Thomson's theorem. It states that charges distribute themselves on surfaces of conductors in such a way that the energy of the resultant electrostatic field is a minimum. This statement is valid, of course, only if the integration in Eq. (11-129) extends over the entire domain of the field. It should be noted, however, that if the integration extends over any sub-region of space which is source-free, then Eq. (11-126) once again reduces to Eq. (11-129); hence the electrostatic energy contained (and we are speaking in rather loose terms here because the concept of energy localization is open to some criticism) in a volume of arbitrary size enclosing no charge is stationary in the variational sense.

We next derive an expression, similar to Eq. (11-129), which expresses mathematically the behavior of fields generated by charges in uniform motion, i.e., by direct currents. We approach this problem in very much the same way as we approached the derivation of Thomson's theorem. In fact, it should be clear a priori that the steps in the two analyses bear a one-to-one relation. We have from Eq. (11-125)

$$\delta \int_V (-\tfrac{1}{2}\mu H^2 + \mathbf{J} \cdot \mathbf{A})\, dv = 0 \tag{11-130}$$

For closed systems, the first term on the left becomes

$$\int_V \tfrac{1}{2}\mu H^2\, dv = \int_V \tfrac{1}{2}(\boldsymbol{\nabla} \times \mathbf{A}) \cdot \mathbf{H}\, dv$$

$$= \int_V \tfrac{1}{2}[\boldsymbol{\nabla} \cdot (\mathbf{A} \times \mathbf{H}) + \mathbf{A} \cdot (\boldsymbol{\nabla} \times \mathbf{H})]\, dv$$

$$= \oint_\Sigma \tfrac{1}{2}(\mathbf{A} \times \mathbf{H}) \cdot \mathbf{n}\, da + \int_V \tfrac{1}{2}\mathbf{A} \cdot \mathbf{J}\, dv \tag{11-131}$$

As in the case of Eq. (11-127), the closed surface integral vanishes as the surface is allowed to expand into a sphere of infinite radius about the origin. So Eq. (11-130) transforms to

$$\delta \int_V \tfrac{1}{2}\mu H^2\, dv = 0 \tag{11-132}$$

which expresses mathematically a property of magnetostatic fields, analogous to Thomson's theorem. Thus Eq. (11-132) states that the distribution of direct currents is such that the total energy of the resultant magnetostatic field is a minimum.

To sum up, electromagnetic fields tend to be in the lowest energy state consistent with the constraints of the system. This property is expressed tacitly by Maxwell's equations and explicitly by Eq. (11-123).

11.8   Field Theory and the Circuit Concept. Let us next turn our attention to circuits and show that Kirchhoff's voltage law can be derived using the variational principle as a point of departure.

To this end, we must first identify the three basic circuit components, that is, capacitors, inductors, and resistors, with terms in the integrand of Eq. (11-123). In the process, we must be sure to note the highly specialized nature of fields that we normally associate with lumped elements. We know that the field produced by a given configuration of primary sources can be confined to a finite region of space only if it is completely surrounded by electric walls and/or magnetic walls, that is, surfaces on which the tangential components of the electric and/or magnetic field intensities vanish. We also know that such walls do not exist in the case of circuit components. Yet our concept of circuit components as electromagnetic structures requires that the associated fields be confined to a limited region, having rather vague boundaries, in the immediate neighborhood of the elements.

Our transition from fields to circuits as disciplines is also contingent upon a second equally fundamental and extremely important assumption. We are referring, of course, to our circuit concept of voltage in which a certain basic limitation is often overlooked. Voltage difference, in the circuit sense, can be precisely defined in terms of field quantities if and only if the field is electrostatic; otherwise the time integral of the electric field intensity is dependent upon the path of integration. That is, the electric field intensity must be derivable from a scalar potential. An extension of this concept to time-varying fields is valid only if the time variations are so slow that the electric field behaves almost as an electrostatic field. This is just the class of fields which will be considered in the present discussion. We have previously called them quasistatic fields.

Consider the field produced in the electromagnetic structure shown in Fig. 11-6. Three distinct types of fields are associated with the conductor

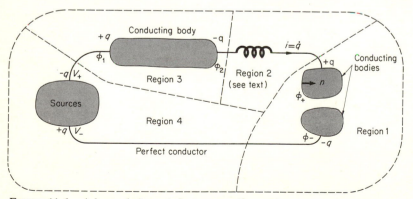

FIGURE 11-6.   A lumped-element electromagnetic structure.

configurations shown.   We assume that the field in each region is confined to
that region.   We further assume that the various physical bodies in the
system are electrically interconnected by very thin perfect conductors and
that no fields are associated with the electrical interconnections.

We begin our examination with region 1.   This portion of the overall
structure consists of two arbitrarily shaped conducting bodies embedded in a
perfect dielectric.   Since only very slow time variations are allowed, the
field in this region is virtually electrostatic; consequently, the variational
principle for this region is Eq. (11-126).   The integral in Eq. (11-126) may
be transformed as in Eq. (11-127), and so Eq. (11-128) applies; hence the net
contribution for region 1 is

$$-\int_V \tfrac{1}{2}\phi\rho \, dv = -[\tfrac{1}{2}\phi_+(+q) + \tfrac{1}{2}\phi_-(-q)] = -\frac{q}{2}(\phi_+ - \phi_-) = -\frac{1}{2C}q^2$$

$$(11\text{-}133)$$

In this expression

$$C = \frac{q}{\phi_+ - \phi_-} \qquad (11\text{-}134)$$

is the usual definition for electrostatic capacitance, and q is the magnitude of
the total charge deposited by the sources on the surface of each conductor.
Needless to say, conservation of charge is assumed throughout this discussion.

Next we turn our attention to region 2.   Here the field is generated by a
current $i = \dot{q}$, where $\dot{q} = dq/dt$, flowing through a perfectly conducting helix.
For relatively slow time variations, the magnetic field predominates, so that
in this region the variational principle is, essentially, Eq. (11-130).   The
transformation (11-131) may again be applied because of the localized nature
of the field, giving

$$\int_V \tfrac{1}{2}\mathbf{A} \cdot \mathbf{J} \, dv = \tfrac{1}{2}\dot{q} \oint_C \mathbf{A} \cdot d\mathbf{l} \qquad (11\text{-}135)$$

In replacing the volume integral by a line integral we tacitly assume that the
current $i = \dot{q}$, flowing through the helical path in region 2, is distributed with
equal density over a filament of infinitesimal cross section.   The closed contour, of course, is formed by the helix, together with a return path along the
region boundary joining the two end points.

The closed-contour integral in Eq. (11-135) may now be transformed,
with the aid of Stokes' theorem, into a surface integral, giving for the contribution of the region 2 field

$$\tfrac{1}{2}\dot{q} \oint_C \mathbf{A} \cdot d\mathbf{l} = \tfrac{1}{2}\dot{q} \int_S (\mathbf{\nabla} \times \mathbf{A}) \cdot \mathbf{n} \, da = \tfrac{1}{2}\dot{q} \int_S \mathbf{B} \cdot \mathbf{n} \, da = \tfrac{1}{2}L\dot{q}^2 \quad (11\text{-}136)$$

In this expression

$$L = \frac{1}{\dot{q}} \int_S \mathbf{B} \cdot \mathbf{n} \, da \qquad (11\text{-}137)$$

is the usual definition of the self-inductance of the configuration. The integration in Eq. (11-137) extends over any two-sided surface bounded by the contour of the circuit, and $\mathbf{n}$ is a unit vector normal to that surface, directed according to the right-hand-screw rule. Hence Eq. (11-135) becomes

$$\int_V \tfrac{1}{2}\mathbf{A} \cdot \mathbf{J} \, dv = \tfrac{1}{2}L\dot{q}^2 \tag{11-138}$$

The field in region 3 exhibits some unusual properties; it does not behave like a pure resistance, as we might suspect. Since the body contained in this region is assumed to have finite conductivity, the field within the conductor is electromagnetostatic. Accordingly, the structure in region 3 must exhibit capacitive as well as resistive characteristics, which can be completely identified by a capacitance C_R and a resistance R such that

$$\dot{q}R = \phi_1 - \phi_2 \tag{11-139}$$

$$q = C_R(\phi_1 - \phi_2) \tag{11-140}$$

Inductive effects, although present, are neglected in a circuit approximation.

In view of the foregoing discussion, the contribution from the region 3 field is given by

$$-\int_V \tfrac{1}{2}\phi\rho \, dv = -\frac{1}{2C_R}q^2 \tag{11-141}$$

In region 4 we have a source whose performance, by definition, is unaffected by the responses and reactions it causes. It is a region such that the charge q undergoes a change in energy level

$$(V_+ - V_-)q = Vq \tag{11-142}$$

in passing through the source. That is, the contribution to the integral of Eq. (11-123) is equal to Vq for region 4.

The contributions to the volume integral from all four regions of interest is now complete. The integral to be extremized is obtained by a simple algebraic addition of the results expressed by Eqs. (11-133), (11-138), (11-141), and (11-142), giving

$$\delta\int_0^t \left(-\frac{1}{2C}q^2 + \frac{1}{2}L\dot{q}^2 - \frac{1}{2C_R}q^2 + Vq\right) dt = 0 \tag{11-143}$$

The Euler equation of the variational problem, Eq. (11-143), is

$$-\frac{q}{C} - L\ddot{q} - \frac{1}{C_R}q + V = 0 \tag{11-144}$$

Using Eqs. (11-139) and (11-140) to express C_R in terms of R transforms Eq. (11-144) to

$$L\ddot{q} + R\dot{q} + \frac{1}{C}q = V \qquad (11\text{-}145)$$

This is just Kirchhoff's voltage law. It was derived strictly from electromagnetic considerations in a manner which clearly shows that a circuit is a special type of electromagnetic structure in which particular types of fields are localized within particular regions of space.

Certain other interesting relaxation properties of circuits may likewise be studied using the variational principle. We first consider another steady-state property of circuits, and follow this discussion with a particular property of circuits operating under transient conditions.

Consider the complex Poynting theorem, Eq. (11-91), in connection with a circuit consisting of sources and lumped elements, all located a finite distance from some arbitrary origin. If again we postulate localization of fields, the left-hand member of Eq. (11-91) is identically equal to zero. Also, by virtue of Eqs. (11-127) and (11-131), the second integral on the right of Eq. (11-91) is exactly equal to the volume integral in Eq. (11-125). It follows that the latter reduces to

$$\delta \int_V \tfrac{1}{2}\mathbf{E} \cdot \mathbf{J}\, dv = 0 \qquad (11\text{-}146)$$

which is the general expression for what has been called the *least power theorem*.[†] It states that the distribution of currents in a network is such as to minimize the power transformation.

In simple illustration, consider a parallel combination of two resistors R_1 and R_2 fed by a current source I, and let I_1 and I_2 denote the currents flowing through R_1 and R_2, respectively. The principle of conservation of charge constrains I_1 and I_2 in such a way that $I = I_1 + I_2$. If we now use this relation to eliminate one of the currents, for example I_2, from the expression for the power consumed by the circuit, we obtain

$$\int_V \tfrac{1}{2}\mathbf{E} \cdot \mathbf{J}\, dv = I_1{}^2 R_1 + I_2{}^2 R_2 = I_1{}^2 R_1 + (I - I_1)^2 R_2 \qquad (11\text{-}147)$$

By taking the derivative of the quantity on the right with respect to I_1 and setting it equal to zero, we obtain

$$I_1 = \frac{R_2}{R_1 + R_2}\, I \qquad (11\text{-}148)$$

[†] Sir James Jeans, "The Mathematical Theory of Electricity and Magnetism," Cambridge University Press, New York, 1963.

This is just the familiar current division rule. Note, however, that it was derived without recourse to Kirchhoff's voltage law.

Now let us consider the principle of conservation of flux linkages. We wish to derive this principle from Eq. (11-123). To this end we note that, by integrating the Euler equation of the variational problem, Eq. (11-143), that is,

$$\left(V - \frac{q}{C} - \frac{q}{C_R}\right) - \frac{d}{dt}(L\dot{q}) = 0 \tag{11-149}$$

over a time interval (t_0, t), we obtain

$$(L\dot{q})_{t_0}^t = \int_{t_0}^t \left(V - \frac{q}{C} - \frac{q}{C_R}\right) dt \tag{11-150}$$

Unless the integrand on the right is an impulse in the interval $[t_0, t]$, the integral itself vanishes in the limit as $t \to t_0$. It follows that

$$\lim_{t \to t_0} (L\dot{q})_{t_0}^t = 0 \tag{11-151}$$

which expresses mathematically the principle of conservation of flux linkages. If, on the other hand, the source voltage is an impulse, Eq. (11-150) can still be used to determine the initial conditions. Suppose, for example, that an initially relaxed series RLC circuit is excited by a unit voltage impulse in the interval $[t_0, t]$. From Eq. (11-150) it is clear that $L\dot{q}(0) = 1$, which simply means that the initial current through the circuit will have a magnitude of $1/L$.

No discussion of the field basis of circuit theory is complete without a look at Kirchhoff's current law and its relation to Maxwell's equations. We saw in Chap. 2 that

$$\oint_\Sigma \mathbf{J} \cdot \mathbf{n} \, da = -\frac{d}{dt}\int_V \rho \, dv \tag{11-152}$$

is an expression (in integral form) for the principle of conservation of charge. Let us apply this equation to a closed surface completely surrounding an arbitrary N-terminal network. Let I_n denote the current carried out of the volume V by the nth conducting lead. Then

$$\oint_\Sigma \mathbf{J} \cdot \mathbf{n} \, da = \sum_{n=1}^N I_n \tag{11-153}$$

If we let

$$Q = \int_V \rho \, dv \tag{11-154}$$

denote the net charge accumulation in V, we can write at once, in place of Eq. (11-152), the simpler relation

$$\sum_{n=1}^N I_n + \frac{dQ}{dt} = 0 \tag{11-155}$$

Evidently, the second term on the left represents a current whose presence is dependent upon the net charge in V. If, in particular, $Q = 0$, then

$$\sum_{n=1}^{N} I_n = 0 \qquad (11\text{-}156)$$

which is the more familiar form of Kirchhoff's current law.

To sum up, we have shown that in a lumped electric circuit the actual distribution of voltages as predicted by the lowest possible state in the energy sense results in Kirchhoff's voltage law.   Moreover, we have given a general proof of the theorem of constant flux linkages.   Finally, we have derived Kirchhoff's current law from Maxwell's equations.

11.9   Time Rate of Change of the Flux Linked by a Moving Contour.

We now switch our attention from electric circuits to electrical and electro-mechanical devices, the main objective being to develop the theory under-lying energy transformations taking place in all such devices.

To this end, let us compute the time rate of change of the flux, Ψ, of a vector $\mathbf{A}$ through a surface S spanning a moving closed contour C (Fig. 11-7). We assume that $\mathbf{A}$ is a divergenceless, but otherwise arbitrary, function of the space coordinates and time.   Then, if the surface is at rest,

$$\frac{d\Psi}{dt} = \int_S \frac{\partial \mathbf{A}}{\partial t} \cdot \mathbf{n}\, da \qquad (11\text{-}157)$$

is the time rate of change of Ψ that is caused by the time rate of change of $\mathbf{A}$ itself.   If, on the other hand, $\partial \mathbf{A}/\partial t = 0$ everywhere on S but the contour C

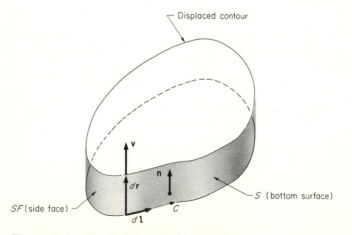

FIGURE 11-7.   The displacement of a contour.

is moving, Ψ will still change because of the change in the orientation of $\mathbf{A}$ relative to S.

Let us suppose that every element of the surface is moving with a velocity $\mathbf{v}$, and that the displacement occurs during a short time interval Δt. During this interval, every element of S is given a displacement $d\mathbf{r} = \mathbf{v}\,\Delta t$, which can be made as small as desired with the proper choice of Δt. We recall that

$$\oint_{\Sigma} \mathbf{A} \cdot \mathbf{n}\,da = 0 \tag{11-158}$$

by assumption. Hence the change of the flux of $\mathbf{A}$ through S is just the amount crossing the side face of the pillbox formed in the course of an infinitesimal displacement $d\mathbf{r}$, which is

$$\Delta \Psi = \int_{SF} \mathbf{A} \cdot (d\mathbf{r} \times d\mathbf{l}) \tag{11-159}$$

By a well-known identity,

$$\mathbf{A} \cdot (d\mathbf{r} \times d\mathbf{l}) = d\mathbf{l} \cdot (\mathbf{A} \times d\mathbf{r})$$

Since $d\mathbf{r} = \mathbf{v}\,\Delta t$, the surface integral

$$\Delta \Psi = \int_{SF} (\mathbf{A} \times \mathbf{v}\,\Delta t) \cdot d\mathbf{l}$$

may be written in the form of a contour integral

$$\Delta \Psi = \Delta t \oint_{C} (\mathbf{A} \times \mathbf{v}) \cdot d\mathbf{l} \tag{11-160}$$

From this we obtain

$$\frac{d\Psi}{dt} = \lim_{\Delta t \to 0} \frac{\Delta \Psi}{\Delta t} = \oint_{C} (\mathbf{A} \times \mathbf{v}) \cdot d\mathbf{l} \tag{11-161}$$

for the time rate of change of Ψ due to the motion of the surface.

If $\mathbf{A}$ is time-dependent *and* the surface S is in motion, then obviously, the time rate of change of Ψ is the sum

$$\frac{d\Psi}{dt} = \int_{S} \frac{\partial \mathbf{A}}{\partial t} \cdot \mathbf{n}\,da + \oint_{C} (\mathbf{A} \times \mathbf{v}) \cdot d\mathbf{l} \tag{11-162}$$

This is the desired result.

Equation (11-162) applies equally well when, instead of moving, the contour is deforming. For, in this case, $d\mathbf{r} \times d\mathbf{l}$ is the vector element of area which in time Δt is spanned by a moving element $d\mathbf{l}$ of the contour C, and therefore $(d\mathbf{r} \times d\mathbf{l}) \cdot \mathbf{A}$ is the change of the flux of $\mathbf{A}$ due to this infinitesimal displacement. The net change can be calculated as the sum of all such infinitesimal changes around the complete contour C. Once again this leads to Eq. (11-160), and the proof of our assertion is now complete.

To sum up, there are two causes for the time rate of change of the flux through a surface: one is the time rate of change of the vector itself; the other is the motion of the contour bounding the surface.   Equation (11-162) is an exact expression for both of them, when $\mathbf{A}$ is divergenceless.

11.10   The emf and *Blv* **Concepts.**   From the Lorentz force density law we can define

$$\mathbf{E}' = \frac{\mathbf{f}}{\rho} = \mathbf{E} + \mathbf{v} \times \mathbf{B} \qquad (11\text{-}163)$$

as the force per unit charge acting on a body which is moving with a velocity $\mathbf{v}$ relative to $\mathbf{B}$.   We assume that $\mathbf{E}$, $\mathbf{B}$, $\mathbf{v}$, and $\mathbf{E}'$ are all measured in the same frame of reference.

Let a and b denote the terminal points of a contour C.   We show presently that *voltage*, defined through

$$V_{ab} = \int_C \mathbf{E}' \cdot d\mathbf{l} = \int_C (\mathbf{E} + \mathbf{v} \times \mathbf{B}) \cdot d\mathbf{l} \qquad (11\text{-}164)$$

is consistent with Faraday's law of induction, and that this expression leads to the more conventional definition of *potential difference* for the electrostatic case when $\partial/\partial t = 0$.

We begin by noting that, in terms of the potentials of the field, the electric field intensity vector is given by

$$\mathbf{E} = -\nabla\phi - \frac{\partial \mathbf{A}}{\partial t} \qquad (11\text{-}165)$$

so that
$$V_{ab} = \int_C \left(-\nabla\phi - \frac{\partial \mathbf{A}}{\partial t} + \mathbf{v} \times \mathbf{B} \right) \cdot d\mathbf{l} \qquad (11\text{-}166)$$

If C is closed, this expression becomes

$$V = \oint_C \left(-\nabla\phi - \frac{\partial \mathbf{A}}{\partial t} + \mathbf{v} \times \mathbf{B} \right) \cdot d\mathbf{l} \qquad (11\text{-}167)$$

and V is called the *induced emf*.   By Stokes' theorem, the first two terms transform to

$$\oint_C \left(-\nabla\phi - \frac{\partial \mathbf{A}}{\partial t} \right) \cdot d\mathbf{l} = -\int_S \left[\nabla \times \nabla\phi + \frac{\partial}{\partial t} (\nabla \times \mathbf{A}) \right] \cdot \mathbf{n} \, da \quad (11\text{-}168)$$

where the surface integration extends over any regular surface bounded by the contour C.   Since the curl of the gradient of a scalar function vanishes identically, and since, by definition, $\mathbf{B} = \nabla \times \mathbf{A}$,

$$\oint_C \left(-\nabla\phi - \frac{\partial \mathbf{A}}{\partial t} + \mathbf{v} \times \mathbf{B} \right) \cdot d\mathbf{l} = -\int_S \frac{\partial \mathbf{B}}{\partial t} \cdot \mathbf{n} \, da + \oint_C (\mathbf{v} \times \mathbf{B}) \cdot d\mathbf{l} \quad (11\text{-}169)$$

By Eq. (11-162), the sum on the right is just the negative of $d\Phi/dt$, where Φ is the flux of $\mathbf{B}$ through S. Thus the induced emf is

$$V = \oint_C (\mathbf{E} + \mathbf{v} \times \mathbf{B}) \cdot d\mathbf{l}$$

$$= -\frac{d\Phi}{dt} = -\frac{d}{dt} \int_S \mathbf{B} \cdot \mathbf{n} \, da$$

$$= -\int_S \frac{\partial \mathbf{B}}{\partial t} \cdot \mathbf{n} \, da + \oint_C (\mathbf{v} \times \mathbf{B}) \cdot d\mathbf{l} \quad (11\text{-}170)$$

The sign of V is completely determined by the vector symbolism. A positive direction for C is chosen arbitrarily (usually counterclockwise). The direction of the unit vector $\mathbf{n}$ is that of advance of a right-hand screw as it turns in the direction in which C is traversed. Then Eq. (11-170) gives the increase in potential (sometimes called *voltage rise*) in the chosen direction. Thus the *induced current* will flow, if permitted, through the loop in such a direction as to oppose the primary field. This is *Lenz's law*.

A few common cases must be singled out for special consideration.

Let us consider an ordinary transformer. We know that for such a device $\mathbf{v} = 0$. Hence the second term on the right of Eq. (11-170) is zero, and thus

$$V = -\int_S \frac{\partial \mathbf{B}}{\partial t} \cdot \mathbf{n} \, da = -\frac{d}{dt} \int_S \mathbf{B} \cdot \mathbf{n} \, da \quad (11\text{-}171)$$

The interchange of the order of differentiation and integration is permissible only when C is fixed in space for all times.

Another simplification results when $\mathbf{B}$ is independent of time. Then

$$V = \oint_C (\mathbf{v} \times \mathbf{B}) \cdot d\mathbf{l} \quad (11\text{-}172)$$

If it happens that C is deforming and only a straight-line segment, of length l, is actually moving in such a way that $\mathbf{v}$, $\mathbf{B}$, and $d\mathbf{l}$ are mutually orthogonal, then Eq. (11-172) reduces to the familiar *flux-cutting relation*

$$V = Blv \quad (11\text{-}173)$$

This is the expression normally used to calculate the induced emfs in rotating machines. Although this approach is perfectly correct, the reasons for doing it are generally not well understood. At power frequencies, ordinary lengths of wires do not carry currents unless they are a part of a completely closed circuit.   This is precisely the case for the conductors which "cut flux" in a rotating machine.   They are actually a part of a complete winding kept in relative motion with respect to a spatially variable magnetic field.   So it is the completely closed loops of rotating conductors which are in effect responsible for the induced emfs predicted by Eq. (11-173).   In fact, Eq. (11-170) provides the basis for explaining why the motional emf may be calculated either from the flux-cutting relation or from the $d\Phi/dt$ relation when $\mathbf{B}$ is independent of time.   For, in this case,

$$-\frac{d\Phi}{dt} = \oint_C (\mathbf{v} \times \mathbf{B}) \cdot d\mathbf{l} \qquad (11\text{-}174)$$

It should be clear by now that, in general, the measurement of induced emfs is always affected by the physical arrangement of the measuring apparatus.   This is so because the instrument itself, together with the leads which connect it to the generator, close the loop which determines the contour C.   Such effects are more pronounced at higher frequencies.   Shorter leads are normally required for that reason.

A most important conclusion to be drawn from the foregoing discussion is that, generally speaking, voltage is not equal to the line integral of $\mathbf{E}$, but, as shown by Eq. (11-170),

$$V = \oint_C (\mathbf{E} + \mathbf{v} \times \mathbf{B}) \cdot d\mathbf{l} = -\frac{d}{dt} \int_S \mathbf{B} \cdot \mathbf{n}\, da \qquad (11\text{-}175)$$

When $\mathbf{v} \times \mathbf{B} = 0$,

$$V_{ab} = \int_C \mathbf{E} \cdot d\mathbf{l} = \int_C \left(-\nabla\phi - \frac{\partial \mathbf{A}}{\partial t} \right) \cdot d\mathbf{l} \qquad (11\text{-}176)$$

for an open contour with terminal points a and b.   From the definition of gradient,

$$\int_C (-\nabla\phi) \cdot d\mathbf{l} = \phi(a) - \phi(b)$$

on the basis of which Eq. (11-176) becomes

$$V_{ab} = \int_C \mathbf{E} \cdot d\mathbf{l} = \phi(a) - \phi(b) - \frac{\partial}{\partial t} \int_C \mathbf{A} \cdot d\mathbf{l} \qquad (11\text{-}177)$$

In the static case the last term is obviously zero, and the voltage V_{ab} is simply the difference in potential.   (If we let b recede to infinity, we obtain the *potential* at a.)   But as can be seen easily from inspection of Eq. (11-177), voltage is not synonymous with potential difference for time-varying fields. In general, V_{ab} *depends upon the path of integration.*

An apparent question that might well be asked at this point concerns the case of a very thin, infinitely conducting dipole antenna.   We know that the tangential component of **E** is zero along the antenna, so that $\mathbf{E} \cdot d\mathbf{l} = 0$. If the antenna is fixed in space, $\mathbf{v} = 0$, and therefore, for any two points along the antenna,

$$V = \int_C (\mathbf{E} + \mathbf{v} \times \mathbf{B}) \cdot d\mathbf{l} = 0 \tag{11-178}$$

This states that the voltage induced in the antenna is zero.   At first glance we have a glaring contradiction.   Everybody knows that dipole antennas "work." Is it true, then, that our theory suffers from serious limitations?   Fortunately, the answer is no!   In our example of the dipole antenna we merely failed to formulate the problem correctly.   What really happens is that, when a wave impinges upon a perfectly conducting structure, it causes charges to be induced and currents to flow on the surface of the structure which give rise to a secondary field.   This field combines with the primary field of the wave to give a zero tangential **E** field at every point on the surface of the structure.   Thus, while the net tangential **E** field is zero on the surface, sources *are* induced in the structure, which, in turn, make the antenna "work."

Example 11-5   Induced emf.   A conducting bar is rolled down a 60° incline (Fig. 11-8) with a velocity $\mathbf{v} = \frac{1}{2}x(\mathbf{a}_x - \sqrt{3}\mathbf{a}_z)$ in the presence of a time-varying magnetic field

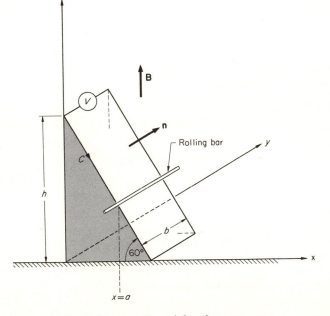

FIGURE 11-8.   Induced emf in a deforming contour.

$\mathbf{B} = B_0 y \sin t \mathbf{a}_z$, where B_0 is a positive constant. What is the instantaneous reading of the voltmeter when the bar is passing the plane $x = a$?

From Eq. (11-170), we have

$$V = -\int_S \frac{\partial \mathbf{B}}{\partial t} \cdot \mathbf{n} \, da + \oint_C (\mathbf{v} \times \mathbf{B}) \cdot d\mathbf{l}$$

The first term represents transformer emf, V_T, the second motional emf, V_m. Thus

$$V_T = -\int_S \frac{\partial}{\partial t} (B_0 y \sin t) \mathbf{a}_z \cdot \left(\frac{\sqrt{3}}{2} \mathbf{a}_x + \frac{1}{2} \mathbf{a}_z \right) da$$

$$= -\int_0^b \int_0^a \tfrac{1}{2} B_0 y \cos t \, dx \, dy = -\tfrac{1}{4} B_0 ab^2 \cos t$$

and

$$\mathbf{v} \times \mathbf{B} = \begin{vmatrix} \mathbf{a}_x & \mathbf{a}_y & \mathbf{a}_z \\ \tfrac{1}{2}x & 0 & -\dfrac{\sqrt{3}}{2} x \\ 0 & 0 & B_0 y \sin t \end{vmatrix} = -(\tfrac{1}{2}x B_0 y \sin t) \mathbf{a}_y$$

The only moving portion of the contour is the bar itself. Since $d\mathbf{l} = dy \mathbf{a}_y$,

$$\oint_C (\mathbf{v} \times \mathbf{B}) \cdot d\mathbf{l} = -\int_0^b (\tfrac{1}{2}x B_0 y \sin t) \, dy = -\tfrac{1}{4} B_0 x b^2 \sin t$$

Therefore, at $x = a$,

$$V_m = -\tfrac{1}{4} B_0 ab^2 \sin t$$

Adding V_T and V_m, we find

$$V = -\tfrac{1}{4} B_0 ab^2 (\sin t + \cos t) = -\frac{1}{2\sqrt{2}} B_0 ab^2 \sin (t + 45°)$$

The minus sign simply indicates that the voltmeter terminal farthest away from the xz plane will be negative relative to the nearest terminal if $0 < (t + 45) < 180°$.

In summary, voltage can be induced in a circuit either as a result of time variations or as a result of motion. The two causes are distinct, and neither of them can be deduced from the other. The emf of a device is the open-circuit potential difference across the output terminals.

11.11 Summary. The integers from 1 to 10 form a *discrete* set of numbers. In contrast, all the real numbers from 1 to 10 form a *continuous* set of numbers. In a similar sense, an electric circuit is a *discrete* structure while a transmission line is a *continuous* structure. This chapter has been devoted mainly to a study of *discrete electromagnetic systems*.

We saw that the character of the electromagnetic field in lumped circuit elements is such that only the zero-order and the first-order fields are needed to define it completely. However, as soon as the frequency becomes high enough so that the wavelength is comparable with the dimensions of the device, the quasistatic approximation to the actual field is no longer adequate to predict the terminal behavior of normal circuit components. "Stray" effects are actually the result of waves set up in the regions of space occupied by the circuit elements themselves, and can be taken into account by equivalent networks.

We saw, furthermore, that the time-honored laws of circuit theory can be traced to Maxwell's equations. Specifically, Kirchhoff's current law is a statement of the principle of conservation of charge, and Kirchhoff's voltage law is a special form of a variational principle which stems directly from the field laws and states that any electromagnetic system is in the lowest energy state consistent with the system constraints. The discussion was completed with some remarks on least-power-circuit theorems, as well as on the principle of conservation of flux, both of which were shown to be specialized expressions of the variational principle.

Finally, the emf and Blv concepts were discussed from the field theory point of view. We saw that

$$V = -\int_S \frac{\partial \mathbf{B}}{\partial t} \cdot \mathbf{n}\, da + \oint_C (\mathbf{v} \times \mathbf{B}) \cdot d\mathbf{l}$$

is the most general expression for voltage induced in a closed circuit, where the first term accounts for transformer action, and the last term accounts for relative motion.

Problems

11-1   Quasistatics of the Inductor. An ideal inductor is formed by winding N turns of a *perfectly conducting wire* on a cylinder. Show that, for this structure,

$$V_1 = \int_C \mathbf{E}^1 \cdot d\mathbf{l} = -j\omega L I_0$$

where $\mathbf{E}^1$ is the quasistatic electric field, and I_0 the zero-order current.

11-2   Variations of an Inductor with Frequency. The perfectly conducting sheets shown in Fig. 11-1 are connected at $z = 0$ by a perfectly conducting sheet. Find the equivalent circuits for the structure in succession through third order, assuming that the current per unit length in the shorting plate is maintained constant at an amplitude K_0 by the sources at $z = -a$.

11-3   Quasistatic Field. The accompanying figure shows three thin, perfectly conducting sheets, all of which extend to infinity in both the positive and negative x directions. A system of sources at $z = -l$ maintains a current

$$\mathrm{Re}\,(\mathbf{K}_0) = \mathrm{Re}\,(K_0 \mathbf{a}_y) = \mathrm{Re}\,(K_{0m} e^{j\omega t} \mathbf{a}_y)$$

in the shorting plate at $z = 0$. Assuming that the resulting electric field is y-directed, determine the quasistatic field at all points within the structure.

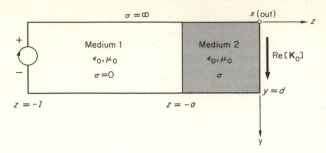

11-4   Quasistatics of a Coaxial Cable. The accompanying figure shows a lossless coaxial cable terminated by a resistive sheet of σ_s ℧/square and driven by sinusoidal sources at $z = -l$, which maintain the load $(z = 0)$ voltage across the cable constant at V_0. Calculate (a) the quasistatic field in the region between the conductors, and (b) the equivalent circuit at the input terminals correct up to first order.

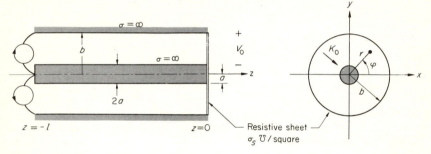

PROBLEM 11-4

11-5   Quasistatic Approximation. A unique method of forming, to rather close tolerances, the inside diameter of a long conducting nonmagnetic tube is depicted in the accompanying figure. A nonmagnetic mandrel is placed inside the tube, and a current

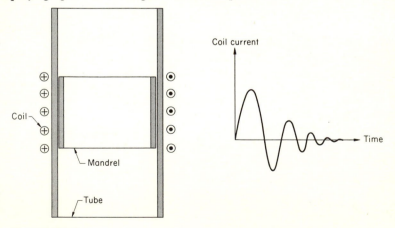

PROBLEM 11-5

impulse is allowed to flow through a coil placed outside the long tube. To form the long tube along its entire length, the sizing mandrel is moved to other positions along the tube, where the procedure is repeated again. The outside diameter of the mandrel is almost equal to the original inside diameter of the tube.

Describe the forces generated in the process, and justify their directions. Then, by making reasonable assumptions, calculate an order of magnitude for the pressure exerted on the long tube.

11-6 Circuit Concept of Complex Power. Based on the field concept of complex power, prove that $VI^*/2$ is the expression for complex power in the circuit sense. A rigorous mathematical development is required.

11-7 Induced emf. A uniform magnetic field **B** is directed normal to the plane of a flat metal plate which is moving at a constant rate **v** normal to **B**. Determine the reading of the voltmeter, and state the polarity of the induced emf.

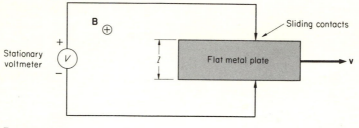

PROBLEM 11-7

11-8 Induced emf. A rectangular conducting loop is rotating about the y axis of a coordinate system with a constant angular velocity ω. A uniform static magnetic field $\mathbf{B} = -B_0\mathbf{a}_z$, where B_0 is a scalar constant, exists everywhere in the immediate space about the loop. Calculate the instantaneous value of the emf induced in the loop as a function of the angle $\theta = \omega t$.

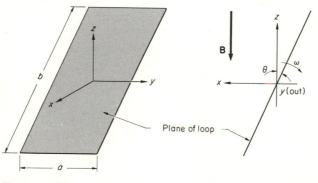

PROBLEM 11-8

11-9 Induced emf. A uniform magnetic field

$$\mathbf{B} = \sin\left(\omega t - \frac{\pi}{2}\right)\mathbf{a}_z \qquad \text{Wb/m}^2$$

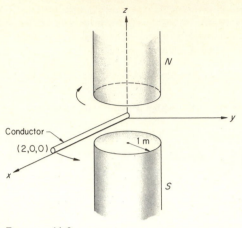

PROBLEM 11-9

exists in the free space between the pole faces of an electromagnet.   The construction is such as to allow the north pole to rotate about the z axis (looking down) in the clockwise direction at 600 rpm, while a 2-m length of rigid conductor is rotating at 600 rpm in the counterclockwise direction.   The conductor is moving in the xy plane with one end fastened at the origin.   If fringing is neglected, what voltage is induced in the conductor at the very instant when the conductor coincides with the x axis?

11-10   Faraday disk.   Faraday found, experimentally, that a metal disk [see part (a) of the figure] functions as a generator as it rotates in the field of a magnet.   For this problem, consider that a thin brass disk is rotating at ω_1 rad/sec with its plane normal to a magnetic field $\mathbf{B} = -(B_0 \cos \omega_2 t)\mathbf{a}_z$.   Find the emf, V, developed at the terminals of a pair of brushes, one at $x = 0$, $y = a$, the other at $x = 0$, $y = b$.

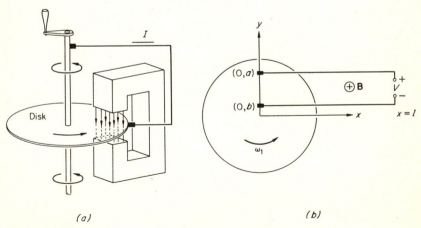

(a)   (b)

PROBLEM 11-10

11-11 Changing Flux and the Lorentz Force. For the magnetic flux density vector **B**, Eq. (11-162) can be transformed by Stokes' theorem to give

$$\frac{d\Phi}{dt} = \frac{d}{dt} \int_S \mathbf{B} \cdot \mathbf{n} \, da = \int_S \left[\frac{\partial \mathbf{B}}{\partial t} + \nabla \times (\mathbf{B} \times \mathbf{v}) \right] \cdot \mathbf{n} \, da$$

Since $\partial \mathbf{B}/\partial t = -\nabla \times \mathbf{E}$, the integral on the right obviously reduces to the contour integral of $\mathbf{E}' = \mathbf{E} + \mathbf{v} \times \mathbf{B}$, the total force per unit charge in a moving body. Prove that the total derivative of **B** is

$$\frac{d\mathbf{B}}{dt} = \frac{\partial \mathbf{B}}{\partial t} + (\mathbf{v} \cdot \nabla)\mathbf{B}$$

and that Faraday's law for a moving medium is

$$\nabla \times \mathbf{E} = -\frac{d\mathbf{B}}{dt}$$

11-12 Electromagnetic Transients. A thin conducting disk is suspended in the middle of a narrow gap cut out of a toroid of circular cross section. Determine the time required for the flux to reach 36.8 percent of its final value after a current step has been applied to the coil. Fringing, though present, may be neglected.

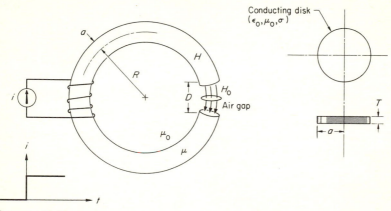

PROBLEM 11-12

APPENDIX

1. THE INTERNATIONAL SYSTEM OF UNITS

(*Abbreviated SI, for Système Internationale d'Unités*)
Multiplied units: symbols plus center dot
Divided units: symbols plus shilling bar

Term	Unit	Abbreviation of unit
Electric current, magnetomotive force	ampere	A
Magnetic field strength	ampere per meter	A/m
Luminous intensity	candela	cd
Luminance	candela per square meter	cd/m²
Electric charge	coulomb	C (A · s)
Volume	cubic meter	m³
Thermodynamic temperature	degree Kelvin	°K
Capacitance	farad	F (A · s/V)
Inductance	henry	H (V · s/A)
Frequency	hertz	Hz (s⁻¹)
Work, energy, quantity of heat	joule	J (N · m)
Mass	kilogram	kg
Density	kilogram per cubic meter	kg/m³
Luminous flux	lumen	lm (cd · sr)
Illumination	lux	lx (lm/m²)
Length	meter	m
Velocity	meter per second	m/s
Acceleration	meter per second squared	m/s²
Force	newton	N (kg · m/s²)
Pressure (stress)	newton per square meter	N/m²
Dynamic viscosity	newton-second per square meter	N · s/m²
Electric resistance	ohm	Ω (V/A)
Plane angle	radian	rad
Angular velocity	radian per second	rad/s
Angular acceleration	radian per second squared	rad/s²
Time	second	s
Area	square meter	m²
Kinematic viscosity	square meter per second	m²/s
Solid angle	steradian	sr
Magnetic flux density	tesla	T (Wb/m²)
Voltage, potential difference, electromotive force	volt	V (W/A)

Term	Unit	Abbreviation of unit
Electric field strength	volt per meter	V/m
Power	watt	W (J/s)
Magnetic flux	weber	Wb (V · s)

Prefixes		
tera-	T	10^{12}
giga-	G	10^9
mega-	M	10^6
kilo-	k	10^3
hecto-	h	10^2
deka-	da	10
deci-	d	10^{-1}
centi-	c	10^{-2}
milli-	m	10^{-3}
micro-	μ	10^{-6}
nano-	n	10^{-9}
pico-	p	10^{-12}
femto-	f	10^{-15}
atto-	a	10^{-18}

2. UNIT CONVERSION TABLES

To convert	Into	Multiply by
Electromagnetic quantities		
abamperes	amperes	10
abfarads	farads	10^9
abhenrys	henrys	10^{-9}
abohms	ohms	10^{-9}
abvolts	volts	10^{-8}
gammas	teslas	10^{-9}
gauss	teslas	10^{-4}
gilberts	amperes	0.79577
maxwells (lines)	webers	10^{-8}
oersteds	amperes per meter	79.577
statamperes	amperes	3.3356×10^{-10}
statfarads	farads	1.1126×10^{-12}
stathenrys	henrys	8.9876×10^{11}
statohms	ohms	8.9876×10^{11}
statvolts	volts	299.79
Length		
angstroms	meters	10^{-10}
feet	meters	0.3048

To convert	Into	Multiply by

Length

inches	centimeters	2.54
miles	meters	1609.3
nautical miles	meters	1852

Area

acres	square meters	4046.9
barns	square meters	10^{-28}
circular mils	square millimeters	5.0671×10^{-4}
hectares	square meters	10^4
square feet	square meters	0.092903
square inches	square centimeters	6.4516

Volume

acre feet	cubic meters	1233.5
barrels (U.S.)	cubic meters	0.15899
cubic feet	cubic meters	0.028317
cubic inches	cubic centimeters	16.387
fluid ounces (U.K.)	cubic centimeters	28.413
fluid ounces (U.S.)	cubic centimeters	29.574
gallons (U.K.)	cubic centimeters	4546.1
gallons (U.S.)	cubic centimeters	3785.4
liters	cubic centimeters	1000

Speed

feet per minute	millimeters per second	5.08
kilometers per hour	meters per second	0.27778
knots	meters per second	0.51444
miles per hour	meters per second	0.44704

Mass

long tons	kilograms	1016.0
ounces (avoirdupois)	grams	28.350
pounds	kilograms	0.45359
short tons	kilograms	907.18
slugs	kilograms	14.594
tonnes	kilograms	1000

Density

| pounds per cubic foot | kilograms per cubic meter | 16.018 |
| pounds per cubic inch | kilograms per cubic meter | 27,680 |

Force

dynes	newtons	10^{-5}
kilograms-force	newtons	9.80665
ounces-force	newtons	0.27801
poundals	newtons	0.13825
pounds-force	newtons	4.4482

Pressure

bars	newtons per square meter	10^5
conventional feet of water	newtons per square meter	2989.1
conventional millimeters of mercury	newtons per square meter	133.32
normal atmospheres (760 torrs)	newtons per square meter	101,325
poundals per square foot	newtons per square meter	1.4882
pounds-force per square foot	newtons per square meter	47.880
pounds-force per square inch	newtons per square meter	6894.8
technical atmospheres (1 kg · f/cm²)	newtons per square meter	98,066.5
torrs	newtons per square meter	133.32

Energy, work

British thermal units (International Table)	joules	1055
British thermal units (thermochemical)	joules	1054
calories (I.T.)	joules	4.1868
calories (thermochemical)	joules	4.184
electronvolts	joules	1.6021×10^{-19}
ergs	joules	10^{-7}
foot poundals	joules	0.042140
foot pounds-force	joules	1.3558

Power

British thermal units (I.T.) per hour	watts	0.2931
ergs per second	watts	10^{-7}
foot pounds-force per second	watts	1.3558
horsepower (British)	watts	745.70
horsepower (electrical)	watts	746
horsepower (metric)	watts	735.50

Quantities of light

foot-candles	lux (lumens per square meter)	10.764
foot-lamberts	candelas per square meter	3.4263

3. ELECTROMAGNETIC PROPERTIES OF MATERIALS

Fundamental constants

$\varepsilon_0 =$ permittivity of free space

$= 8.854 \times 10^{-12} \approx \dfrac{1}{36\pi} \times 10^{-9} \text{ F/m}$

$\mu_0 =$ permeability of free space

$= 4\pi \times 10^{-7} = 1.257 \times 10^{-6} \text{ H/m}$

$c =$ velocity of light in free space

$= 2.997925 \times 10^8 \text{ m/s}$

$m =$ electron mass

$= 9.1091 \times 10^{-31} \text{ kg}$

$e =$ electronic charge

$= 1.6021 \times 10^{-19} \text{ C}$

$e/m =$ specific electronic charge

$= 1.7588 \times 10^{11} \text{ C/k}$

$h =$ Planck's constant

$= 6.6256 \times 10^{-34} \text{ J·s}$

Properties of dielectrics

Material	Low-frequency dielectric constants	Conductivity, ℧/m at 20°C
Glass	4–7	10^{-12}
Lucite	3.4	10^{-13}
Marble	8.3	10^{-7}–10^{-9}
Mica	4.5–7.5	10^{-11}–10^{-15}
Paraffin	2.1–2.5	10^{-14}–10^{-16}
Porcelain	5.7	3×10^{-13}
Rubber, hard	2.3–4.0	10^{-14}–10^{-16}
Shellac	2.3–4.0	10^{-14}
Water, distilled	81	2×10^{-4}
Wood	2.5–7.7	10^{-8}–10^{-11}

Typical variations of dielectric constants with frequency

Material	Frequency	
	100 MHz	10 GHz
Bakelite	4.40	3.52
Formica	3.77	3.36
Micarta	3.98	3.62
Neoprene	4.50	4.00
Polystyrene	3.83	3.46

Properties of conductors

Material	Conductivity, $\mho$/m at 20°C
Aluminum	3.54×10^7
Brass	1.50×10^7
Copper, annealed	5.80×10^7
Gold	4.10×10^7
Iron, pure	1.00×10^7
Lead	0.48×10^7
Mercury	0.104×10^7
Nickel	1.28×10^7
Silver, pure	6.14×10^7
Tin	0.869×10^7
Tungsten	1.81×10^7
Zinc	1.74×10^7

Conduction properties of the earth

Substance	Conductivity, $\mho$/m
Ground, dry	10^{-4}–10^{-5}
Ground, wet	10^{-2}–10^{-3}
Water, fresh	10^{-3}
Water, sea	4

ANSWERS TO PROBLEMS

CHAPTER 1

1-1 (a) $\mathbf{A} = \mathbf{a}_x + 4\mathbf{a}_y + 12\mathbf{a}_z$ $\mathbf{B} = 2\mathbf{a}_x + 4\mathbf{a}_y + 12\mathbf{a}_z$.

(b) $3\mathbf{a}_x + 8\mathbf{a}_y$

(c) $xy + y^4 + 3zt^2$

(d) 27

(e) $xy + y^4 + 9t^2$

(g) $(y^2zt - 3y^2t)\mathbf{a}_x + (3ty - xzt)\mathbf{a}_y + (xy^2 - y^3)\mathbf{a}_z$

(h) $3\mathbf{a}_y - 4\mathbf{a}_z$

(i) $(4zt - 12t)\mathbf{a}_x + (6t - xzt)\mathbf{a}_y + (4x - 8)\mathbf{a}_z$

1-2 (a) $3(x^2 + y) + x(4 + x) + 4y = 45$

(b) $\mathbf{A} \cdot (\mathbf{B} \times \mathbf{C}) = 48 - 4x^3 + y(x^3 - 4y) = -\mathbf{A} \cdot (\mathbf{C} \times \mathbf{B})$

$\mathbf{C} \cdot (\mathbf{A} \times \mathbf{B}) = -4y^2 + x^3(y - 4) + 48 = -(\mathbf{B} \times \mathbf{A}) \cdot \mathbf{C}$

$\mathbf{B} \cdot (\mathbf{C} \times \mathbf{A}) = x^3(y - 4) + 4(12 - y^2) = -\mathbf{A} \cdot (\mathbf{C} \times \mathbf{B})$

(c) $(4x - 4y - xy)\mathbf{a}_x + (x^2y + y^2 - 12)\mathbf{a}_y + (12 + 3x - x^3 - xy)\mathbf{a}_z$

1-3 (a) $AB\mathbf{n}$ with $\mathbf{n}$ undefined

(b) 0

1-4 (a) $\mathbf{A} = r(\cos^2 \varphi + r \sin^3 \varphi)\mathbf{a}_r + r \sin \varphi \cos \varphi(r \sin \varphi - 1)\mathbf{a}_\varphi + 3t\mathbf{a}_z$

$\mathbf{B} = r \sin \varphi(\cos \varphi + r \sin^2 \varphi)\mathbf{a}_r + r \sin^2 \varphi(r \cos \varphi - 1)\mathbf{a}_\varphi + zt\mathbf{a}_z$

(b) $\mathbf{A} = \dfrac{9}{\sqrt{5}} \mathbf{a}_r + \dfrac{2}{\sqrt{5}} \mathbf{a}_\varphi + 12\mathbf{a}_z$

$\mathbf{B} = \dfrac{10}{\sqrt{5}} \mathbf{a}_r + 12\mathbf{a}_z$

1-5 (a) $\mathbf{A} \cdot \mathbf{B} = r^2 \sin \varphi \cos \varphi + r^4 \sin^4 \varphi + 3zt^2$

(b) 27

1-6 (a) $[(z - 3)tr^2 \sin^2 \varphi \cos \varphi - tr \sin \varphi(z \cos \varphi - 3 \sin \varphi)]\mathbf{a}_r$

$+ [(3 - z)tr^2 \sin^3 \varphi + tr \cos \varphi(3 \sin \varphi - z \cos \varphi)]\mathbf{a}_\varphi$

$+ r^3 \sin^2 \varphi(\cos \varphi - \sin \varphi)\mathbf{a}_z$

(b) $\dfrac{6}{\sqrt{5}} \mathbf{a}_r + \dfrac{3}{\sqrt{5}} \mathbf{a}_\varphi - 4\mathbf{a}_z$

1-7 10.03

1-8 $A = \left(\dfrac{zx^2}{x^2 + y^2} - y^2 \right) a_x + \left(\dfrac{xyz}{x^2 + y^2} + xy \right) a_y + 16\sqrt{x^2 + y^2}\, a_z$

1-9 (a) $B = ra_r$ (b) $5a_r$

1-10 Answers are given by Eqs. (1-13) and (1-14).

1-11 25.1°

1-12 $A = a_x + a_y - a_z$ or $A = a_x - \frac{17}{13}a_y + \frac{7}{13}a_z$

1-13 (a) $r = $ constant; circles

(b) $(1,0,0),\ (0,1,0),\ (-1,0,0),\ (0,-1,0)$

1-14 (a) $a^2x^2 + b^2y^2 = $ constant; ellipses

(b) $x^2 + y^2 = $ constant; circles

1-15 $B_x = 29\frac{1}{2}$ $B_z = -5\frac{1}{2}$ $C_y = -9$

1-16 $\dfrac{\pi}{4}$

1-17 (a) $-125\frac{2}{3}$ (b) $-241\frac{1}{6}$

1-18 $\frac{3}{2}$

1-19 $128\frac{7}{15}$

1-20 9

1-21 $10\frac{2}{3}$

1-22 -9

1-23 $-\frac{1}{3}$

1-24 -8 in the direction of $n = \dfrac{1}{\sqrt{5}} a_y - \dfrac{2}{\sqrt{5}} a_z$

1-25 $32\frac{2}{3}$

1-26 18 in the direction of $n = \dfrac{1}{\sqrt{2}} a_y + \dfrac{1}{\sqrt{2}} a_z$

1-27 $27\frac{1}{4}$ in the direction of $n = a_z$

1-28 16π in the direction of $n = a_z$

1-29 $\frac{8}{5}$ in the direction of $n = \dfrac{2x}{\sqrt{5 - 4z}} a_x + \dfrac{2y}{\sqrt{5 - 4z}} a_y + \dfrac{1}{\sqrt{5 - 4z}} a_z$

1-30 0

1-31 $\dfrac{\pi}{2}$ in the direction of the outward radial vector

1-32 $\dfrac{\pi}{6}(5\sqrt{5} - 1)$

1-33 $\frac{352}{3}\pi$

1-34 4π

1-35 $\pm 2312 a_z$

1-37 $-\frac{3}{2}$

1-39 $\nabla \cdot \mathbf{A} = 0$ $\qquad$ $\nabla \cdot \mathbf{B} = 2x$ $\qquad$ $\nabla \cdot \mathbf{C} = 0$

$\nabla \times \mathbf{A} = \mathbf{a}_x + \mathbf{a}_z$ $\qquad$ $\nabla \times \mathbf{B} = 0$ $\qquad$ $\nabla \times \mathbf{C} = 0$

1-40 (a) $2x\mathbf{a}_x + 2y\mathbf{a}_y + 2z\mathbf{a}_z$

(b) $2r\mathbf{a}_r + 2z\mathbf{a}_z$

(c) $2r\mathbf{a}_r$

1-41 (a) $\mathbf{a}_x + \mathbf{a}_y + 2z\mathbf{a}_z$

(b) $(\cos \varphi + \sin \varphi)\mathbf{a}_r + (\cos \varphi - \sin \varphi)\mathbf{a}_\varphi + 2z\mathbf{a}_z$

(c) $[\sin \theta(\cos \varphi + \sin \varphi) + 2r \cos^2 \theta]\mathbf{a}_r + \cos \theta(\cos \varphi + \sin \varphi - 2r \sin \theta)\mathbf{a}_\theta$
$\quad + (\cos \varphi - \sin \varphi)\mathbf{a}_\varphi$

CHAPTER 2

2-2 No

2-3 180,000 C

2-4 $\mathbf{J} = -[\rho_+(v_+ + v_0) + \rho_-(v_0 - v_-)]\mathbf{a}_x$

2-5 No

2-6 On the same plate, 3.94 m away

2-8 10 eV

2-9 Surrounding medium has zero electric conductivity but nonzero thermal conductivity.

2-10 (a) No $\qquad$ (b) Yes $\qquad$ (c) Yes

2-12 $\mathbf{H} = 0$ $\qquad$ $\mathbf{D} = \dfrac{Q}{4\pi R^2} \mathbf{a}_r$, Q total charge, $R > r_1 + r_0$

2-13 $J = 5.7$ A/m²

2-14 (a) $\mathbf{D} = -\dfrac{Q}{4\pi r^2} \mathbf{a}_r$ $\qquad$ (b) $\phi = \dfrac{Q}{4\pi\varepsilon}\left(\dfrac{1}{a} - \dfrac{1}{b}\right)$

2-17 (a) $\mathbf{J} = -11.3e^{-0.113 \times 10^6 t}\mathbf{a}_x$ A/m²

(b) Zero

2-19 $\theta_2 = 16.1°$ $\qquad$ $E_2 = 18$ V/m

2-21 $[F] = ITVL^{-1}$ $\qquad$ $[Q] = IT$ $\qquad$ $[m] = IT^3VL^{-2}$ $\qquad$ $[W] = ITV$ $\qquad$ $[\epsilon] = ITV^{-1}L^{-1}$

2-22 $B = 1.63 \times 10^{-5}$ Wb/m² $\qquad$ $\mu_r = 1.3$ $\qquad$ $\chi_m = 0.3$

2-23 11.79 percent

2-24 Blood would be subject to both types of effects.

2-25 Aluminum paramagnetic
Copper diamagnetic
Gold diamagnetic
Magnesium paramagnetic
Mercury diamagnetic
Silver diamagnetic
Tungsten paramagnetic

2-27 $\mathbf{J}_1 = \dfrac{\rho_0 \sigma_1}{3\epsilon_1} re^{-(\sigma_1/\epsilon_1)t}\mathbf{a}_r$

$\mathbf{J}_2 = \dfrac{\rho_0 \sigma_2}{3\epsilon_2} re^{-(\sigma_2/\epsilon_2)t}\mathbf{a}_r$

$\mathbf{D} = \dfrac{\rho_0}{3} \dfrac{r_2{}^3}{r^2}\mathbf{a}_r \qquad r > r_2$

CHAPTER 3

3-2 (a) 2.5 μC in $\mathbf{a}_y$ direction

(b) 2V

3-3 100 V

3-4 6 V

3-5 (a) 11.1 pF (b) 3.33 μC

3-6 $\dfrac{6\pi\epsilon_0}{d}\left(1 - \dfrac{a^2}{2}\right)$

3-7 $\phi = \dfrac{V}{\sinh\left(\pi\dfrac{b}{a}\right)} \sin\left(\dfrac{\pi}{a}x\right)\sinh\left(\dfrac{\pi}{a}y\right) + \dfrac{V}{\sinh\left(\pi\dfrac{a}{b}\right)}\sinh\left(\dfrac{\pi}{b}x\right)\sin\left(\dfrac{\pi}{b}y\right)$

3-8 $\phi = \dfrac{xV}{d + (\epsilon/\epsilon_0)(l - d)}$ (dielectric)

$\phi_0 = \dfrac{[d(\epsilon_0/\epsilon - 1) + x]V}{d(\epsilon_0/\epsilon - 1) + l}$ (air space)

3-10 $Ka \ln \dfrac{b}{a}$ V

3-11 (a) $\varphi = E_0\left(\dfrac{a^2}{r} - r\right)\cos\varphi$ (b) $\mathbf{E} \approx E_0\mathbf{a}_x$

3-12 $\varphi = E_0\left(r + \dfrac{a^2}{r}\dfrac{\epsilon - \epsilon_0}{\epsilon + \epsilon_0}\right)\sin\varphi$ (dielectric)

$\phi_0 = E_0\left(\dfrac{2\epsilon}{\epsilon + \epsilon_0}\right)r\sin\varphi$ (cavity)

3-13 $Q_2 = -\dfrac{Q_1}{2}$

3-15 (a) $\phi = V$ $\qquad\qquad$ $\mathbf{E} = 0$ $\qquad$ $0 \le r < a$

$$\phi = \frac{aV}{b-a}\left(\frac{b}{r} - 1\right) \qquad \mathbf{E} = \frac{abV}{(b-a)r^2}\,\mathbf{a}_r \qquad a < r < b$$

$\phi = 0$ $\qquad\qquad\qquad$ $\mathbf{E} = 0$ $\qquad$ $b \le r$

(b) $\rho_s = \begin{cases} \dfrac{\epsilon V}{b-a}\dfrac{b}{a} & \text{(inner sphere)} \\[3mm] -\dfrac{\epsilon V}{b-a}\dfrac{a}{b} & \text{(outer sphere)} \end{cases}$

3-16 $C = \dfrac{4\pi\epsilon_0\epsilon ab(a+b)}{(b-a)(a\epsilon + b\epsilon_0)}$

3-17 (b) Potential at all points exterior to the sphere may be calculated by super-imposing the fields of three point charges: the charge q_1, an image charge $q_i = -aq_1/b$ at $R = a^2/b$, and a third charge $q_2 - q_i$ at the center of the sphere. The potential at all interior points is $\phi = (q_2 - q_i)/4\pi\epsilon_0 a$.

3-18 For points above the plate,

$$\phi = \frac{\rho_l}{2\pi\epsilon}\ln\sqrt{\frac{(h+b)^2 + a^2}{(h-b)^2 + a^2}}$$

where b is the height of P above the plane, and a the lateral distance from the line charge. For all points on the flat plane, or below, $\phi = 0$.

3-19 0, 1.69, 8.16, 11 cm

3-20 Maps the entire upper half of the Z plane into the interior of a semi-infinite horizontal trough bounded by the lines $v = 0$, $u = 0$, $v = j\pi$.

3-22 $\phi = \dfrac{\rho_0}{\epsilon}(a - r) + \left[V + \dfrac{\rho_0}{\epsilon}(b - a)\right]\dfrac{\ln(r/a)}{\ln(b/a)}$

3-23 $\rho = \epsilon\,\dfrac{\cos r}{r}$

$$\rho_s = \frac{-\epsilon(R\sin R + \cos R)}{R^2}$$

$q = 4\pi\epsilon$ $\qquad$ at origin

3-24 (b) Two-dimensional dipole

3-26 42 pF/m

3-30 Continuous charge distribution

$$\rho = -a^2\epsilon\,\frac{e^{-ar}}{r}$$

throughout space and a point charge $q = 4\pi\epsilon$ at the origin. The system as a whole is neutral.

3-32 $\mathbf{E} = \begin{cases} 0 & \text{for } r < b \\[2mm] E_0\left(\dfrac{b^2}{r^2} + 1\right)\cos\varphi\,\mathbf{a}_r + E_0\left(\dfrac{b^2}{r^2} - 1\right)\sin\varphi\,\mathbf{a}_\varphi & \text{for } r \ge b \end{cases}$

3-33 (a) $Q_1 = Q_2 \dfrac{R_1(R_2 - R_3)}{R_2(R_3 - R_1)}$

  (b) $Q_3 = Q_2 \dfrac{R_3(R_1 - R_2)}{R_2(R_3 - R_1)}$

  (c) $\phi_2 = \dfrac{Q_2}{4\pi\epsilon_0} \dfrac{(R_2 - R_1)(R_2 - R_3)}{R_2^2(R_3 - R_1)}$

3-37 $\rho_s = \dfrac{\epsilon V}{\cosh^{-1}\dfrac{d^2 - 2a^2}{2a^2}} \left[\dfrac{a - c(\cos \varphi)}{a^2 + c^2 - 2ac(\cos \varphi)} + \dfrac{(b - c)\cos \varphi - a}{(b - c)^2 + a^2 - 2a(b - c)\cos \varphi} \right]$

where $c = (b - d)/2$, and φ is a cylindrical angle centered about the axis of the conductor on the left in Fig. 3-19c.

3-38

$L = 100a$					

$N = 10$			$N = 20$			
K	Y		K	Y	K	Y
1	0.1451		1	0.1762	11	0.1053
2	0.1150		2	0.1241	12	0.1055
3	0.1094		3	0.1177	13	0.1061
4	0.1067		4	0.1131	14	0.1070
5	0.1056		5	0.1102	15	0.1083
6	0.1056		6	0.1082	16	0.1102
7	0.1067		7	0.1070	17	0.1131
8	0.1094		8	0.1062	18	0.1178
9	0.1150		9	0.1055	19	0.1241
10	0.1451		10	0.1053	20	0.1761

$N = 50$									
K	Y	K	Y	K	Y	K	Y	K	Y
1	0.2686	11	0.1108	21	0.1054	31	0.1057	41	0.1117
2	0.1263	12	0.1098	22	0.1053	32	0.1059	42	0.1131
3	0.1366	13	0.1087	23	0.1053	33	0.1064	43	0.1145
4	0.1258	14	0.1082	24	0.1050	34	0.1065	44	0.1165
5	0.1224	15	0.1079	25	0.1051	35	0.1071	45	0.1189
6	0.1187	16	0.1067	26	0.1050	36	0.1077	46	0.1222
7	0.1165	17	0.1068	27	0.1051	37	0.1081	47	0.1258
8	0.1148	18	0.1065	28	0.1051	38	0.1088	48	0.1363
9	0.1128	19	0.1057	29	0.1054	39	0.1097	49	0.1268
10	0.1117	20	0.1058	30	0.1055	40	0.1107	50	0.2683

K = interval number along wire axis
Y = relative charge density

CHAPTER 4

4-1 $B = \dfrac{\mu a d}{4\pi \sqrt{h^2 + (a^2 + d^2)/4}} \left(\dfrac{1}{h^2 + a^2/4} + \dfrac{1}{h^2 + d^2/4} \right)$

4-3 0.101 A; yes

4-4 $\mathbf{H} = \dfrac{bkt}{2} \mathbf{a}_z$

4-5 $\mathbf{H} = \dfrac{K}{6} \left(\dfrac{a}{r} \right)^3 (3 \cos^2 \theta - 1) \mathbf{a}_z$

where K is the surface current density.

4-6 A second, infinitely long wire located in the xy plane at $y = 2$ and carrying a current of 10 A in the positive x direction.

4-7 $\mathbf{H} = \frac{1}{2} J_0 D \mathbf{a}_y$; center of hole at $x = D$; current flow in the positive z direction; conductor axis coincident with the z axis

4-8 Integration of Poisson's equation is contingent upon a *bounded* distribution of static sources.

4-9 $F = \dfrac{\mu_0 I I'}{\pi} \left[\dfrac{a}{R} - \ln\left(1 + \dfrac{a}{R} \right) \right]$ attractive

4-10 $\lambda = \dfrac{1}{2\pi} \mu_0 I \ln \dfrac{b}{a}$

4-11 $2.78 \cos 400t$ mV/m

4-12 Downward

4-14 (*a*) 0.054 H   (*b*) 0.0202 H

4-15 605 kilolines

4-16 1.656 A

4-17 97 kilolines

4-20 $\mathbf{H}_1 = -0.0103 H_0 \mathbf{a}_y$

$\mathbf{H}_2 = -0.0515 H_0 \left[\left(1 - \dfrac{25 \times 10^{-4}}{r^2} \right) \sin \varphi \, \mathbf{a}_r + \left(1 + \dfrac{25 \times 10^{-4}}{r} \right) \cos \varphi \, \mathbf{a}_\varphi \right]$

$\mathbf{H}_3 = -H_0 \left[\mathbf{a}_y + \dfrac{26.01 \times 10^{-4}}{r^2} (\sin \varphi \, \mathbf{a}_r - \cos \varphi \, \mathbf{a}_\varphi) \right]$

4-21 If μ is infinite, $\mathbf{H}_2 = 0$, and

$\mathbf{B}_2 = -\dfrac{2\mu_0 H_0}{1 - a^2/b^2} \left[\left(1 - \dfrac{a^2}{r^2} \right) \sin \varphi \, \mathbf{a}_r + \left(1 + \dfrac{a^2}{r^2} \right) \cos \varphi \, \mathbf{a}_\varphi \right]$

If μ is large but finite:

$$\mathbf{H}_2 = -\frac{2(\mu_0/\mu)H_0}{1 - a^2/b^2}\left[\left(1 - \frac{a^2}{r^2}\right)\sin\varphi\,\mathbf{a}_r + \left(1 + \frac{a^2}{r^2}\right)\cos\varphi\,\mathbf{a}_\varphi\right]$$

$$\mathbf{H}_1 = -\frac{4(\mu_0/\mu)H_0}{1 - a^2/b^2}\,\mathbf{a}_y$$

4-22 Inside the sphere

$$B = 9B_0\frac{\mu_r}{(2\mu_r + 1)(\mu_r + 2) - 2(\mu_r - 1)^2(a/b)^3}$$

4-24 $\displaystyle \mathbf{H} = K_0 e^{-\pi b/2a}\left[\sin\left(\frac{\pi}{a}x\right)\cosh\left(\frac{\pi}{a}y\right)\mathbf{a}_x - \cos\left(\frac{\pi}{a}x\right)\sinh\left(\frac{\pi}{a}y\right)\mathbf{a}_y\right]$

4-25 $\displaystyle \mathbf{H}_1 = -H_0\left(\frac{2\mu_0}{\mu_0 + \mu}\right)(\sin\varphi\,\mathbf{a}_r + \cos\varphi\,\mathbf{a}_\varphi)$

$$\mathbf{H}_2 = -H_0\left[\left(1 - \frac{a^2}{r^2}\frac{\mu_0 - \mu}{\mu_0 + \mu}\right)\sin\varphi\,\mathbf{a}_r + \left(1 + \frac{a^2}{r^2}\frac{\mu_0 - \mu}{\mu_0 + \mu}\right)\cos\varphi\,\mathbf{a}_\varphi\right]$$

$$\mathbf{M}_1 = \left(\frac{\mu}{\mu_0} - 1\right)\mathbf{H}_1$$

4-27 Air gap: $\displaystyle \mathbf{H}_0 = -\left(A - \frac{G}{r^2}\right)\cos\varphi\,\mathbf{a}_r + \left(A + \frac{G}{r^2}\right)\sin\varphi\,\mathbf{a}_\varphi$

Rotor: $\mathbf{H} = C(-\cos\varphi\,\mathbf{a}_r + \sin\varphi\,\mathbf{a}_\varphi)$

where $\displaystyle A = \frac{-\mu K_0 a^2}{a^2(\mu_0 - \mu) + (a + t)^2(\mu_0 + \mu)}$

$$G = \frac{\mu K_0 a^2(a + t)^2}{a^2(\mu_0 - \mu) + (a + t)^2(\mu_0 + \mu)}$$

$$C = \frac{-\mu_0 K_0[a^2 + (a + t)^2]}{a^2(\mu_0 - \mu) + (a + t)^2(\mu_0 + \mu)}$$

4-28 $\displaystyle \mathbf{K} = -\frac{2\cos\varphi}{R^2}\,\mathbf{a}_z$

4-29 The residual magnetism in the substance corresponds to a point, such as P, on the magnetization curve, where it intersects the straight line

$$B_{\varphi i} = -\frac{\mu_0 H_{\varphi i}(2\pi r - t)}{t}$$

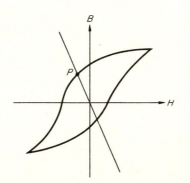

CHAPTER 5

5-3 $\rho_s = \begin{cases} \dfrac{\epsilon V}{b} & \text{(top)} \\[2ex] \dfrac{-\epsilon V}{b} & \text{(bottom)} \end{cases}$

5-4   22.12 s

5-6 $\phi = \dfrac{Vb^2(r^2 - a^2)}{r^2(b^2 - a^2)}$

5-8 $\mathbf{J} = J_0 \sin \varphi \left(1 - \dfrac{a^2}{r^2}\right)\mathbf{a}_r + J_0 \cos \varphi \left(1 + \dfrac{a^2}{r^2}\right)\mathbf{a}_\varphi$

        $J_{\max} = 2J_0$ at $r = a$ and $\varphi = 0°, 180°$

5-9 $\dfrac{\varphi_0}{\sigma_s \ln (b/a)}$

5-11 $\phi = -\dfrac{2I}{\pi \sigma t R} \displaystyle\sum_{n=0}^{\infty} \left(\dfrac{r}{R}\right)^{2n+1} \dfrac{\sin (2n + 1)\,\Delta\varphi}{(2n + 1)\,\Delta\varphi} \dfrac{\sin (2n + 1)\varphi}{(2n + 1)}$

where t is the thickness, and σ is the conductivity of the coin.

5-12 $\mathbf{E} = \dfrac{I}{2\pi\sigma r^2}\mathbf{a}_r$ $\mathbf{H} = -\dfrac{I(1 + \cos \theta)}{2\pi r \sin \theta}\mathbf{a}_\varphi$ (in metal)

        $\mathbf{E} = 0$ $\mathbf{H} = \dfrac{I}{2\pi r}\mathbf{a}_\varphi$ (in air)

5-14   With $a = 2$, $b = 1.5$, $x_0 = 0.4a$, $y_0 = 0.3b$, a tolerance limit of 0.1, and for a grid size of 20 by 15, $R\sigma_s = 1.234$, where R is the resistance from point P to the silver-painted electrode.

CHAPTER 6

6-1 $\dfrac{Q^2}{8\pi\epsilon}\left(\dfrac{1}{a} - \dfrac{1}{b}\right)$

6-2 $\dfrac{\mu I^2}{3\pi a^4}(a^3 - r^3)$

6-6 $\approx \sqrt{\dfrac{e}{2mV_0}}\, B_0 l L$

6-8 $F = \dfrac{1}{4\pi\epsilon}\left[\dfrac{Qq}{d^2} - \dfrac{q^2 R^3}{d^3}\dfrac{2d^2 - R^2}{(d^2 - R^2)^2}\right]$

6-9 $\dfrac{\mu_0(\mu - \mu_0)I^2}{4\pi h(\mu + \mu_0)}$ toward the iron slab

6-10 Away from the source

6-11 $\epsilon_0 E^2 A / 2$, where E is the strength of the field, and A is the area of the plates.

6-13 $\dfrac{Q^2 d}{2b} \dfrac{\epsilon - \epsilon_0}{[x(\epsilon - \epsilon_0) + L\epsilon_0]^2}$; field pulls dielectric in.

6-15 17.708×10^{-10} J

6-16 $\theta = \dfrac{\epsilon_0 V^2 R^2}{4K} \dfrac{t}{2d(2d + t)}$ $K = $ spring constant

6-18 $\dfrac{9\pi\epsilon_0 a^2 E_0^2}{4} \left(\dfrac{\epsilon_r - 1}{\epsilon_r + 2} \right)^2$ (attractive)

CHAPTER 7

7-2 $f = 88$ MHz, $\lambda = 3.41$ m, elliptical polarization, cw rotation

7-3 (a) $E_x = 3 \cos(\omega t - \beta z)$

$$E_y = 4 \cos\left(\omega t - \beta z + \frac{\pi}{4}\right)$$

Equation of ellipse: $\left(\dfrac{E_x}{3}\right)^2 - \dfrac{\sqrt{2}}{12} E_x E_y + \left(\dfrac{E_y}{4}\right)^2 = \dfrac{1}{2}$

7-7 $v_p = \dfrac{\omega}{\sqrt{\omega^2 \mu\epsilon - (\pi/a)^2}}$ $v_g = \dfrac{1}{\mu\epsilon} \dfrac{\sqrt{\omega^2 \mu\epsilon - (\pi/a)^2}}{\omega}$

7-8 (a) $\alpha = 6.28 \times 10^3$ Np/m $\beta = 6.28 \times 10^3$ rad/m $v_p = 10^5$ m/s
$\quad\quad\quad \lambda = 10^{-3}$ m
$\quad\quad$ (b) $\alpha = 1.167 \times 10^{-8}$ Np/m $\beta = 3.37$ rad/m $v_p = 1.86 \times 10^8$ m/s
$\quad\quad\quad \lambda = 1.86$ m

7-10 0.053 mW/m²

7-11 (a) $\lambda = 3$ m $\beta = 2.093$ rad/m
$\quad\quad$ (b) $\langle W_e \rangle = \langle W_m \rangle = 2.22 \times 10^{-18}$ J/m³
$\quad\quad$ (c) 1.325×10^{-9} W/m² 2.65×10^{-9} W/m² (peak)

7-12 $\dfrac{20}{3} \ln\left(\cot \dfrac{\theta_0}{2}\right)$

7-14 Orthogonality in the time domain always implies orthogonality in the frequency domain.   If either G_1 or G_2 is linearly polarized, orthogonality in the time domain implies orthogonality in the frequency domain, and conversely.   The same is true if G_1 and G_2 are out of phase by an odd multiple of $\pi/2$; this includes circular polarization as a special case.

7-15 $v_1 = \left[\mu_0\epsilon_0\left(1 - \dfrac{\omega_0{}^2}{\omega^2 - \omega_g{}^2}\right)\right]^{-1/2}$ $\qquad$ $v_2 = \left[\mu_0\epsilon_0\left(1 - \dfrac{\omega_0{}^2}{\omega^2}\right)\right]^{-1/2}$

7-16 (b) Isotropic

(c) $\gamma\mathbf{n}\cdot\gamma\mathbf{n} = -\omega^2\mu_0\epsilon_0\left[1 + \dfrac{\omega_0{}^2}{j\omega(f_c + j\omega)}\right]$

(d) $\epsilon = \epsilon_0\left(1 - \dfrac{\omega_0{}^2}{\omega^2 + f_c{}^2}\right)$

(e) $\sigma = \dfrac{\epsilon_0\omega_0{}^2 f_c}{\omega^2 + f_c{}^2}$

CHAPTER 8

8-1 $\Gamma_\perp = \dfrac{\sqrt{1 - \sin^2\theta_1} - 2\sqrt{1 - (\sin^2\theta_1)/4}}{\sqrt{1 - \sin^2\theta_1} + 2\sqrt{1 - (\sin^2\theta_1)/4}}$

$\Gamma_\parallel = \dfrac{2\sqrt{1 - \sin^2\theta_1} - \sqrt{1 - (\sin^2\theta_1)/4}}{2\sqrt{1 - \sin^2\theta_1} + \sqrt{1 - (\sin^2\theta_1)/4}}$

8-2 $15.1e^{-j2.95}$ $\qquad$ $\mu\text{V/m}$

8-3

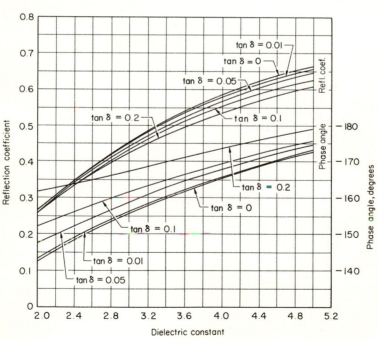

8-4 $\epsilon_r = \left(1 + \dfrac{\Delta}{\beta_0 d}\right)^2$ $\tan \delta = -\dfrac{\lambda_0 \ln \sqrt{P/P_0}}{\pi d \sqrt{\epsilon_r}}$

where Δ is the measured phase shift caused by sample insertion, and P and P_0 are the amounts of power transmitted with and without the sample; also, β_0 is the usual phase-shift constant in free space.   It is assumed that the dielectric constant of the sample is not significantly larger than 1.

8-6   Both slabs must be integral multiples of a half wavelength thick.

8-9   Sun pressure $= 9.33 \times 10^{-6}$ N/m^2

8-12 $\epsilon_r = 1 + 2\dfrac{\Delta}{\beta_0 d}\cos\theta_1 + \left(\dfrac{\Delta}{\beta_0 d}\right)^2$ $\tan \delta = -\dfrac{\lambda_0 \ln \sqrt{P/P_0}\sqrt{\epsilon_r - \sin^2\theta_1}}{\pi d \epsilon_r}$

where the symbolism is the same as in Prob. 8-4; θ_1 is the angle of incidence.

8-14   No particles in vacuum to scatter light in our direction.

8-18

θ_0, deg	$T_\perp^2$	$R_\perp^2$	$T_\parallel^2$	$R_\parallel^2$
0	0.926	0.00440	0.926	0.00440
10	0.925	0.00457	0.927	0.00442
20	0.921	0.00668	0.928	0.00643
30	0.908	0.0164	0.926	0.0126
40	0.875	0.0459	0.926	0.0183
50	0.799	0.118	0.935	0.0130
60	0.650	0.268	0.951	0.000259
70	0.410	0.518	0.898	0.060
80	0.134	0.821	0.512	0.460
88	0.006	0.984	0.032	0.962

For $\theta_0 = 30 + j0$, the angles of refraction are

$\theta_1 = 14.0 + j0.106$

$\theta_2 = 28.0 + j0.073$

$\theta_3 = 14.0 + j0.106$

$\theta_4 = 16.4 + j0.292$

$\theta_5 = 10.4 + j0.139$

$\theta_6 = 30.0 + j0$

8-19   (a)   28.73 m   (b)   68.3 kHz

CHAPTER 9

9-1   50 Ω

9-2   50 $\underline{/-53.2°}$ Ω

9-3 $Z = 7.36 \underline{/33°}\ \Omega$

$3.6 \cos(\omega t - 73.9°)$ A

9-4 1.64

9-5 Distance $= 0.091\lambda$ $Z_L = 38 + j92.5$ Ω

9-7 Distance to stub point $= 0.095\lambda$ Length of shorted stub $= 0.153\lambda$

9-8 (a) $Z_L = 30 - j40$ Ω
    (b)   Stub length $= 4.56$ cm at 1.68 cm from load

9-9 $d = 0.042\lambda$ $l_1 = 0.446\lambda$ $l_2 = 0.375\lambda$

9-10 $l_1 = 6.04$ cm $l_2 = 14.64$ cm

9-11 $d = 0.363\lambda$ $l = 0.161\lambda$

9-12 $1.59\ \mho$

9-13 133Ω (inductive), $298\ \Omega$ (capacitive)

9-14 $R = 8.66\ \Omega/\text{m}$ $G = 8.66 \times 10^{-4}\ \mho/\text{m}$
      $C = \frac{1}{3} \times 10^{-10}$ F/m $L = \frac{1}{3} \times 10^{-6}$ H/m

9-15 (a) $37.5 \underline{/-20°}\ \Omega$
    (b) 2.94×10^7 m/s
    (c)   85 percent

9-16 (a) $\alpha = 0.866$ Np/m $\beta = 0.5$ rad/m

    (b) $13.6 \sin\left(\omega t - \frac{\pi}{6} + \frac{1}{\sqrt{3}}\right)$
    (c) $0.05 \underline{/-60°}$
    (d) $50 \sin \omega t - 2.5 \sin(\omega t - 60°)$
    (e) $R = 1\ \Omega$ $X = 1.732\ \Omega$ $G = 0.5\ \mho$ $B = 0\ \mho$
    (f) $4.69 \underline{/63.7°}\ \Omega$

9-17

9-18 (a)   Voltage maximum
    (b) 0.25λ least distance; 0.2812λ greatest distance
    (c) 0.082λ

9-19 $V(z, s) = F(s)\dfrac{Z_0(s)}{Z_0(s) + Z_s(s)}[e^{-z\gamma(s)} + \Gamma_0(s)e^{(z-2d)\gamma(s)} + \Gamma_s(s)\Gamma_0(s)e^{-(z+2d)\gamma(s)} + \cdots]$

where $F(s)$ is the Laplace transform of $f(t)$.

9-23

	0–2T	2T–4T	4T–6T	∞
$V(0, t)$	0.1	0.28	0.424	1
$I(0, t)$	$\dfrac{0.1}{Z_0}$	$\dfrac{0.08}{Z_0}$	$\dfrac{0.064}{Z_0}$	0

T = time required for a wave to travel length of line

9-24 $V(z,t) = \dfrac{1}{2} u_{-1}\left(t - \dfrac{z}{v}\right) + u_{-1}\left(t - 2T + \dfrac{z}{v}\right)\left[\exp\left(-\dfrac{t - 2T + z/v}{LY_0}\right) - \dfrac{1}{2}\right]$

9-25 $f_c = 6.56$ GHz   maximum frequency = 13.12 GHz

9-26

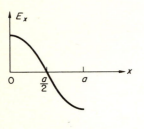

(a)

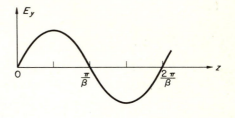

(b)

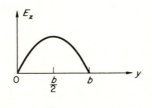

(c)

9-27   Yes, 6.56 to 9.28 GHz

9-28 $0.485 \, \underline{/-63°}$

9-29   452.4 m

9-30   max $E_y = 60 \times 10^3$ V/m   at $x = a/2$

         max $H = 127$ A/m   at $x = 0$ and $x = a$

9-31 $C = $ a constant proportional to strength of the field

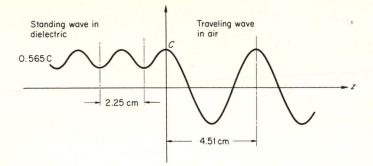

Standing wave in dielectric

Traveling wave in air

0.565 C

C

2.25 cm

4.51 cm

z

9-32 $0.989 P_t$

9-33 Inside diameter $= 1.954$ cm

TE_{11}	9 GHz
TM_{01}	11.77 GHz
TE_{21}	14.92 GHz
TE_{01} and TM_{11}	18.72 GHz

9-34 $\mathbf{K} = j\dfrac{\omega\epsilon}{h} CJ_0'(ha)e^{-j\beta z}\mathbf{a}_z$

$P_L = \pi a R_s \left[\dfrac{\omega\epsilon}{h} CJ_0'(ha) \right]^2$

$ha = 2.405$

9-36 $a = 2.24$ cm, $l = 3.95$ cm

9-37

l, cm	1	2	3	4	5
f_r, GHz	15.6	8.7	6.65	5.78	5.33

CHAPTER 10

10-2 $\rho_l = -j\dfrac{\beta}{\omega} I_m \sin \beta z \, e^{j\omega t}$; 90° out of phase

10-5 Circular polarization; orientation in a plane parallel to xy plane is immaterial.

10-7 Array factor $= 1 + 2 \cos (\pi \cos \varphi)$

10-8 Field intensity $= 4.24 \times 10^{-6}$ V/m

10-9 (*a*) 83.6° for main beam

(*b*) 141° for main beam

10-10 (*a*) 34.8° (*b*) 91°

10-11 In either plane

$$\theta_{01} = \sin^{-1} \frac{1}{20} = 2°52'$$

$$\theta_{02} = \sin^{-1} \frac{1}{10} = 5°44'$$

$$\theta_{03} = \sin^{-1} \frac{3}{20} = 8°38'$$

$$\theta_{04} = \sin^{-1} \frac{1}{5} = 11°32'$$

. . . .

10-12

$$\frac{je^{-j\beta R}}{\lambda R} (ab) E_0 \frac{\sin\left[\frac{\beta a}{2}(\sin\theta_0 + \sin\theta\cos\varphi)\right]}{\frac{\beta a}{2}(\sin\theta_0 + \sin\theta\cos\varphi)} \frac{\sin\left(\frac{\beta b}{2}\sin\theta\sin\varphi\right)}{\frac{\beta b}{2}\sin\theta\sin\varphi}$$

10-13

$$\frac{je^{-j\beta R}}{\lambda R} (ab) \frac{E_0}{2} \left[\frac{(2\pi/a)^2}{(2\pi/a)^2 - \beta^2 \sin^2\theta} \frac{\sin\left(\frac{\beta a}{2}\sin\theta\right)}{\frac{\beta a}{2}\sin\theta} \right]$$

$$\sin\theta_{3\,DB} \approx 0.716 \frac{\lambda}{a}$$

10-14 (*a*) $f = 120$ MHz

10-17 Fourier coefficients:

$$a_0 = \frac{1}{2}$$

$$a_k = \frac{1}{k^2\pi^2} [1 - (-1)^k] \qquad k = 1, 2, 3, \ldots$$

$$b_k = 0$$

10-18 (*a*) In general, no.

(*b*) In general, no.

(*c*) Must consider coupling effects.

(*d*) In general, no.

(*e*) Yes.

(*f*) Reduced area of loop, and hence R_{rad}.

(*g*) To some extent, yes.

(*h*) Doppler radar.

(*i*) Ionospheric backscatter; airborne repeater.

CHAPTER 11

11-2 $Z^0 = 0$, short circuit

$$Z^1 = j\omega L \qquad \text{where } L = \frac{\mu a d}{b} \text{, a pure inductance}$$

$$Z^2 = \frac{j\omega L}{1 - \omega^2 LC} \qquad \text{where } C = \frac{\epsilon ab}{2d} \text{ , parallel } LC \text{ combination}$$

$$Z^3 = \frac{j\omega L(1 - \omega^2 LC/3)}{1 - \omega^2 LC} \qquad L \text{ in parallel with a series combination of } \frac{2C}{3} \text{ and } \frac{L}{2}$$

11-3   Media 1 and 2: $\mathbf{E}_0 = 0$ $\qquad \mathbf{H}_0 = -\text{Re}(K_0)\mathbf{a}_x$

Medium 1: $\mathbf{E}_1 = -\text{Re}(j\mu_0 z K_0)\mathbf{a}_y$ $\qquad \mathbf{H}_1 = -\text{Re}\left(j\frac{\sigma\mu_0 a^2}{2} K_0\right)\mathbf{a}_x$

Medium 2: $\mathbf{E}_1 = -\text{Re}(j\mu_0 z K_0)\mathbf{a}_y$ $\qquad \mathbf{H}_1 = -\text{Re}\left(j\frac{\sigma\mu_0 z^2}{2} K_0\right)\mathbf{a}_x$

11-4   (a) $\mathbf{E}_0 = -\dfrac{V_0}{r \ln (b/a)}\,\mathbf{a}_r$ $\qquad \mathbf{H}_0 = -\dfrac{\sigma_s V_0}{r \ln (b/a)}\,\mathbf{a}_\varphi$

$\mathbf{E}_1 = \dfrac{j\,\mu\sigma_s V_0}{\ln (b/a)}\dfrac{z}{r}\,\mathbf{a}_r$ $\qquad \mathbf{H}_1 = \dfrac{j\epsilon V_0}{\ln (b/a)}\dfrac{z}{r}\,\mathbf{a}_\varphi$

(b)   If $\sigma_s < \sqrt{\epsilon/\mu}$: $Y^1 = \dfrac{1}{R} + j\omega C$ $\qquad$ (parallel combination of R and C)

where $R = \dfrac{\ln (b/a)}{2\pi\sigma_s}$ $\qquad$ and $\qquad C = \dfrac{2l\pi(\epsilon - \mu\sigma_s^2)}{\ln (b/a)}$

If $\sigma_s > \sqrt{\epsilon/\mu}$: $Z^1 = R + j\omega L$ $\qquad$ (series combination of R and L)

where $L = \dfrac{\ln (b/a)}{2\pi} l\left(\mu\sigma_s - \dfrac{\epsilon}{\sigma_s}\right)$

If $\sigma_s = \sqrt{\dfrac{\epsilon}{\mu}}$: $Z^1 = R$ $\qquad$ (pure resistance)

11-5 $\frac{1}{2}\mu_0 N^2 I^2$

11-7 Blv

11-8 $ab\omega B_0 \cos \theta$

11-9 $10\pi \sin\left(\omega t - \dfrac{\pi}{2}\right)$ $\quad$ V

11-10 $V = \omega_2 l(b - a)B_0 \sin \omega_2 t - \omega_1 \dfrac{b^2 - a^2}{2} B_0 \cos \omega_2 t$

11-12 $\dfrac{\mu\mu_0\sigma a^2 T}{8(\mu D + 2\pi\mu_0 R)}$

INDEX